Plant–Animal Interactions
in the
Marine Benthos

The Systematics Association
Special Volume No. 46

Plant–Animal Interactions in the Marine Benthos

Edited by

D.M. JOHN
Department of Botany
The Natural History Museum, London

S.J. HAWKINS
Port Erin Marine Laboratory
University of Liverpool, Isle of Man

and

J.H. PRICE
Department of Botany
The Natural History Museum, London

Published for the SYSTEMATICS ASSOCIATION by
CLARENDON PRESS · OXFORD
1992

Oxford University Press, Walton Street, Oxford OX2 6DP
Oxford New York Toronto
Delhi Bombay Calcutta Madras Karachi
Petaling Jaya Singapore Hong Kong Tokyo
Nairobi Dar es Salaam Cape Town
Melbourne Auckland

and associated companies in
Berlin Ibadan

Oxford is a trade mark of Oxford University Press

Published in the United States
by Oxford University Press, New York

© The Systematics Association, 1992

All rights reserved. No part of this publication may be reproduced,
stored in a retrieval system, or transmitted, in any form or by any means,
electronic, mechanical, photocopying, recording, or otherwise, without
the prior permission of Oxford University Press

A catalogue record for this book is available from the British Library

Library of Congress Cataloging in Publication Data
Plant–animal interactions in the marine benthos / edited by D. M. John,
S. J. Hawkins, and J. H. Price.
p. cm. — (The Systematics Association special volume ; no. 46)
Includes index.
1. Benthos. 2. Animal–plant relationships. 3. Marine ecology.
I. John, D. M. (David M.) II. Hawkins, S. J. (Stephen J.) III. Price, J. H.
(James H.) IV. Series.
QH541.5.S3P53 1992   574.5'2636—dc20   91-25977

ISBN 0-19-857754-0

Set by
Colset Pte. Ltd., Singapore

Printed in Great Britain by
St. Edmundsbury Press,
Bury St. Edmunds, Suffolk

Preface

Traditional compartmentalization of biological sciences is increasingly breaking down or being considered irrelevant by new generations of biological scientists. Marine biologists have generally been ahead of such trends, as is evident from much recent research on plant–animal interactions in the marine benthos. This research has both contributed to a greater understanding of coastal ecosystems and made considerable contributions to general ecological concepts. The symposium from which the present volume derives was generated by the desire to integrate and consolidate the large body of work published over the last two decades. This four-day symposium held in Liverpool, 18–21 September 1990, provided a useful forum for the almost 140 participants (botanists, zoologists, marine biologists, behaviourists, and so on) to consider current problems in, and future approaches to, the study of plant–animal interactions. The symposium programme consisted of 'keynote' or invited papers, with contributed papers[1] and posters. Of the 60 plus papers presented, the majority tended to focus on the more obvious form of biotic interaction—the relationship between herbivorous animals and their food plants—reflecting current interest in trophic interactions.

It was appropriate to hold the symposium at Liverpool University, for much of the early work on experimental shore ecology was carried out on the Isle of Man from The Port Erin Marine Laboratory of Liverpool University. The logo adopted for the symposium was the famous Liverbird, heraldic symbol of the City of Liverpool. The name Liverpool is a corruption of 'laver pool'. Laver is a red seaweed (*Porphyra*) eaten as a delicacy in some parts of the British Isles (South Wales, south-west England, Ireland) and elsewhere (e.g., Japan, China). Some heraldic licence has to be permitted in the form of the plant diet as the bird (supposedly a cormorant) is a carnivore.

The symposium was opened by Professor A.J. Southward who gave an account of some of the early work on plant–animal interactions with an unashamedly Liverpudlian emphasis. The chapters mostly fall

---

[1] Some (ed. S.J. Hawkins and G.A. Williams) to be published in Vol. 71 of the *Journal of the Marine Biological Association of the UK*.

under two major headings. Those under the first deal with habitats mainly from a regional perspective and thus enable the thought-provoking question 'do assemblages have different ecologies or are there different types of ecologist?' to be addressed. We felt there to be a need to bring together authoritatively written reviews to facilitate the necessary inter-regional comparisons. The omission of some regional or detailed habitat accounts was due to the existence of a recent review (e.g., coral reefs: see papers in *Proceedings of the Sixth International Coral Reef Symposium*, 3 vols. (1988)), the difficulty of finding a suitable authority prepared to write a review, or the complete lack of any information on interactions. The second heading deals primarily with trophic interactions: chapters cover types of grazer, feeding mechanisms, foraging behaviour, succession, competition, co-evolution, and structural, chemical, and behavioural adaptations of plants to grazing.

We have attempted to ensure as complete a coverage as possible of plant–animal interactions from an ecological perspective. To broaden coverage, however, the penultimate chapter reviews our knowledge of endosymbiosis in cnidarians. Habitats where conditions are not strictly marine have been deliberately omitted. Mangrove swamps and salt marshes are therefore outside the scope of the volume. Many chapters look to the directions of future studies and focus attention on areas where further research is needed. It is our hope that the volume will stimulate research effort in a fascinating field where much still remains to be discovered.

We would like to thank Professor Brian Moss for providing symposium facilities in the Department of Environmental and Evolutionary Biology (of which The Port Erin Marine Laboratory is part) and for making the welcoming address. The Symposium Dinner was held at the Adelphi Hotel, Liverpool, and we would like to extend thanks to Professor Trevor Norton (Director of The Port Erin Marine Laboratory) for his after-dinner speech. Thanks go also to all those who chaired the formal sessions. We are very much aware that our task of producing this volume to a reasonable time-scale would have been impossible without the full cooperation of all authors and those who kindly refereed the chapters.

The organizers would like to thank certain organizations for financial or scientific support given to the meeting: The Natural History Museum, London; University of Liverpool; The Systematics Association; The Marine Biological Association (UK); The Royal Society; The British Phycological Society; The Malacological Society; The Estuarine and Coastal Science Association. Thanks go also to the following for their generous sponsorship: Royal Skandia Life Assurance; American

Airlines; Cambridge University Press; Blackwell Scientific Publications; Kall-Kwik Printers.

We are especially grateful to others whose often less obvious contributions to aspects of the organization were extremely valuable: Mrs Jenny Moore (The Natural History Museum, London), who edited the Book of Abstracts and whose efforts as the Meeting Secretary did much to ensure the smooth running of the symposium; Mrs B. Brereton (Port Erin), who had to handle the accounts and much of the registration. The meeting would not have functioned without students and technicians of Liverpool and Port Erin, who worked tirelessly to resolve individual problems, produced T-shirts and operated the slide projector; deserving of special thanks are Janette Allen, Mark Davies, Andy Hill, Tim Hill, Stuart Johnson, Marcus Perrin, Helena Corte-Real, and Susan Spence.

London, Port Erin, and Hong Kong D. M. J.
January 1991 S. J. H.
  J. H. P.
  G. A. W.

Contents

List of Contributors

Habitats and regional perspectives

1. Plant–animal interactions on hard substrata in the north-east Atlantic — 1
   S. J. Hawkins, R. G. Hartnoll, J. M. Kain, and T. A. Norton

2. Plant–animal interactions in the north-west Atlantic — 33
   R. L. Vadas and R. W. Elner

3. Plant–animal interactions in the north-east Pacific: the importance of grazing to seaweed evolution and assemblage structure — 61
   M. S. Foster

4. Tropical east Atlantic and islands: plant–animal interactions on tropical shores free of biotic reefs — 87
   D. M. John, J. H. Price, and G. W. Lawson

5. Ecology of tropical rocky shores: plant–animal interactions in tropical and temperate latitudes — 101
   Deborah M. Brosnan

6. Grazing on sediment shores — 133
   K. Reise

7. A perspective on plant–animal interactions in seagrasses: physical and biological determinants influencing plant and animal abundance — 147
   R. J. Orth

8. Plant–animal trophic relationships in the *Posidonia oceanica* ecosystem of the Mediterranean Sea: a review — 165
   L. Mazzella, M. C. Buia, M. C. Gambi, M. Lorenti, G. F. Russo, M. B. Scipione, and V. Zupo

9. Interactions between macrofaunal epiphytes and their host algae ... 189
   G. A. Williams and R. Seed

10. A trophodynamic model of fish production on a windward reef tract ... 213
    N. V. C. Polunin and D. W. Klumpp

Processes and types of plant–animal interaction

11. Mesoherbivores ... 235
    Susan H. Brawley

12. Herbivory in benthic suspension-feeding molluscs ... 265
    B. L. Bayne and A. J. S. Hawkins

13. Foraging behaviour of marine benthic grazers ... 289
    M. G. Chapman and A. J. Underwood

14. Chemical mediation of seaweed–herbivore interactions ... 319
    M. E. Hay and W. Fenical

15. Herbivorous fishes: feeding and digestive mechanisms ... 339
    M. H. Horn

16. Digestion survival in seaweeds: an overview ... 363
    B. Santelices

17. Role of algae in the recruitment of marine invertebrate larvae ... 385
    A. N. C. Morse

18. Algal 'gardening' by marine grazers: a comparison of the ecological effects of territorial fish and limpets ... 405
    G. M. Branch, J. M. Harris, C. Parkins, R. H. Bustamante, and S. Eekhout

19. Grazing and succession ... 425
    W. P. Sousa and J. H. Connell

20. Competition and marine plant–animal interactions ... 443
    A. J. Underwood

21. Plant–herbivore coevolution: a reappraisal from the marine realm and its fossil record ... 477
    R. S. Steneck

22.	Bioerosion and biogeomorphology T. Spencer	492

Endosymbiosis

23.	Endosymbiosis in marine cnidarians P. Spencer Davies	511

Summary and future view

24. Summary and future prospects for plant–animal interactions

541

A.J. Underwood

Index

List of Systematics Association Publications 545

Contributors

B.L. BAYNE (Dr)
The Plymouth Marine Laboratory, Prospect Place, West Hoe, Plymouth PL1 3DH, England, UK.

G.M. BRANCH (Professor)
Department of Zoology and Marine Biology Research Institute, University of Cape Town, Rondebosch 7700, South Africa.

SUSAN BRAWLEY (Dr)
Department of Plant Ecology and Pathology, University of Maine, Orono, ME 04469, USA.

DEBORAH M. BROSNAN (Dr)
Department of Zoology, Oregon State University, Corvallis, 97331, USA.

M.C. BUIA (Dr)
Laboratorio di Ecologia del Benthos, Stazione Zoologica di Napoli, Punta S. Pietro, 80077, Ischia Porto (NA), Italy.

R.H. BUSTAMANTE
Department of Zoology and Marine Biology Research Institute, University of Cape Town, Rondebosch 7700, South Africa.

M.G. CHAPMAN (Ms)
Institute of Marine Ecology, University of Sydney, Zoology Building, A08, NSW 2009, Australia.

J.H. CONNELL (Professor)
Department of Biological Sciences, University of California, Santa Barbara, CA 93106, USA.

S. EEKHOUT
Department of Zoology and Marine Biology Research Institute, University of Cape Town, Rondebosch 7700, South Africa.

R.W. ELNER (Dr)
Department of Fisheries and Oceans, Pacific Biological Station, Nanaimo, British Columbia, V9R 5K6, Canada.

W. FENICAL (Professor)
Scripps Institution of Oceanography, University of California, San Diego, La Jolla, CA 92093-0227, USA.

M.S. FOSTER (Professor)
Moss Landing Marine Laboratories and San Jose State University, P.O. Box 450, Moss Landing, CA 95039-0450, USA.

M.C. GAMBI (Dr)
Laboratorio di Ecologia del Benthos, Stazione Zoologica di Napoli, Punta S. Pietro, 80077, Ischia Porto (NA), Italy.

J.M. HARRIS (Ms)
Department of Zoology and Marine Biology Research Institute, University of Cape Town, Rondebosch 7700, South Africa.

R.G. HARTNOLL (Dr)
Port Erin Marine Laboratory, University of Liverpool, Port Erin, Isle of Man.

A.J.S. HAWKINS (Dr)
The Plymouth Marine Laboratory, Prospect Place, West Hoe, Plymouth PL1 3DH, England, UK.

S.J. HAWKINS (Dr)
Port Erin Marine Laboratory, University of Liverpool, Port Erin, Isle of Man.

M.E. HAY (Dr)
University of North Carolina at Chapel Hill, Institute of Marine Sciences, Morehead City, NC 28557, USA.

M.H. HORN (Professor)
Department of Biological Science, California State University, Fullerton, CA 92634, USA.

D.M. JOHN (Dr)
Department of Botany, The Natural History Museum, Cromwell Road, London SW7 5BD, England, UK.

J.M. KAIN (Dr)
Port Erin Marine Laboratory, University of Liverpool, Port Erin, Isle of Man.

D.W. KLUMPP (Dr)
Australian Institute of Marine Science, PMB No. 3, Townsville MC, Queensland 4810, Australia.

G.W. LAWSON (Dr)
Department of Botany, The Natural History Museum, Cromwell Road, London SW7 5BD, England, UK.

M. LORENTI (Dr)
Laboratorio di Ecologia del Benthos, Stazione Zoologica di Napoli, Punta S. Pietro, 80077, Ischia Porto (NA), Italy.

LUCIA MAZZELLA (Dr)
Laboratorio di Ecologia del Benthos, Stazione Zoologica di Napoli, Punta S. Pietro, 80077, Ischia Porto (NA), Italy.

AILEEN N.C. MORSE (Dr)
University of California, Marine Science Institute, Santa Barbara, CA 93106, USA.

T.A. NORTON (Professor)
Port Erin Marine Laboratory, University of Liverpool, Port Erin, Isle of Man.

R.J. ORTH (Dr)
School of Marine Science, Virginia Institute of Marine Science, College of William and Mary, Gloucester Point, Virginia 23062, USA.

C. PARKINS (Dr)
Department of Zoology and Marine Biology Research Institute, University of Cape Town, Rondebosch 7700, South Africa.

N.V.C. POLUNIN (Dr)
Department of Marine Sciences and Coastal Management, University of Newcastle upon Tyne, Newcastle upon Tyne NE1 7RU, England, UK.

J.H. PRICE
Department of Botany, The Natural History Museum, Cromwell Road, London SW7 5BD, England, UK.

K. REISE (Dr)
Wadden Sea Institute Sylt of Biologische Anstalt Helgoland, D-2282 List, Germany.

G.F. RUSSO (Dr)
Laboratorio di Ecologia del Benthos, Stazione Zoologica di Napoli, Punta S. Pietro, 80077, Ischia Porto (NA), Italy.

B. SANTELICES (Professor)
Facultad de Ciencias Biologicas, Pontificia Universidad Catolica de Chile, Casilla 114-D, Santiago, Chile.

M.B. SCIPIONE (Dr)
Laboratorio di Ecologia del Benthos, Stazione Zoologica di Napoli, Punta S. Pietro, 80077, Ischia Porto (NA), Italy.

R. SEED (Dr)
School of Ocean Sciences, University of Wales Bangor, Menai Bridge, Gwynedd LL59 5EY, Wales, UK.

W.P. SOUSA (Dr)
Department of Integrative Biology, University of California, Berkeley, CA 94720, USA.

T. SPENCER (Dr)
Department of Geography, University of Cambridge, Downing Place, Cambridge CB2 3EN, England, UK.

P. SPENCER DAVIES (Dr)
Department of Zoology, The University, Glasgow G12 8QQ, Scotland, UK.

R.S. STENECK (Professor)
Department of Botany and Plant Pathology and Department of Oceanography, University of Maine, Darling Marine Centre, Walpole, Me 04573, USA.

A.J. UNDERWOOD (Dr)
Institute of Marine Ecology, University of Sydney, Zoology Building, A08, NSW 2009, Australia.

R.L. VADAS (Professor)
Department of Plant Biology and Pathology, University of Maine, 209 Deering Hall, Orono, Maine 04469, USA.

G.A. WILLIAMS (Dr)
Swire Marine Laboratory, University of Hong Kong, Cape d'Aguilar, Hong Kong.

1. Plant–animal interactions on hard substrata in the north-east Atlantic

S.J. HAWKINS, R.G. HARTNOLL, J.M. KAIN (JONES), and T.A. NORTON
Port Erin Marine Laboratory, University of Liverpool, Isle of Man, UK

Abstract

The composition of intertidal and subtidal communities within this area is briefly outlined with reference to the British Isles. A major feature is a general trend of decreasing plant dominance and greater numbers of grazers and filter feeders from shelter to wave-exposure in the eulittoral. This pattern changes from north to south, with animal-dominated communities extending further into shelter at more southerly locations. Similarly, southwards, dominance by kelp beds declines low on the shore and in the subtidal, and red algal turfs become more important and grazer diversity increases. The role of plant–animal interactions in energy flow within the mid-intertidal is discussed. Sheltered-shore fucoid beds are net exporters of energy, primarily to the detrital pathway. More exposed shores are net importers of energy, although most *in situ* microalgal production is consumed within the community. Shores of intermediate exposure are probably modest net exporters. Comparatively little macroalgal production passes through the grazing pathway. The way in which plant–animal interactions structure communities on the environmental gradients of wave-exposure and tidal height/depth in the British Isles is considered. *Patella* grazing prevents establishment of fucoids on more exposed shores. On moderately exposed shores, escapes of fucoids from grazing help to generate patchiness. Within these patchy communities, a complex suite of both positive and negative interactions occurs between plants and animals. On the vertical gradient, grazing is a major structuring agency in the midshore region and just below the kelp canopy in the subtidal. The balance of the interactions is altered by latitude, and grazing becomes less effective further north.

Plant–Animal Interactions in the Marine Benthos, (ed. D.M. John, S.J. Hawkins, and J.H. Price), Systematics Association Special Volume No. 46, pp. 1–32. Clarendon Press, Oxford, 1992. © The Systematics Association, 1992.

Introduction

The north-east Atlantic has a long history of studies of the distribution of benthic plants and animals (e.g., Audouin and Milne-Edwards 1832). It was also here that some of the pioneering field experiments on plant–animal interactions were undertaken (e.g., Hatton 1938; Jones 1946, 1948; Lodge 1948; Jones and Kain 1967), including the long series of investigations at Lough Ine, south-west Eire (reviewed by Kitching and Ebling 1967). Unfortunately, in recent years it has not been fashionable to pursue field experimental studies here as vigorously as in other parts of the world.

In this review, attention is concentrated on the British Isles, since it is situated in the centre of the region, and because the vast majority of experimental work has been done there. We set the scene by briefly summarizing the distribution patterns of the major plants and animals in the intertidal and subtidal zones. We then select two major themes. Firstly, we outline the energetics of plant–animal interactions in the intertidal and secondly, focus on the plant–animal interactions involved in setting distributions and structuring communities along the three main environmental gradients of wave exposure, tidal height/depth, and latitude. Finally, after some conclusions are drawn, suggestions for further work are proposed.

Distribution patterns

Figure 1.1 provides a schematic summary of the distribution patterns of intertidal and subtidal communities on gently sloping rocks on the south-western coasts of the British Isles. It is based on the large body of early qualitative descriptions of the intertidal zone (see Southward 1958; Ballantine 1961a, b; Lewis 1964, for reviews) combined with similar subtidal studies since the advent of scuba diving (e.g., Hiscock 1983).

In the British Isles the high shore (littoral fringe of Lewis 1964) has low diversity and biomass, being characterized by large numbers of littorinids, lichens, and blue-greens (Cyanobacteria). Various ephemeral algae occur only from late autumn to spring.

The eulittoral (*sensu* Lewis 1964) of sheltered shores is dominated by large fucoids, usually forming clearly defined bands. As exposure increases, *Fucus* cover becomes patchy and interspersed with barnacles and limpets on moderately exposed shores. Limpets become more numerous in fully exposed conditions, where mussels and barnacles cover the rocks. Small patches of ephemeral algae and small clumps of

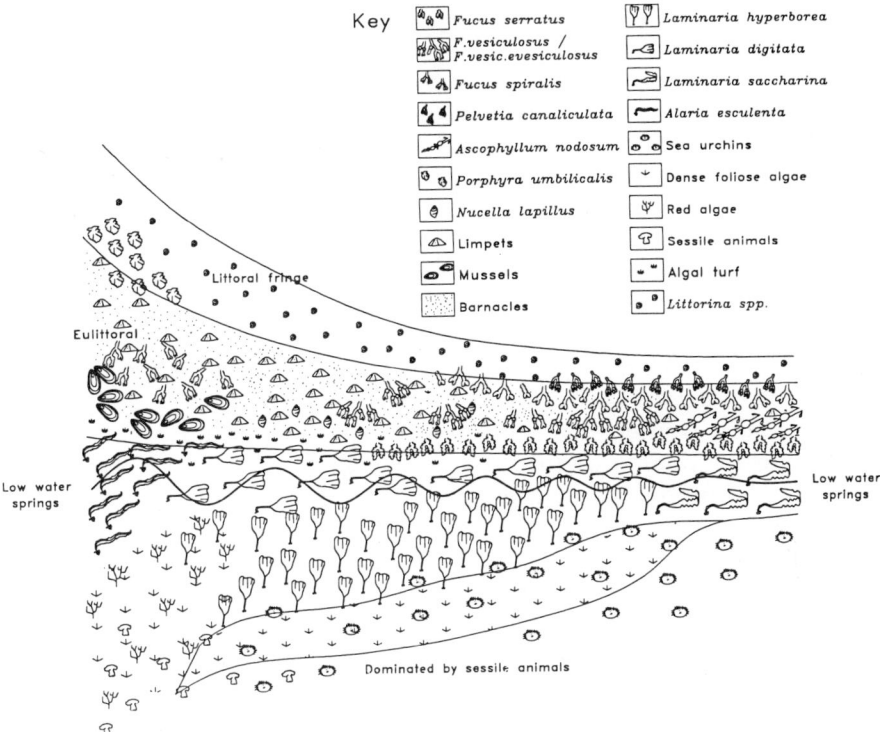

Fig. 1.1. Schematic diagram of distribution patterns on open seaward-facing rock on the south-western coasts of the British Isles, for the intertidal and subtidal combined: left side very exposed, right side very sheltered (modified from Ballantine 1961a; Lewis 1964; Hiscock 1983).

the stunted bladderless form of *Fucus vesiculosus* (var. *evesiculosus*) are found on exposed shores.

The region immediately above and below low water is dominated by macroalgae: *Fucus serratus*, a red algal turf and *Himanthalia* give way below to various kelps, which form a dense canopy in the sublittoral fringe. A dense forest of *Laminaria hyperborea* dominates subtidally in all but the extremes of shelter and exposure. This forest thins out with depth, giving way to an assemblage of small foliose algae with many grazing urchins. Deeper still, sessile animals become increasingly common.

This general pattern is modified by small-scale topographic influences (geology, shore profile, and crevices) and changes with latitude.

Energetics of plant–animal interactions

It should be stated at the outset that this aspect of the review will be based on limited evidence, since both the production of the plants, and the consumption of the grazers, have been poorly documented in energetic terms. Consideration will be limited to intertidal rocky shores — the very important subtidal *Laminaria* communities have provided even less data for the north-east Atlantic, although quite detailed studies have been made on the Atlantic coast of Canada (Miller et al. 1971).

The best-studied community is that around midtide level (MTL), where there is a relatively simple set of plant–animal interactions. The primary production consists of two components which are normally fairly discrete: a microbial film on the rock surface and a macroalgal population dominated by fucoids. On the shores studied on the Isle of Man, these components are grazed respectively by the limpet *Patella* and by the winkle *Littorina*, other grazers playing very minor roles. The major energetic pathways in such a community are shown in Fig. 1.2. Rather tentative energy budgets for this community were proposed by Hartnoll (1983a). How do these energy budgets stand up in light of more recent data?

The budget for *Patella vulgata* (from Wright and Hartnoll 1981) requires modification following recent measurements of mucus production (Davies et al. 1990), which turns out to be much greater than was previously thought. The revised energy budget (using standard International Biological Programme (IBP)) terminology, with an additional term M for mucus production) in kJ m^{-2} yr^{-1} becomes:

$$C = P_g + P_r + R + F + U + M$$
$$>3082 = 68 + 96 + 498 + 1698 + 2 + >720$$

The main impact of this change on the community energy budget is that the estimate of microalgal consumption by the limpets is almost doubled. However, the budgets for *Littorina obtusata* and *Nucella lapillus* remain unchanged (Table 1.1). No further data have been produced to supplement those in Wright (1977).

The budget for *Semibalanus balanoides* had been based on studies of *Balanus glandula* (Wu and Levings 1978), an unsatisfactory situation. The values for consumption and production are in any event certain to be substantial overestimates for *B. glandula* since, as the authors point out, they are based only on fast-growing first-year specimens ignoring mortality effects. Also, the annual temperature regime differs. Therefore, in preference, use was made of data on the respiration of *Semi-*

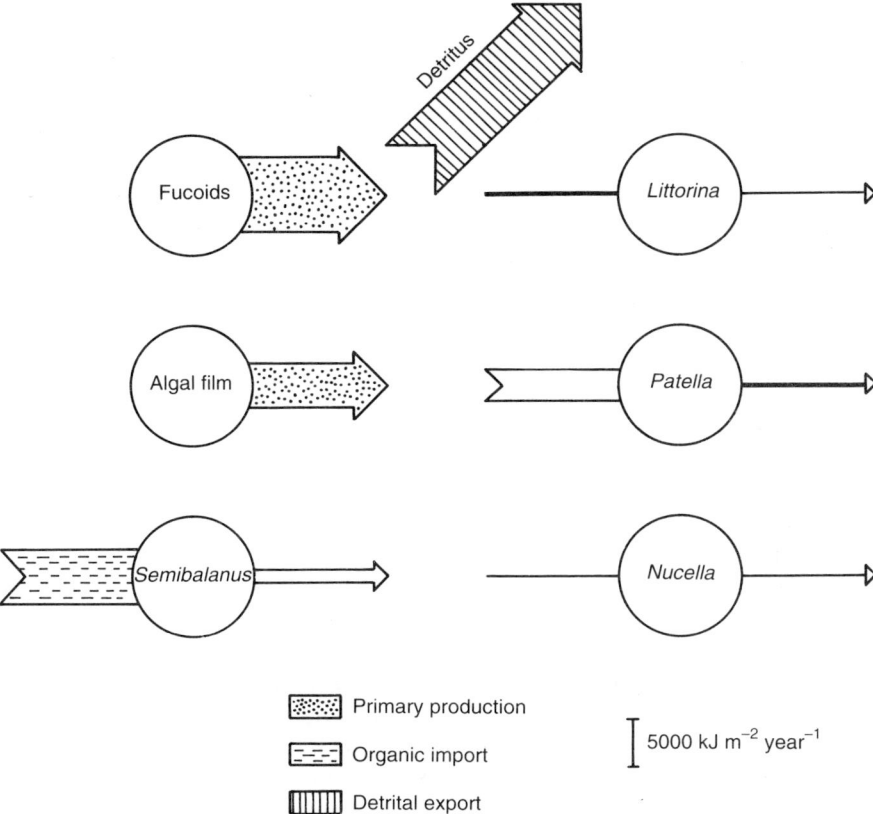

Fig. 1.2. Energy-flow patterns in a midshore community on a moderately exposed rocky shore in the Isle of Man. Production and consumption are indicated by arrows whose widths are in proportion to the energy flux (as detailed in Table 1.1).

balanus balanoides, from Barnes *et al.* (1963). Based on a series of rather gross assumptions, an annual energy expenditure for respiration of the studied population of 3900 kJ m^{-2} yr^{-1} was derived, very different from the earlier estimate of 12 370 kJ m^{-2} yr^{-1}. If the energy partitioning is assumed to be as in *B. glandula*, then a revised budget can be calculated (Table 1.1).

At the time of the previous study (Wright and Hartnoll 1981), lack of data made it impossible to estimate either biomass or production of the microalgal film. The value given for production was based solely on the assumption that the limpets grazed effectively all microbial production since the film did not proliferate. It is now possible to

Table 1.1. Components of the energy budgets for the major elements in the Derbyhaven MTL community (values derived as explained in the text). All energy values in kJ m^{-2} yr^{-1}, biomass in kJ m^{-2}.

	B	C	P	R	F	U + E	M
Algal film	700		4900				
Fucoids	5860		8000				
Patella	346	> 3082	164	498	1698	2	> 720
Semibalanus	865	5780	1318	3900	431	130	
Littorina	40	390	92	193	101	1	
Nucella	24	123	25	76	22		

Biomass (B), Consumption (C), Production (P), Respiration (R), Faeces (F), Urine and Excretion (U + E), Mucus (M)

estimate both biomass and productivity of the microalgal film, and to compare the latter with consumption by limpets. Workman (1983) estimated microbial production by the difference in biomass between grazed and ungrazed treatments. His lower station is more comparable with our midtide level Manx (Isle of Man) study site; his estimate of annual microalgal production, around 4900 kJ m^{-2} yr^{-1}, was sufficient to provide the limpet consumption postulated above. His estimate of limpet consumption was 1930 kJ m^{-2} yr^{-1}, less than half of production. However, the technique of brushing the rock clean at the start of the experiment may well produce artefactual effects by interfering with grazing efficiency. This consumption rate of 1930 kJ m^{-2} yr^{-1}, for a biomass of around 100 kJ m^{-2}, is substantially higher in proportion than the value for Manx populations. However, his population of many small limpets, as opposed to the Manx one of a smaller number of large individuals, could well have much higher energy needs.

Direct measurements of microbial film biomass based on chlorophyll extraction have been made by Hill and Hawkins (1990). They obtained a mean value of around 8 µg chlorophyll cm^{-2}. Conversion of this to weight or energy values is a rather problematical procedure, since the chlorophyll content varies between algae and not all the film may have been of living algae. Taking a value of 500 µg chlorophyll g^{-1} fresh weight as found for fucoids by Ramus et al. (1977), converting this to dry weight using experimental data and using a calorific value of 14·3 kJ g^{-1} dry weight (Paine and Vadas 1969), gives a biomass of 700 kJ m^{-2}. A high production/biomass (P/B) ratio would be expected in grazed microalgae, and a not unreasonable value of 7 would give a productivity similar to that of Workman. While there is still uncertainty, the independently derived values for limpet consumption, microalgal production, and microalgal biomass are in reasonable agreement.

There is a surprising dearth of relevant information on macroalgal production. The value used previously was based on a P/B ratio of 1:1, an estimate taken from Blinks (1955) for Californian *Fucus*. Niell (1977), working in north-western Spain, calculated rather higher values: 1·97 for *Fucus spiralis* and 3·02 for *Bifurcaria bifurcata*. With a biomass of 5860 kJ m^{-2}, these various ratios would give productivities ranging from 6250 to 17 580 kJ m^{-2} yr^{-1}. Recent studies in the Isle of Man, involving measuring the change in weight of individual *Fucus vesiculosus* clumps, indicate a productivity of 5700 kJ m^{-2} yr^{-1} (A. Hill, pers. comm.). Inability to determine the amount of tissue lost by fragmentation of older clumps renders this a minimal value. For fucoids (unspecified) on the Atlantic coast of Canada, Mann (1972) gave a productivity of *c.* 700 g C m^{-2} yr^{-1}, which is equivalent to about 26 000 kJ m^{-2} yr^{-1}. There is no indication of the biomass involved, but it presumably relates to 100 per cent fucoid cover, in contrast to the 50 per cent cover of the shore on the Isle of Man. A best guess from this restricted and conflicting evidence is that fucoid production in the community under analysis is in the range 6000–13 000 kJ m^{-2} yr^{-1}, and a conservative value of 8000 kJ m^{-2} yr^{-1} will be adopted here.

The outcome of these revised estimates is that the balance of the community energy budget (Table 1.1; Fig. 1.2) for this midtide area on a semi-exposed shore has changed from the earlier version (Hartnoll 1983a). Previously, this community was regarded as a modest net consumer of energy, but increased estimates for micro- and macroalgal production, and reduced estimates for barnacle consumption, now suggest that it is a net producer to the extent of about 3500 kJ m^{-2} yr^{-1}. The conclusion remains unchanged that microalgal production is largely consumed *in situ*, but that macroalgal production is almost all exported. However, perhaps only two-thirds of microalgal production is consumed by limpets, so other agencies may be involved in preventing the proliferation of this film. These may include grazing by the tiny *Littorina neglecta*, but other larger grazers such as *Littorina littorea* are absent. The reduced estimate for barnacle production is still an order of magnitude greater than consumption by *Nucella*.

The above energy budget is for a semi-exposed shore on the Isle of Man. Using appropriate biomass values, and assuming that the energy budget ratios are unchanged (a possibly dubious assumption), some very approximate estimates can be derived for energy flow on exposed and very sheltered shores (Table 1.2). These estimates suggest that an exposed shore is a net consumer of energy by more than 20 000 kJ m^{-2} yr^{-1} and that a sheltered shore has, by contrast, a very high net production of > 55 000 kJ m^{-2} yr^{-1}.

There are few other studies of energetics in grazer–plant

Table 1.2. Estimated values for biomass, consumption and production on exposed and very sheltered shores. All energy values in kJ m^{-2} yr^{-1}, biomass in kJ m^{-2}.

	EXPOSED			SHELTERED		
	B	C	P	B	C	P
Algal film	700		4900	430		3000
Fucoids	600		820	45000		62000
Patella	430	3800	200	170	1500	80
Semibalanus	3450	23000	5250	215	1440	330
Littorina	0	0	0	640	6240	1470
Nucella	24	120	25	70	370	75

interactions for the area. The study by Workman (1983) in north-east England has already been mentioned. At MHWN there was a microalgal production of around 3000 kJ m^{-2} yr^{-1}, of which some 600 kJ m^{-2} yr^{-1} were consumed by limpets, and some 800 kJ m^{-2} yr^{-1} by littorinids, more abundant at this location. At a lower site between MLWN and MLWS, microalgal production was higher at around 5000 kJ m^{-2} yr^{-1}, of which 1930 kJ m^{-2} yr^{-1} was consumed by limpets. These figures suggest that the assumption that grazers are consuming virtually all microalgal production is in need of re-examination.

Jansson et al. (1982) have studied the energetics of the rather specialized sublittoral *Fucus vesiculosus* community in the Baltic. They produced an energy budget for the community, based on detailed short-term measurements. The macrophytes had a biomass of 4826 kJ m^{-2} and a net production of 44·2 kJ m^{-2} day^{-1}. A group of 'browsers' (various gastropods and isopods) provided the main consumers of the macrophytes. They had a biomass of 206 kJ m^{-2} and, while their consumption was not measured, their respiration rate of 7·0 kJ m^{-2} day^{-1} indicated that they could consume a significant proportion of algal production.

Wolff (1977) produced a budget for the benthos of the Grevelingen Estuary in the Netherlands. Production of the microphytobenthos was estimated at 25–57 g C m^{-2} yr^{-1} and production of the grazer *Hydrobia ulvae* at 2·4 g C m^{-2} yr^{-1}. This probably means a consumption of at least 10 g C m^{-2} yr^{-1} — a substantial part of the production of the microphytobenthos.

Despite the paucity of information, it is possible to draw some general conclusions. On rocky shores, the balance between algal production and the consumption by primary consumers varies within wide limits. At the mid-tide level, exposed shores are net consumers and sheltered shores are net producers; shores of intermediate wave-exposure are near a trophic balance. There will certainly also be variation with tidal height: lower on the shore the increase in algal biomass and reduction of barnacles will together produce an increasing trend towards net production by the community. On subtidal rocky areas, the *Laminaria* beds also display even higher levels of production (see Savidge and Kain 1990), and will probably have a net community productivity of up to $22\,000\,\text{kJ}\,\text{m}^{-2}\,\text{yr}^{-1}$ with limited *in situ* consumption. This surplus production is largely exported from the immediate ecosystem in the form of algal detritus, making an important contribution to the food input of those intertidal and subtidal communities which function as net consumers.

Interactions on the horizontal exposure gradient

To keep this review succinct, we have concentrated on the role of plant–animal interactions in structuring communities along the wave-exposure gradient in the midshore region (eulittoral of Lewis 1964).

1. Broad-scale changes between exposure and shelter

The underlying causes of the change from fucoid domination on sheltered shores to domination by barnacles, mussels, and limpets in exposed conditions (see Fig. 1.1) were first investigated by experimental work on the Isle of Man (Jones 1946, 1948; Lodge 1948; Burrows and Lodge 1950; Southward 1956, 1964). This work demonstrated that *Fucus* spp., particularly mid-shore *Fucus vesiculosus*, were directly prevented from extending in quantity on to shores of greater wave-exposure by limpet (*Patella*) grazing. Removal of limpets from large strips (10 m wide) or squares, rapidly resulted in a dense sward of various ephemeral or opportunist species (diatoms, *Enteromorpha*, *Blidingia*, and *Porphyra*) followed after six months by a growth of *Fucus* spp. forming virtually 100 per cent cover. These swathes of fucoids persisted for 3 to 5 years depending on the size of the original removal area. No attempts were made continually to exclude or remove encroaching limpets during these classic experiments. The fucoid cover eventually decreased as losses caused by old age and insecure anchorage on barnacles (Barnes and Topinka 1969) were not matched by new recruitment. Fucoid recruitment was inhibited initially by the shading and canopy effects of the adult plants (see Schonbeck and Norton 1980) and, more importantly, by the grazing pressure of large numbers of limpets which eventually aggregated under the canopy. These limpets included both adult migrants from surrounding areas and newly recruited juveniles (see also Hartnoll and Hawkins 1985). The adult plants prevented settlement of barnacles by sweeping (see Hatton 1938; Hawkins 1983). Loss of plants also resulted in reduced barnacle cover as both plants and barnacles were plucked from the surface. The experimental consequence of the aggregation of limpets was reduced grazing pressure on either side of the strip, resulting eventually in a bare zone recolonized by barnacles in the position of the original strip, with enhanced fucoid growth on either side (Southward 1964). This also happened in a repeat experiment by Hawkins (1981a; Hawkins, unpublished data).

Confirmation of the importance of limpet grazing followed the *Torrey Canyon* oil spill. Dense growth of fucoids occurred at very exposed sites in West Cornwall, following widespread killing of limpets by indiscriminate applications of dispersants to clean up the oil (Southward and Southward 1978; Southward 1979). This dramatically confirmed that, even in the most wave-exposed conditions, establishment of fucoids is prevented directly by grazing rather than by wave action. Once established, fucoids do not live for as long on exposed shores as in sheltered conditions (Southward and Southward 1978). This suggests that their persistence may be governed by their inability to resist wave action, although wave-resistant forms may be selected for (Russell and Fielding 1981). Interactions between mussels and the small tufted *Fucus vesiculosus* (*evesiculosus* form) on exposed north-east Atlantic coasts have not been explored by experimental manipulation. This contrasts with the situation on the New England coast, where mussels are thought to outcompete fucoids in wave-beaten conditions (Menge 1976; Menge and Sutherland 1976).

'Red tides' of toxic dinoflagellates, particularly *Gyrodinium aureolum*, in the inshore phytoplankton can lead to catastrophic effects on littoral ecosystems. In extreme cases they severely disrupt littoral community structure by selectively killing space-occupying animals and important grazers and predators, although having little effect on plants (Cross and Southgate 1980; Southgate et al. 1984). After the death of sessile animals, predators, and particularly grazers (e.g., *Patella*), a pattern of recolonization ensues similar to that described after oil spills.

Grazing by *Littorina littorea* has also been shown to be important on those shores where this species is numerous, although its distribution is very variable in the north-east Atlantic region (see comments by Norton et al. 1990). In the more sheltered waters of Norwegian fjords (Lein 1984) and the Firth of Clyde in Scotland (Clokie and Boney 1980), their grazing has been shown locally to reduce algal vegetation. *Littorina* spp. do not seem, however, to exert the same control as *Patella* on broad-scale patterns of algal distribution.

While the importance of grazing in preventing the extension of fucoids into more exposed conditions is indisputable, the factors which prevent barnacles and limpets from extending into shelter are less clear. Studies on the effects of fucoid clumps and canopies on *Semibalanus balanoides* settlement on various shores around the Isle of Man have been made by Hawkins (1983). On moderately exposed shores the sweeping effect of quite small clumps of fucoids reduced *S. balanoides* settlement markedly (see also Hatton 1938; Lewis 1964). High on a sheltered shore, recruitment was greater under canopies of *F. spiralis* and *F. vesiculosus* compared with cleared areas, presumably because post-settlement survival was enhanced (see also Dayton 1971). *Fucus serratus* prevented settlement by its sweeping action, on both moderately exposed and sheltered shores. Barnacles settled well in clearings amongst *F. serratus* despite the absence of conspecific adults. Very few barnacles were found or could be induced to settle under *Ascophyllum*, except on transplanted boulders carrying barnacles.

The factors preventing establishment of limpets on very sheltered shores dominated by *Fucus* spp. and particularly *Ascophyllum*, are also not clearly understood. Lewis and Bowman (1975) suggest that fucoids act as a barrier to limpet recruitment. The silt deposited under fucoids on sheltered shores, and trapped by understorey algal turfs, may clog the gills of adults and provide an unsuitable surface for recruitment. Turfs of understorey algae under *Ascophyllum* limit the space available for limpets. Larval input may be less on sheltered shores. Certainly, pools lined with calcified red algae ('lithothamnia', *Corallina*) are rare on sheltered shores. Limpet larvae are thought to settle preferentially in these pools (Bowman 1981; see also Morse, Chapter 17, this volume). Ballantine (1961*b*) showed the precarious nature of clearings amongst fucoids containing limpets in the lower eulittoral of sheltered shores; thinning of the limpets resulted in rapid colonization by algae which swamped the limpets. Preliminary experiments suggest that removal of the few limpets present on sheltered shores has much less effect on algal recruitment than removal of the canopy (Hawkins 1979). *Nucella* foraging activity is enhanced under canopies (Menge 1978), and its predation may be important in reducing numbers of both limpets and barnacles on sheltered shores. Further work is needed to clarify the processes causing low limpet densities on sheltered shores, particularly prevention of recruitment to the *Ascophyllum* zone.

2. Patchiness on moderately exposed shores

In the British Isles and northern France, shores of intermediate exposure support a eulittoral community consisting of patches of *Fucus*, barnacles (*Chthamalus* spp., but primarily *Semibalanus balanoides*), and bare rock interspersed with grazing *Patella* and predatory *Nucella* (see Lewis 1964 for further details). Good examples of this kind of shore are found on the Isle of Man and were described in detail by Southward (1953). Various large-scale (see previous section) and small-scale experiments (Hawkins 1981*a*, 1981*b*, 1983; Hawkins and Hartnoll 1982*a*, 1982*b*), coupled with detailed monitoring of fixed areas (Hartnoll and Hawkins 1980, 1985; Hawkins and Hartnoll 1983*a*), have resulted in this community being well understood (qualitative models proposed in Hawkins and Hartnoll 1983*b*, refined in Hartnoll and Hawkins 1985). The patches form mosaics which show considerable temporal variation in the abundance of fucoids, barnacles, and limpets. A complex suite of both positive and negative interactions occurs between *Fucus vesiculosus*, *Patella vulgata*, *Semibalanus balanoides*, and various opportunist ephemeral algae (see Fig. 1.3). *Nucella* is not numerous on this shore, probably because of lack of refuges. Mussels are almost completely absent. The virtual absence of *Littorina littorea* is perhaps due to lack of larval supply (see preliminary data in Norton *et al.* 1990). Studies on the Orkney Islands (Baxter *et al.* 1985) showed similar but less rapid fluctuations, with an ordered sequence of events and an indeterminate time-scale.

Patchy moderately exposed shores

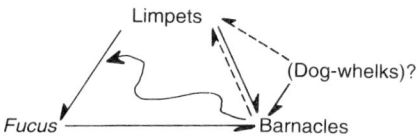

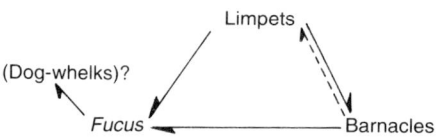

Fig. 1.3. Directions of positive (e.g., provision of shelter, removal of competitors) and negative interactions (feeding, sweeping) in the fucoid–barnacle–limpet mosaic on a moderately exposed rocky shore on the Isle of Man. Strong interactions: solid line; weak interactions: dotted line. Wavy line is where one interaction modifies another. The role of dog-whelks is unclear and under investigation.

Clumps of *Fucus* can be induced in barnacle-dominated areas by experimental exclusion of *Patella* (Hawkins 1981a). If limpets are experimentally removed from patchy areas, virtually complete cover of fucoids can result (Hawkins 1981b), although algal recruitment is slowed within the radius of sweep of fucoid clumps. Patches of ephemeral algae and clumps of fucoids arise naturally in localized areas of low or zero limpet density. The causes of these areas of reduced grazing intensity probably include mortality due to sea-bird predation (see Feare and Summers 1985) and possibly physical disturbance from storm-strewn boulders and pebbles (Southward 1956, 1964). The natural tendency of limpets to aggregate also leaves low intensity grazing areas between groups (e.g., Hartnoll and Hawkins 1985). Escapes by algae from *Patella* grazing are more likely amongst dense clusters of barnacles (Burrows and Lodge 1950; Hawkins 1981a, 1981b; see Lubchenco 1983 and Petraitis 1983 for similar work with *Littorina* in New England). New germlings are safe from larger grazers amongst the barnacle matrix, although the tiny but numerous *Littorina neglecta* which dwell in dead barnacle shells are thought to be important grazers (Hawkins 1981a). The complex surface may also modify the physical environment, which facilitates settlement and early survival of *Fucus*. Thus escapes of fucoids are more likely to occur in years following heavy barnacle settlement, or in areas with localized heavy settlement.

The composition and sequence of algae which colonize limpet exclusion areas or naturally lightly grazed areas vary with season (Hawkins 1981a). On barnacle-covered rock, diatoms may be initial colonizers in autumn to late spring, followed by various ephemeral algae such as *Enteromorpha, Blidingia*, and *Porphyra*, which are eventually outgrown by fucoids. In the summer, fucoids can directly colonize barnacles without any preliminary stages, although sequences of colonization vary between years (Hawkins and Hartnoll, unpublished observations). Hawkins (1981a) also recorded marked qualitative and quantitative differences in the algae colonizing barnacle-covered rock and scraped-bare areas from which limpets had been excluded. Ephemeral algae always grew and persisted on the initially bare rock, and subsequent increases in *Fucus* cover were slow. On the barnacles, some very fine ephemerals such as *Ulothrix* were not seen, and other ephemerals were less dense because they were quickly replaced by *Fucus* which colonized the barnacles very rapidly. This indicates the importance of the role of barnacles in modifying grazing interactions by either providing a rough substrate or a habitat for vast numbers of *Littorina neglecta* (see Hawkins 1981b; Hawkins and Hartnoll 1983b for detailed discussion). This work has been rightly criticized by Jernakoff (1985) because of lack of replication, but has been recently verified by

Hill (1990), with a replicated experimental design.

The microbial communities within patchy mosaics have also been recently examined (Hill 1990; Hill and Hawkins 1990, 1991). The microbial community is also highly patchy, and there are differences between areas covered by barnacles and rock bare of macrobiota. *Patella* grazing has a marked effect on microbial biomass and community composition, and helps to maintain a lawn of microscopic diatoms. Escapes of diatoms from grazing are enhanced by barnacles, particularly during the summer and autumn. Examination of diatoms in the *Patella* gut contents compared with adjacent epilithic communities suggest that *Patella* is a much more selective grazer than previously thought.

Patches of ephemeral algae do not usually persist long in the face of encroaching limpets. In contrast, once fucoids reach 5 cm or so, they are little affected by *Patella* grazing (Burrows and Lodge 1950; Hawkins and Hartnoll 1983a) and eventually clumps of *Fucus* become evident. Observations suggest that both adult and juvenile limpets aggregate rapidly under *Fucus* clumps and disperse much more slowly when the clumps disappear after two years or so. This generates areas of low grazing intensity nearby and can restart the cycle (Hawkins and Hartnoll 1983a; Hartnoll and Hawkins 1985, unpublished observations).

Extrinsically generated fluctuations in recruitment of important species such as *Patella* (e.g., Bowman and Lewis 1977), barnacles (e.g., Hawkins and Hartnoll 1982a; Kendall *et al*. 1985), and fucoids (e.g., Jones *et al*. 1979) can initiate and drive the cycles seen on these shores. Good recruitment of fucoids and barnacles will tend to promote escapes, as will poor limpet recruitment. Certain recruitment conditions, for example continued good limpet recruitment, can probably halt the cycle for a time at a particular stage. Thompson (1980) suggested that low and variable recruitment of *Patella vulgata* helps to initiate cyclical changes in fucoid cover on shores in Bantry Bay, Eire. The variable and sometimes low recruitment of barnacles to the isolated populations on the Isle of Man (see discussion of Hawkins and Hartnoll 1982a) and also of limpets (compare Hawkins and Hartnoll 1983b with Lewis and Bowman 1975) may amplify the fluctuations as compared with studies in the Orkneys and elsewhere.

Fluctuations in recruitment are particularly marked in species with a planktonic larval phase. Drifting larvae are subject to unpredictable stochastic factors, which subsequently affect settlement and ultimately recruitment to the breeding population. Winds can affect settlement of *Semibalanus balanoides* larvae in different ways on different coasts (reviewed by Kendall *et al*. 1985). Onshore winds can enhance (Hawkins and Hartnoll 1982a, Isle of Man) or reduce settlement

(Bennell 1981, Anglesey). No clear relationship has been found in other studies (e.g., Kendall *et al.* 1985). Poor development of the spring phytoplankton can badly affect *S. balanoides* recruitment (Barnes 1956). No doubt similar events occur with the less easily studied veliger larvae of gastropods and bivalves.

Since the above ideas were proposed in the mid-1980s, it has become apparent that *Nucella* has markedly declined in areas near ports and harbours, due to leachates from antifouling paints, which induce 'imposex' (Bryan *et al.* 1986): female dog-whelks develop male reproductive organs which block the female genital duct, leading to sterility (Gibbs *et al.* 1987), reduced recruitment, and low numbers (Spence *et al.* 1990). It has been suggested that this could lead to increased cover of *Fucus* on moderately sheltered shores, since denser barnacles resulting from reduced predation would promote escapes (Spence *et al.* 1990). No firm evidence exists for this assertion, but it is currently under investigation.

In summary, work on the Isle of Man and elsewhere has demonstrated that mosaic communities on moderately exposed shores are intrinsically unstable compared with those on more exposed or sheltered ones (see Southward and Southward 1978). Both intrinsic and extrinsic factors interact to affect the balance of the community. Generally, the extrinsic factors promote change and intrinsic factors keep fluctuations within broad but contained limits. A dynamic equilibrium exists along the wave-exposure gradient, from barnacles, mussels, and limpets, in exposed conditions, to fucoids in sheltered conditions. At the point of balance on patchy intermediate shores, fluctuations occur prompted by irregular predation, and, more importantly, by variation in recruitment.

Vertical gradients: high shore to subtidal

To examine interactions between plants and animals along the vertical environmental gradient, we consider here a hypothetical band on a moderately exposed shore, from the splash zone down to the limit of light penetration in the subtidal zone. Information from more sheltered locations is also included where relevant.

1. The intertidal zone

There is virtually no published experimental work on plant–animal interactions high on the shore in the littoral fringe in the north-east Atlantic. The comments below are therefore speculative. The high-shore flora has a marked seasonal pattern, thriving in the winter and being bleached and killed each spring or summer. At higher latitudes it may persist for much of the summer. Competitive interactions between plants and animals are few, as space-occupying animals are virtually absent. Observations and preliminary experiments (in Hawkins and Hartnoll 1983b), plus extrapolations from other regions (see Norton et al. 1990 for review), suggest that grazing by littorinids can be important. High-shore littorinids are, however, often restricted to pits and crevices. Localized bare haloes around these refuges are often seen. In areas with many refuges, numbers of littorinids are higher (Raffaelli and Hughes 1978), and grazing pressure would be expected to be greater. There is a clear need for experimental manipulation of this system, particularly to explore the interaction between seasonally varying activity patterns of littorinids and the ability of algae to proliferate. Our current hypothesis is that the direct effects of physical factors are of overriding importance at this high-shore level.

A quite dense band of *Fucus spiralis* occurs at the top of the eulittoral in sheltered conditions, and can even extend on to moderately exposed shores (Ballantine 1961a; Lewis 1964). Removal of the canopy high on the shore leads to a reduction in limpet numbers (Hawkins and Hartnoll 1983b). Higher recruitment of barnacles occurs under canopies of high-shore fucoids than in nearby cleared areas on sheltered shores (Hawkins 1983). Therefore the canopy can have beneficial sheltering effects on sub-canopy animals, presumably by reducing physical extremes.

Interactions in the mid-eulittoral of more exposed shores have been discussed previously. The grazing activity of limpets is probably the major interaction structuring such shores. Limpets prevent algal growth which would otherwise outcompete the barnacles. *Fucus* clumps that arise by escapes from grazing, reduce barnacle settlement (Hawkins 1983), but seem to enhance recruitment of *Patella vulgata*, *Nucella lapillus*, and *Actinia equina* (Hartnoll and Hawkins 1985). It has been suggested (Hawkins and Hartnoll 1983b) that the patchy and diffuse zonation on many moderately exposed shores of *Fucus spiralis*, *F. vesiculosus*, and *F. serratus* is the result of differential ability to escape limpet-grazing, due to interspecific differences in the growth rate of young fucoids (see Schonbeck and Norton 1978, 1980). The mixing of algal zones after the early large-scale removal experiments on the Isle of Man (see Southward 1964), and experimental induction of *Fucus serratus* above its normal level (Hawkins 1981b), strengthen this suggestion, but it needs to be formally tested in the field.

In the lower eulittoral, cover of canopy- and turf-forming algae increases and the influence of limpet-grazing wanes as limpets become less common (Hawkins 1981*b*). The canopy and sweeping effects of *Fucus serratus* can exert considerable influence at this level on barnacle and algal recruitment (see previous comments; also Hawkins and Harkin 1985). On some shores (e.g., Peveril Point, Swanage) there is a sharp cut-off point where limpets and barnacles stop and where begins a dense red algal turf that forms a virtually complete cover. Removal of limpets above this turf allowed it to occur higher on the shore, while repeated scraping of the turf allowed a slight downward extension of limpets (Hawkins, unpublished observations). Similarly, after large-scale kills of limpets by dispersant application following the *Torrey Canyon* oil spill, many low-shore algae were observed to extend higher up the shore (Southward and Southward 1978). Algal turves may also prevent low-shore extension of barnacles (Hawkins 1983). These experiments echo the work in Australia (Underwood 1980; Underwood and Jernakoff 1981) which first showed that there is a point low on the shore where the ability of algae to grow and cover the space exceeds the ability of gastropods to graze. A series of photographs stretching back over 30 years (Southward, unpublished data) and a resurvey in 1985 of Grubb's (1936) transects suggest that this point is remarkably stable, despite being an equilibrium point between biological interactions.

A recent comprehensive study by Janke (1990) has examined interactions in the rather atypical rocky intertidal of Helgoland, where *Patella* is virtually absent. High on the shore, littorinid grazers did not prevent the growth of ephemeral algae. Physical factors were considered the most important. In the mid-shore region, *Fucus* proliferated from 20 per cent cover to complete cover within two years in the absence of grazing. Competition with *Mytilus* (originally 50 per cent cover) prevented growth of ephemerals and fucoids directly on to the rock surface, but they grew upon the mussels. Low on the shore, in a region completely dominated by *Fucus serratus*, *Littorina littorea* grazing at normal densities could not prevent growth of *Fucus* germlings. When the numbers of these winkles were experimentally doubled they prevented *Fucus* recruitment. Predation by crabs also aided establishment of the fucoid canopy by removing competing mussels low on the shore. Even in the absence of limpets, it seems that physical factors exert considerable control high on the shore, giving way to grazing interactions at mid levels; with canopy interactions and the indirect effects of predation being of greatest importance low down.

The sublittoral fringe has a very dense canopy of the flexibly-stiped *Laminaria digitata*, which prevents colonization by ephemeral algae by sweeping the rock surface (Hawkins and Harkin 1985). Grazers are rare on the rock surface here, and we suspect that canopy interactions are the most important structuring agency.

2. The subtidal zone

That sessile animals are often in direct competition with macroalgae in the subtidal was first noticed by Kitching *et al.* (1934) in their classic study of a shallow rocky gully at Wembury Bay, near Plymouth, England. They identified two 'associations', one dominated by plants, on upward-facing surfaces, and one by animals, on overhangs. While this separation was true of the dominant organisms, some animals (such as *Pomatoceros*) did form part of the undergrowth on upward-facing rocks. Similarly, some red algae successfully competed on vertical or slightly overhanging surfaces.

It is now generally accepted that upward-facing rocks in shallow, clear water are dominated by algae which occur in several different layers, with a considerable total biomass per unit area. Many of the interactions in this zone will be between plants, with canopy effects being particularly important (Kain 1975; Hawkins and Harkin 1985). However, the undergrowth can include animals that are successful in their occupation of space. The *Laminaria hyperborea* forest at Carrigathorna in south-west Eire has 17 species of sessile animals in the understorey (Norton *et al.* 1977). Vertical surfaces in shallow water can support algae, but animals often assume a greater importance (Forster 1958; Hiscock and Hoare 1975), while, as noted by Kitching *et al.* (1934), overhangs are populated almost exclusively by animals (Forster 1958; Hiscock 1979). The quantitative change-over from plant to animal domination, with increasing angle, from upward-facing horizontal surfaces, has been shown by Chappell (in Hartnoll 1983*b*). Some animal biomass was present on surfaces at all angles, but plants were absent from near-horizontal, downward-facing surfaces.

The most obvious explanation for these differences in the success of plants and animals is the influence of light, essential for plant growth. It can be assumed that the amount of light reaching an overhang in shallow water can be as little as that reaching an upward-facing surface in much deeper water. This may not be the only explanation, however, as before algae can grow they first have to settle. Some animal larvae have been shown to possess behavioural repertoires which could take them to the undersides of rock (Crisp 1976, for review). Few species of algae have spores that exhibit a taxic response and those best fitted to survive in a shaded environment belong to the red algae, which are entirely devoid of motile spores. Sinking alone is likely to carry algal spores largely to upward-facing surfaces, but water motion seems to be far more important for propagule dispersal (Norton and Fetter 1981). Such turbulence might distribute propagules on to overhangs where they might adhere. Nonetheless, it seems that animal larvae may have a tactical colonizing advantage.

The most important grazers of subtidal algae in the region are the echinoids. The commonest species in Great Britain, *Echinus esculentus*, is less common in very shallow water, but becomes more abundant with depth and shelter. Forster (1959) was the first to realize that *Echinus* could destroy both sessile animals and algae. The quantitative effect on the plants must depend partly on grazer density. In many areas the density of *Echinus* generally does not exceed 1 animal m^{-2} (Forster 1959; Larsson 1968; Comely and Ansell 1988) and even at a little above this density there is little obvious effect on the vegetation, which in shallow water may include *Laminaria* forests (Vost 1985; Comely and Ansell 1988). However, Vost (1985) demonstrated that even at this low intensity, *Echinus* grazing can have some effects. She maintained the end of the ruined breakwater off Port Erin, Isle of Man, clear of *Echinus* for two years. After a year, the mean biomass of understorey plants was significantly greater than in the control area even in shallow water (1–4 m below LAT), where the *Echinus* density was 1·4 animals m^{-2}.

Higher densities of 4 to 5 *Echinus* m^{-2}, achieved at certain depths (Jones and Kain 1967; Vost 1985; Comely and Ansell 1988), can have a devastating effect on the plants, preventing the establishment of fleshy algae. The lower blocks and boulders of Port Erin breakwater are normally devoid of any but encrusting algae but after three years of repeated clearance of *Echinus* they became colonized by *Laminaria hyperborea* (Jones and Kain 1967). An intermediate density of 3·4 animals m^{-2} at around 5 m depth prevents a full *L. hyperborea* forest but allows opportunistic *Saccorhiza polyschides* to colonize (Jones and Kain 1967; Norton and Burrows 1969; Vost 1985). In deeper water (17–20 m below LAT), at another Isle of Man site, a similar density of 3·5 animals m^{-2} completely inhibited laminarian colonization until the animals were removed (Vost 1985). This agrees with the suggestion (Jones and Kain 1967) that the effect of grazing on algae is linked to depth: slower growth in weaker light makes the plants more vulnerable. Off the Isle of Man, population densities of *Echinus* have remained remarkably stable over nearly 20 years (Kain, unpublished observation). As a result, the lower limit of the *Laminaria* forest has changed little.

Other echinoids, although mainly smaller, can be just as destructive. Kitching and Ebling (1961) realized that *Paracentrotus lividus* could be controlling shallow subtidal algae in Lough Ine, south-west Ireland, and confirmed this by clearing an area and observing rapid algal colonization. The urchin is common in the Lough, occupying grazed patches at densities of 15 to 61 animals m^{-2}, but can vary very considerably from year to year (Kitching and Thain 1983).

Off the Norwegian coast, population densities of *Strongylocentrotus droebachiensis* have been by no means stable and increased dramatically about 10 years ago (Hagen 1983) in some northern areas. Extensive kelp beds were devastated, flourishing forests of *Laminaria hyperborea* being replaced by barren areas in which only crustose corallines survived. A mean density of 48 animals m^{-2} was recorded for one of these areas (Hagen 1983). This parallels events in the north-west Atlantic (see Vadas, Chapter 2, this volume).

While *Echinus esculentus* will eat calcareous crustose algae only if no other food is available, the gastropod *Acmaea virginea* feeds only on such crusts (although closely related species ignore them: Clokie and Norton 1974). *Acmaea virginea* also scrapes away at bivalve shells, feeding on the conchocelis phase of *Porphyra* when these filaments grow within the shells (Farrow and Clokie 1979).

More obvious grazing of fleshy subtidal algae is carried out by gastropods inhabiting their thalli. *Helcion* (as *Patina*) *pellucidum* is common on species of *Laminaria* (Graham and Fretter 1947) and *Saccorhiza* (Ebling *et al.* 1948; Norton 1971). The animals start sedentary life on the blades, eating into the surface and causing shallow cavities. On *L. hyperborea* they may later migrate down the stipe, crawl into the holdfast, and gouge out cavities larger than themselves. In some sites in the British Isles, around half the older plants can be infested in this way, endangering their attachment (Kain 1963, 1977). In Norway, however, although the animals migrate proximally (Vahl 1972), they remain on the blades, which may also be badly damaged (Kain and Svendsen 1969). Seasonally, the population density is maximal when the *L. hyperborea* blades have their maximum dry-matter content (Vahl 1971). Although grazing by *Helcion* is likely to affect the population dynamics of the kelp, it is unlikely to be a major influence on community structure.

In deeper water, animals gradually become more important than plants (Forster 1958; Castric-Fey *et al.* 1973; Hiscock 1979). With increasing depth, plants first disappear from steeper angles, and later from horizontal surfaces (Forster 1955; 1961).

Latitudinal changes

It is difficult to assess how interactions between plants and animals vary throughout the region. Very few workers have related the distribution of herbivores to that of seaweeds on a broad geographical scale, although correlations are apparent on less extensive stretches of coastline (Arrontes and Anadon 1990). No regionally integrated experimental studies have been attempted.

In the north-east Atlantic, even allowing for variations in the intensity of collecting, several trends in the algal flora are apparent (Table 1.3). From north to south there is a marked change in species richness. The preponderance of brown algae gives way to an abundance of red algae. The proportion that are calcified and/or crustose increases. Changes in the flora are even more dramatic if only the dominant species are considered.

Ephemeral green algae, such as *Ulva* and *Enteromorpha*, are ubiquitous. To the north, several species of fucoids (*Fucus* spp., *Pelvetia*, and *Ascophyllum*) blanket shores. These become increasingly restricted to shelter at lower latitudes (Ballantine 1961), with only *Fucus spiralis* occurring in the far south. From southern Portugal southwards, the majority of seaweeds are confined to low on the shore (e.g., Lawson and Norton 1971) or to the subtidal (e.g., Saldanha 1974). Although species of *Sargassum*, *Cystoseira*, and *Saccorhiza polyschides* may form zones, the most striking feature is a dense turf of low-shore red algae on either side of low water (e.g., Lawson and Norton 1971; Saldanha 1974; Hawkins et al. 1990).

Diversity and biomass of grazers has not been studied at some locations throughout the region. From the faunistic literature and early biogeographical studies (e.g., Fischer-Piette 1935; Crisp and Fischer-Piette 1959), it is clear that the diversity of major grazers is greater in southern Europe and North Africa than in the far north. It seems likely that the balance between limpet grazing and fucoid growth along the wave-exposure gradient changes with latitude (Ballantine 1961a). Whereas limpets are effective grazers in most of the region, recent work by Kupfer (unpublished observation) suggests that limpets are less effective in northern Norway. Further south, the presence of a multiplicity of grazers suggests that escapes of seaweeds from grazing are much less likely to occur. We hypothesize that if recruitment of one species of limpet or trochid fails or is poor, then others might take its place to maintain grazing pressure, particularly as the diet of a suite of grazers has been shown to be very similar (Hawkins et al. 1989).

Plant growth is mainly light-limited, and grazer foraging activity is mainly temperature dependent. In the far north, both factors will be highly seasonal and largely synchronized. In the south of the region, algal production and grazing activity will continue throughout the year. We believe that at intermediate latitudes (e.g., the British Isles) there will be some asynchrony. Plant growth is light-dependent and increases in late winter–early spring. Foraging activity, particularly of tide-in foragers (such as limpets), will be temperature dependent and lag

Table 1.3. The seaweed flora of various sites in the north-east Atlantic (data from various sources).

Site	Latitude	Total number of species recorded	Ratio Rhodophyta Phaeophyta	Percentage calcareous	Percentage crustose
Spitzbergen	76°–80°N	67	0.6	1.5	6.0
Finnmark, N. Norway	70°–71°N	201	0.7	4.5	7.5
Faeroes	62°N	213	1.1	5.5	8.0
British Isles	50°–60°N	679	1.8	5.0	9.0
Roscoff, N.W. France	48° 40′N	547	1.8	5.3	7.1
Portugal	37° 30′–42°N	404	2.5	6.9	9.4
Canary Islands	27°–29°N	340	3.7	13.3	10.9

behind the plants. This could account for the flush of ephemeral algae seen on shores in the late winter–early spring.

Regional changes in the vegetation may have far more influence on grazers than simply a change in menu. They also entail a qualitative and quantitative change in the anti-herbivore defences of the dominant algae. The brown algae, especially the fucoids, contain phlorotannins. These polyphenolic compounds effectively discourage many benthic grazers (Ragan and Glombitza 1986). Northern fucoids, such as *Ascophyllum nodosum*, are particularly rich in the more astringent polyphenols and are most repellent to grazing littorinids (Norton *et al.* 1990; Norton and Manley 1990). The red algae that dominate the vegetation in the south are protected by acetylene-containing lipids (Paul and Fenical 1980) and a variety of halogenated compounds, especially brominated phenols (e.g., Fenical 1975). While there are some differences in the suites of grazing species between the north and the south of the region, broadly distributed species (e.g., *Patella vulgata*, *Littorina littorea*) will encounter differing algal defences in different parts of their range. For a review of biochemical defence mechanisms in marine algae, see Hay and Fenical (Chapter 14 this volume).

Concluding remarks

Very little of the macroalgal production in the north-east Atlantic flows through herbivores. The consumption of the microbial film, as well as being the main energy source for many grazers, is a major influence on algal recruitment, distribution patterns, and hence community structure.

The balance between the ability of plants or animals to dominate an area is determined by the major environmental gradients. In exposed conditions and in southern Europe, grazers and sessile animals dominate most of the eulittoral zone. The balance is tilted in favour of plants, with increasing shelter and at higher latitudes in the mid-shore region. The region immediately above and below low water seems to be structured by the competitive interactions of either algal turfs dominating space on the rock surface or large kelps forming canopies. Deeper down, grazing can be important, but is variable. With attenuating light, animals eventually dominate at greater depths. We consider that although biological interactions directly structure many communities, the intensities and directions of these interactions are shaped by underlying physical factors. The types of interaction also depend on the species present, which in turn reflects the biogeography of the area.

Much work still needs to be done: many of the experiments described (including our own work) do not conform to current methodological standards (see comments by Underwood, Chapter 24 this volume). There is the need to repeat some of the work using properly replicated factorial designs, to determine interactions between the various components of the system under study. There are still some glaring gaps in our knowledge that require specific studies if they are to be bridged. For example, productivity of the microbial film needs to be determined; experiments are needed on high-shore littorinid grazing; manipulative experiments are required to explore interactions on exposed mussel-dominated shores; much more experimentation is required on subtidal communities; the effects of microhabitat variation in locally mediating interactions need to be examined. Of greatest importance is the need for an integrated programme throughout the region to examine experimentally the importance of plant–animal interactions in structuring communities. This needs to be co-ordinated to ensure that similar methodologies are used throughout. The almost continuous nature of rocky coastline from the Arctic to the Straits of Gibraltar is ideal for work of this nature. The results would be even more interesting if the other side of the Atlantic could be incorporated into the programme.

Acknowledgements

A large proportion of this work has been funded by NERC in various ways. Paul Fernandes produced Figs. 1.1 and 1.3.

References

Arrontes, J. and Anadon, R. (1990). Distribution of intertidal isopods in relation to geographical changes in macroalgal cover in the Bay of Biscay. *Journal of the Marine Biological Association of the United Kingdom*, **70**, 283–94.

Audouin, J.V. and Milne-Edwards, H. (1832). *Recherches pour servir à l'histoire naturelle du littoral de la France* **1**. Crochard, Paris.

Ballantine, W.J. (1961a). A biologically-defined exposure scale for the comparative description of rocky shores. *Field Studies*, **1**, 1–17.

Ballantine, W.J. (1961b). The population dynamics of *Patella vulgata* and other limpets. Unpublished Ph.D Thesis. Queen Mary College, University of London.

Barnes, H. (1956). *Balanus balanoides* (L.) in the Firth of Clyde: the development and annual variation in the larval population and the causative factors. *Journal of Animal Ecology*, **25**, 72–84.

Barnes, H. and Topinka, J.A. (1969). Effect of the nature of the substratum on the force required to detach a common littoral alga. *American Zoologist*, **9**, 753–8.

Barnes, H., Barnes, M., and Finlayson, D. M. (1963). The seasonal changes in body weight, biochemical composition and oxygen uptake of two common boreo-arctic cirripedes, *Balanus balanoides* and *B. balanus*. *Journal of the Marine Biological Association of the United Kingdom*, **43**, 185–211.

Baxter, J. M., Jones, A. M., and Simpson, J. A. (1985). A study of long-term changes in some rocky shore communities in Orkney. *Proceedings of the Royal Society of Edinburgh*, **B75**, 47–63.

Bennell, S. J. (1981). Some observations on the littoral barnacle populations of North Wales. *Marine Environmental Research*, **5**, 227–40.

Blinks, L. R. (1955). Photosynthesis and productivity of littoral marine algae. *Journal of Marine Research*, **14**, 363–73.

Bowman, R. S. (1981). The morphology of *Patella* spp. juveniles in Britain, and some phylogenetic inferences. *Journal of the Marine Biological Association of the United Kingdom*, **61**, 647–66.

Bowman, R. S. and Lewis, J. R. (1977). Annual fluctuations in the recruitment of *Patella vulgata* L. *Journal of the Marine Biological Association of the United Kingdom*, **57**, 793–815.

Bryan, G. W., Gibbs, P. E., Hummerstone, L. G., and Burt, G. R. (1986). The decline of the gastropod *Nucella lapillus* around south-west England: evidence for the effect of tributyltin from antifouling paints. *Journal of the Marine Biological Association of the United Kingdom*, **67**, 525–44.

Burrows, E. M. and Lodge, S. M. (1950). Note on the interrelationships of *Patella*, *Balanus* and *Fucus* on a semi-exposed coast. *Report of the Marine Biological Station, Port Erin*, **62**, 30–34.

Castric-Fey, A. Girard-Descatoire, A., Lafargue, F., and L'Hardy-Halos, M.-T. (1973). Etagement des algues et des invertébrés sessiles dans l'Archipel de Glénan. *Helgoländer Wissenschaftlichen Meeresuntersuchungen*, **24**, 490–509.

Clokie, J. J. P. and Boney, A. D. (1980). The assessment of changes in intertidal ecosystems following major reclamation work: framework for interpretation of algal-dominated biota and the use and misuse of data. In *The shore environment* Vol. 2: *Ecosystems*. Systematics Association Special Volume No. 17, (ed. J. H. Price, D. E. G. Irvine, and W. F. Farnham), pp. 609–75. Academic Press, London and New York.

Clokie, J. J. P. and Norton, T. A. (1974). The effects of grazing on the algal vegetation of pebbles from the Firth of Clyde. *British Phycological Journal*, **9**, 216.

Comely, C. A. and Ansell, A. D. (1988). Population density and growth of *Echinus esculentus* L. on the Scottish west coast. *Estuarine, Coastal and Shelf Science*, **27**, 311–34.

Crisp, D. J. (1976). Settlement responses in marine organisms. In *Adaptations to environment: essays on the physiology of marine animals*, (ed. R. C. Newell), pp. 83–124. Butterworths, London.

Crisp, D.J. and Fischer-Piette, E. (1959). Reparation des principales espèces intercotidals de la côte atlantique française en 1954-55. *Annales Institute d'Oceanographique Monaco*, **36**, 275-388.

Cross, T.F. and Southgate, T. (1980). Mortalities of fauna of rocky substrates in South-west Ireland associated with the occurrence of *Gyrodinium aureolum* blooms during autumn 1979. *Journal of the Marine Biological Association of the United Kingdom*, **60**, 1071-3.

Davies, M.S., Hawkins, S.J., and Jones, H.D. (1990). Mucus production and physiological energetics in *Patella vulgata* L. *Journal of Molluscan Studies*, **56**, 499-503.

Dayton, P.K. (1971). Competition, disturbance and community organization: the provision and subsequent utilization of space in a rocky intertidal community. *Ecological Monographs*, **41**, 351-89.

Ebling, F.J., Kitching, J.A., Purchon, R.D., and Bassindale, R. (1948). The ecology of the Lough Ine rapids with special reference to water currents 2. The fauna of the *Saccorhiza* canopy. *Journal of Animal Ecology*, **17**, 223-44.

Farrow, G.E. and Clokie, J. (1979). Molluscan grazing of sublittoral algal-bored shells and the production of carbonate mud in the Firth of Clyde, Scotland. *Transactions of the Royal Society of Edinburgh*, **70**, 139-48.

Feare, C.J. and Summers, R. (1985). Birds as predators on rocky shores. In *The ecology of rocky coasts*, (ed. P.G. Moore and R. Seed), pp. 249-64. Hodder and Stoughton, Sevenoaks, Kent.

Fenical, W. (1975). Halogenation in the Rhodophyta: a review. *Journal of Phycology*, **11**, 243-59.

Fischer-Piette, E. (1935). Systématique et biogéographie. Les Patelles d'Europe et d'Afrique du Nord. *Journal de Conchyliologie*, **79**, 5-66.

Forster, G.R. (1955). Underwater observations on rocks off Stoke Point and Dartmouth. *Journal of the Marine Biological Association of the United Kingdom*, **34**, 197-9.

Forster, G.R. (1958). Underwater observations on the fauna of shallow rocky areas in the neighbourhood of Plymouth. *Journal of the Marine Biological Association of the United Kingdom*, **37**, 473-82.

Forster, G.R. (1959). The ecology of *Echinus esculentus* L. Quantitative distribution and rate of feeding. *Journal of the Marine Biological Association of the United Kingdom*, **38**, 361-7.

Forster, G.R. (1961). An underwater survey on the Lulworth Banks. *Journal of the Marine Biological Association of the United Kingdom*, **41**, 157-60.

Gibbs, P.E., Bryan, G.W., Pascoe, P.L., and Burt, G.R. (1987). The use of the dogwhelk, *Nucella lapillus*, as an indicator of Tributyltin (TBT) contamination. *Journal of the Marine Biological Association of the United Kingdom*, **67**, 507-23.

Graham, A. and Fretter, V. (1947). The life history of *Patina pellucida* (L.).

Journal of the Marine Biological Association of the United Kingdom, **26**, 590-601.

Grubb, V. M. (1936). Marine algal ecology and the exposure factor at Peveril Point, Dorset. *Journal of Ecology*, **24**, 392-423.

Hagen, N. T. (1983). Destructive grazing of kelp beds by sea urchins in Vestfjorden, northern Norway. *Sarsia*, **68**, 177-90.

Hartnoll, R. G. (1983a). Bioenergetics of a limpet-grazed intertidal community. *South African Journal of Science*, **79**, 166-7.

Hartnoll, R. G. (1983b). Substratum. In *Sublittoral ecology. The ecology of the shallow sublittoral benthos*, (ed. R. Earll and D. G. Erwin), pp. 97-124. Oxford University Press, Oxford.

Hartnoll, R. G. and Hawkins, S. J. (1980). Monitoring rocky shore communities: a critical look at spatial and temporal variation. *Helgoländer Wissenschaftlichen Meeresuntersuchungen*, **33**, 484-95.

Hartnoll, R. G. and Hawkins, S. J. (1985). Patchiness and fluctuations on moderately exposed rocky shores. *Ophelia*, **24**, 53-63.

Hatton, H. (1938). Essais de bionomie explicative sur quelques espéces intercotidals d'algues et d'animaux. *Annales Institute d'Oceanographique, Monaco*, **17**, 241-348.

Hawkins, S. J. (1979). Field studies on Manx rocky shore communities. Unpublished Ph.D. Thesis. University of Liverpool.

Hawkins, S. J. (1981a). The influence of *Patella* on the fucoid/barnacle mosaic on moderately exposed rocky shores. *Kieler Meeresforschungen*, **5**, 537-43.

Hawkins, S. J. (1981b). The influence of season and barnacles on the algal colonisation of *Patella vulgata* exclusion areas. *Journal of the Marine Biological Association of the United Kingdom*, **61**, 1-15.

Hawkins, S. J. (1983). Interaction of *Patella* and macroalgae with settling *Semibalanus balanoides* (L.). *Journal of Experimental Marine Biology and Ecology*, **71**, 55-72.

Hawkins, S. J. and Harkin, E. (1985). Experimental canopy removal in algal dominated communities low on the shore and in the shallow subtidal. *Botanica Marina*, **28**, 223-30.

Hawkins, S. J. and Hartnoll, R. G. (1982a). Settlement patterns of *Semibalanus balanoides* (L.) in the Isle of Man. *Journal of Experimental Marine Biology and Ecology*, **62**, 271-83.

Hawkins, S. J. and Hartnoll, R. G. (1982b). The influence of barnacle cover on the numbers, growth and behaviour of *Patella vulgata* L. on an artificial pier. *Journal of the Marine Biological Association of the United Kingdom*, **62**, 855-67.

Hawkins, S. J. and Hartnoll, R. G. (1983a). Changes in a rocky shore community: an evolution of monitoring. *Marine Environmental Research*, **9**, 131-81.

Hawkins, S. J. and Hartnoll, R. G. (1983b). Grazing of intertidal algae by marine invertebrates. *Oceanography and Marine Biology Annual Review*, **21**, 195-282.

Hawkins, S.J., Burnay, L.P., Neto, A.I., Tristão da Cunha, R., and Frias Martins, A.M. (1990). A description of the zonation patterns of molluscs and other important biota on the South coast of São Miguel, Azores. *Açoreana, Supplement*, 21-38.

Hawkins, S.J., Watson, D.C., Hill, A.S., Harding, S.P., Kyriakides, M.A., Hutchinson, S., and Norton, T.A. (1989). A comparison of feeding mechanisms in microphagous herbivorous intertidal prosobranchs in relation to resource partitioning. *Journal of Molluscan Studies*, **55**, 151-65.

Hill, A.S. (1990). The grazing of microbial films on moderately exposed shores on the Isle of Man. Unpublished Ph.D. Thesis. University of Liverpool.

Hill, A.S. and Hawkins, S.J. (1990). An investigation of methods for sampling microbial films on rocky shores. *Journal of the Marine Biological Association of the United Kingdom*, **70**, 77-88.

Hill, A.S. and Hawkins, S.J. (1991). Seasonal and spatial variation of epilithic microalgal distribution and abundance and its ingestion by *Patella vulgata* on a moderately exposed rocky shore. *Journal of the Marine Biological Association of the United Kingdom*, **71**, 403-25.

Hiscock, K. (1979). A survey of sublittoral habitats and species in the region of Padstow, Cornwall. *Progress in Underwater Science*, **4**, 121-30.

Hiscock, K. (1983). Water movement. In *Sublittoral ecology. The ecology of the shallow sublittoral benthos*, (ed. R. Earll and D.G. Erwin), pp. 58-96. Oxford University Press, Oxford.

Hiscock, K. and Hoare, R. (1975). The ecology of sublittoral communities at Abereiddy Quarry, Pembrokeshire. *Journal of the Marine Biological Association of the United Kingdom*, **55**, 833-64.

Janke, K. (1990). Biological interactions and their role in the community structure in the rocky intertidal of Helgoland (German Bight, North Sea). *Helgoländer Meeresuntersuchungen*, **44**, 219-63.

Jansson, A.-M., Kautsky, N., Oertzen, J.-A. von, Schramm, W., Sjöstedt, B., Wachenfeldt, T. von, and Wallentinus, I. (1982). Structural and functional relationships in a southern Baltic *Fucus* ecosystem. *Contributions of the Askö Laboratory*, **28**, 1-95.

Jernakoff, P. (1985). An experimental evaluation of the influence of barnacles, crevices and seasonal patterns of grazing and algal diversity and cover in an intertidal zone. *Journal of Experimental Marine Biology and Ecology*, **88**, 287-302.

Jones, N.S. (1946). Browsing of *Patella. Nature*, **158**, p. 557.

Jones, N.S. (1948). Observations and experiments on the biology of *Patella vulgata* at Port St. Mary, Isle of Man. *Proceedings and Transactions of the Liverpool Biological Society*, **56**, 60-77.

Jones, N. S. and Kain, J. M. (1967). Subtidal algal colonization following the removal of *Echinus*. *Helgoländer Wissenschaftlichen Meeresuntersuchungen*, **15**, 160–6.

Jones, W. E., Fletcher, A., Bennell, S. J., McConnell, B. J., and Mack-Smith, S. (1979). Changes in littoral populations as recorded by long term surveillance. In *Cyclic phenomena in marine plants and animals*, (ed. E. Naylor and R. G. Hartnoll), pp. 93–100. Pergamon Press, Oxford.

Kain, J. M. (1963). Aspects of the biology of *Laminaria hyperborea* II. Age, weight and length. *Journal of the Marine Biological Association of the United Kingdom*, **43**, 129–51.

Kain, J. M. (1975). Algal recolonization of some cleared subtidal areas. *Journal of Ecology*, **73**, 739–65.

Kain, J. M. (1977). Aspects of the biology of *Laminaria hyperborea* X. The effect of depth on some populations. *Journal of the Marine Biological Association of the United Kingdom*, **57**, 587–607.

Kain, J. M. and Svendsen, P. (1969). A note on the behaviour of *Patina pellucida* in Britain and Norway. *Sarsia*, **38**, 25–30.

Kendall, M. A., Bowman, R. S., Williamson, P., and Lewis, J. R. (1985). Annual variation in the recruitment of *Semibalanus balanoides* on the North Yorkshire coast 1969–1981. *Journal of the Marine Biological Association of the United Kingdom*, **65**, 1009–30.

Kitching, J. A. and Ebling, F. J. (1961). The ecology of Lough Inc XI. The control of algae by *Paracentrotus lividus* (Echinoidea). *Journal of Animal Ecology*, **30**, 373–83.

Kitching, J. A. and Ebling, F. J. (1967). Ecological studies at Lough Ine. *Advanced Ecological Research*, **4**, 198–291.

Kitching, J. A. and Thain, V. M. (1983). The ecological impact of the sea urchin *Paracentrotus lividus* (Lamarck) in Lough Ine, Ireland. *Philosophical Transactions of the Royal Society of London*, Series B, **300**, 513–52.

Kitching, J. A., Macan, T. T., and Gilson, H. C. (1934). Studies in sublittoral ecology. I. A submarine gully in Wembury Bay, South Devon. *Journal of the Marine Biological Association of the United Kingdom*, **19**, 677–705.

Larsson, B. A. S. (1968). SCUBA-studies on vertical distribution of Swedish rocky-bottom echinoderms. A methodological study. *Ophelia*, **5**, 137–56.

Lawson, G. W. and Norton, T. A. (1971). Some observations on littoral and sublittoral zonation at Teneriffe (Canary Isles). *Botanica Marina*, **14**, 116–20.

Lein, T. E. (1984). A method for the experimental exclusion of *Littorina littorea* L. (Gastropoda) and the establishment of fucoid germlings in the field. *Sarsia*, **69**, 83–6.

Lewis, J. R. (1964). *The ecology of rocky shores*. The English Universities Press, London.

Lewis, J. R. and Bowman, R. S. (1975). Local habitat-induced variations in the population dynamics of *Patella vulgata* L. *Journal of Experimental Marine Biology and Ecology*, **17**, 165-204.

Lodge, S. M. (1948). Algal growth in the absence of *Patella* on an experimental strip of foreshore, Port St. Mary, Isle of Man. *Proceedings and Transactions of the Liverpool Biological Society*, **56**, 78-83.

Lubchenco, J. (1983). *Littorina* and *Fucus*: effects of herbivores, substratum heterogeneity and plant escapes during succession. *Ecology*, **64**, 1116-23.

Menge, B. A. (1976). Organization of the New England rocky intertidal community: role of predation, competition and environmental heterogeneity. *Ecological Monographs*, **46**, 355-93.

Menge, B. A. (1978). Predation intensity in a rocky intertidal community: effect of an algal canopy, wave action and desiccation on predator foraging rates. *Oecologia* (Berlin), **34**, 17-35.

Menge, B. A. and Sutherland J. P. (1976). Species diversity gradients: synthesis of the roles of predation, competition, and temporal heterogeneity. *American Naturalist*, **110**, 351-69.

Miller, R. J., Mann, K. H., and Scarratt, D. J. (1971). Production potential of a seaweed-lobster community in eastern Canada. *Journal of the Fisheries Research Board of Canada*, **28**, 1733-8.

Niell, F. X. (1977). Rocky intertidal benthic systems in temperate seas: a synthesis of their functional performances. *Helgoländer Wissenschaftlichen Meeresuntersuchungen*, **30**, 315-33.

Norton, T. A. (1971). An ecological study of the fauna inhabiting the sublittoral marine alga *Saccorhiza polyschides*. *Hydrobiologia*, **37**, 215-31.

Norton, T. A. and Burrows, E. M. (1969). Studies on marine algae of the British Isles. 7. *Saccorhiza polyschides* (Lightf.) Batt. *British Phycological Journal*, **4**, 19-53.

Norton, T. A. and Fetter, R. (1981). The settlement of *Sargassum muticum* propagules in stationary and flowing water. *Journal of the Marine Biological Association of the United Kingdom*, **61**, 929-40.

Norton, T. A. and Manley, N. L. (1990). The characteristics of algae in relation to their vulnerability to grazing snails. In *Behavioural mechanisms of food selection*, (ed. R. N. Hughes) NATO Advanced Study Institute Series, **G20**, pp. 461-78. Springer-Verlag, Berlin.

Norton, T. A., Hiscock, K., and Kitching, J. A. (1977). The ecology of Lough Ine XX. The *Laminaria* forest at Carrigathorna. *Journal of Ecology*, **65**, 919-41.

Norton, T. A., Hawkins, S. J., Manley, N. L., Williams, G. A., and Watson, D. C. (1990). Scraping a living: a review of littorinid grazing. *Hydrobiologia*, **193**, 117-38.

Paine, R. T. and Vadas, R. L. (1969). Calorific values of benthic marine algae and their postulated relation to invertebrate food preference. *Marine Biology*, **4**, 79-86.

Paul, V. and Fenical, W. (1980). Toxic acetylene containing lipids from the red alga *Liagora farinosa* Lamouroux. *Tetrahedron Letters*, **21**, 3327-30.

Petraitis, P. S. (1983). Grazing patterns of the periwinkle and their effect on sessile intertidal organisms. *Ecology*, **64**, 522-33.

Raffaelli, D. G. and Hughes, R. N. (1978). The effect of crevice size and availability on populations of *Littorina rudis* and *Littorina neritoides*. *Journal of Animal Ecology*, **47**, 71-83.

Ragan, M. A. and Glombitza, K.-W. (1986). Phlorotannins, brown alga polyphenols. *Progress in Phycological Research*, **4**, 129-241.

Ramus, J., Lemons, F., and Zimmerman, C. (1977). Adaptation of light-harvesting pigments to downwelling light and the consequent photosynthetic performance of the eulittoral rockweeds *Ascophyllum nodosum* and *Fucus vesiculosus*. *Marine Biology*, **42**, 293-303.

Russell, G. and Fielding, A. J. (1981). Individuals, populations and communities. In *Biology of seaweeds*, (ed. C. S. Lobban and M. J. Wynne), Botanical Monograph 17, pp. 393-420. Blackwell, London.

Saldanha, L. (1974). Estudo do povoamento dos horizontes superiores da rocha litoral da Costa da Arrábida (Portugal). *Arquivos do museu Bocage*, Series 2, Vol. **V** no. 1.

Savidge, C. and Kain, J. M. (1990). Productivity of the Irish Sea. In *The Irish Sea. Part 3. Exploitable living resources*, (ed. T. A. Norton and A. J. Geffen), pp. 9-43.

Schonbeck, M. W. and Norton, T. A. (1978). Factors controlling the upper limits of fucoid algae. *Journal of Experimental Marine Biology and Ecology*, **31**, 303-13.

Schonbeck, M. W. and Norton, T. A. (1980). Factors controlling the lower limits of fucoid algae on the shore. *Journal of Experimental Marine Biology and Ecology*, **43**, 131-50.

Southgate, T., Wilson, K., Cross, T. F., and Myers, A. A. (1984). Recolonisation of a rocky shore in South-west Ireland following a toxic bloom of the Dinoflagellate *Gyrodinium aureolum*. *Journal of the Marine Biological Association of the United Kingdom*, **64**, 485-92.

Southward, A. J. (1953). The ecology of some rocky shores in the Isle of Man. *Proceedings and Transactions of the Liverpool Biology Society*, **59**, 1-50.

Southward, A. J. (1956). The population balance between limpets and seaweeds on wave-beaten rocky shores. *Report of the Marine Biological Station, Port Erin*, **67**, 20-29.

Southward, A. J. (1958). The zonation of animals and plants on rocky sea shores. *Biological Reviews*, **33**, 137-77.

Southward, A. J. (1964). Limpet grazing and the control of vegetation on rocky shores. In *Grazing in terrestrial and marine environments*, (ed. D. J. Crisp), pp. 265-73. Blackwell, Oxford.

Southward, A. J. (1979). Cyclic fluctuations in population density during eleven years recolonisation of rocky shores in West Cornwall following the 'Torrey Canyon' oil-spill in 1967. In *Cyclic phenomena in marine plants and animals*, (ed. E. Naylor and R. G. Hartnoll), pp. 85-92. Pergamon Press, Oxford.

Southward, A.J. and Southward, E. C. (1978). Recolonization of rocky shores in Cornwall after use of toxic dispersants to clean up the Torrey Canyon spill. *Journal of the Fisheries Research Board of Canada*, **35**, 682-706.

Spence, S. K., Bryan, G. W., Gibbs, P. E., Masters, D., Morris, L., and Hawkins, S.J. (1990). Effects of TBT contamination on *Nucella* populations. *Functional Ecology*, **4**, 425-32.

Thompson, G. B. (1980). Population dynamics of the limpet *Patella vulgata* L. in Bantry Bay. *Journal of Experimental Marine Biology and Ecology*, **45**, 173-217.

Underwood, A.J. (1980). The effects of grazing by gastropods and physical factors on the upper limits of distribution of intertidal macroalgae. *Oecologia* (Berlin), **46**, 201-13.

Underwood, A.J. and Jernakoff, P. (1981). Interactions between algae and grazing gastropods in the structure of a low-shore algal community. *Oecologia* (Berlin), **48**, 221-33.

Vahl, O. (1971). Growth and density of *Patina pellucida* (L.) (Gastropoda: Prosobranchia) on *Laminaria hyperborea* (Gunnerus) from Western Norway. *Ophelia*, **9**, 31-50.

Vahl, O. (1972). On the position of *Patina pellucida* (L.) (Gastropoda) on the frond of *Laminaria hyperborea*. *Ophelia*, **10**, 1-9.

Vost, L. M. (1985). The influence of grazing by the sea urchin *Echinus esculentus* L. on subtidal algal communities. Unpublished Ph.D. Thesis. University of Liverpool.

Wolff, W.J. (1977). A benthic food budget for the Grevelingen Estuary, the Netherlands, and a consideration of the mechanisms causing high benthic secondary production in estuaries. In *Ecology of marine benthos*, (ed. B.C. Coull), pp. 267-80. University of South Carolina Press.

Workman, C. (1983). Comparisons of energy partitioning in contrasting age-structured populations of the limpet *Patella vulgata* L. *Journal of the Marine Biological Association of the United Kingdom*, **68**, 81-103.

Wright, J. R. (1977). The construction of energy budgets for three intertidal rocky shore gastropods, *Patella vulgata*, *Littorina littoralis* and *Nucella lapillus*. Unpublished Ph.D. Thesis. University of Liverpool.

Wright, J. R. and Hartnoll, R. G. (1981). An energy budget for a population of the limpet *Patella vulgata*. *Journal of the Marine Biological Association of the United Kingdom*, **61**, 627-46.

Wu, R. S. S. and Levings, C. D. (1978). An energy budget for individual barnacles (*Balanus glandula*). *Marine Biology*, **45**, 225-35.

2. Plant–animal interactions in the north-west Atlantic

ROBERT L. VADAS, Sr.* and
ROBERT W. ELNER[†]

*Department of Plant Biology and Pathology, University of Maine, Orono, USA
[†]Department of Fisheries and Oceans, Pacific Biological Station, Nanaimo, British Columbia, Canada

Abstract

A feature of the north-west Atlantic is low macroalgal diversity, which may be related in part to grazing. The major herbivores, the sea urchin *Strongylocentrotus droebachiensis* and the gastropod *Littorina littorea*, affect distribution and abundance patterns of macroalgae, especially annual forms. Persistent subtidal and intertidal algae are mainly perennials and defended species. Grazing by herbivores appears to influence but not control algal recruitment. Temporal and spatial refuges provide variability for algal colonization. Algae serve as habitat and food for a variety of subtidal and intertidal animals, including crustaceans. Most algal production appears to pass into detrital rather than into herbivore food webs. The systems and interactions are influenced by historical and stochastic events. The introduction of *L. littorea* in the early nineteenth century displaced *Nassarius obsoletus* and probably catalysed large-scale changes in the intertidal. Population explosions, followed by massive mortalities of urchins off Nova Scotia, have resulted in an alternation between subtidal macroalgal beds and coralline algal-dominated barrens. Dense populations of littorinids and dogwhelks recycle patches of free space in the intertidal by removing ephemeral algae and barnacles. Feeding by *L. littorea* on drift algae may have indirectly led *S. droebachiensis* to increase grazing pressure on living plants. Although much remains unknown about the roles of recruitment, abiotic disturbance and stress, and episodic events, herbivores clearly have a significant impact on benthic algal assemblages.

Plant-Animal Interactions in the Marine Benthos, (ed. D.M. John, S.J. Hawkins, and J.H. Price), Systematics Association Special Volume No. 46, pp. 33–60. Clarendon Press, Oxford, 1992. © The Systematics Association, 1992.

'The components of any given vegetational belt may, of course, change markedly from century to century.'

(Johnson and Skutch, 1928, p. 334)

Introduction

Research on nearshore interactions in the north-west Atlantic during the last two decades has emphasized non-equilibrium processes. In fact, studies on kelp–sea urchin interactions and community structure sought verification for the 'keystone predator' paradigm (Elner and Vadas 1990). Intertidal studies have also invoked predation by a limited number of consumers as one of the major forces structuring communities (Menge 1976, 1978; Vadas *et al.* 1977; Lubchenco and Menge 1978). Competition was invoked only where predation was reduced or made ineffective by wave disturbance (Menge 1978; Lubchenco 1986) or reduced salinity (Keser and Larson 1984a). Abiotic factors have been neglected, but disturbance by waves is important in shallow subtidal communities (Witman 1987), and affects recruitment of a dominant macroalga in intertidal communities (Vadas *et al.* 1990). Herbivore recruitment pulses may also influence algal assemblages (Keats *et al.* 1984; Himmelman 1986). As Underwood and Fairweather (1989) note, studies that lack an understanding of recruitment may not be able to provide a clear statement of processes structuring communities. Although several intertidal studies were dismissed by Chapman and Johnson (1990) because of flaws in experimental design, such works have provided useful information and ideas for testing.

Here we review important works on herbivores and algae and the interactions between them. Since space is a critical necessity for both algae and sessile invertebrates, biotic and abiotic disturbances influencing space are integrated where appropriate. We partition our review into two broad categories (intertidal vs subtidal habitats) and provide general remarks on each. Within each, we deal with:

(1) herbivore effects on the distribution of macroalgae, including both horizontal and vertical gradients;
(2) macroalgal effects on animals, including algae as food and habitat.

Complicating effects, such as historical, physical, and stochastic factors, are considered throughout. Our review focuses on hard substrates and includes Newfoundland, Nova Scotia, Gulf of St Lawrence, Bay of Fundy, Gulf of Maine, and Long Island Sound.

Littoral systems

1. Regional description and patterns

Intertidal habitats in the north-west Atlantic are strongly affected by the availability of hard substrata and by ice scour. Most striking are the overall changes in the physical gradient from north to south. Rocky shores in Newfoundland, Cape Breton, and the Gulf of St Lawrence are regularly scoured by ice (Bourget *et al.* 1985). Consequently, there are fewer reports of herbivore effects, except for periodic sea urchin excursions into littoral areas (Himmelman *et al.* 1983; Keats *et al.* 1985). Rocky coasts dominate north of Cape Cod, whereas more southernly regions consist of a mixture of ledge, boulder, cobble, sand flats, mudflats, and salt-marshes. Luxuriant intertidal algal assemblages develop from central Nova Scotia to Cape Cod. Beyond these points (to the Gulf of Maine) the flora is relatively impoverished. This cursory survey does not mean that herbivores are unimportant in marginal habitats. In fact, several studies (e.g., Bertness 1984) demonstrate the significant effects of herbivores in floristically impoverished areas.

Several herbivores inhabit north-west Atlantic shores but only one, *L. littorea*, has made a significant impact on its intertidal flora (Petraitis 1983). *Littorina littorea* is a recent invader (Vermeij 1982) and is generally important throughout the region, although it may be absent or sparse in the St Lawrence Estuary (Himmelman and Lavergne 1985) and in parts of Newfoundland. The recognized disproportionate effects that an introduced species can have on potential gastropod competitors (e.g., *Nassarius obsoletus* and prey) are apparent with *L. littorea* (Brenchley and Carlton 1983).

Summaries of horizontal and vertical intertidal distribution patterns of major taxa or occupiers of space, important consumers, and major processes are provided in Figs 2.1 and 2.2.

2. Herbivore effects on macroalgae

(a) Horizontal gradients Two types of spatial pattern influencing herbivory are considered: wave-exposure and salinity gradients. The former effect is often confused by ice abrading sheltered shores (Bourget *et al.* 1985). Himmelman and Lavergne (1985) and Keats *et al.* (1985) document disturbance by ice scour in the north, whereas Wethey (1985) suggests that sea ice is a frequent catastrophic source of mortality to barnacles and other species in New England. Wave-exposure is a conspicuous process influencing algal distribution patterns (Figs 2.1 and 2.2). Considerable morphological plasticity is exhibited by fucoids along salinity/wave-exposure gradients (Jordan and Vadas 1972;

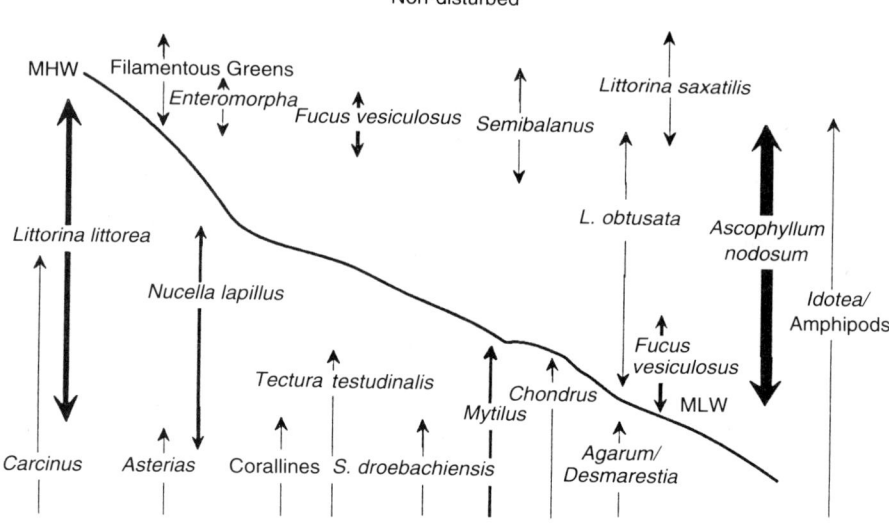
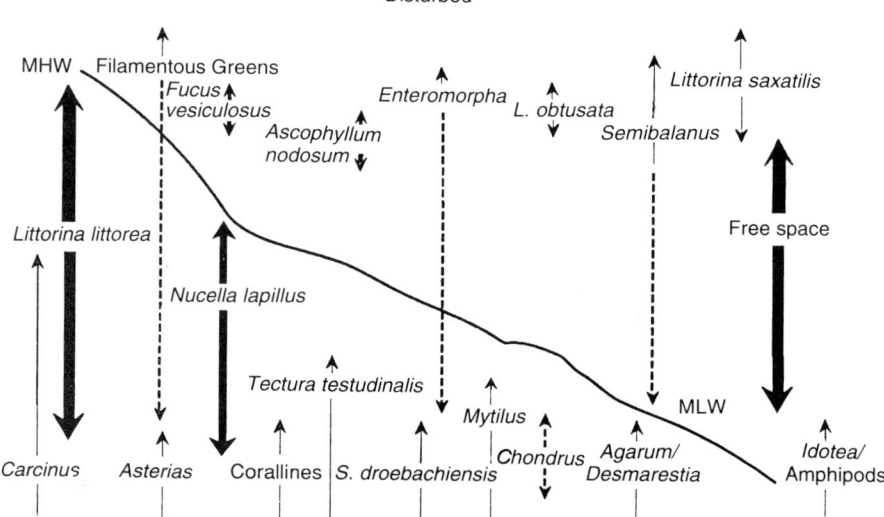

Fig. 2.1. Schematic diagram showing horizontal and vertical distribution patterns of major taxa and processes for moderately sheltered/sheltered intertidal shores (non-disturbed and disturbed) in the north-west Atlantic. The vertical distribution is shown by the length of the arrows, while the width depicts the relative abundance or functional importance of a taxon or process.

A dashed line denotes a changing or ephemeral, seasonal pattern.

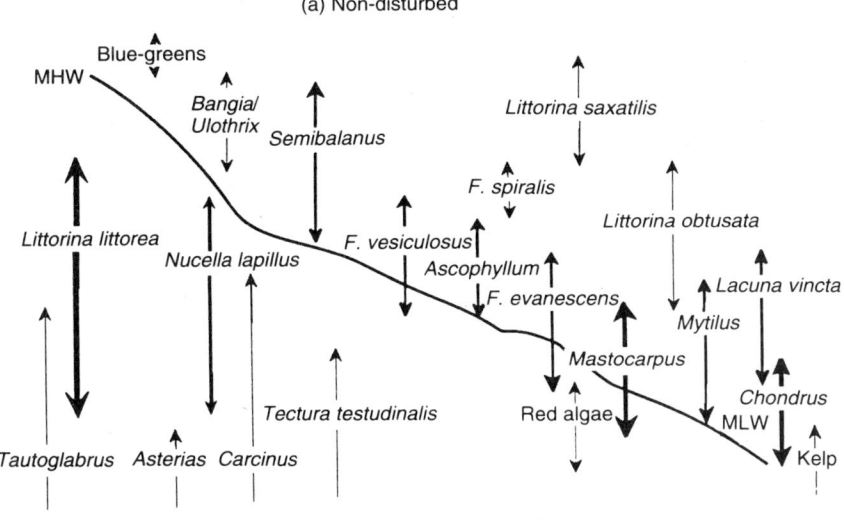

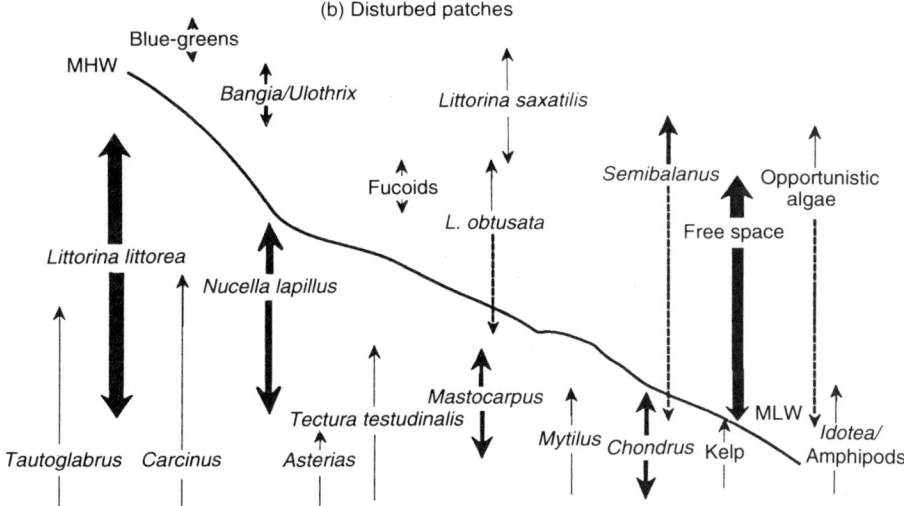

Fig. 2.2. Schematic diagram (as in Fig. 2.1) showing horizontal and vertical distribution patterns of major taxa and processes for relatively exposed intertidal shores (non-disturbed and disturbed) in the north-west Atlantic.

Rice *et al.* 1985). For example, in *Fucus vesiculosus*, vesiculate thalli are replaced by evesiculate forms on exposed coasts. Species replacements along this gradient are also evident; *Ascophyllum nodosum* and *F. vesiculosus* are gradually replaced by *F. distichus* subspp. *evanescens* and *edentatus* (= *F. evanescens*) on exposed shores (Lubchenco 1980;

Mathieson and Penniman 1986). *Fucus spiralis* occurs on intermediate to relatively exposed sites. Barker and Chapman (1990) have shown differential susceptibility among fucoids to grazing by different littorinids, but there is no evidence linking any horizontal pattern to herbivory. In the field, *F. vesiculosus* was found to be susceptible to grazers, whereas *F. distichus* was grazed least. With *F. spiralis*, littorinids had no effect on juveniles but did affect the cover of adults (Chapman 1989). Littorinids can have a significant impact on both stages of *F. vesiculosus* (Cheney and Mathieson 1978; Keser *et al.* 1981; Petraitis 1987). These works indicate that the interactions between a grazer and a particular fucoid species cannot be extrapolated to life history stage or other fucoid species.

Impoverished river-estuarine sites are common in the north-west Atlantic, but most contain fucoids (*F. vesiculosus*, *Ascophyllum*) and a few opportunistic species such as *Enteromorpha* spp. Herbivores such as *Littorina saxatilis* and *L. obtusata* are uncommon or absent (Himmelman and Lavergne 1985). Similarly, tide pools with fluctuating salinities lack littorinids (Sze 1980; Wolfe and Harlin 1988a, 1988b) and are dominated by green algae (Lubchenco 1978). At more exposed estuarine sites *L. littorea* can be present at high densities but its effects on algae vary. It has been reported to have a negligible influence on *Ulothrix flacca* at low-salinity sites (Archambault and Bourget 1983) but grazed all ephemeral algae at a site with seasonally reduced salinities (Vadas, unpublished observations).

More attention has been directed towards understanding interactions along exposure gradients. In protected low-energy environments in southern New England, high densities of *L. littorea* exclude all erect algae (Bertness *et al.* 1983; Petraitis 1983) and in some areas extensively modify the physical habitat. Within 10 days of removal of *L. littorea*, sediment accumulation is noticeable. With time and continued sediment deposition, soft-bottom organisms become more numerous and *Spartina alterniflora* expands its range into the removal areas (Bertness 1984). At sheltered sites, grazing by *L. littorea* reduces the diversity of ephemeral algal species and promotes perennial forms (Lubchenco 1978, 1983; Lubchenco and Menge 1978). However, other work indicates that *L. littorea* prevents the establishment of perennials on sheltered shores (Petraitis 1987; Keser and Larson 1984a; Vadas 1991), or promotes the enhancement of ephemerals (Chapman 1990b). There are also contradictory reports about the abundance of algae on sheltered shores. Most workers (Keser *et al.* 1981; Topinka et al. 1981; Vadas and Wright 1986) conclude that sheltered shores, unless influenced by ice, contain few species but a large biomass. Petraitis (1987), however, reports that algae are rare at sheltered sites. Both assertions may be correct and could relate to grazing and prior disturbance (Fig. 2.1). Sheltered shores can be partitioned on the basis of disturbance: undisturbed shores conforming to the *Ascophyllum*-dominated pattern (Fig. 2.1a), and disturbed shores being controlled by consumers and containing abundant free space (Fig. 2.1b; see below, Vadas 1991).

Intermediate to moderately exposed shores in Maine and Nova Scotia contain some of the highest standing stocks of algae recorded in the North-west Atlantic (Mann 1973; Keser *et al.* 1981; Cousens 1984). These shores contain nearly pure stands of *Ascophyllum*, with narrow zones of *Fucus spiralis* and *F. vesiculosus* above, and *Mastocarpus stellatus* and *Chondrus crispus* below. The algae coexist with high densities of *L. littorea*, *L. obtusata*, and *L. saxatilis*. Seasonally there are blooms of ephemeral, mostly epiphytic, green and brown algae. The subtidal fringe contains a moderately diverse assemblage of filamentous and foliose red algae. Competition, rather than herbivory, is probably the major structuring force in these communities (Cousens 1985; Lubchenco 1986; Chapman 1990a). Long-term monitoring of *Ascophyllum* populations indicate that, even at high grazer densities, herbivore effects alone on this alga are negligible (Vadas, unpublished observation).

At relatively exposed locations, the band of fucoids is compressed. Barnacles (*Semibalanus balanoides*), blue-green algae and ephemeral greens dominate the upper littoral and perennial red algae, primarily *Mastocarpus* (above) and *Chondrus* (below), dominate the lower littoral (Fig. 2.2a). Both species are abundant in the Gulf of Maine, and *Mastocarpus* often comprises over 50 per cent of the biomass in the lower littoral (Dudgeon *et al.* 1989; Vadas, unpublished observation). Studies by Lubchenco and Menge reported only *Chondrus* in the low intertidal of Maine. Neither species, however, is exploited by *L. littorea* (Cheny and Mathieson 1978; Lubchenco 1986; Dudgeon, unpublished observation). Densities of *L. littorea* can vary over short distances on these shores because of slight variations in the direct force of the waves. Subtle variations in exposure, due to offshore ledges and islands, and also seasonal foraging activities, can create 'thresholds' for predation by gastropods, thereby reducing grazing intensity and permiting the development of algae (Vadas 1991). Algal diversity in the Gulf of Maine may be highest on relatively exposed shores (Mathieson and Penniman 1986).

Relatively exposed habitats are zoned and dominated by:
(1) barnacles, blue-green and green algae, with *L. saxatilis* in the upper zone;
(2) barnacles in the next zone;
(3) fucoids (*Fucus* spp. and *Ascophyllum*), barnacles, *Littorina obtusata*, and *L. littorea*;
(4) *Mastocarpus*, *Chondrus*, and *L. littorea*, with limpets (*Tectura testudinalis*) in the lowest zone.

Amphipods and isopods are common seasonally in the canopy of perennial red algae, in moist depressions and in tide pools, but rare on emergent substrata (Chapman 1989; Hacker and Steneck 1990). In addition, there are isolated patches of bare space in the algal matrix of the mid and low littoral (Fig. 2.2b). The patches are predictably colonized by r-selected species and barnacles each spring, and contain seasonally 10 to 15 species of algae (mostly filamentous greens). Patches routinely surveyed on the shore have been recycled annually for over 15 years by 'multiphasic patch recycling', which involves intensive sequential predation by *L. littorea* and *N. lapillus* (Vadas 1991). Clearly, consumers maintain these patches, but the origin of patches in the midst of homogeneous stands of *Mastocarpus* and *Chondrus* is unknown since neither alga is utilized by *L. littorea*. The patches increase the heterogeneity and diversity of these habitats.

Exposed shores contain few herbivores and predators, and are usually reported to be dominated by barnacles and mussels (Lubchenco and Menge 1978). Infra-red aerial photographs and observations of exposed shores in Maine, however, reveal high coverage by algae in mid and low zones (Vadas and Manzer 1971). Because wave-action reduces biotic disturbance, competition becomes an important structuring force on these shores. Barnacles and algae are outcompeted by mussels (Grant 1977; Lubchenco and Menge 1978), whereas *Chondrus* outcompetes *Mastocarpus* (S. Dudgeon and Vadas, unpublished observations).

(b) Vertical gradients Vertical bands of algae are conspicuous on emergent surfaces and also in microhabitats (e.g., tide pools). There are few documented studies concerning the effects of grazers on specific patterns of algal zonation. Most reports suggest that competition for light and space or pre-emptive effects determine vertical limits. Lubchenco (1980) demonstrated the competitive exclusion of fucoids by perennial red algae in the low intertidal of relatively exposed shores. However, in patches or gaps within this zone, fucoids were capable of establishing when grazers were excluded (Vadas 1991, unpublished observations). On protected shores, *Ascophyllum* outcompetes *Fucus* through its longevity, and influences vertical ranges (Keser and Larson 1984b). The limited vertical migrations of *L. littorea* suggest that grazing could temporally influence zonation patterns. On Rhode Island, persistent intertidal populations control ephemeral algae in the low- and mid-intertidal zones (Bertness *et al.* 1983). In Maine, intense grazing produces a marked lower limit on ephemeral algae at lower levels, but these and perennial forms persist into autumn in exclusion cages (Vadas *et al.* 1977; Keser and Larson 1984a). In Nova Scotia, *L. littorea* does not forage in the upper intertidal (Chapman 1989).

Little is known about herbivore effects on the zonation of perennial algae. Chapman (1990a) showed that the distribution of *F. spiralis* was set by competition rather than herbivory. Vadas *et al.* (1977) suggest that herbivory controls zonation in fucoids, but wave disturbance of attaching zygotes may influence distribution patterns (Vadas *et al.* 1990). It is possible that zonation could be determined by systematic grazing of sporeling stages (Lubchenco 1978; Keser *et al.* 1981), but this does not occur with *F. spiralis* (Chapman 1989). Episodic invasions by herbivores may also influence zonation patterns. Low intertidal populations of adult *Fucus distichus* subsp. *edentatus* (= *F. evanescens*) were nearly destroyed by *Lacuna vincta* in the Bay of Fundy (Thomas and Page 1983). Intertidal migrations of sea-urchins can alter the lower limits of littoral or littoral fringe species (Arnold 1976). At moderately protected sites, the lower limits of *Chondrus* are controlled by foraging movements of *S. droebachiensis* (Lubchenco 1980). Lower limits on algae may also be determined indirectly by higher-order predators such as the green crab, *Carcinus maenas*, acting on herbivorous gastropods (Elner and Raffaelli 1980).

(c) Tide pools Tide-pool studies have provided information on herbivore effects on littoral algae along both vertical and horizontal gradients. Although some experimental studies (Lubchenco 1978) have been criticized (cf. Chapman and Johnson 1990), the patterns derived from both experimental and correlative approaches have yielded similar conclusions. Exposed-shore and high-tide pools typically have low densities of *L. littorea* and high abundances of *Enteromorpha* spp. In contrast, mid- and low-tide pools in protected habitats have higher densities of *L. littorea* and contain perennial algae such as *Chondrus* and *Fucus distichus*. Juvenile densities of the latter are reduced by grazers (Chapman 1990b). Temporal patterns of algal abundance are also evident, due mainly to life history and climatic effects, or to herbivores (Lubchenco 1978; Sze 1980; Thomas 1983; Wolfe and Harlin 1988a, 1988b).

3. Macroalgal effects on animals

(a) Algae as habitat Herbivores such as *L. littorea* can have important consequences for algae and invertebrates. Species that rely on littoral algae for protection from desiccation and predation or other functions (e.g., egg-laying, substrate provision) are at risk when algal turfs or canopies are destroyed. Numerous algae and sessile invertebrates rely on *Ascophyllum* as a substrate (Thomas *et al.* 1983). On a sunny day *Ascophyllum* canopies can reduce temperatures by 10°C, increase humidity to 100 per cent and decrease light intensities by 99 per cent

(Vadas and Wright, unpublished data). The presence of an extensive algal cover also affects mobile species, such as amphipods, isopods, crabs and some gastropods, that forage in the intertidal during high tide. By utilizing fronds as refuges (Menge 1978: Hacker and Steneck 1990), these species can migrate higher and remain protected from predation and desiccation.

Algae can also have negative effects on animals, especially on sedimentary shores. Algal blooms have occurred in the Gulf of Maine, causing anoxia as the mats accumulate and decompose (Vadas and Beal 1987). The blooms may represent a release from herbivory (due to periwinkle harvesting) or nutrient enhancement. These mats affect the survival of soft-shelled clams and provide shelter for predators such as green crabs and moon snails. They also provide insights on how some shores might have appeared prior to the introduction of *L. littorea*.

(b) Algae as food   Intertidal shores contain productive algal assemblages (Mann 1973), but less than 10 per cent of this production goes through the two major herbivore (sea urchins, periwinkles) food chains (Mann 1972). Considerable algal biomass is lost during winter (Vadas 1979) and ends up as drift or stranded wrack (Josselyn and Mathieson 1980). Much is apparently fragmented by small snails, isopods, and amphipods (Mann 1973). However, the wrack consists mainly of fucoids, *Chondrus*, and seagrasses, species that are not preferred by periwinkles or sea urchins. Interestingly, many of these same forms are not preferred by, and do not fulfil the energetic requirements of, intertidal amphipods (Pederson and Capuzzo 1984). Nonetheless, micrograzers may be important in energy-flow on exposed shores where they and non-defended algae have a refuge from consumers. Although dominance by a defended algal flora supports the notion that algal production passes through decomposer food chains, there is little information on energy-flow from non-defended forms.

4.   Overview of littoral

Littoral shores in the north-west Atlantic are characterized by a defended (perennial) algal flora, barnacles, and mussels. Free space within this assemblage is colonized by ephemeral algae and barnacles. *Littorina littorea* is present in high densities in most habitats, but mainly controls ephemerals on disturbed sites (however, see Petraitis 1987; Chapman 1990*b*). Coupled with *Nucella lapillus*, which removes most low-shore barnacles, these consumers annually recycle free space in disturbed areas. Periwinkles are reported both to enhance (Lubchenco 1986) and to prevent succession (Petraitis 1987; Vadas 1991), but these observations may depend on whether or not dog-whelks are present. In the absence of whelks, barnacles provide a refuge for fucoid algae (Lubchenco 1983) and jointly pre-empt space. The ecological importance of these processes, however, is problematical. Since *L. littorea* was only recently introduced, it is likely that generalizations regarding herbivore interactions can apply only to the last century. Despite the apparent stability of *Ascophyllum*, long-term interactions on these shores may be quite variable. Green crabs are also recent invaders in the north-west Atlantic (Vermeij 1982), whereas the large intertidal sea mink (*Mustela macrodon*) disappeared in recent times (D. Sanger, pers. comm.). Such changes in consumer groups, coupled with catastrophic climatic shifts (Wethey 1985), suggest that the patterns and processes on these shores are potentially highly variable. Johnson and Skutch (1928) recognized the instability of the north-west Atlantic littoral over 60 years ago.

Sublittoral Systems

1. Regional description and patterns

The extensive sublittoral hard substrate along the north-west Atlantic supports a diverse epifauna and epiflora (Logan *et al.* 1983). For example, Mathieson (1979) identified 100 taxa of plants from open coastal sites in northern New England. Dominant fleshy macrophytes are the kelps *Laminaria longicruris*, *L. digitata*, *L. saccharina*, *Saccorhiza dermatodea*, *Alaria esculenta*, and *Agarum cribrosum*. However, other algae such as *Desmarestia* can form extensive beds, and species such as *Chondrus* dominate either ledges or understories of kelps. Crustose corallines such as species of *Lithothamnion*, *Clathromorphum*, and *Phymatolithon* are ubiquitous and can build up rhodoliths several metres deep where other macrophytes are absent (Hooper 1980; Steneck 1983, 1986). Grazers include sea urchins, *S. droebachiensis*, molluscs and crustaceans. The principal predators of sublittoral grazers are lobsters, *Homarus americanus*, *Cancer* (crab) spp., green crabs, seastars, *Asterias* spp., and numerous fish species (Fig. 2.3; see Keats *et al.* 1987*a*).

Linkages between herbivory and community structure in the sublittoral are dramatically apparent. Two alternate community states exist, depending on the population density of *S. droebachiensis*. When urchins are rare, communities of kelp and other fleshy macrophytes flourish. Where corallines dominate, the system is maintained through intensive grazing by sea urchins (Fig. 2.3). While large kelp beds do occur, hard bottoms around the Gulf of St Lawrence, Newfoundland, and Labrador are covered by corallines (Hooper 1980; Himmelman *et al.* 1983). Conversely, kelp beds flourish in the Bay of Fundy and offshore islands and ledges in the Gulf of Maine (Vadas and Steneck 1988). Nearshore areas are dominated by corallines (Steneck 1986), but

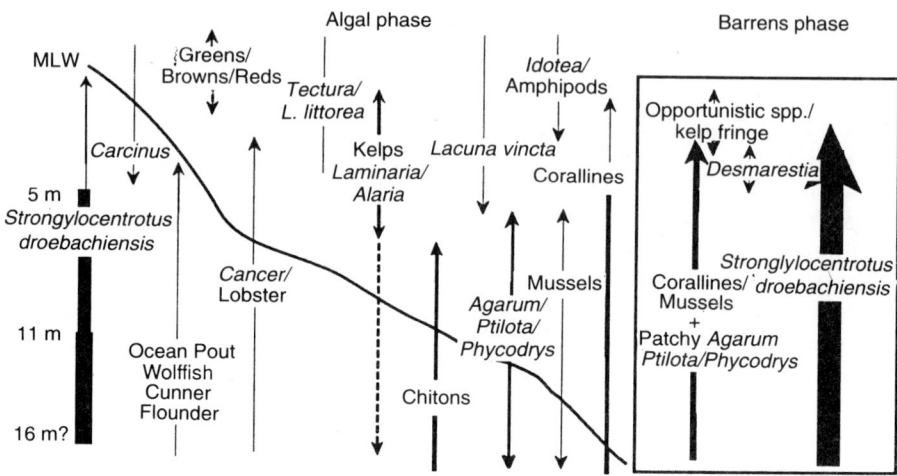

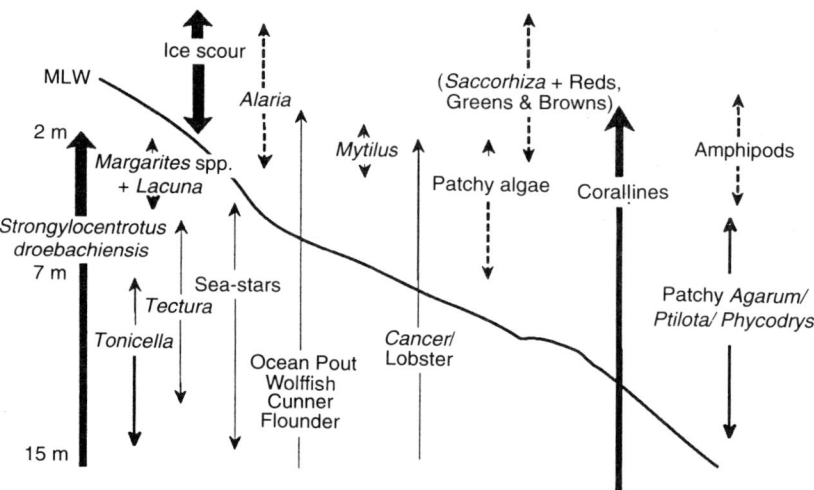

Fig. 2.3. Schematic diagram (as in Fig. 2.1) showing common sublittoral patterns and processes in the northern and central regions of the north-west Atlantic.

contain numerous patchy (refuge) kelp beds. In some areas, extensive beds of mussel and kelp cover the bottom (Witman 1987) and sponge or bryozoan communities may cover vertical walls (Logan 1988; Sebens 1986). Sea urchins also play an important role in these communities (Briscoe and Sebens 1988). Urchins can increase locally in abundance and switch from feeding on drift algae to grazing live plants (Arnold 1976; Hooper 1980), thereby producing holes in the canopy. Isolated stands of algae can survive in coralline barrens where water motion provides refuges from urchin grazing (Himmelman and Lavergne 1985). On a geographical scale, eastern Nova Scotia appears to be a fulcrum zone where community states alternate between kelp and sea urchin barrens, depending on the presence and behaviour of urchins (Miller 1985a). Population explosions, followed by massive mortalities of urchins, have resulted in an alternation between assemblages (Mann 1985; Chapman and Johnson 1990).

2. Herbivore effects on the distribution of macroalgae

(a) Vertical gradients Community structure and various schemes of zonation of sublittoral sites in the north-west Atlantic have been described (Lamb and Zimmerman 1964; Sears and Wilce 1975; Himmelman *et al.* 1983). Mechanisms invoked for the patterns have included abiotic and biotic factors, depending on the particular zone and site characteristics. One of the earliest studies (Johnson and Skutch 1928) surveyed algae to 1.2 m (4 feet) below low water and suggested that three species of macroalgae characteristic of the lower littoral were being outcompeted for space by kelps and *Devalerea* (*Halosaccion*). Detailed surveys of sublittoral macroalgae on the Atlantic coast of Nova Scotia recognized three (Edelstein *et al.* 1969) or eight (Mann 1972) vertical zones. St Margaret's Bay was regarded by Mann as representative of the sublittoral of northern New England, Nova Scotia, and Newfoundland. In northern New England, Mathieson (1979) recorded a consistent vertical patterning with diversity and ratios of annuals and perennials, Chlorophyceae, and Rhodophyceae decreasing with increase in depth. Mathieson (1979) stressed the role of light interaction with the substrate as being responsible for zonation. Meanwhile, other workers (Himmelman and Steele 1971; Breen and Mann 1976a; Hooper 1980; Himmelman and Lavergne 1985) documented the importance of urchin grazing in controlling the distribution of sublittoral algae. The destructive grazing along the coast of Nova Scotia during the 1970s and the maintenance of the coralline assemblage (Steneck 1986) attracted numerous investigations (Miller 1985a; Vadas *et al.* 1986). Subsequently, sea urchin die-offs in the early 1980s

prompted various surveys to track recolonization (Miller 1985b; Moore and Miller 1983; Scheibling 1984). Novaczek and McLachlan (1986) found that the recolonized communities resembled those documented at the same sites prior to destructive grazing (Edelstein et al. 1969). Vadas and Steneck (1988) described three algal zones on a deep-water rock pinnacle in the central Gulf of Maine. The occurrence of algae at record depths for the Gulf of Maine was related to the absence of large herbivores and the high productivity potential of the benthos. However, zonation was consistent with nearshore patterns and with other regions. Himmelman and Lavergne (1985) described two major subtidal zones in the St Lawrence Estuary, a shallow macroalgae-dominated zone and a deeper urchin-dominated zone. The former included the sublittoral fringe, where large phaeophytes were conspicuous due to refuge effects from urchins. Only defended algae, such as *Agarum* and *Ptilota serrata*, were able to coexist with urchins in the deepest zone (Fig. 2.3; see Himmelman and Nedelec 1990).

Witman (1987) studied an exposed, sublittoral site in the Gulf of Maine where the horse mussel *Modiolus modiolus* dominated hard bottoms at intermediate depths (11–18 m). At shallower depths (4–8 m), kelps (*Laminaria digitata*, *L. saccharina*) predominated. Storm-generated dislodgement of mussels overgrown by kelp was determined to be the mechanism preventing *Modiolus* from establishing in the shallow zone. An urchin-removal experiment in the *Modiolus* zone led to kelp overgrowth on the mussels and a 30-fold increase in dislodgement. Also, as higher urchin densities were found inside mussel beds than outside, possibly due to increased survivorship, the *Modiolus–Strongylocentrotus* interaction was postulated as a facultative mutualism. Removal of urchins from the lower edge of the kelp zone resulted in a downward extension of kelps to 12.5 m. In an earlier study of the functional ecology of *Modiolus* beds, Witman (1985) found three major invertebrate communities between 8 and 30 m. The *Modiolus* beds provided both a refuge for *Strongylocentrotus* and other infauna and protection from severe disturbance due to urchin grazing. Overall, the lower level of disturbance inside the mussel beds influenced the abundance and spatial distribution of *Modiolus* communities.

(b) Horizontal gradients Debate on the role of lobsters as keystone predators in the north-west Atlantic has continued for almost 20 years. The argument in the early 1970s developed from the observations that commercial landings of lobster had declined dramatically, and that kelp beds along Nova Scotia had disappeared due to overgrazing by urchins. The original hypothesis (Mann and Breen 1972; Breen and Mann 1976b) to account for increased sea urchin densities was that the abun-

dance of lobster, purportedly a major predator on urchins, had declined sufficiently to allow a population explosion of urchins. However, data on the control of urchins by lobsters were lacking; feeding and behavioural studies (Carter and Steele 1982; Vadas *et al.* 1986; Elner and Campbell 1987) showed that urchins were not the preferred prey of lobster. Subsequently, the lobster predation hypothesis was modified to include a suite of predators, and aspects of predator behaviour, and eventually developed into an agglomerative complex. In the early 1980s, massive mortalities of urchins due to disease allowed macroalgae to re-establish (Miller and Colodey 1983; Scheibling and Stephenson 1984; Scheibling 1986). Much remains unknown concerning the triggering mechanisms for succession from algae to barrens and back to macroalgae. The four major unresolved problems are:

(1) the mechanism responsible for population explosions of urchins;
(2) the conditions necessary for the large-scale die-offs of urchins;
(3) the comparative ecological value of the two community phases;
(4) the mechanism for formation of urchin fronts, a precursor to destructive grazing.

Johnson and Mann (1988) postulated that low urchin-grazing pressure allows understorey species such as *Chondrus* and *Phyllophora truncata* to form extensive, slow-growing turfs that limit kelp recruitment on the south coast of Nova Scotia. For the east coast of Nova Scotia, where understorey guilds are mainly ephemeral plants, they hypothesized that disturbance from intensive urchin-grazing prevented turf formation. East coast sites, where *Chondrus* dominates, tend to be shallow, exposed habitats where wave-action provides a refuge from grazing.

3. Macroalgal effects on animals

(a) Algae as food Production and energy pathways in seaweed and barren communities along the Atlantic coast of Nova Scotia have been of particular interest due to possible impacts on commercial species, such as lobster and on seaweed harvesting (Mann 1985). Miller *et al.* (1971) estimated that the production of seaweeds (>7000 kcal m^{-2} yr^{-1}) in St Margaret's Bay, Nova Scotia, far exceeded the ingestion rate by herbivores (663 kcal m^{-2} yr^{-1}) and reasoned that most of the seaweed production was exported out of the coastal system in the form of suspended particulate matter. As the production of lobster prey species (83.9 kcal m^{-2} yr^{-1}) was estimated to be well above the ingestion rate of the lobster population (6.4 kcal m^{-2} yr^{-1}), they suggested that lobster production might be increased by reducing predation and competition

for food. The barrens created along the coast of Nova Scotia in the 1970s were assessed as being markedly less productive than kelp beds for lobster and other commercial species. Chapman (1981) concluded that more than 60 per cent of the primary productivity in St Margaret's Bay had been lost with the demise of kelp beds. Such assertions, coupled with the postulated habitat-related importance of kelp to lobster (Breen and Mann 1976b; see also Johns and Mann 1987, for extension to juveniles), led Wharton and Mann (1981) to hypothesize that the removal of kelp had triggered further decreases in lobster abundance. Despite the abundance of urchins, the perception was that lobster and other species were now food-limited and there was a general downward and irreversible spiral of the whole system (Wharton and Mann 1981). With the re-establishment of kelp in the early 1980s, lobster landings increased to record levels (Elner and Cambell 1991). Miller (1985a) argued that evidence for enhancement of lobster abundance, food, growth, and survival by seaweeds was equivocal. Moreover, he noted that the strong upturn in landings was based on lobsters that were raised on barrens. Michaud and Elner (unpublished data) determined that coralline barrens in Nova Scotia off Ingomar had higher invertebrate biomass than nearby kelp beds off Pubnico. Production of lobster prey species at shallow Pubnico sites, however, was significantly higher than at any depth for Ingomar. Elner and Campbell (1987) found that the natural diets of lobsters collected from the same barren-ground and kelp habitats off Nova Scotia were similar in terms of bulk and spectrum, despite variations in availability of some prey. Thus, they argued that the lobster's food supply was not limiting on the barrens. Clearly, benthic biomass and production for Nova Scotia are poor predictors of habitat quality, and the question of bioavailability of primary and secondary production needs further research.

Mohn and Miller (1987) developed a ration-based model of a seaweed–sea urchin community and suggested that the community was driven by seaweed abundance and catastrophic expressions of the sea urchin pathogen. Increasing the abundance of urchin predators in the model did not decrease the mean level of sea urchin abundance.

(c) Algae as a habitat The ecological significance of sublittoral algae as shelter and refuge for fish and larger benthic invertebrates, such as lobster and crab, appears to be conjectural. While some fish species such as juvenile cod (Keats *et al.* 1987b) may be more abundant among algae, there has been only limited testing of whether they have a survival advantage over conspecifics on the outside. Surveys to compare biomass of invertebrate species inside and outside beds are problematic because of difficulties in quantifying the abundance of animals in

macroalgal beds. One clear, habitat-related use of macroalgae is as a substrate for numerous epifauna. Macroalgae can also serve as egg-deposition sites for fish such as herring (Cooper et al. 1975). There have been various claims that seaweeds provide important cover for lobsters, especially juveniles (Botero and Atema 1982; Wharton and Mann 1981). Laboratory experiments by Johns and Mann (1987) indicate that Irish moss and habitat complexity attract juvenile lobster, and reduce predation by cunner. However, field manipulations to test the hypothesis that seaweed cover is beneficial to lobsters, have yet to be carried out. As argued persuasively by Miller (1985a), the relationship has only weak support, and measures of relatively greater food, and/or survival advantage inside macroalgal beds, do not mean that more lobsters will necessarily use the habitat. Destructive grazing of subtidal algal beds has reduced or eliminated other invertebrates that use macroalgae as a habitat or food (Himmelman et al. 1983; Himmelman and Lavergne 1985).

4. Overview of the sublittoral

Nearshore sublittoral areas in the north-west Atlantic are characterized by the presence of sea urchins, defended and perennial algae, crustose corallines, and mussels. Annuals, kelps and non-defended algal species are common in wave-exposed refuges and in areas with low urchin densities. Disturbance is primarily biological (but see Witman 1987) and results from intensive grazing by sea urchins. In protected waters, urchins enhance the development of coralline barrens.

Anecdotal accounts collected by Miller (1985a, 1985b) suggest that urchin populations on the Atlantic coast of Nova Scotia have exploded, then collapsed, several times over the last 100 years. Sea urchins were reported in great abundance in eastern Canada during the 1860s by Dawson (1868), who referred to them as 'submarine rodents'. Considered overall, the population dynamics of urchins as a group appear to be prone to periodic massive recruitment episodes, possibly as a result of high potential fecundity coupled with favourable large-scale oceanographic or climatic events (Scheibling 1986). Subsequently, the urchins tend to dominate benthic habitats until massive mortalities are induced by disease.

General discussion

Although the littoral and sublittoral are considered separately, parallels appear between them that may yield insights into processes influencing north-western Atlantic systems. For example, grazing intensity in both of these habitats is predominantly related to the abundance of a single herbivore, *Littorina littorea* and *Strongylocentrotus droebachiensis*, respectively (see also Johnson and Mann 1986). Controls on the abundance of each herbivore remain largely unresolved, although the role of predation seems to be low or ambiguous in both cases. The role played by disease in the systems is more certain. A sea urchin pathogen is the critical mechanism for the switch from barrens to macroalgal phases in the sublittoral (Scheibling 1984). Similarly, parasites may influence the distribution and behaviour of littorinids (Elner and Raffaelli 1980). Both herbivores form destructive grazing fronts at critical densities (Bernstein and Mann 1982; Vadas 1991) and can maintain barren areas devoid of algae. The resultant cleared space in both systems is temporally richer in ephemeral algae, but more variable than the ungrazed states that they replace. However, while *S. droebachiensis* removes annuals and perennials, *L. littorea* primarily grazes ephemerals and lightly corticated forms (Steneck and Watling 1982). In terms of ecological impact, *L. littorea* and *S. droebachiensis* appear to be out-of-scale to the other herbivores in their respective habitats. Interestingly, *L. littorea* is a recent invader (Vermeij 1982), whereas the urchin is subject to population explosions: both events are ecologically destabilizing.

As both *L. littorea* and *S. droebachiensis* exist at high densities, they graze intensively, and the chances for algal escapes are limited. Patches and zones of algae growing in habitats with high densities of herbivores are invariably comprised of defended species. Because both herbivores feed on drift, there is probably less drift available in nearshore systems than before the introduction of *L. littorea*, and most likely more pressure on the standing flora. It is plausible that a lack of drift, due to *L. littorea*, has been a contributing factor to destructive grazing episodes by urchins.

The recent invasion of *L. littorea* also has implications for the structure of intertidal communities and the persistence of *Ascophyllum*. Based on the longevity of current stands and the species inability to readily colonize (Vadas *et al.* 1990), successful recruitment pulses probably occur naturally only once in several decades. The widespread abundance of *L. littorea* probably means that there is less opportunity for *Ascophyllum* recruitment success. Coupling between intensive grazing by *L. littorea*, and the presumed reduction in recruitment success by *Ascophyllum*, suggests that more 'free space' is being made available in the intertidal. Where *Nucella* coexists with *L. littorea*, these consumers are capable of destructively grazing and maintaining areas free of algae (Lubchenco 1978; Bertness *et al.* 1983; Petraitis 1987; Vadas 1991).

There are conflicting views on the perceived stability of these communities. Mann and colleagues (Mann 1982; Chapman and Johnson 1990) have argued that the different community states in the subtidal represent stable patterns, and that the alternation between them is caused by abundance patterns of sea urchins. Similarly, there also appear to be different community states intertidally: disturbed and non-disturbed patches or sites. The longevity and widespread dominance of *Ascophyllum* augments the notion of stability on non-disturbed shores, but abundance patterns of this alga may also be changing. The cover on a small island in Frenchman's Bay, Maine, has been decreasing over the last decade, apparently due to periodic ice-flows, recruitment failure, lack of refuges, and dense *L. littorea* populations (Vadas, unpublished data). This site may be a microcosm of what is occurring on a regional basis.

Indeed, numerous studies document the widespread occurrence of anomalous events and species changes during the last few hundred years. For example, lobster (common intertidally in the eighteenth century) and cod have largely been overfished and are now represented by smaller stocks. Introductions of exotic species, such as *L. littorea* and *Carcinus*, appear to be commonplace, as are post-glacial intrusions of warm-water species into the north-west Atlantic (Bousfield 1967). Similarly, the loss of the sea mink may have altered shallow-water communities. The regular occurrence of sea ice in New England (Wethey 1985) suggests that stability in intertidal communities may only consist of phased events between bouts of ice. Apparently irregular blooms of red algae encouraged Johnson and Skutch (1928) to suggest that vegetation zones in Maine change markedly from century to century. In addition, our experience with episodic recruitment on bare substrates in lower intertidal communities suggests that the search for stability in the north-west Atlantic may be illusory or misleading. Clearly, we understand little about the effects of less-structured events on the stability of either intertidal or subtidal communities. If stability involves short temporal scales: a decade or two, then we may have witnessed periods of stability. Alternatively, if stability connotes decades, centuries or millenia, then it seems premature to ascribe that feature to these communities.

Our understanding of the linkages in plant–animal interactions in the north-west Atlantic appears to be sharply asymmetrical. We have a good appreciation of the effects of grazers on algal species and, to some extent, on algal assemblages. With the converse, however, we know relatively little about the role or effects of algae on animals. Algae are also purported to serve as habitat and nursery areas for numerous fish and invertebrates, but the limited data on the subject are equivocal at best (R. Black and R. Miller, pers. comm.).

Lastly, we believe that the major, perhaps only, creditable mark of success in understanding plant–animal interactions is an ability to predict ecological outcomes of specific interactions. We need to understand the fine-scale behaviour of species in the interaction or system. Ecologists have done this to some extent with a few herbivores and predators in both intertidal and subtidal systems. However, we have little understanding of the broader-scale events controlling these systems, e.g., recruitment phenomena (Underwood and Fairweather 1989). What controls the population size of sea urchins, or the processes involved in successful recruitment of *Ascophyllum*, for example? Until we understand these, there is little likelihood of real predictability. We will know short-term patterns and be able to predict in limited temporal and spatial scales or states, but not between states.

Acknowledgements

We thank Wes Wright, Steve Dudgeon, Mike Eagles, and Donna Wilbur for assistance with the manuscript. RLV acknowledges the support of the NOAA Sea Grant Program (NA-81aa-D-00035), the Maine Agricultural Experiment Station, and a Faculty Research Award from the University of Maine.

References

Archambault, D. and Bourget, E. (1983). Importance du régime de dénudation sur la structure et la succession des communautés intertidales de substrat rocheux in milieu subarctique. *Canadian Journal of Fisheries and Aquatic Sciences*, **40**, 1278–92.

Arnold, D.C. (1976). Local denudation of the sublittoral fringe by the green sea-urchin *Strongylocentrotus droebachiensis* (O.F. Müller). *Canadian Field-Naturalist*, **90**, 186–7.

Barker, K.M. and Chapman, A.R.O. (1990). Feeding preferences of periwinkles among four species of *Fucus*. *Marine Biology*, **106**, 113–18.

Bernstein, B.B. and Mann, K.H. (1982). Changes in the nearshore ecosystem of the Atlantic coast of Nova Scotia, 1968–1981. *Northwest Atlantic Fisheries Organization Science Council Study*, **5**, 101–5.

Bertness, M.D. (1984). Habitat and community modification by an introduced herbivorous snail. *Ecology*, **65**, 370–81.

Bertness, M.D., Yund, P.O., and Brown, A.F. (1983). Snail grazing and the abundance of algal crusts on a sheltered New England rocky beach. *Journal of Experimental Marine Biology and Ecology*, **71**, 147–64.

Botero, L. and Atema, J. (1982). Behavior and substrate selection during larval settling in the lobster *Homarus americanus*. *Journal of Crustacean Biology*, **2**, 59-69.

Bourget, E., Archambault, D., and Bergeron, P. (1985). Effet des propriétés hivernales sur les peuplements epibenthiques intertidaux dans un milieu subarctique, l'estuaire du Saint-Laurent. *Le Naturaliste canadien, Revue d'Écologie et de Systematique*, **112**, 131-42.

Bousfield, E. L. (1967). Studies on littoral marine arthropods from the Bay of Fundy region. *National Museum of Canada Bulletin*, **183**, 46-62.

Breen, P. A. and Mann, K. H. (1976a). Destructive grazing of kelp by sea-urchins in eastern Canada. *Journal of the Fisheries Research Board of Canada*, **33**, 1278-83.

Breen, P. A. and Mann, K. H. (1976b). Changing lobster abundance and the destruction of kelp beds by sea-urchins. *Marine Biology*, **34**, 137-42.

Brenchley, G. A. and Carlton, J. T. (1983). Competitive displacement of native mud snails by introduced periwinkles in the New England intertidal zone. *Biological Bulletin*, **165**, 543-58.

Briscoe, C. S. and Sebens, K. P. (1988). Omnivory in *Strongylocentrotus droebachiensis* (Müller) (Echinodermata: Echinoea): predation on subtidal mussels. *Journal of Experimental Marine Biology and Ecology*, **115**, 1-24.

Carter, J. A. and Steele, D. H. (1982). Stomach contents of immature lobsters (*Homarus americanus*) from Placentia Bay, Newfoundland. *Canadian Journal of Zoology*, **60**, 337-47.

Chapman, A. R. O. (1981). Stability of sea-urchin dominated barren grounds following destructive grazing of kelp in St. Margaret's Bay, eastern Canada. *Marine Biology*, **62**, 307-11.

Chapman, A. R. O. (1989). Abundance of *Fucus spiralis* and ephemeral seaweeds in a high eulittoral zone: effects of grazers, canopy and substratum type. *Marine Biology*, **102**, 565-72.

Chapman, A. R. O. (1990a). Competitive interactions between *Fucus spiralis* L. and *F. vesiculosus* L. (Fucales, Phaeophyta). *Hydrobiologia*, **204/205**, 205-9.

Chapman, A. R. O. (1990b). Effects of grazing, canopy cover and substratum type on the abundance of common species of seaweeds inhabiting littoral fringe tide pools. *Botanica Marina*, **33**, 319-26.

Chapman, A. R. O. and Johnson, C. R. (1990). Disturbance and organization of macroalgal assemblages in the Northwest Atlantic. *Hydrobiologia*, **192**, 77-121.

Cheney, D. P. and Mathieson, A. C. (1978). On the ecological and evolutionary significance of vegetative reproduction in seaweeds. *Journal of Phycology*, **14 (suppl.)**, p. 27.

Cooper, R. A., Uzmann, J. R., Clifford, R. A., and Pecci, K. J. (1975). *Direct observations of herring* (Clupea harengus harengus *L.) egg beds on Jeffreys*

Ledge, *Gulf of Maine in 1974*. International Commission for the Northwest Atlantic Fisheries Research Document 75/93.

Cousens, R. (1984). Estimation of annual production by the intertidal brown alga *Ascophyllum nodosum* (L.) Le Jolis. *Botanica Marina*, **27**, 217-27.

Cousens, R. (1985). Frond size distributions and the effects of the algal canopy on the behavior of *Ascophyllum nodosum* (L.) Le Jolis. *Journal of Experimental Marine Biology and Ecology*, **92**, 231-49.

Dawson, J. W. (1868). The food of the common sea urchin. *American Naturalist*, **1**, 124-5.

Dudgeon, S. R., Davison, I. R., and Vadas, R. L. (1989). Effect of freezing on photosynthesis of intertidal macroalgae: relative tolerance of *Chondrus crispus* and *Mastocarpus stellatus* (Rhodophyta). *Marine Biology*, **101**, 107-14.

Edelstein, T., Craigie, J. S., and McLachlan, J. (1969). Preliminary survey of the sublittoral flora of Halifax county. *Journal of the Fisheries Research Board of Canada*, **26**, 2703-13.

Elner, R. W. and Campbell, A. (1987). Natural diets of lobster *Homarus americanus* from barren ground and macroalgal habitats off southwestern Nova Scotia, Canada. *Marine Ecology Progress Series*, **37**, 131-40.

Elner, R. W. and Campbell, A. (1991). Spatial and temporal patterns in recruitment for the American lobster *Homarus americanus*, in the northwest Atlantic. In *Proceedings of the International Crustacean Conference, Brisbane, Australia, July 1990*. Memoirs of the Queensland Museum. (In press).

Elner, R. W. and Raffaelli, D. G. (1980). Interactions between two marine snails, *Littorina rudis* Maton and *Littorina nigrolineata* Gray, a predator, *Carcinus maenas* (L.), and a parasite, *Microphallus similis* Jagerskiold. *Journal of Experimental Marine Biology and Ecology*, **43**, 151-60.

Elner, R. W. and Vadas, R. L. (1990). Inference in ecology: the sea urchin phenomenon in the North-west Atlantic. *American Naturalist*, **136**, 108-25.

Grant, W. S. (1977). High intertidal community organization on a rocky headland in Maine, USA. *Marine Biology*, **44**, 15-25.

Hacker, S. D. and Steneck, R. S. (1990). Habitat architecture and the abundance and body-size dependent habitat selection of a phytal amphipod. *Ecology*, **71**, 2269-85.

Himmelman, J. H. (1986). Population biology of green sea urchins on rocky barrens. *Marine Ecology Progress Series*, **33**, 295-306.

Himmelman, J. H. and Lavergne, Y. (1985). Organization of rocky subtidal communities on the St. Lawrence Estuary. *Le Naturaliste canadien, Revue d'Ecologie et de Systématique*, **112**, 143-54.

Himmelman, J. H. and Nedelec, H. (1990). Urchin foraging and algal survival strategies in intensely grazed communities in eastern Canada.

Canadian Journal of Fisheries and Aquatic Sciences, **47**, 1011–26.

Himmelman, J. H. and Steele, D. H. (1971). Foods and predators of the green sea-urchin *Strongylocentrotus droebachiensis* in Newfoundland waters. *Marine Biology*, **9**, 315–22.

Himmelman, J. H., Cardinal, A., and Bourget, E. (1983). Community development following removal of urchins, *Strongylocentrotus droebachiensis*, from the rocky subtidal zone of the St. Lawrence Estuary, eastern Canada. *Oecologia* (Berlin), **59**, 27–39.

Hooper, R. (1980). Observations on algal–grazer interactions in Newfoundland and Labrador. *Canadian Technical Report of Fisheries and Aquatic Sciences*, **954**, 120–24.

Johns, P. M. and Mann, K. H. (1987). An experimental investigation of juvenile lobster habitat preference and mortality among habitats of varying structural complexity. *Journal of Experimental Marine Biology and Ecology*, **109**, 275–85.

Johnson, C. R. and Mann, K. H. (1986). The importance of plant defence abilities to the structure of subtidal seaweed communities: the kelp *Laminaria longicruris* de la Pylaie survives grazing by the snail *Lacuna vincta* (Montagu) at high population densities. *Journal of Experimental Marine Biology and Ecology*, **97**, 231–67.

Johnson, C. R. and Mann, K. H. (1988). Diversity, patterns of adaptation and stability of Nova Scotian kelp beds. *Ecological Monographs*, **58**, 129–54.

Johnson, D. S. and Skutch, A. F. (1928). Littoral vegetation on a headland of Mt. Desert Island, Maine. II. Tide-pools and the environment and classification of submersible plant communities. *Ecology*, **9**, 307–38.

Jordan, A. J. and Vadas, R. L. (1972). Influence of environmental parameters on intraspecific variation in *Fucus vesiculosus*. *Marine Biology*, **14**, 248–52.

Josselyn, M. N. and Mathieson, A. C. (1980). Seasonal influx and decomposition of autochthonous macrophyte litter in a north temperate estuary. *Hydrobiologia*, **71**, 197–208.

Keats, D. W., South, G. R., and Steele, D. H. (1984). Ecology of juvenile green sea-urchins (*Strongylocentrotus droebachiensis*) at an urchin dominated sublittoral site in eastern Newfoundland. In *Echinodermata, Proceedings of the Fifth International Conference, Galway, 24–29 September 1984*, (ed. B. F. Keegan and B. D. S. O'Connor), pp. 295–302. Balkema, Rotterdam.

Keats, D. W., South, G. R., and Steele, D. H. (1985). Algal biomass and diversity in the upper subtidal at a pack-ice disturbed site in eastern Newfoundland. *Marine Ecology Progress Series*, **25**, 151–8.

Keats, D. W., Steele, D. H., and South, G. R. (1987*a*). Ocean pout (*Macrozoarces americanus* (Bloch and Schneider) (Pisces: Zoarcidae)) predation on green sea urchins *Strongylocentrotus droebachiensis* (O. F. Müll.) (Echinodermata: Echinoidea) in eastern Newfoundland. *Canadian Journal of Zoology*, **65**, 1515–21.

Keats, D.W., Steele, D.H., and South, G.R. (1987b). The role of fleshy macroalgae in the ecology of juvenile cod (*Gadus morhua* L.) in inshore waters off eastern Newfoundland. *Canadian Journal of Zoology*, **65**, 49-53.

Keser, M. and Larson, B.R. (1984a). Colonization and growth dynamics of three species of *Fucus*. *Marine Ecology Progress Series*, **15**, 125-34.

Keser, M. and Larson, B.R. (1984b). Colonization and growth of *Ascophyllum nodosum* (Phaeophyta) in Maine. *Journal of Phycology*, **20**, 83-7.

Keser, M., Vadas, R.L., and Larson, B.R. (1981). Regrowth of *Ascophyllum nodosum* and *Fucus vesiculosus* under various harvesting regimes in Maine. *Botanica Marina*, **24**, 29-38.

Lamb, I.M. and Zimmerman, M.H. (1964). Marine vegetation of Cape Ann, Essex county, Massachusetts. *Rhodora*, **66**, 217-54.

Logan, A. (1988). A sublittoral hard substrate epibenthic community below 30 m in Head Harbour Passage, New Brunswick, Canada. *Estuarine, Coastal and Shelf Science*, **27**, 445-59.

Logan, A., McKay, A.W., and Noble, J.P.A. (1983). Sublittoral hard substrates. In *Marine and coastal Systems of the Quoddy Region*, Canadian Special Publication of Fisheries and Aquatic Sciences, No. 64, (ed. M.L.H. Thomas), pp. 119-39. Department of Fisheries and Oceans, Ottawa.

Lubchenco, J. (1978). Plant species diversity in a marine intertidal community: importance of herbivore food preference and algal competitive abilities. *American Naturalist*, **112**, 23-39.

Lubchenco, J. (1980). Algal zonation in the New England rocky intertidal community: an experimental analysis. *Ecology*, **61**, 333-44.

Lubchenco, J. (1983). *Littorina* and *Fucus*: effects of herbivores, substratum heterogeneity, and plant escapes during succession. *Ecology*, **64**, 1116-23.

Lubchenco, J. (1986). Relative importance of competition and predation: early colonization by seaweeds in New England. In *Community ecology*, (ed. J. Diamond and T.J. Case), pp. 537-55. Harper and Row, New York.

Lubchenco, J. and Menge, B.A. (1978). Community development and persistence in a low rocky intertidal zone. *Ecological Monographs*, **59**, 67-94.

Mann, K.H. (1972). Ecological energetics of the seaweed zone in a marine bay on the Atlantic coast of Canada. I. Zonation and biomass of seaweeds. *Marine Biology*, **12**, 1-10.

Mann, K.H. (1973). Seaweeds: their productivity and strategy for growth. *Science*, **182**, 975-81.

Mann, K.H. (1982). Kelp, sea urchins and predators: a review of strong interactions in rocky subtidal systems of eastern Canada, 1970-1980. *Netherlands Journal of Sea Research*, **16**, 414-23.

Mann, K.H. (1985). Invertebrate behaviour and the structure of marine benthic communities. In *Behavioural ecology: ecological consequences of adaptive behaviour*, (ed. R. M. Sibley and R. H. Smith), pp. 227-46. Blackwell, London.

Mann, K. H. and Breen, P. A. (1972). The relation between lobster abundance, sea-urchins, and kelp beds. *Journal of the Fisheries Research Board of Canada*, **29**, 603-9.

Mathieson, A. C. (1979). Vertical distribution and longevity of subtidal seaweeds in northern New England, U.S.A. *Botanica Marina*, **22**, 511-20.

Mathieson, A. C. and Penniman, C. A. (1986). A phytogeographic interpretation of the marine flora from the Isles of Shoales, U.S.A. *Botanica Marina*, **29**, 413-34.

Menge, B. A. (1976). Organization of the New England rocky intertidal community: role of predation, competition, and environmental heterogeneity. *Ecological Monographs*, **46**, 355-93.

Menge, B. A. (1978). Predation intensity in a rocky intertidal community: effect of an algal canopy, wave action and desiccation on predator feeding rates. *Oecologia* (Berlin), **34**, 17-35.

Miller, R. J. (1985a). Seaweeds, sea urchins, and lobsters: a reappraisal. *Canadian Journal of Fisheries and Aquatic Sciences*, **42**, 2061-72.

Miller, R. J. (1985b). Succession in sea urchin and seaweed abundance in Nova Scotia, Canada. *Marine Biology*, **84**, 275-86.

Miller, R. J. and Colodey, A. G. (1983). Widespread mass mortalities of the green sea urchin in Nova Scotia, Canada. *Marine Biology*, **73**, 263-7.

Miller, R. J., Mann, K. H., and Scarratt, D. J. (1971). Production potential of a seaweed-lobster community in eastern Canada. *Journal of the Fisheries Research Board of Canada*, **28**, 1733-8.

Mohn, R. K. and Miller, R. J. (1987). A ration-based model of a seaweed-sea urchin community. *Ecological Modelling*, **37**, 249-67.

Moore, D. S. and Miller, R. J. (1983). Recovery of macroalgae following widespread sea-urchin mortality with a description of the nearshore hard-bottom habitat on the Atlantic coast of Nova Scotia. *Canadian Technical Report of Fisheries and Aquatic Sciences*, **1230**, 94 pp.

Novaczek, I. and McLachlan, J. (1986). Recolonization by algae of the sublittoral habitat of Halifax County, Nova Scotia, following the demise of sea-urchins. *Botanica Marina*, **29**, 69-73.

Pederson, J. B. and Capuzzo, J. M. (1984). Energy budget of an omnivorous rocky shore amphipod, *Calliopius laeviusculus* (Kroyer). *Journal of Experimental Marine Biology and Ecology*, **76**, 277-91.

Petraitis, P. S. (1983). Grazing patterns of the periwinkle and their effect on sessile intertidal organisms. *Ecology*, **64**, 522-33.

Petraitis, P. S. (1987). Factors organizing rocky intertidal communities of New England: herbivory and predation in sheltered bays. *Journal of Experimental Marine Biology and Ecology*, **109**, 117-36.

Rice, E. L., Kenchington, T. J., and Chapman, A. R. O. (1985). Intraspecific geographic–morphological variation patterns in *Fucus distichus* and *F. evanescens*. *Marine Biology*, **88**, 207–15.

Scheibling, R. E. (1984). Echinoids, epizootics and ecological stability in the rocky subtidal off Nova Scotia, Canada. *Helgoländer Meeresuntersuchungen*, **37**, 233–42.

Scheibling, R. E. (1986). Increased macroalgal abundance following mass mortalities of sea-urchins *Strongylocentrotus droebachiensis* along the Atlantic coast of Nova Scotia. *Oecologia* (Berlin), **68**, 186–98.

Scheibling, R. E. and Stephenson, R. L. (1984). Mass mortality of *Strongylocentrotus droebachiensis* (Echinodermata: Echinoidea) off Nova Scotia, Canada. *Marine Biology*, **78**, 153–64.

Sears, J. R. and Wilce, R. T. (1975). Sublittoral benthic marine algae of southern Cape Cod and adjacent islands: seasonal periodicity, associations, diversity, and floristic composition. *Ecological Monographs*, **45**, 337–65.

Sebens, K. P. (1986). Community ecology of vertical rock walls in the Gulf of Maine, U.S.A.: small-scale processes and alternative community states. In *The ecology of rocky coasts*, (ed. P. G. Moore and R. Seed), pp. 346–71. Columbia University Press, New York.

Steneck, R. S. (1983). Escalating herbivory and resulting adaptive trends in calcareous algal crusts. *Paleobiology*, **9**, 44–61.

Steneck, R. S. (1986). The ecology of coralline algal crusts: convergent patterns and adaptive strategies. *Annual Review of Ecology and Systematics*, **17**, 273–303.

Steneck, R. S. and Watling, L. (1982). Feeding capabilities and limitation of herbivorous molluscs: a functional group approach. *Marine Biology*, **68**, 299–319.

Sze, P. (1980). Aspects of the ecology of macrophytic algae in high rockpools at the Isles of Shoals (U.S.A.). *Botanica Marina*, **23**, 313–18.

Thomas, M. L. H. (1983). Tide pool systems. In *Marine and coastal systems of the Quoddy Region*, (ed. M. L. H. Thomas), Canadian Special Publication of Fisheries and Aquatic Sciences, No. 64, pp. 95–106. Department of Fisheries and Oceans, Ottawa.

Thomas, M. L. H. and Page, F. H. (1983). Grazing by the gastropod, *Lacuna vincta*, in the lower intertidal area at Musquash Head, New Brunswick, Canada. *Journal of the Marine Biological Association of the United Kingdom*, **63**, 725–36.

Thomas, M. L. H., Arnold, D. C., and Taylor, A. R. A. (1983). Rocky intertidal communities. In *Marine and coastal systems of the Quoddy Region*, (ed. M. L. H. Thomas), Canadian Special Publication of Fisheries and Aquatic Sciences, No. 64, pp. 35–73. Department of Fisheries and Oceans, Ottawa.

Topinka, J., Tucker, L., and Korjeff, W. (1981). The distribution of fucoid macroalgal biomass along central coastal Maine. *Botanica Marina*, **24**, 311-19.
Underwood, A.J. and Fairweather, P.G. (1989). Supply-side ecology and benthic marine assemblages. *Trends in Ecology and Evolution*, **4**, 16-20.
Vadas, R.L. (1979). Seaweeds: an overview; ecological and economic importance. *Experientia*, **35**, 429-32.
Vadas, R.L. (1991). Littorinid grazing and algal patch dynamics in Maine. *Journal of Molluscan Studies*. (In press.)
Vadas, R.L. and Beal, B. (1987). Green algal ropes: a novel estuarine phenomenon in the Gulf of Maine. *Estuaries*, **10**, 171-6.
Vadas, R.L. and Manzer, F. (1971). The use of aerial color photography for studies on rocky intertidal benthic marine algae. In *Proceedings of the Third Biennial Workshop on aerial color Photography in the plant Sciences and related Fields*, (ed. A. Anson), pp. 255-66. American Society of Photogrammetry.
Vadas, R.L. and Steneck, R.S. (1988). Zonation of deep water benthic algae in the Gulf of Maine. *Journal of Phycology*, **24**, 338-46.
Vadas, R.L. and Wright, W.A. (1986). Recruitment, growth, and management of *Ascophyllum nodosum*. In *Actas Segundo Congreso nacional sobre Algas marinas Chilenas*, pp. 101-13. Universidad Austral de Chile.
Vadas, R.L., Kesser, M., Larson, B., and Grant, W.S. (1977). Influence of *Littorina littorea* on algal zonation. *Proceedings of the International Seaweed Symposium*, **12**, 84.
Vadas, R.L., Elner, R.W., Garwood, P.E., and Babb, I.G. (1986). Experimental evaluation of aggregation behavior in the sea urchin *Strongylocentrotus droebachiensis*: a reinterpretation. *Marine Biology*, **90**, 433-48.
Vadas, R.L., Wright, W.A., and Miller, S.L. (1990). Recruitment of *Ascophyllum nodosum*: wave action as a source of mortality. *Marine Ecology Progress Series*, **61**, 263-72.
Vermeij, G.J. (1982). Environmental change and the evolutionary history of the periwinkle (*Littorina littorea*) in North America. *Evolution*, **36**, 561-80.
Wethey, D.S. (1985). Catastrophe, extinction, and species diversity: a rocky intertidal example. *Ecology*, **66**, 445-56.
Wharton, W.G. and Mann, K.H. (1981). Relationship between destructive grazing by the sea urchin, *Strongylocentrotus droebachiensis*, and the abundance of the American lobster, *Homarus americanus*, on the Atlantic coast of Nova Scotia. *Canadian Journal of Fisheries and Aquatic Science*, **38**, 1339-49.
Witman, J.D. (1985). Refuges, biological disturbance, and rocky subtidal community structure in New England. *Ecological Monographs*, **55**, 421-45.
Witman, J.D. (1987). Subtidal coexistence: storms, grazing, mutualism, and the zonation of kelps and mussels. *Ecological Monographs*, **57**, 167-87.

Wolfe, J. M. and Harlin, M. M. (1988a). Tidepools in southern Rhode Island, U.S.A. I. Distribution and seasonality of macroalgae. *Botanica Marina*, **31**, 525-36.

Wolfe, J. M. and Harlin, M. M. (1988b). Tidepools in southern Rhode Island, U.S.A. II. Species diversity and similarity analysis of macroalgal communities. *Botanica Marina*, **31**, 537-46.

3. How important is grazing to seaweed evolution and assemblage structure in the north-east Pacific?

MICHAEL S. FOSTER

Moss Landing Marine Laboratories and San Jose State University, Moss Landing, California, USA

Abstract

Numerous studies have shown that grazing by large invertebrates can be important in shaping seaweed assemblage structure, and that such grazing may have been important to macroalgal evolution. The questions of how important this interaction is to assemblage structure along the temperate shores of the north-east Pacific, and how important it may have been to evolution in this geographic region are reviewed. The reported effects of grazing on foliose algal cover at spatial scales from geographic to small patches suggest that grazing is generally (at most of the sites studied) important only in high intertidal–splash zone assemblages, where limpets, littorines, and other herbivores prevent the establishment of most seaweeds. While sea urchins can greatly reduce the cover of foliose plants in subtidal assemblages, this effect is highly variable in space and time. The importance of grazing to patch structure at scales of centimetres to metres is potentially high but remains largely unexplored. While grazers may have large effects in cleared areas during the early stages of intertidal succession, these effects are short-lived and grazer densities are only weakly correlated with recovery rate.

Contrasts between assemblages over large areas in the geographic region, and among assemblages within local sites, suggest that the importance of grazing increases when intertidal desiccation and subtidal physical disturbance reduce the availability of plant biomass. This simple relationship is, however, confounded by differences in the biology of the seaweeds and the grazers.

Plant–Animal Interactions in the Marine Benthos, (ed. D. M. John, S. J. Hawkins, and J. H. Price), Systematics Association Special Volume No. 46, pp. 61–85. Clarendon Press, Oxford, 1992. © The Systematics Association, 1992.

There is little strong evidence for most of the suggested seaweed 'defences' against grazers in the north-east Pacific, or for grazing as an important agent of selection in the evolution of these plants. It is proposed that seaweeds in this region are, in fact, weeds; that their biological attributes reflect the primary importance of seasonal and episodic physical disturbance.

Introduction

Of all the possible interactions between plants and animals on rocky intertidal shores and subtidal reefs in the north-east Pacific, grazing has probably received the most attention. This interaction has been summarized and discussed in a number of recent regional (Dayton 1985; Foster and Schiel 1985; Foster et al. 1988) as well as general (Branch 1981; Lubchenco and Gaines 1981; Gaines and Lubchenco 1982; Hawkins and Hartnoll 1983; Schiel and Foster 1986; Harrold and Pearse 1987; Hay and Fenical 1988; Horn 1989) reviews. Numerous grazing interactions in the north-east Pacific can have significant effects on local seaweed assemblages, and it has been suggested that a variety of macroalgal traits evolved as a result of such interactions.

In this paper the literature and some unpublished data have been used to evaluate the importance of grazing to seaweed evolution and assemblage structure in this geographical region. In its most rigorous form, this evaluation would require determining the proportion of the variance in assemblage structure and plant traits accounted for by grazing relative to other factors (*sensu* Welden and Slauson 1986). Such rigour is impossible given the limitations of available information, but I hope that a review of what is known about the relative importance of factors that affect seaweeds in the region will at least contribute to a more scientific evaluation of the evidence, narrow the boundaries of speculation, and perhaps stimulate attempts to better understand these plants.

The review focusses on grazing by macroinvertebrates along the coasts of California, Oregon, and Washington (ca. 32°–50°N), and particularly coasts north of Point Conception in California (>35°N). Most of the relevant studies have been done in these areas and on invertebrate grazers. The number and activity of herbivorous fishes are low, especially north of Point Conception (see Horn 1989). The geographical generality of the conclusions will be affected both by these biases and because the massive, protected and semi-protected, rocky reefs which have been most studied may not be the most common rocky habitats in the region (Foster et al. 1988; Foster 1990).

Grazing and seaweed assemblages

Factors and their interactions may directly (e.g., a sea urchin eats a seaweed) or indirectly (e.g., high-water motion prevents a sea-star from eating a sea urchin, and that sea urchin eats a seaweed) affect organisms. I will focus primarily on direct interactions. Ultimately, indirect effects and history must be known in order to understand why grazing varies in importance (Lubchenco and Gaines 1981; Underwood 1986; Menge and Sutherland 1987; Foster 1990), but most of the available evidence for evaluating the importance of factors concerns direct effects and even at this level there is considerable debate over factor primacy (e.g., Connell 1983; Sih et al. 1985). Sorting out what we do know about direct effects and their variation is necessary to evaluate the reality of higher-level syntheses. It also avoids confusion of causal levels (e.g., is water motion, predation, or grazing more important to seaweed mortality in the example above?) before we have the information to confront them rigorously.

1. Geographical patterns

There are great differences in the biota at the latitudinal extremes of the north-east Pacific, but changes in between are gradual, particularly in cold-water habitats north of Point Conception (Scagel 1963; Foster et al. 1988). Geographical differences in algal assemblages are generally attributed directly to changes in temperature and other water-mass characteristics (Murray et al. 1980; Druehl 1981), but may also be related to increases in desiccation south of Point Conception (Seapy and Littler 1982; Barry 1988). Latitudinal changes in grazers may influence assemblage structure (Gaines and Lubchenco 1982), and Sousa et al. (1981) suggest that reduced grazing at more northern latitudes may be one reason why, in the low intertidal zone, there are more kelp assemblages in the north and more algal turf assemblages in the south. Such differences, however, may also be related to increased disturbance by sand in the south (Seapy and Littler 1982; Taylor and Littler 1982; Stewart 1989). There are too few data to evaluate the various alternatives and, as Dethier and Duggins (1988) have shown, even if community composition is similar, the effect of grazers can be quite different in different geographical regions.

2. Patterns among local sites

In local areas where a common species pool is likely, there may be great differences in algal assemblage structure between nearby sites. These differences are commonly associated with variation in local oceanographic conditions, substratum characteristics, and water motion (e.g., Neushul 1965; Stephenson and Stephenson 1972; Foster 1982a; Schiel and Foster 1986; Foster et al. 1988; Harrold et al. 1988; Littler et al. 1991). In one of the few studies that have examined intertidal algal assemblage organization along a wave-exposure gradient, Dayton (1975) found that, with the exception of patches caused by sea urchin grazing, competition between algae was most important in affecting structure. Littler (1980) concluded that variation in local sites in southern California was due primarily to local differences in physical characteristics such as wave-action, substratum stability, and sand inundation.

Based on work at three kelp-forest sites, Dayton et al. (1984) indicated that differences in patch stability between sites were attributable to physical differences (e.g., removal of plants by water motion), and the importance of such direct physical effects on overall kelp-forest assemblage structure has been documented in other kelp forests (Cowen et al. 1982; Foster 1982a; Harrold et al. 1988). Sea-urchin grazing can cause great differences between subtidal sites in the Aleutian Islands (Estes and Palmisano 1974), and has been shown to affect algal abundance at additional subtidal sites in south-east Alaska (Duggins 1980) and British Columbia (Foreman 1977). However, as I have argued elsewhere (Foster and Schiel 1988), without broad scale surveys such as those available for California, it is impossible to know the extent of grazing effects on local sites in these areas. While there are numerous examples of sea urchin grazing within local sites in California (see below), there are relatively few examples of deforestation over large (>100 m) areas (Foster and Schiel 1988). Thus, with few exceptions, the available evidence indicates that physical factors are generally most important at the scale of local sites.

3. Patterns within sites

Of all the spatial variation evident in algal assemblages, changes with tidal height and depth probably are the best described and most thoroughly investigated, particularly on semi-protected rocky bottoms (Table 3.1). The algal assemblages in these habitats range from the splash zone characterized by ephemeral micro- and macroalgae to the deep subtidal dominated by perennial, non-geniculate coralline algae. There is generally a peak in total cover, species richness (Aleem 1956; Neushul 1967) and, no doubt, productivity in the low intertidal to mid-subtidal zones, and declines in all these attributes both up the shore and down into deeper water. While almost all kinds of grazers can be found throughout this depth range, relatively small, sedentary limpets and littorines are most common high on the shore, larger and more mobile gastropods are most conspicuous in lower intertidal assemblages, and sea urchins (primarily *Strongylocentrotus* spp.) are frequent in the subtidal zone.

Studies, particularly those involving experimental removal, suggest that grazing is the most important factor affecting assemblage structure in the splash zone. In this assemblage in the north-east Pacific (and often elsewhere: Branch 1981; Hawkins and Hartnoll 1983) limpets, littorines, insect larvae, and grapsid crabs often keep long stretches of coast free of macroalgae for most of the year (citations in Table 3.1). At progressively lower tidal levels, foliose plant cover increases, and changes in distribution from small (cm) patches surrounded by bare rock (common in the high intertidal zone) to more uniform cover interrupted by smaller and smaller patches of bare rock and/or crustose algae (mid-to low intertidal: Foster 1990 and personal observations). The more uniform cover is usually composed of patches of various foliose species. Foliose algal cover may be reduced or absent at some low intertidal sites where sea urchins are aggregated in pools or burrowed into the rock, or where large, mobile chitons occur (Paine and Vadas 1969; Gaines 1985).

Shores in southern California (south of Point Conception) may differ from this general pattern. Murray (pers. comm.) and Barry (1988) indicate that foliose algae are rare year-around in the splash zone in this region, and that bare rock and compact algal turfs commonly occur down to the low intertidal zone. Relative to coasts north of Point Conception, this lack of algal growth may be a result of increased desiccation caused by reduced wave-action and higher air temperatures (particularly during winter when low tides occur during the day). Barry (1988) has found that grazing by the chiton *Nuttallina kata* can greatly affect high and mid-intertidal algal assemblages under such circumstances.

In the shallow and mid-subtidal zones, foliose plant cover often exceeds 100 per cent if all canopies are considered (e.g., Reed and Foster 1984). This high cover is the norm in California kelp forests, but sea-urchin grazing can produce areas nearly devoid of foliose algal cover, on scales of metres to hundreds of metres, in these forests (Foster and Schiel 1988) and elsewhere in the north-east Pacific (e.g., Pace 1981; reviewed in Harrold and Pearse 1987). Seaweed assemblages may vary temporally in the splash and high intertidal zones, owing in part to grazing (Cubit 1984), but I am not aware of any reports of grazers causing such variation lower in the intertidal zone. There are, however, numbers of reports of temporal variation in algal cover in kelp forests caused directly by grazing (North and Pearse 1970; Foreman 1977; Pearse and Hines 1979; Dean *et al.* 1984; Harrold and Reed 1985; Ebeling *et al.* 1985).

Table 3.1. A summary of algal and grazer assemblages, and grazer effects, on semi-protected shores and subtidal reefs in the central northeast Pacific. Based on reviews (Dawson *et al.* 1960; Ricketts *et al.* 1985; Foster and Schiel 1985; Foster *et al.* 1988) as well as the references listed and personal observations.

Zones	Common algae	Common grazers	Grazing effects	References
Intertidal				
Splash	Diatoms *Ulva Urospora Porphyra*	Limpets Littorines Insect larvae Crabs (*Pachygrapsus*)	Reduce flora to diatoms and juvenile seaweeds, especially in summer	Castenholtz 1961 Cubit 1984 Robles 1982
High	*Endocladia Mastocarpus Fucus Pelvetia*	Limpets Littorines Turbans (*Tegula*) Crabs (*Pachygrapsus*)	Patches in flora (rock or crusts) of varying size	Stimpson 1970 Dayton 1971
Mid	*Gigartina Iridaea Gelidium* Coralline algae	Limpets Turbans Chitons	Patches, but less frequent than in high zone	Foster 1982*b* Gaines 1985
Low	*Gigartina* Coralline algae	Limpets Turbans Chitons	Urchins can produce large cleared patches	Paine and Vadas 1969 Dayton 1975 Duggins and Dethier 1985
Subtidal Shallow	*Cryptopleura Laminaria Egregia Phyllospadix*	Sea urchins Crabs (*Pugettia*)	Limpets and chitons produce patches of varying size	

Information for the deep subtidal is scant, but suggests that sea urchins can have effects similar to those in shallower subtidal areas or, presumably as a result of reduced light, very little effect even if they are present. After a mass mortality of sea urchins at the deep edge of a kelp forest, Pearse and Hines (1979) found that foliose algae extended

Table 3.1. contd

Zones	Common algae	Common grazers	Grazing effects	References
Mid	Corallines *Gigartina* *Macrocystis* *Nereocystis* *Pterygophora* *Desmarestia*	Limpets Chitons Sea urchins	Urchins clear areas around crevices; can reduce flora to crusts over large areas Limpets/chitons produce small (cm) patches	Estes and Palmisano 1974 Pace 1981 Cowen *et al.* 1982 Ebeling *et al.* 1985 Harrold and Reed 1985 Foster and Schiel 1988
Deep	Non-geniculate corallines	Sea urchins	Urchin effects variable with depth?	Pearse and Hines 1979 (and see text)

into deeper water but did not occupy hard substrata below about 12 m depth. Agegian, Foster, and Heine (unpublished data) removed sea urchins from a 12 m wide strip between depths of 6–18 m at a site near that of Pearse and Hines (1979). Before removal, the area to be cleared and a control strip of the same size were dominated by non-geniculate coralline algae and sessile animals. Ten months after removal, foliose algae on the removal strip had increased, but only above 12 m depth even though sea urchins were still rare below 12 m. Assemblages at all depths on the control strip remained unchanged.

Numerous grazers can be found in many of the above assemblages where high foliose algal cover is common. Rather than affecting cover, grazers might alter species composition in these assemblages. Effects on cover are usually obvious from the presence of cleared patches with grazers inside; patches that appear to be physically suitable for foliose algal growth. In contrast, possible effects on species composition are not often obvious from simple observation. Data from long-term grazer removal and succession experiments, however, suggest that grazers do not cause widespread changes in species composition in the north-east Pacific. Duggins and Dethier (1985) kept low intertidal areas, with little algal cover, free of large chitons for four years. Foliose algal biomass increased dramatically and various kelps, that were previously very rare or absent, invaded the area. However, these kelps were eventually killed by 'stressful summer conditions' and winter storms maintained patchiness in the post-removal assemblage. Dayton (1975) found that changes in algal species composition after grazer removal in lower intertidal areas did not persist. Similarly, after mid-shore removals of both grazers and algae, Foster (1982b) found that the prior algal dominant quickly re-established even when grazers were excluded by fences.

Dean et al. (1988) showed that subtidal grazing by white sea urchins (*Lytechinus anamesus*) enhanced the survivorship of juvenile *Cystoseira osmundacea* relative to juvenile Laminariales, resulting in an increase of *C. osmundacea*. These authors suggested that this change might persist regardless of the fate of the sea urchins. Persistence may be unlikely, however, because *C. osmundacea* does not grow or reproduce well in low light intensities (Schiel 1985). It is likely that, with time, *Macrocystis pyrifera* plants would invade the patches, or that the surface canopies of surrounding *M. pyrifera* would shade the patches, inhibiting the growth and reproduction of *C. osmundacea*.

These observations suggest that the species composition of foliose algae, in areas where their cover is high, may not be greatly influenced by grazers. Instead, as found by Underwood and Jernakoff (1981), with good growing conditions foliose algae may outcompete some grazers for space.

Long-term studies of succession in high and mid-intertidal clearings also suggest that grazing may be relatively unimportant to assemblage structure except early in the successional process. We (Foster and colleagues at Kinnetic Labs) have been following succession in the high intertidal *Endocladia muricata–Mastocarpus papillatus* assemblage at six widely separated sites in central and northern California. Three plots (1 × 2 m each) were completely cleared and burned at each site in both spring and autumn. Recovery has been assessed at six-month intervals for over four years, by comparing species composition and abundance in randomly selected areas within the cleared plots and control plots using Bray–Curtis similarity indices. Grazers re-established after clearing, and highest grazer densities in the areas within clearings ($0 \cdot 18$ m^2) varied between 34 and 242. These densities are in grazer units; different species of grazers were scaled to a common grazer unit based on algal consumption experiments (e.g., for the three most common grazers: one *Tegula* sp. $= 1$; one limpet $= 0 \cdot 34$; one *Littorina* sp. $= 0 \cdot 07$). The rate at which the similarity of a cleared plot approached the range of similarities among replicate control plots (percentage recovery/month) was used as a measure of recovery rate. In both seasons the correlation between maximum grazer density and recovery rate was negative but weak, and it was only significant for plots cleared in the spring (spring: $n = 18$, $r = -0.49$, $p = 0.04$; autumn: $n = 18$, $r = -0.43$, $p = 0.08$).

Related studies in a mid-intertidal assemblage of mussels and algae examined edge effects due primarily to grazing. De Vogelaere (1991) found by observation and field experiments that initial floristic differences (primarily in ephemeral algal abundance) between borders and centres of clearings (Fig. 3.1b) were primarily due to differences in the abundance of the grazing chiton *Nuttallina californica* (Fig. 3.1a). However, as found by Sousa (1984), these differences did not persist. The floras in the two regions of the clearings became quite similar after 18 months (Fig. 3.1b–d). Because chitons eventually reinvaded all areas of the clearings, one could argue that chitons (or some other grazer such as crabs) removed the ephemerals, allowing later successional species to invade. While Sousa (1979) found that certain crabs did this in his experiments, he also pointed out that ephemerals can be removed by desiccation stress. In De Vogelaere's experiments, desiccation and removal by waves appeared most important in removing *Ulva* both during the first year after clearing and seasonally thereafter (De Vogelaere, personal observations, Fig. 3.1b). Moreover, there were early differences in chiton and *Ulva* abundance within the clearings (Fig. 3.1a, b), but these had little effect on later assemblage composition (Fig. 3.1c, d).

4. Patterns within assemblages

While the above evidence suggests that relative to other processes, grazing on rocky substrata north of Point Conception may not be important to the overall assemblage structure except in the splash zone and, to a much lesser degree, the mid-subtidal zone algal assemblages, it may be of great importance to patch structure within all assemblages. Centimetre- to metre-sized patches of bare rock or crustose algae are commonly observed among foliose algal stands. While these can arise as a direct result of physical disturbance (Sousa 1985) and modification of the physical environment by topographical variation (e.g., tide pools, crevices), they also can result from the localized effects of grazers such as limpets (Stimpson 1970). Close inspection of such patches in the intertidal and subtidal zones frequently reveals the presence of limpets, chitons, or sea urchins (Foster and Schiel 1988; Foster, personal observations).

These patches could originate in a number of ways, including immigration of grazers into small clearings made by some physical disturbance, or as a result of retreat into smaller areas as perennial algae

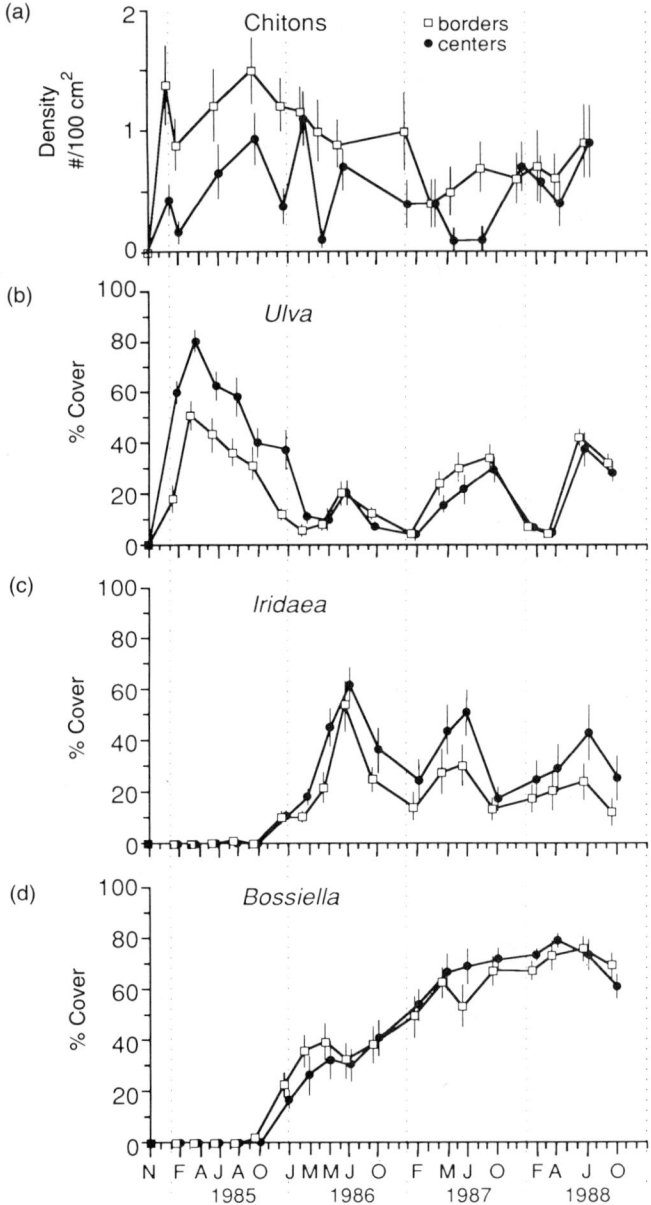

gradually re-occupy the larger areas after the latter have been disturbed. In either case, perhaps grazers (other than *Lottia gigantea*, which maintains a territory: Stimpson 1970) can maintain small clearings better than large clearings (e.g., due to contraction into most favourable micro-sites where their activities can prevent algal colonization, differences in behavioural effects at different spatial scales, protection from predators by surrounding algae, etc.). This will all be largely speculation (maybe the patches are maintained by other processes that simply also make the area suitable for grazers?) until we have a better understanding of patch structure and dynamics, and the effects of grazer removals on them. At present, there are very few published data on even the size distribution of patches of bare rock or crustose algae within assemblages dominated by foliose algae (Sousa 1985).

5. What indirect factors may account for variation in the importance of grazing?

While information on grazer effects is meager for many of the spatial scales discussed above, most of the available evidence suggests that the importance of grazing declines with decreasing tidal height into the mid-subtidal zone where, at least north of Point Conception in California, its importance is generally low but highly variable (Fig. 3.2*b*). Importance declines again in the deep subtidal zone.

This summary differs slightly from that proposed by Hawkins and Hartnoll (1983) for north-east Atlantic shores. These authors indicate that grazing is most important in the high and mid-intertidal zone (= eulittoral) and in the mid- to deep subtidal zone (Fig. 3.2*c*), and that the algal assemblage in the splash zone (= littoral fringe) is limited directly by physiological stress. The latter difference may be due to the presence of limpets in the north-east Pacific but not the north-east Atlantic (Figure 25*a* in Harkins and Hartnoll 1983). It also may be due to semantics; their littoral fringe may be above what I consider to be the splash zone, and the curve in Fig. 3.2*c* is shifted to the far right relative to Fig. 3.2*b*. This same sort of difficulty may apply when interpreting the results (relative to tidal height) of studies in the north-east Pacific. For example, while Robles (1982) described his work as being in the

---

Fig. 3.1. Changes in the abundance of the chiton, *Nuttallina californica* (a), the ephemeral alga, *Ulva* spp. (b), and the perennial algae, *Iridaea splendens* (c) and *Bossiella plumosa* (d) in borders (outer 10 cm) and centers of cleared plots (De Vogelaere 1991). $N = 12$ randomly located plots that varied in size from 50×50 to 150×150 cm. Plots were randomly located in a mid intertidal zone *Mytilus*/perennial algal assemblage, and completely cleared by scraping and burning in November 1984. Data are means ± 1 SE (standard error).

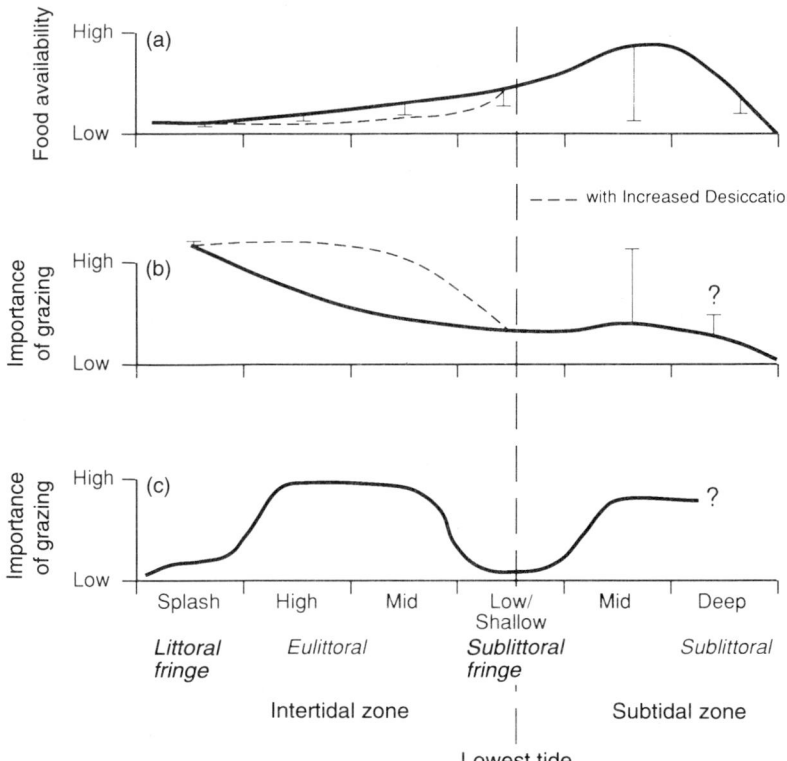

Fig. 3.2. A graphical summary or model of variation in grazer importance and its relationship to food availability on semi-protected shores in the Northeast Pacific, and a comparison with the model of Hawkins and Hartnoll (1983). (a) Changes in food availability (solid line: maximum availability with bars representing variation; dashed line: effect of increased desiccation). (b) Changes in grazing importance at the spatial scale of assemblage (solid line: minimum importance with bars representing variation; dashed line: effect of reduction in algal availability due to desiccation). (c) Hawkins and Hartnoll's (1983) model of grazer importance. See Table 3.1 for common algae and grazers for (a) and (b); ? = lack of information.

high-intertidal zone, some of his manipulations were as high as +3.3m, well above +1.5m, the generally accepted upper limit of the high-intertidal zone in central California (Ricketts *et al.* 1985). Because of such variation, and because different investigators use different zonation schemes, differences in grazing relative to exact zonal boundaries may be difficult to resolve. However, most investigators seem to be in agreement on general locations (high, mid, low), and site-specific information can thus be used to examine general trends.

It is interesting that in central California in areas above the splash zone (above the reach of the tide and all but exceptionally high waves) there are some littorines but few limpets (Ricketts et al. 1985). *Prasiola meridionalis* is a small, fleshy, green alga found above the splash zone around rocks occupied by birds in California (Abbott and Hollenberg 1976). Anderson (1987) found that the lower limit of this alga at one site was determined by grazers in the splash zone, and its horizontal distribution and seasonality by physiological tolerance, both findings in agreement with the summary of Hawkins and Hartnoll (1983) if one assumes that their littoral fringe is above my splash zone. The causes of suggested differences in grazer importance within the high to mid-intertidal zone (declining vs consistently high, Fig. 3.2b vs 3.2c) are unknown, but may be related to differences in desiccation discussed below.

Rather than a low but variable importance, Hawkins and Hartnoll (1983) suggest that grazing is consistently important in the mid- to deep subtidal (= sublittoral) of the north-east Atlantic. Possible reasons for this difference are discussed below.

Assuming that these relationships on semi-protected shores in the north-east Pacific reflect reality, what indirect effects may cause the variation in grazer importance? In general, I suggest that variation in food availability produces the strongest overall effect on the ability of grazers to modify algal assemblages. I agree with Hawkins and Hartnoll (1983) that physical factors are ultimately most important at high levels on the shore (physiological stress) and in deep water (insufficient light for plant growth). However, physiological stress appears to have a relatively greater impact on the seaweeds than on grazers in the splash zone (Cubit 1984) and, therefore, grazers can consume foliose algae before the algae reach adult size. At progressively lower intertidal levels, climatic and oceanographic conditions rapidly shift towards those more favourable for algal growth, and this increased growth apparently exceeds the grazers' ability to consume it.

Underwood and Jernakoff (1981) have shown that, in New South Wales (Australia), such dense growth in the lower intertidal zone increases grazer mortality by reducing the solid substrata available for attachment, especially during periods of high water motion (for other examples, see Branch 1981). The large chiton *Katharina tunicata* is an exception in the north-east Pacific, and can maintain high abundances in the low intertidal zone (Dethier and Duggins 1988). *Strongylocentrotus purpuratus* can also maintain high densities in this habitat, but usually only in patches where rock burrows or tide pools (Paine and Vadas 1969; Dayton 1975) presumably provide a refuge from removal by waves.

Hawkins and Hartnoll (1983) have suggested that competition among plants probably replaces grazing as the most important assemblage organizer lower on the shore, and this appears to be true in the north-east Pacific (Dayton 1975; Hruby 1976; Hodgson 1980; Foster 1982b).

As previously discussed, studies in southern California suggest that increased desiccation reduces algal growth in lower intertidal zones, and the importance of grazing may also extend lower in the intertidal zone. Reduced algal growth would tend to lower food availability (dashed line in Fig. 3.2a) and, based on the above arguments, shift the importance of grazing to the right (dashed line in Fig. 3.2b). Perhaps desiccation on north-east Atlantic shores has a greater effect on algal growth than on shores north of Point Conception? Alternatively, perhaps limpets in the north-east Atlantic are capable of removing more algal growth, as some limpets are in South Africa (*Patella cochlear*; Branch 1981).

In the mid subtidal in much of the north-east Pacific, the availability of food (in the form of drift algae) is high and grazing importance is low (Fig. 3.2a, b). In contrast to the north-east Atlantic, surface canopy plants commonly supply most of this drift (Harrold and Reed 1985), but are vulnerable to episodic removal by storms (Zobell 1971; Rosenthal et al. 1974; Foster 1982a; Tegner and Dayton 1987). When this occurs, food availability declines and sea urchins may alter their behaviour from passively consuming drift vegetation to active grazing on attached plants, reducing the algal assemblage to one dominated by non-geniculate coralline algae (Dean et al. 1984; Ebeling et al. 1985; Harrold and Reed 1985). The resulting temporal variability in importance at this depth can, therefore, be considerable (Fig. 3.2b). In addition to removal of kelps by storms, this dynamic range of assemblage structure can also be driven by the removal of sea urchins during storms (Ebeling et al. 1985) and by disease (Pearse et al. 1977), variation in algal and sea urchin recruitment (Harrold and Reed 1985; Ebert and Russell 1988; Reed et al. 1988), variation in temperature and nutrients that affect algal growth (Jackson 1977), and by variation in the availability of alternate food sources for sea urchins (Duggins 1981).

Numerous aspects of the biology of seaweeds and grazers further confound any simple relationship between food availability and grazing importance. As adults, seaweeds may not be affected by associated grazers because of their size, toughness, etc. Thus, although present, the resulting food is not available except as drift. In the subtidal zone, the importance of grazing can vary with sea urchin behaviour as well as density (Schiel 1982; Harrold and Reed 1985; Harrold and Pearse 1987). Other aspects, such as the ability of sea urchins (relative to gastropods) to starve but not die (Andrew 1989), no doubt add further complexity.

Grazing and seaweed functional morphology and chemistry

The diverse biological attributes of seaweeds are ignored when these plants are categorized as available food. For example, there is considerable potential food for grazing chitons in an attached, mature giant kelp plant, but, while they may scrape small bits off the holdfast, the rest of the plant is essentially unavailable as food. Seaweeds clearly have numerous life history, morphological, and chemical attributes that contribute to variation in their consumption by grazers. However, it is equally obvious that these attributes may have other functions unrelated to grazing (e.g., giant kelp surface-canopies place most of the biomass out of reach of benthic grazers, but also where there is more light).

There are numerous observations and experiments showing that seaweeds vary in their vulnerability or resistance to grazers in the north-east Pacific (e.g., some non-geniculate coralline algae vs fleshy algae: reviewed in Harrold and Pearse 1987; crustose vs foliose phases of the same species: Lubchenco and Cubit 1980; Slocum 1980; different crustose species: Kitting 1980). Moreover, a number of morphological (e.g., Gaines 1985; Steneck 1985; Van Alstyne 1989) and chemical (Steinberg 1985; Van Alstyne 1988) characteristics that may contribute to this variation have been studied. However, even in these cases it is not always clear whether this variation is a function of algal resistance, grazer preference, or some combination of the two (Kitting 1980), or if the characteristic was selected for by grazing.

The evidence for chemical resistance in the north-east Pacific is particularly weak. With the exception of Van Alstyne (1988), who showed that littorines removed less tissue from *Fucus* that had higher levels of phenolics, the field experiments necessary to determine if certain algal chemicals reduce consumption in natural field settings have not been done. While it may be interesting to identify grazer preferences in the laboratory (Vadas 1977; Steinberg 1984, 1985), such preferences may not be relevant to interactions under natural field conditions (Schiel 1982). Furthermore, experiments done in the field that do not duplicate the species composition and structural arrangement of natural algal–grazer associations (e.g., grazing by *Tegula funebralis* on *Alaria marginata*; Steinberg 1984) are insufficient. *Desmarestia* spp. is a particularly good example of the difficulties. While various observations (e.g., high acid content, association with sea urchins in the field, erosion of sea urchin teeth when fed *Desmarestia* in the laboratory) have led to the suggestion that it is chemically resistant to sea urchin grazing (reviewed in Hay and Fenical 1988), there is no clear field evidence that this alga is avoided by sea urchins in the north-east Pacific, and the observational evidence suggests it is not (e.g., Foster 1982*a*). *Desmarestia* spp. do occur in disturbed areas such as those created by sea urchins, but this may be a plant response to increased light (Reed and Foster 1984), not a result of avoidance by grazers.

Grazing and seaweed evolution

Are differences in algal resistance to grazing a result of algal evolution in response to grazers? If the function of a particular algal characteristic was clearly only to deter grazing, then negative correlations between the presence of the characteristic and the amount of grazing might be reasonably convincing. However, I am not aware of any characteristics of seaweeds in the north-east Pacific that have such an obvious anti-grazer function; seaweed thalli are generally less specialized for particular functions than terrestrial plants (e.g., blades, unlike most leaves, are the site of nutrient uptake as well as photosynthesis). Therefore, the possible agents of selection for a particular seaweed trait are probably even more diverse than those of terrestrial plants. For example, the morphological characteristics of non-geniculate coralline algae may have evolved in response to benthic grazers (Steneck 1983, 1985), but they also may have evolved in response to competition for space (e.g., Woelkerling and Foster 1989), or the often high water motion, sand abrasion and burial, fouling, and/or low light conditions typical of many of habitats in which they occur. The adaptive value of crustose and foliose phases in some seaweed life histories may also be unrelated to grazing (Dethier 1981), but perhaps associated with temporal variability in the physical environment (e.g., Mathieson 1982). Grazers may induce the formation of branches and grazing wounds (these may be higher in compounds that inhibit grazing), but similar branching may be produced by physical damage (Van Alstyne 1989) and the compounds thus induced may have evolved to help heal or reduce infection of damaged tissues.

There are numerous other examples of correlations between seaweed characteristics and variation in the physical environment (reviews in Neushul 1972; Littler and Littler 1980; Norton *et al.* 1981). Chemicals that deter some grazers may also deter parasites, epiphytes, endophytes, and/or fungal and bacterial pathogens (Levin 1976; Hay and Fenical 1988; Bernays *et al.* 1989). While these and other possible alternatives are often mentioned in the seaweed 'chemical defense' literature (e.g., Van Alstyne 1988), they are rarely investigated. Any putative 'adaptations to herbivory' (Duffy and Hay 1990), at least in the temperate waters of the north-east Pacific, may be adaptations to some other aspect(s) of the environment. They may be unrelated to grazing or 'pre-adaptations' to grazing (Schiel 1982; see general discussion of evidence for 'adaptive stories' in Gould and Lewontin 1979).

Given these alternative evolutionary interpretations, the lack of strong evidence for choosing among them, and the growing awareness that, even in ecological time, factor importance is highly variable, it seems at best premature to refer to any adaption of a seaweed in the north-east Pacific as a 'defense' against grazers. While seaweeds may resist grazing in various ways, they may not be doing so as a result of prior attacks by grazers. Speculations about higher order effects of grazing such as the evolution of kelps in the North Pacific (Estes and Steinberg 1988) are little more than 'just so' stories in this context (Jarvinen et al. 1986).

Carefully designed studies that evaluate multiple hypotheses concerning the adaptive value of particular seaweed traits under realistic field conditions are essential if these questions are to be addressed with rigour. In the north-east Pacific, species such as *Mastocarpus papillatus* (Slocum 1980; Zupan and West 1990) may be particularly good candidates for such studies, especially if the studies include contrasts between plants in habitats where grazing appears to vary greatly in importance (e.g., splash/high vs mid-intertidal zone).

An alternative hypothesis: seaweeds are weeds

In the interest of providing a general alternative hypothesis concerning seaweed evolution in the north-east Pacific, and a hypothesis for which there is considerable correlative evidence as well as evidence from the biology of terrestrial plants, I suggest that seaweeds are basically weeds. Relative to many of their vascularized, flowering, terrestrial counterparts, seaweeds generally recruit quickly, grow rapidly (less than one year to reproductive maturity), have a high fecundity (DeWreede and Klinger 1988), and are commonly self-compatible. Fertilization occurs via water transport and chemotaxis, not through animal vectors. Many perennial algae (perhaps all perennial red algae?) can reproduce vegetatively or regenerate from fragments. These characteristics are common among weedy species of flowering plants (Baker 1974; ruderals in Grime's (1979) classification). Although there are exceptions (Paine et al. 1979), seaweed life-spans generally appear to be similar to or shorter than those of the more common macro-invertebrates that graze on them, not longer. (In contrast, the turnover of potential seaweed pathogens is probably much quicker.) The few studies available suggest that seaweed dispersal may often be quite short for many species (e.g., Sousa 1984), perhaps because physical disturbances occur more frequently on smaller spatial scales than in terrestrial habitats. Dispersal also may be of the order of kilometres, especially during storms (Reed *et al.* 1988).

I suggest that most of these characteristics, like those of land weeds (Baker 1974), are adaptations for life in disturbed habitats, and that the wave- and surge-swept rocky shores and subtidal reefs of the north-east Pacific are highly disturbed. In addition to life history features, many morphological characteristics of seaweeds can be interpreted as adaptations to physical processes (examples above and Denny 1988). Thus, in the absence of better evidence to the contrary, I speculate that the direct effects of the physical environment have been and are most important in shaping seaweed evolution and assemblage structure in this region.

Acknowledgements

I thank J. Barry, A. De Vogelaere, M. Kingsford, and S. Murray for stimulating discussions on various topics in this paper; D. Hardin, L. Kiguchi, E. Nigg, and numerous others associated with Kinnetic Labs for their generous help and collaboration with the succession studies; A. De Vogelaere and J. Tarpley for providing some of their unpublished data, and J. Barry, A. De Vogelaere, S. Murray, M. Neushul, D. Reed, and D. Schiel for helpful comments on the manuscript. The succession research was supported by the US Minerals Management Service, Contract 14-12-0001-30057.

References

Abbott, I. A. and Hollenberg, G. J. (1976). *Marine algae of California.* Stanford University Press, California.

Aleem, A. A. (1956). Quantitative underwater study of benthic communities inhabiting kelp beds off California. *Science*, **123**, 183.

Anderson, B. S. (1987). Factors controlling the distribution of the high intertidal green alga, *Prasiola meridionalis*. Unpublished M. Sc. Thesis. San Jose State University, California.

Andrew, N. L. (1989). Contrasting ecological implications of food limitation in sea-urchins and herbivorous gastropods. *Marine Ecology Progress Series*, **51**, 189-93.

Baker, H. G. (1974). The evolution of weeds. *Annual Review of Ecology and Systematics*, **5**, 1-24.

Barry, J. P. (1988). Pattern and process: patch dynamics in a rocky intertidal community in southern California. Unpublished Ph.D. Thesis. University of California, San Diego.

Berneys, E. A., Driver, G. C., and Bilgener, M. (1989). Herbivores and plant tannins. *Advances in Ecological Research*, **19**, 263-302.

Branch, G. M. (1981). The biology of limpets: physical factors, energy flow, and ecological interactions. *Oceanography and Marine Biology Annual Review*, **19**, 235-380.

Castenholz, R. W. (1961). The effect of grazing on marine littoral diatom populations. *Ecology*, **42**, 783-94.

Connell, J. H. (1983). On the prevalence and relative importance of interspecific competition: evidence from field experiments. *American Naturalist*, **122**, 661-96.

Cowen, R. K., Agegian, C. R., and Foster, M. S. (1982). The maintenance of community structure in a central California kelp forest. *Journal of Experimental Marine Biology and Ecology*, **64**, 189-201.

Cubit, J. D. (1984). Herbivory and the seasonal abundance of algae on a high intertidal rocky shore. *Ecology*, **65**, 1904-17.

Dawson, E. Y., Neushul, M., and Wildman, R. D. (1960). Seaweeds associated with kelp beds along southern California and northwestern Mexico. *Pacific Naturalist*, **1**, 1-81.

Dayton, P. K. (1971). Competition, disturbance, and community organization: the provision and subsequent utilization of space in a rocky intertidal community. *Ecological Monographs*, **41**, 351-89.

Dayton, P. K. (1975). Experimental evaluation of ecological dominance in a rocky intertidal algal community. *Ecological Monographs*, **45**, 137-59.

Dayton, P. K. (1985). Ecology of kelp communities. *Annual Review of Ecology and Systematics*, **16**, 215-45.

Dayton, P. K., Currie, V., Gerrodette, T., Keller, B. D., Rosenthal, R., and VenTresca, D. (1984). Patch dynamics and stability of some California kelp communities. *Ecological Monographs*, **54**, 253-89.

Dean, T. A., Jacobsen, F. R., Thies, K., and Lagos, S. L. (1988). Differential effects of grazing by white sea-urchins on recruitment of brown algae. *Marine Ecology Progress Series*, **48**, 99-102.

Dean, T. A., Schroeter, S. C., and Dixon, J. D. (1984). Effects of grazing by two species of sea urchins (*Strongylocentrotus franciscanus* and *Lytechinus anamesus*) on recruitment and survival of two species of kelp (*Macrocystis pyrifera* and *Pterygophora californica*). *Marine Biology*, **78**, 301-13.

Denny, M. W. (1988). *Biology and the mechanics of the wave-swept environment*. Princeton University Press.

Dethier, M. N (1981). Heteromorphic alga life histories: the seasonal pattern and response to herbivory of the brown crust, *Ralfsia californica*. *Oecologia* (Berlin), **48**, 333-39.

Dethier, M. N., and Duggins, D. O. (1988). Variation in strong interactions in the intertidal zone along a geographical gradient: a Washington-Alaska comparison. *Marine Ecology Progress Series*, **50**, 97-105.

De Vogelaere, A. P. (1991). Disturbance, succession and distribution patterns in rocky intertidal communities of central California. Unpublished Ph.D. Thesis. University of California, Santa Cruz.

DeWreede, R. E., and Klinger, T. (1988). Reproductive strategies in algae. In *Plant reproductive ecology*, (ed. J. L. Doust and L. L. Lovett), pp. 267-84. Oxford University Press.

Druehl, L. D. (1981). Geographical distribution. In *The biology of seaweeds*, (ed. C. S. Lobban and M. J. Wynne), pp. 306-25. University of California Press, Berkeley.

Duffy, J. E. and Hay, M. E. (1990). Seaweed adaptations to herbivory. *BioScience*, **40**, 368-75.

Duggins, D. O. (1980). Kelp beds and sea urchins: an experimental approach. *Ecology*, **61**, 447-53.

Duggins, D. O. (1981). Sea-urchins and kelp: the effects of short-term changes in urchin diet. *Limnology and Oceanography*, **26**, 391-4.

Duggins, D. O. and Dethier, M. N. (1985). Experimental studies of herbivory and algal competition in a low intertidal habitat. *Oecologia* (Berlin), **67**, 183-91.

Ebeling, A. W., Laur, D. R., and Rowley, R. J. (1985). Severe storm disturbances and reversal of community structure in a southern California kelp forest. *Marine Biology*, **84**, 287-94.

Ebert, T. A. and Russell, M. P. (1988). Latitudinal variation in size structure of the west coast purple sea urchin: a correlation with headlands. *Limnology and Oceanography*, **33**, 286-94.

Estes, J. A. and Palmisano, J. F. (1974). Sea otters: Their role in structuring nearshore communities. *Science*, **185**, 1058-60.

Estes, J. A. and Steinberg, P. D. (1988). Predation, herbivory, and kelp evolution. *Paleobiology*, **14**, 19-36.

Foreman, R. E. (1977). Benthic community modification and recovery following intensive grazing by *Strongylocentrotus drobachiensis*. *Helgoländer Wissenschaftlichen Meeresuntersuchungen*, **30**, 468-84.

Foster, M. S. (1982a). The regulation of macroalgal associations in kelp forests. In *Synthetic and degradative processes in marine macrophytes*, (ed. L. Srivastava), pp. 185-205. Walter de Gruyter, Berlin.

Foster, M. S. (1982b). Factors controlling the intertidal zonation of *Iridaea flaccida* (Rhodophyta). *Journal of Phycology*, **18**, 285-94.

Foster, M. S. (1990). Organization of macroalgal assemblages in the Northeast Pacific: the assumption of homogeneity and the illusion of generality. *Hydrobiologia*, **192**, 21-33.

Foster, M. S. and Schiel, D. R. (1985). *The ecology of giant kelp forests in California: a community profile*. US Fish and Wildlife Service Biological Report 85. US Fish and Wildlife Service, Washington, DC.

Foster, M. S. and Schiel, D. R. (1988). Kelp communities and sea otters: keystone species or just another brick in the wall? In *The community ecology of sea otters*, (ed. G. R. VanBlaricom and J. A. Estes), pp. 92–115. Springer-Verlag, Berlin.

Foster, M. S., De Vogelaere, A. P., Harrold, C., Pearse, J. S., and Thum, A. B. (1988). Causes of spatial and temporal patterns in rocky intertidal communities of central and northern California. *Memoirs of the California Academy of Sciences*, **9**, 1–45.

Gaines, S. D. (1985). Herbivory and between-habitat diversity: the differential effectiveness of defenses in a marine plant. *Ecology*, **66**, 473–85.

Gaines, S. D. and Lubchenco, J. (1982). A unified approach to marine plant–herbivore interactions. II. Biogeography. *Annual Review of Ecology and Systematics*, **13**, 111–38.

Gould, S. J. and Lewontin, R. C. (1979). The spandrels of San Marco and the Panglossian paradigm: a critique of the adaptationist programme. *Proceedings of the Royal Society of London Series B*, **205**, 581–98.

Grime, J. P. (1979). *Plant strategies and vegetation processes*. John Wiley, New York.

Harrold, C. and Pearse, J. S. (1987). The ecological role of echinoderms in kelp forests. In *Echinoderm studies vol. II*, (ed. M. Jangoux and J. M. Lawrence), pp. 137–233. Balkema, Rotterdam.

Harrold, C. and Reed, D. C. (1985). Food availability, sea urchin grazing, and kelp forest community structure. *Ecology*, **66**, 1160–9.

Harrold, C., Watanabe, J., and Lisin, S. (1988). Spatial variation in the structure of kelp forest communities along a wave exposure gradient. *P.S.Z.N.I.: Marine Ecology*, **9**, 131–56.

Hawkins, S. J. and Hartnoll, P. G. (1983). Grazing of intertidal algae by marine invertebrates. *Oceanography and Marine Biology Annual Review*, **21**, 195–282.

Hay, M. E. and Fenical, W. (1988). Marine plant-herbivore interactions: the ecology of chemical defence. *Annual Review of Ecology and Systematics*, **19**, 111–45.

Hodgson, L. (1980). Control of the intertidal distribution of *Gastroclonium coulteri* in Monterey Bay, California, USA. *Marine Biology*, **57**, 121–26.

Horn, M. H. (1989). Biology of marine herbivorous fishes. *Oceanography and Marine Biology Annual Review*, **27**, 167–272.

Hruby, T. (1976). Observations of algal zonation resulting from competition. *Estuarine and Coastal Marine Science*, **4**, 231–3.

Jackson, G. A. (1977). Nutrients and production of the giant kelp *Macrocystis pyrifera* off southern California. *Limnology and Oceanography*, **22**, 979–95.

Jarvinen, O., Babin, C., Bambach, R. K., Flugel, E., Fursich, F. T., Futuyma, D. J. et al. (1986). The neontologico-paleontological interface of community evolution: how do the pieces of the kaleidoscopic biosphere move? In *Patterns and processes in the history of life*, (ed. D. M. Raup and D. Jablonski), pp. 331–50. Springer-Verlag, Berlin.

Kitting, C. L. (1980). Herbivore–plant interactions of individual limpets maintaining a mixed diet of intertidal marine algae. *Ecological Monographs*, **50**, 527–50.

Levin, D. A. (1976). The chemical defenses of plants to pathogens and herbivores. *Annual Review of Ecology and Systematics*, **7**, 121–59.

Littler, M. M. (1980) Overview of rocky intertidal systems of southern California. In *The California Islands: proceedings of a multidisciplinarty symposium*, (ed. D. M. Power), pp. 265–306. Santa Barbara Museum of Natural History.

Littler, M. M. and Littler, D. S. (1980). The evolution of thallus form and survival strategies in benthic marine macroalgae: field and laboratory test of a functional form model. *American Naturalist*, **116**, 25–44.

Littler, M. M., Littler, D. S., Murray, S. N., and Seapy, R. R. Southern California intertidal ecosystems. In *Ecosystems of the world*, vol. 24, (ed. P. Nienhuis and A. C. Mathieson), Elsevier, Amsterdam. (In press).

Lubchenco, J. and Cubit, J. (1980). Heteromorphic life histories of certain marine algae as adaptations to variations in herbivory. *Ecology*, **61**, 676–87.

Lubchenco, J. and Gaines, S. D. (1981). A unified approach to marine plant–herbivore interactions. I. Populations and communities. *Annual Review of Ecology and Systematics*, **12**, 405–37.

Mathieson, A. C. (1982). Field ecology of the brown alga *Phaeostrophion irregulare* Setchell et Gardner. *Botanica Marina*, **25**, 67–85.

Menge, B. A. and Sutherland, J. P. (1987). Community regulation: variation in disturbance, competition, and predation in relation to environmental stress and recruitment. *American Naturalist*, **130**, 730–57.

Murray, S. N., Littler, M. M., and Abbott, I. A. (1980) Biogeography of the California marine algae with emphasis on the southern California islands. In *The California Islands: proceedings of a multidisciplinary symposium*, (ed. D. M. Power), pp. 325–39. Santa Barbara Museum of Natural History.

Neushul, M. (1965). SCUBA diving studies of the vertical distribution of benthic marine plants. In *Proceedings of the Fifth Marine Biology Symposium*, pp. 161–76. Acta Universitatis, Göteborg, Sweden.

Neushul, M. (1967). Studies of subtidal marine vegetation in western Washington. *Ecology*, **48**, 83–94.

Neushul, M. (1972). Functional interpretation of benthic marine algal morphology. In *Contributions to the systematics of benthic marine algae of the North Pacific*, (ed. I. A. Abbott and M. Kurogi), pp. 47–74. Japanese Phycology Society, Kobe.

North, W. J. and Pearse, J. S. (1970). Sea urchin explosion in southern California coastal waters. *Science*, **167**, 209.

Norton, T. A., Mathieson, A. C., and Neushul, M. (1981). Morphology and environment. In *The biology of seaweeds*, (ed. M.J. Wynne and C.S. Lobban), pp. 421–51. University of California Press, Berkeley.

Pace, D. (1981). Kelp community development in Barkley Sound, British Columbia following sea urchin removal. *Proceedings of the International Seaweed Symposium*, **8**, 457–63.

Paine, R.T. and Vadas, R.L. (1969). The effects of grazing by sea urchins, *Strongylocentrotus* spp., on benthic algal populations. *Limnology and Oceanography*, **14**, 710–19.

Paine, R.T., Slocum, C.J., and Duggins, D.O. (1979). Growth and longevity in the crustose red alga *Petrocelis middendorffii*. *Marine Biology*, **51**, 185–92.

Pearse, J.S. and Hines, A.H. (1979). Expansion of a central California kelp forest following mass mortality of sea urchins. *Marine Biology*, **51**, 83–91.

Pearse, J.S., Costa, D.P., Yellin, M.B., and Agegian, C.R. (1977). Localized mass mortality of red sea urchins, *Strongylocentrotus franciscanus*, near Santa Cruz, California. *Fishery Bulletin*, **75**, 645–8.

Reed, D.C. and Foster, M.S. (1984). The effects of canopy shading on algal recruitment and growth in a giant kelp forest. *Ecology*, **65**, 937–48.

Reed, D.C., Laur, D.R., and Ebeling, A.W. (1988). Variation in algal dispersal and recruitment: the importance of episodic events. *Ecological Monographs*, **58**, 321–35.

Ricketts, E.F., Calvin J., Hedgpeth, J.W., and Phillips, D.W. (1985). *Between pacific tides* (5th edn.). Stanford University Press.

Robles, C.D. (1982). Disturbance and predation in an assemblage of herbivorous Diptera and algae on rocky shores. *Oecologia*, **54**, 23–31.

Rosenthal, R.J., Clarke, W.D., and Dayton, P.K. (1974). Ecology and natural history of a stand of giant kelp, *Macrocystis pyrifera*, off Del Mar, California. *Fishery Bulletin*, **72**, 670–84.

Scagel, R.F. (1963). Distribution of attached marine algae in relation to oceanographic conditions in the northeast Pacific. In *Marine distributions*, (ed. M.J. Dunbar), pp. 37–50. Royal Society of Canada Special Publication No. 5. University of Toronto Press.

Schiel, D.R. (1982). Selective feeding by the echinoid, *Evechinus chloroticus*, and the removal of plants from subtidal algal stands in northern New Zealand. *Oecologia* (Berlin), **54**, 379–88.

Schiel, D.R. (1985). A short-term demographic study of *Cystoseira osmundacea* (Fucales: Cystoseiraceae) in central California. *Journal of Phycology*, **21**, 99–106.

Schiel, D.R. and Foster, M.S. (1986). The structure of subtidal algal stands in temperate waters. *Oceanography and Marine Biology Annual Review*, **24**, 265–307.

Seapy, R.R. and Littler, M.M. (1982). Population and species diversity fluctuations in a rocky intertidal community relative to severe aerial exposure and sediment burial. *Marine Biology*, **71**, 87–96.

Sih, A., Crowley, P., McPeek, M., Petranka, J., and Strohmeier, K. (1985). Predation, competition, and prey communities: a review of field experiments. *Annual Review of Ecology and Systematics*, **16**, 269-311.

Slocum, C.J. (1980). Differential susceptibility to grazers in two phases of an intertidal alga: advantages of heteromorphic generations. *Journal of Experimental Marine Biology and Ecology*, **46**, 99-110.

Sousa, W.P. (1979). Experimental investigations of disturbance and ecological succession in a rocky intertidal algal community. *Ecological Monographs*, **49**, 227-54.

Sousa, W.P. (1984). Intertidal mosaics: patch size, propagule availability, and spatially variable patterns of succession. *Ecology*, **65**, 1918-35.

Sousa, W.P. (1985). Disturbance and patch dynamics on rocky intertidal shores. In *The ecology of natural disturbance and patch dynamics*, (ed. S.T. Pickett and P.S. White), pp. 101-24. Academic Press, New York.

Sousa, W.P., Schroeter, S.C., and Gaines, S.D. (1981). Latitudinal variation in intertidal algal community structure: the influence of grazing and vegetative propagation. *Oecologia* (Berlin), **48**, 297-307.

Steinberg, P.D. (1984). Algal chemical defense against herbivores: allocation of phenolic compounds in the kelp *Alaria marginata*. *Science*, **223**, 405-7.

Steinberg, P.D. (1985). Feeding preferences of *Tegula funebralis* and chemical defenses of marine brown algae. *Ecological Monographs*, **55**, 333-49.

Steneck, R.S. (1983). Escalating herbivory and resulting adaptive trends in calcareous algal crusts. *Paleobiology*, **9**, 44-61.

Steneck, R.S. (1985). Adaptations of crustose coralline algae to herbivory: patterns in space and time. In *Paleoalgology: contemporary research and applications*, (ed. D.F. Toomey and M.H. Nitecki), pp. 352-66. Springer-Verlag, Berlin.

Stephenson, T.A. and Stephenson, A. (1972). *Life between tidemarks on rocky shores*. W.H. Freeman and Company, San Francisco.

Stewart, J.G. (1989). Establishment, presistence and dominance of *Corallina* (Rhodophyta) in algal turf. *Journal of Phycology*, **25**, 436-46.

Stimpson, J. (1970). Territorial behavior of the owl limpet *Lottia gigantea*. *Ecology*, **51**, 113-18.

Taylor, P.R. and Littler, M.M. (1982). The roles of compensatory mortality, physical disturbance, and substrate retention in the development and organization of a sand-influenced, rocky-intertidal community. *Ecology*, **63**, 135-46.

Tegner, M.J. and Dayton, P.K. (1987). El Niño effects on southern California kelp forest communities. *Advances in Ecological Research*, **17**, 243-79.

Underwood, A.J. (1986). Physical factors and biological interactions: the necessity and nature of ecological experiments. In *The ecology of rocky* coasts, (ed. P. G. Moore and R. Seed), pp. 372-90. Columbia University Press, New York.

Underwood, A.J. and Jernakoff, P. (1981). Effects of interactions between algae and grazing gastropods on the stucture of a low-shore intertidal algal community. *Oecologia* (Berlin), **48**, 221-33.

Vadas, R. L. (1977). Preferential feeding: an optimization strategy in sea-urchins. *Ecological Monographs*, **47**, 337-71.

Van Alstyne, K. L. (1988). Herbivore grazing increases polyphenolic defenses in the intertidal brown algal *Fucus distichus. Ecology*, **69**, 655-63.

Van Alstyne, K. L. (1989). Adventitious branching as a herbivore-induced defense in the intertidal brown alga *Fucus distichus. Marine Ecology Progress Series*, **56**, 169-76.

Welden, C. W. and Slauson, W. L. (1986). The intensity of competition versus its importance: an overlooked distinction and some implications. *Quarterly Review of Biology*, **61**, 23-44.

Woelkerling, W.J. and Foster, M. S. (1989). A systematic and ecographic account of *Synarthrophyton schielianum* sp. nov. (Corallinaceae, Rhodophyta) from the Chatham Islands. *Phycologia*, **28**, 39-60.

Zobell, C. E. (1971). Drift seaweeds on San Diego County beaches. *Nova Hedwigia*, **32**, 269-314.

Zupan, J. R. and West, J. A. (1990). Photosynthetic responses to light and temperature of the heteromorphic marine alga *Mastocarpus papillatus* (Rhodophyta). *Journal of Phycology*, **26**, 232-9.

4. Tropical east Atlantic and islands: plant–animal interactions on shores free of biotic reefs

DAVID M. JOHN, JAMES H. PRICE, and GEORGE W. LAWSON

Department of Botany, The Natural History Museum, Cromwell Road, London, UK

Abstract

The western coast of Africa and offshore islands lying between the Tropics contrast with most other tropical regions in the absence of biotic reefs. There is, however, a remarkable biotic resemblance between rocky shores of similar wave-exposure and type within the region, despite its vast size. Animals and non-geniculate coralline algae invariably dominate wave-exposed shores, whilst foliose and other soft algae dominate those protected from wave-action provided that algivorous fish are present in low numbers. Molluscs are unquestionably important grazers on many shores, although little information exists other than the observation that removal of *Siphonaria* from the upper eulittoral leads to a lush growth of blue-green algae. Unexplored experimentally is the grazing of seaweeds and microalgae by the larger limpet *Patella safiana*, littorinids, other molluscan genera and amphipods. Some experimental evidence exists that sea urchin grazing and seasonal changes in the physical environment are responsible for the absence or patchiness of non-calcareous algae at lower levels on unprotected shores. Shore organisms are affected by the climatic seasons: the stormy, wetter period favours intertidal algae by reducing desiccation while adversely affecting subtidal algae by causing instability of cobble banks and increasing turbidity; the calmer drier period has the reverse effect, with conditions more severe for shore algae and favourable to those growing subtidally. This seasonality affects biotic relationships as exemplified by green algae (in

Plant–Animal Interactions in the Marine Benthos, (ed. D. M. John, S. J. Hawkins, and J. H. Price), Systematics Association Special Volume No. 46, pp. 87–99. Clarendon Press, Oxford, 1992. © The Systematics Association, 1992.

the wetter period) out-competing barnacles that settled on primary space vacated by the death (during the drier period) of algae growing towards the uppermost limit of their vertical distribution. Many of the fish associated with rocky reefs showing much physical relief are algivorous and known to occur in tropical regions characterized by coral reefs. What little information exists concerning trophic relationships points to affinities of the tropical eastern Atlantic with the Caribbean — where knowledge of plant–animal interactions is more detailed.

Introduction

The western coast of Africa is remarkably free of indentations although it is frequently interrupted by the mouths of rivers flowing through sometimes wide estuaries or large deltas. Extensive lagoon systems are often to be found behind the sandy beaches that otherwise dominate long stretches of the coastline. Rocky headlands are present at intervals, the most striking being those forming the high cliffs on the Cap Blanc Peninsula and Cap Vert in Sénégal. Along most of the coastline, the restricted rocky areas take the form of low cliffs, wave-cut platforms, or boulder beaches. Harbour systems constructed of breakwaters of rough stone or concrete triads provide relatively recent *refugia* for shore organisms requiring hard and stable surfaces for attachment. Suitable surfaces are also present offshore and include, as well as rocky reefs, seasonally stable banks of cobbles or nodules, often referred to as rhodoliths, formed by non-geniculate coralline algae.

The Atlantic coast of Africa resembles the western shores of other continents (e.g., America) in lacking both coral reefs and the diversified life associated with these biogenic structures. The absence of coral reefs only partly accounts for the impoverishment of marine plant and animal life when compared to the biological diversity associated with coral reefs in tropical East Africa or the Caribbean (with which latter region West Africa shows its greatest biogeographical affinities; Lawson and John 1987). Also lacking from over 80 per cent of the West African coast considered here are the seagrass meadows that elsewhere in the tropics are a characteristic feature of protected lagoons lying landward of fringing reefs. Plant–animal interactions in seagrass ecosystems are reviewed by Orth (Chapter 7, present volume).

The majority of the ecological investigations of littoral assemblages in tropical West Africa have been descriptive, and the focus has tended to be on their composition, structure, and distribution (see John and Lawson 1991, for review). The few experimental studies or longer-term observations designed to obtain a better understanding of the functioning of these assemblages are confined to those associated with rocky reefs or offshore cobble banks — habitats where conditions are fully marine. These was a tendency in earlier investigations to seek explanations for distribution patterns in terms of physical, chemical, or historical factors, rather than biological ones. With the realization of the importance of plant–animal interactions, through experimental studies and following man-made disasters, it is to be hoped that proper consideration will now be given to biological factors.

Biological factors are many and include grazing pressure, fungal and microbial activity, competition for substratum, presence of a protective cover against desiccation during emersion, shading due to overgrowth, and availability of host plants or animals to obligate epiphytes, endophytes, epizoans, endozoans, and parasitic algae. In West Africa, interest has been directed only towards two types of biological interaction: *interspecific competition*, where two or more species utilize a finite resource which is limiting (e.g., available space), and *herbivory* or *algivory*.

The review covers the approximately 8000 mile (12 800 km) long coastline of mainland Africa lying between the tropics and gives some consideration to the islands of volcanic origin lying to the east of the Mid-Atlantic Ridge (Fig. 4.1). Briefly described are some of the more important features of intertidal and subtidal assemblages of benthic organisms associated with hard and relatively stable surfaces. In addition, the role of plant–animal interactions in determining spatial and temporal variations in the composition and structure of such assemblages is evaluated as far as possible. Experiments to test hypotheses concerning importance of grazing to seaweed assemblage structure have been confined to Ghana, a country bordering the Gulf of Guinea and lying about 400 miles (644 km) north of the Equator (see Fig. 4.1). Generalizations based on the findings are predicted to apply to most of the region, due to the remarkable biotic similarity when shores of similar type and exposure to wave-action are compared (John and Lawson 1991). The most noticeable changes in the flora and fauna of the mainland coast occur immediately to the south of the Angola–Namibia border and north of Cap Blanc. Although knowledge of biotic interactions here is limited, we believe there is sufficient to enable comparisons to be made with other tropical regions.

Spatial distributions and plant–animal interrelationships

1. The littoral zone

The distribution of organisms growing on the more accessible littoral zone of most West African shores has been investigated by John and Lawson (see Lawson and John 1987; John and Lawson 1991; for reviews). In common with rocky shores elsewhere, groups of plants and

Fig. 4.1. Map of the tropical Atlantic showing the African mainland and positions of islands. Islands and other geographical features mentioned in the text are indicated.

animals are restricted to definite belts or zones that form convenient bases for describing and comparing shores. These zones possess the universal features recognized and described earlier by Stephenson and Stephenson (1949) and later by Lewis (1961) in a modified scheme. Wave-exposure is recognized as the single most important factor modifying the zonation pattern. It is an especially significant factor in West Africa, where the tidal range is normally less than a metre. Undoubtedly, on any shore, the patterns of distribution at different spatial scales and the relative dominances in terms of cover-abundance of the benthic organisms are reflections of both physical factors (principally wave-action and desiccation mediated by tidal patterns) and biological interactions. These latter involve both intraspecific and interspecific competition for available space and, more directly, complete removal of, or extreme damage to, components of the ecosystem, caused during grazing or browsing.

Unquestionably, grazing is an important factor, as evidenced by the frequent presence in quantity of animals known to be grazers, especially of microalgae. Despite the lack of experimental evidence, patchiness or absence of algae, except the non-geniculate corallines, on more wave-exposed shores is very likely a consequence both of grazing by the larger numbers of algal feeders in such situations and the physical effects of wave action. The only experimental study designed to ascertain the importance of mollusc-grazing in West Africa was carried out by Gauld and Buchanan (1959) on a Ghanaian shore. They observed a rapid and prolific growth of blue-green algae following the removal of *Siphonaria pectinata*. Unexplored experimentally is the importance of other molluscan grazers, including larger limpets or littorinids (e.g., *Littorina punctata, L. granosa*); these latter are often very abundant and characterize the littoral fringe or splash zone. Similarly lacking is information on highly mobile grazers such as the amphipod *Ligia gracilipes* or the crab *Grapsus grapsus* that ranges over the entire shore, especially where there is much wave-action. Amphipods are occasionally found in large numbers in mats or clumps of algae (John and Lawson, personal observations), although shelter as much as feeding may also be involved (Price, personal observations, Antarctic *Desmarestia* spp.). The importance of the night-time feeding activities of active mesograzers such as amphipods has only become appreciated within the last few years (see Brawley, Chapter 11 present volume).

2. The sublittoral fringe

In much of West Africa and offshore islands the most obvious grazers of macroalgae on the lower shore are sea urchins. In this region they may be used to define the upper limit of the sublittoral fringe, since where they are present in quantity algae, other than the crusts of non-geniculate corallines, are virtually absent. No doubt the grazing activities of urchins profoundly influence the distribution of algae in both lower shore and the shallow subtidal. The interpretation of the results of the only experimental study on effects of sea urchin grazing on algal vegetation were complicated by the activities of algivorous fish. Following the experimental removal of the sea urchin *Echinometra lucunter* from a large pool in Ghana, there was significant increase in the ground cover of fleshy and filamentous algae (see John 1986, fig. 9.6).

This increase was principally due to algal gardens developing in territories of a damselfish whose numbers increased in the pool immediately following removal of the sea urchins. The grazing role of sea urchins in these pools seems to have been quickly taken over by non-territorial algivorous and omnivorous fish, as there was no increase in growth of macroscopic algae other than in the newly-formed damselfish territories. As in other tropical regions, algivorous fishes are common (Sanusi 1980) and influence the distribution of lower-shore algae on sheltered shores, since they follow the rising tide to feed actively on intertidal algae.

3. The sublittoral

(a) Grazing interactions Algivorous or omnivorous fish often congregate in considerable numbers around highly structured rocky areas (see Sanusi 1980) where the physical relief is believed to afford them protection from marauding carnivores. The majority of these fish are also known from the Caribbean, where they occur associated with coral rather than with rocky reefs. In West Africa, diversity of these reef-inhabiting fish is lower and the number of endemics significantly less when compared with the Caribbean (see Briggs 1974). The principal Pan-Atlantic non-territorial grazers in terms of population density and the amount of algae consumed belong to two families, the Acanthuridae (surgeonfishes) and Scaridae (parrotfishes). Another important family of generalist herbivores or omnivores is the reef-inhabiting Pomacentridae (damselfish). The adults of this family occupy a distinct territory, and the very aggressive behaviour that they display enables them to maintain algal gardens in their territories by preventing the grazing of other fish. Any conspicuous algae growing outside these damselfish territories might well possess adaptations that enable them to survive persistent grazing (see below).

The important effect of fish-grazing on the distribution of subtidal algae was first demonstrated experimentally in West Africa by John and Pople (1973). They followed the rapid disappearance of different algae transplanted into a habitat believed to be subject to chronic fish-grazing. Loss of tissue from algae placed a short distance away from the roughstone breakwater was found to be significantly less when compared with plants placed on it. This finding gives further support to the widely-held belief that habitats possessing much physical relief provide small reef-inhabiting demersal fish with exploitable long-term protection from marauding carnivores. The shores of the tropical eastern Atlantic island of Ascension are among the more unusual habitats thus far investigated. Virtually all subtidal habitats and accessible intertidal pools or surfaces on this island are intensely grazed by one omnivorous fish, the black triggerfish *Melichthys niger* (Price and John 1980). Its distribution on the island does not seems to be dependent on a prey–predator interaction but rather on physical factors such as severe water movement. Only in the few habitats unavailable to this fish are there appreciable growths of algae (e.g., *Sargassum*) other than the non-geniculate calcareous red algae and blue-green algae that appear to be resistant to its attentions.

An assemblage of algae dominated by *Sargassum* develops in the West African sublittoral fringe or shallow subtidal, provided only that grazing by sea urchins and/or algivorous fish is not too intense. This *Sargassum*-dominated algal community sometimes extends to a depth of 10 m, and is probably the long-term climax vegetation on rocky reefs in West Africa. It is rarely very extensive, as the conditions necessary for its development are uncommon, namely wave-protected to moderately exposed rocky areas where the bottom provides too few *refugia* to support large populations of algivorous fish. If numerous holes, crevices, ledges, and gullies are present then *Sargassum* is absent or else survives only as widely-scattered individuals growing in the form of basal rosettes of foliar appendages closely adpressed to the substratum (Price and John 1980). Such highly structured rocky subtidal areas are characterized by the following:

(1) patches of filamentous algae growing within damselfish territories;
(2) low species diversity and algal biomass;
(3) domination by non-geniculate coralline algae, gelatinous forms, turf-forming species, and clumps of foliose algae that are believed to owe their survival to possession of life history or structural adaptations (e.g., calcareous and crustose; see Steneck, this volume Chapter 21), and/or to biochemical defences against fish (see Hay and Fenical, Chapter 14 this volume).

The type of algal vegetation found outside damselfish territories might be regarded as representing a climax that has been shifted away from domination by *Sargassum* to one composed of species that are well-defended against grazing.

The influence of fish-grazing on the structure and composition of algal vegetation is most dramatically demonstrated when habitats showing much physical relief are compared with nearby areas of seabed where there are few *refugia* for algivorous fish (Lieberman *et al.* 1984). In West Africa, rocky reefs lying immediately offshore have much surface relief and sometimes abut low-lying banks of calcareous cobbles or rhodoliths that cover extensive areas of bottom especially in the Gulf of Guinea (see Lawson and John 1987). Often these banks are misleadingly referred to in the literature as 'coral banks'. During the calmer months of the dry season, the surface swell is insufficient to cause tumbling and abrasion of the surface of these cobbles. On the cessation of tumbling, a highly diverse community of filamentous and foliose algae develops rapidly on these seasonally unstable cobbles, forming what has been termed an 'algal plain' (Lieberman et al. 1979, 1984). Over these banks of cobbles, there are no resident fish or sea urchin populations despite the occurrence of large numbers of such grazers on rocky reefs near-by.

Another important subtidal grazer is likely to be the sea urchin *Diadema antillarum*. Off the Ghanaian coast, this urchin is commonly associated with the cliff areas of rocky reefs where sessile animals are common and the only conspicuous algae are non-geniculate corallines. It is likely that dominance of crustose forms owes much to the nocturnal activities of this grazer. On these reefs the principal grazers of crustose algae are suggested to be parrotfish. No experimental information exists concerning the grazing effects of either *Diadema* or parrotfish in West Africa. It has been suggested (John 1986) that the common occurrence of oysters and barnacles in areas where there are high concentrations of *Diadema* is due to sea urchin grazing that actually promotes establishment of these sessile animals, which are normally competitively inferior to ground-covering filamentous and fleshy algae.

Information on the feeding behaviour of other possible algivorous grazers is almost entirely lacking. Often evidence of the importance of other grazers comes from casual observations or the analysis of stomach contents. For example, Edmunds and Edmunds (1973) mention observing the sea slug *Aplysia winneba* feeding on bushy clumps of the red alga *Laurencia*.

(b) Non-trophic interactions   An interesting non-trophic interaction is the relationship existing between red algae growing on the turret shell *Turritella* and a small pagurid crab. Off the Ghanaian coast, the seabed from about 11 to 36 m depth is characterized by large numbers of *Turritella*, with dead shells very common at the upper and lower depth limits (Buchanan 1958). The original occupants of these shells seem to have been removed by carnivorous gastropods, and most are now occupied by hermit crabs. Shells still occupied by *Turritella* are rarely colonized by algae or bryozoans — its burrowing habit inhibits successful algal settlement and subsequent development. Empty shells also become buried by deposition and so similarly are unavailable. Thus, only those shells kept at the surface by the occupancy of surface-feeding hermit crabs are available for colonization.

Temporal change and plant–animal interactions

1. Seasonality

The various interactions taking place on any shore are constantly adjusting, especially to a significant seasonal component. This was first demonstrated in West Africa by Lawson (1957, 1966) who followed changes in the cover-abundance of algae and shore animals along transects placed on an intertidal reef at Christiansborg, Ghana. He demonstrated that there were regular seasonal invasions of the upper shore by many algae, followed later by their apparent retreat and subsequent replacement by certain animals. The barnacle *Chthamalus dentatus* reached its maximum cover between November and April and colonized lower-shore levels that are generally occupied by zone-forming algae; this is the period when conditions for the growth of shore algae are least favourable. From June to October the lower limit of *Chthamalus* moved up the shore and algae recolonized the higher shore levels (see Lawson 1957, fig. 4). These seasonal changes in distribution were found to correlate with desiccation operating in parallel with the pattern of the tides. During September to March (dry season months) the lowest of the two low waters of each diurnal period occurs during the hours of daylight, whereas the position for the remainder of the year (the rainy season) is reversed. Severe desiccation during the former period was considered to be responsible for bleaching and disappearance of algae that had earlier invaded the upper shore. It was also demonstrated by Lawson (1957) that algae are able to grow at levels on the shore above those where they are normally to be found, and competition was suggested as the limiting factor. In transect observations conducted in 1952–1954, consideration was given by Lawson to changes in abundance of the principal animals but not to whether they play a role, directly or indirectly, in determining the distribution of the algae. There is, however, some circumstantial evidence for the interactions of plants and animals. Thus, although attention was initially focused on seasonal changes described above, there were also some indications of long-term population trends. Most obvious of these was the considerable increase in numbers of *Littorina punctata* over a three-year period, while at the same time there appeared to be a reduction in numbers of *Siphonaria pectinata*. Again, over the same period, certain seaweeds (e.g., as *Chaetomorpha antennina* and *Wrangelia argus*) appeared to show a trend towards increasing cover values. It seems possible that this was due to decreased grazing by *Siphonaria*, since these algae were abundant when numbers of the molluscs were low and were much less abundant when *Siphonaria* was more common (Fig. 4.2). It should be

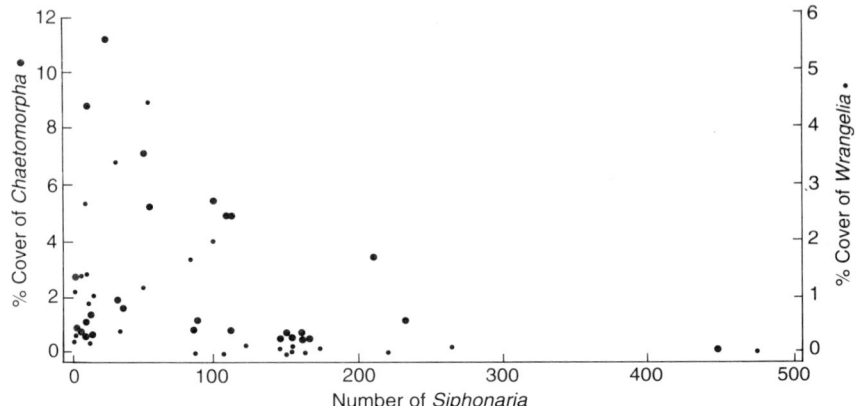

Fig. 4.2. Numbers of *Siphonaria pectinata* plotted against estimated percentage cover of *Chaetomorpha antennina* and *Wrangelia argus* on 28 observation dates over a three-year period. Data from Lawson (1957).

noted that there were also many times when both algae and animals were in low quantity on the transect. Another inference to be drawn from these figures is that none of the algae appeared to be affected by the considerable increase in *Littorina* that took place during the period when observations were being made.

2. Recolonization

The primary intention of long-term studies carried out by one of us (GWL) at Christiansborg, Ghana, was to follow recolonization of denuded intertidal surfaces but the observations also throw some light on plant–animal interrelationships. Initially, denuded rock surfaces were colonized by the green algae *Enteromorpha* and *Ulva*, the latter becoming dominant and forming an almost complete cover for two to three months. This green algal carpet was then largely replaced by nongeniculate corallines which also persisted for several months until the rock was covered by a mixture of algae with no single species becoming completely dominant. The surface was cleared during the month of November, a time when barnacles tend to dominate the upper shore and, in fact, an appreciable settlement of new *Chthamalus* took place within two weeks of the beginning of the experiment and before the green algae had covered more than one-fifth of the upper quadrats. These barnacles were displaced or completely overgrown by *Ulva fasciata* within two weeks and made a further brief appearance about two months later but did not reappear in any quantity until nearly one year later — during the period when lower low waters occur at night and algae retreat from the upper shore (see above). Thus, the normal seasonal cycles were superimposed on the recolonization sequence. These can be detected by comparing events in the recolonization areas with those on the untouched parts of the shore that serve as control areas.

Apart from seasonal effects, long-term observations also reveal changes, which appear to indicate that cleared areas may not always revert to what they were initially. In the case of a quadrat which originally contained an estimated 60 per cent cover of *Chthamalus*, that cover was only again achieved after about one year. Thereafter, however, cover again decreased and, in the second year, the brown filamentous alga *Bachelotia antillarum*, which had not been in the area at all prior to denudation, appeared in small quantities and after six years completely dominated the quadrat (Fig. 4.3). The prostrate filaments and rhizoids of *Bachelotia* tend to bind sand, and this may lead to alteration of the substrate so as to make it unsuitable for barnacle settlement. Such changes taking place over very long periods can hardly be regarded as normal stages in succession; they may be better regarded as something more akin to the mosaic theory which Aubreville (1959) invoked to interpret long-term changes in the tropical rainforest.

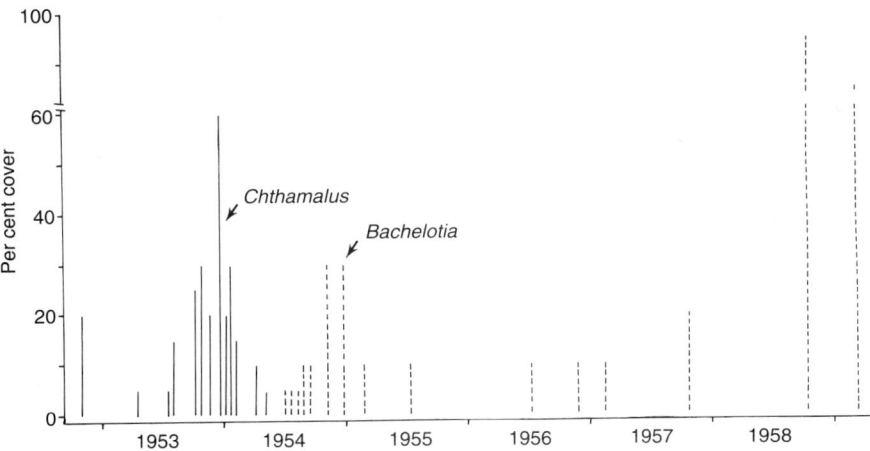

Fig. 4.3. Long-term observations in a recolonization quadrat on the coast of Ghana, showing how *Bachelotia antillarum* (broken vertical lines) gradually took over domination from *Chthamalus dentatus* (unbroken vertical lines). Data from Lawson (1966) and unpublished observations.

Conclusions

This review outlines in very general terms what little is definitively known about some of the apparently more evident and important plant-animal interrelationships believed to be of widespread occurrence in the region. Some of the information is confirmed through experimental studies carried out in Ghana, but much of the evidence is circumstantial or simply speculative and is based on extrapolation from the findings of studies in other tropical regions. Most of the review concerns trophic interactions, since the few investigations performed so far in the region have dealt with this type of interaction. There is tremendous scope for further research on interactions, with much to be achieved through simple and well-designed manipulative experiments. Knowledge of interactions is an important next step in understanding the dynamics of processes operating in the inshore environment of West Africa and in other regions of the tropics. The concerns expressed here regarding the inadequacy of information on plant-animal interactions apply with equal validity to most parts of the tropics outside the Caribbean and, possibly, Australia. What little is known of such interactions in the vast Indo-Pacific region is covered in general reviews (e.g., Hawkins and Hartnoll 1983; John and Lawson 1990).

References

Aubreville, A. (1959). *La flore forestière de la Côte d'Ivoire*. Centre Technique Forestière Tropicale, Nogent-sur-Marne.

Briggs. C. L. (1974). *Marine zoogeography*. McGraw-Hill, New York.

Buchanan, J. B. (1958). The bottom fauna communities across the continental shelf off Accra, Ghana (Gold Coast). *Proceedings of the Zoological Society of London*, **130**, 1–56.

Edmunds, J. and Edmunds, M. (1973). Preliminary report on the mollusca of the benthic communities off Tema, Ghana. *Malacologia*, **14**, 371–2.

Gauld, D. T. and Buchanan, J. B. (1959). The principal features of rock shore fauna in Ghana. *Oikos*, **10**, 121–32.

Hawkins, S. J. and Hartnoll, P. G. (1983). Grazing of intertidal algae by marine invertebrates. *Oceanography and Marine Biology Annual Review*, **21**, 195–282.

John, D. M. (1986). Littoral and sub-littoral vegetation. In *Plant ecology in West Africa*, (ed. G. W. Lawson), pp. 215–46. John Wiley, New York.

John, D. M. and Lawson, G. W. (1990). The effects of grazing animals on algal vegetation. In *Introduction to applied phycology*, (ed. I. Akatsuka), pp. 307–45. SPB Academic Publishing, The Hague.

John, D. M. and Lawson, G. W. (1991). Littoral ecosystems of tropical western Africa. In *Ecosystems of the world*: Vol. 24. *Intertidal and littoral ecosystems*, (ed. P. H. Nienhuis and A. C. Mathieson), Elsevier, Amsterdam, Oxford, and New York. (In press).

John, D. M. and Pople, W. (1973). The fish grazing of rocky shore algae in the Gulf of Guinea. *Journal of Experimental Marine Biology and Ecology*, **11**, 81–90.

Lawson, G. W. (1954). Studies in the intertidal ecology of rocky shores in West Africa. Unpublished Ph.D. Thesis. University of London.

Lawson, G. W. (1957). Seasonal variation of intertidal zonation on the coast of Ghana in relation to tidal factors. *Journal of Ecology*, **45**, 831–60.

Lawson, G. W. (1966). The littoral ecology of West Africa. *Oceanography and Marine Biology Annual Review*, **4**, 405–48.

Lawson, G. W. and John, D. M. (1987). *The marine algae and coastal environment of Tropical West Africa* (2nd edn.). Cramer, Berlin and Stuttgart.

Lewis, J. R. (1961). The littoral zone on rocky shores — a biological or physical entity? *Oikos*, **12**, 281–301.

Lieberman, M., John, D. M., and Lieberman, D. (1979). Ecology of subtidal algae on seasonally devastated cobble substrates off Ghana. *Ecology*, **60**, 1151–61.

Lieberman, M., John, D. M., and Lieberman, D. (1984). Factors influencing algal species assemblages on reef and cobble substrata off Ghana. *Journal of Experimental Marine Biology and Ecology*, **75**, 129–43.

Price, J. H. and John, D. M. (1980). Ascension Island, South Atlantic: a survey of inshore macroorganisms, communities and interactions. *Aquatic Botany*, **9**, 251–78.

Sanusi, S. S. (1980). A study on grazing as a factor influencing the distribution of benthic littoral algae. Unpublished M.Sc. Thesis. University of Ghana, Legon, Ghana.

Stephenson, T. A. and Stephenson, A. (1949). The universal features of zonation between tide-marks on rocky coasts. *Journal of Ecology*, **37**, 289–305.

5. Ecology of tropical rocky shores: plant–animal interactions in tropical and temperate latitudes

DEBORAH M. BROSNAN
Department of Zoology, Oregon State University, Corvallis, Oregon, USA

Abstract

It is only the relative importance of various ecological processes that varies between tropical and temperate shores, not the processes themselves. A comparison of physical and biological features of tropical shores is presented. Overall, herbivorous fish are more common in the tropics, while invertebrate grazers dominate temperate regions; this difference is responsible for much of the variation in algal composition and abundance between tropical and temperate shores. A model predicting how different herbivores influence plant composition is presented. Non-trophic interactions are important at all latitudes, but may be more important in temperate zones. Three-dimensional refuges are more important in the tropics. Models of the importance of species interactions in relation to environmental stress may allow us to predict when plant–animal interactions will be dominant factors on rocky shores, regardless of latitude.

Introduction

We know little of the processes affecting tropical rocky intertidal shores in comparison with the wealth of information available for temperate shores. Recent reviews on marine herbivory (Lubchenco and Gaines 1981; Gaines and Lubchenco 1982; Hawkins and Hartnoll 1983) and the biology and ecology of gastropod grazers (Underwood 1979; Branch 1981, 1986), have emphasized temperate studies (but see Gaines and

Plant–Animal Interactions in the Marine Benthos, (ed. D.M. John, S.J. Hawkins, and J.H. Price), Systematics Association Special Volume No. 46, pp. 101–31. Clarendon Press, Oxford, 1992. © The Systematics Association 1992.

Lubchenco 1982, for some discussion of tropical shores). Much of our knowledge of patterns on tropical shores is based on descriptive studies (e.g., Gauld and Buchanan 1959; Stephenson and Stephenson 1972; Rao and Sundaram 1974; Jones *et al.* 1987), although there has recently been experimental research in Central America (Garrity and Levings 1981; Menge and Lubchenco 1981; Levings and Garrity 1983, 1984; Lubchenco *et al.* 1984; Ortega 1985, 1986; Menge *et al.* 1986a, b; Sutherland and Ortega 1986; Sutherland 1990). Such paucity of information may limit the generalizations we can make about interactions on tropical shores. Here I summarize available evidence and highlight areas for further research.

There are no ecological processes that are uniquely tropical; instead the relative importance of interactions varies between shores (Menge and Lubchenco 1981; Lubchenco *et al.* 1984; Menge and Sutherland 1976, 1987; Sutherland 1990). There may be more similarities between an exposed tropical and an exposed temperate shore than between either of these two and a protected shore at any latitude (e.g., Menge and Farrell 1989). We do not need new conceptual frameworks to describe the ecology of tropical shores. Existing theoretical concepts are applicable to all latitudes. Throughout this review, I shall compare the ecological interactions on tropical shores with those on temperate shores and will use the framework of Menge and Sutherland (1976, 1987) to discuss and predict factors which may be most important in the tropics. Plant–animal interactions in tropical and temperate shores cover a broad spectrum of ecology. Not all areas can be given the coverage they deserve in a single chapter.

Tropical rocky intertidal shores

1. Physical factors affecting tropical rocky shores

Rocky shores do not always spring to mind when considering tropical coasts. Coral reefs and mangroves, generally associated with the tropics, extend over much of these shorelines. Rocky intertidal shores are by no means rare, however. For example, the east coast of India is predominantly rocky (Rao and Sundaram 1974) and large stretches of rocky shoreline occur on the Guinea Coast of West Africa, particularly around the Cap Vert Peninsula (Lawson 1966).

There is a popular belief that the tidal range of most tropical shores is small, but this is not always so. Longhurst and Pauly (1987) found no systematic difference in relative distribution of microtidal (< 2m), mesotidal (2–4 m), and macrotidal (> 4 m) coastal environments between high and low latitudes. As elsewhere, tidal ranges in the tropics are greatest where shelf morphology and coastline shape

Table 5.1. A comparison of the main zonation patterns on tropical and temperate rocky shores along a wave-exposure gradient[a].

Intertidal zone	Exposed	Intermediate	Protected
Tropical shores			
Upper	Barnacles	Barnacles and bivalves	Barnacles and bivalves (cover is less abundant)
Mid	Algal crusts and lithothamnia	Mixed algal assemblage	Bare space
Low	Filamentous reds and *Sargassum*	Filamentous reds, lithothamnia and foliose algae	Algal turf and crusts, filamentous reds, and foliose algae
Temperate shores			
Upper	Barnacles	Barnacles, fucoids and crustose algae	Bare space Algal crusts
Mid-	Mussels	Fucoid canopy	Fucoid canopy, crustose algae, and bare space
Low	Kelps	Kelps	Red algal turf, free space, and occasional kelps

[a] Data from Lawson (1966); Stephenson and Stephenson (1972); John and Pople (1973); Rao and Sundaram (1974); Menge and Lubchenco (1981); Hawkins and Hartnoll (1983); Hodgkiss (1984); Ortega (1985); Menge et al. (1986a, 1988b); Menge and Farrell (1989).

constrict oceanic waters. Tides of >4 m are widespread in the tropical Indo-Pacific and reach 11 m in the north-west Arabian Sea. Tidal ranges of 4–6 m occur in the Bay of Bengal, on both coasts in the straits between Madagascar and Africa, from north-west Australia to southern New Guinea, and in the Gulf of Panama. However, in the Atlantic, only on both sides of the mouth of the Amazon does the range exceed 4 m. The entire west coast of Africa has a tidal range of < 2 m.

Small tidal ranges of about 1 m occur in the Red Sea, the Caribbean–Antilles region, and the Hawaiian archipelago.

Most of the work on tropical shores (except for that on the Pacific coast of Panama where tidal range is 3–6 m), has been carried out on coasts where the tidal range is small, so that the problems associated with small tidal ranges deserve some special mention. In these regions, weather conditions may determine the tidal range. High pressure and calm seas can produce extreme high and low waters for a two-week period in the Red Sea (Jones *et al.* 1987). Not surprisingly, these conditions have been associated with high mortalities of marine organisms (Hay *et al.* 1983). Sea-level can also vary appreciably, dropping by as much as 0·5 m in the dry season. During that season, when other physical factors combine to produce maximum stress, sea-levels are at their minimum and expose areas of the wet season sublittoral to both sun and air. In some parts of the tropics, the dry season coincides with daytime low tides, thereby adding to the already intense stress.

On many tropical coasts, seasonality is expressed by the state of the sea, and the amount of freshwater discharged by rivers, lagoons, and runoff. High seasonal rainfall leads to substantial reductions in salinity over some shores. In the Gulf of Mannar, on the east coast of India, salinities range from 22·88 per cent to 35·82 per cent between wet and dry seasons (Rao and Sundaram 1974). Hodgkiss (1984) reported that seasonal reductions in salinity were partially responsible for the decline in nine species of algae in Hong Kong.

2. Zonation on tropical rocky shores

Many researchers have commented on the bare appearance of tropical rocky shores (e.g., Stephenson and Searles 1960; Gaines and Lubchenco 1982). This seems to be due to the lack of large foliose algae such as the fucoids and kelps which characterize temperate coasts. In fact, many rocky shores are not barren but dominated by low-lying forms such as crustose algae and species richness can be high (Lubchenco *et al.* 1984).

Typical zonation patterns along an exposure gradient differ between tropical and temperate shores (Table 5.1). Barnacles dominate the upper shore in temperate and tropical exposed shores; they are less common in warm temperate compared with cold temperate shores (Menge and Farrell 1989). But while there is a rich mussel band on most mid-zone cold temperate shores, this is not so in the tropics. Instead, the mid-zone is dominated by crustose algae, and there is usually a distinct *Lithothamnion* band in the lower mid-zone. Kelps are entirely absent from the tropics. Lush kelp beds are an important feature of temperate shores, but are less prevalent in warm temperate than in cold temperate regions (Menge and Farrell 1989). Filamentous red algae, and species such as *Sargassum* and *Dictyota*, are common in the tropical lower intertidal zone.

On intermediate (less-exposed) temperate shores, bivalve cover begins to wane as the fucoid canopy becomes more common (Menge and Farrell 1989). However, it is on moderately exposed tropical shores that bivalves reach their peak abundance and oysters often form a well-defined bed just below the barnacle zone. A mixed algal assemblage prevails on intermediate tropical shores. Lithothamnia are displaced to a lower level, where they become more abundant and mingle with filamentous red algae, other crustose algae, and foliose species.

A striking feature of many protected tropical shores is a bare midzone between the upper barnacle zone and the lower algal turf zone. This bare zone does not occur on protected temperate shores, although bare space is more prevalent on protected than on more exposed shores. At high and low latitudes, gastropod grazers are most abundant in the mid- and upper shores.

Within the general pattern described above, the relative abundances and distributions of individual species vary greatly. For instance, Menge *et al.* (1986) found that sessile invertebrates were rare on the Pacific coast of Panama (covering < 10 per cent of available space), whereas Lawson (1966) reported 60 per cent cover of *Chthamalus stellatus* on shores in Ghana. In Costa Rica, sessile invertebrates are more abundant than in Panama (Sutherland and Ortega 1986).

Trophic plant–animal interactions: direct effects

The importance of herbivore–algal interactions on any shore depends on many factors, but primarily on the nature and density of the herbivore, susceptibility of the algae to herbivory, and the growth rates of the algae.

1. Type of herbivore

One striking difference which emerges from a comparison of temperate and tropical herbivores is that fish play an enormous role in structuring tropical rocky shores (Lubchenco and Gaines 1981; Gaines and Lubchenco 1982; Hixon 1986) but seem to have little effect on temperate shores (Menge 1982). Molluscs and sea urchins are the leading herbivores in temperate areas, while molluscs, sea urchins, and fish predominate in the tropics. For instance, in Central America, and the Red Sea, surgeonfish (Acanthuridae) and parrotfish (Scaridae) commonly forage in the intertidal region (Menge and Lubchenco 1981; Lubchenco *et al.* 1984; Hay *et al.* 1983; Walker 1987).

Different herbivores have different feeding habits and structures, with consequent varying effects on the algae. Fish, for example, are fast-moving to visual predators and many have strong calcareous beaks. Parrotfish and sea urchins both feed by scraping algae and often remove pieces of underlying rock (Walker 1987). In contrast, surgeon fish remove individual algal filaments. Slower-moving gastropods, such as limpets, scrape the surface with their radulae, but even these differ in their effects. Siphonarian limpets bite off macroalgal fragments; patellacean limpets remove all algal sporelings by rasping down to bare rock (Branch 1986). While both fish and molluscs have limited feeding periods in the intertidal zone, fast-moving fish can cover a much greater area in a single tidal cycle (Menge and Lubchenco 1981). The presence of herbivorous fish causes dramatic large-scale effects on the appearance of tropical shores by determining the types of algae that occur. Fish are thought to be at least partly responsible for the lack of foliose algae and for the bare appearance of many tropical shores (Menge et al. 1986a), although desiccation may also play a role. The Pacific shores of Panama are dominated by crustose algae, while sessile invertebrates and foliose algae are rare (Menge and Lubchenco 1981; Menge et al. 1986a). Four types of algal crust persist in the low zone (Lubchenco et al. 1984), and consumers are diverse and abundant (including large and small fish, crabs, predatory and herbivorous gastropods). In an experimental separation of the effects of these consumers, Menge et al. (1986a) found that small and large fish, as well as crabs, reduced the cover of sessile animals and foliose algae, although they were unable to graze foliose algae down to the rock surface. Molluscan herbivores grazed foliose algae down to grazer-resistant crusts. Menge et al. concluded that the combined effects of this diverse group of consumers maintain the dominance of crustose algae on the shore. Slow-moving herbivores, primarily molluscs, tended to have short-term (zero- to six-month) effects on algal abundances by decreasing the cover of foliose species; they were also responsible for longer-term (six months to three years) changes in abundances of crustose algae, of *Enteromorpha*, and of *Ulva*. Fish and crabs were more efficient foragers and had greater long-term effects on foliose algae and the rock oyster *Chama*.

Other studies on tropical shores support these findings in Panama. For instance, Hay et al. (1983) (Caribbean coast of Panama), Stephenson and Searles (1960) (Heron Island, Australia), and Walker (1987) (Red Sea), concluded that fish are primarily responsible for patterns of algal distribution and abundance.

There are no temperate shores where fish have a similar effect, or interact with other grazers to the extent described in the Panama study. However, it is interesting that Gaines and Lubchenco (1982) noted that those temperate shores with intense sea urchin grazing resemble the shores of Panama.

2. *Herbivores set limits on algal abundance and distribution*

Herbivores can restrict the distributions of algae on the shore, at times confining them to sub-optimal habitats. This may be a feature of many rocky shores, regardless of latitude. However, because fish are an effective ecological feature of tropical shores only, the extent to which herbivores limit algal distribution and abundance may vary across latitudes.

Hay *et al.* (1983) recently studied the impact of herbivory on spatial and temporal algal distributions on the Caribbean coast of Panama. The study area consisted of a reef-flat, exposed during low tides and calm seas, a reef-slope, and a sandplain. Composed of hard substrate, the reef-flat supports an abundance of upright fleshy algae. Corallines are the most abundant algae on the reef-slope. Herbivorous fish (mainly surgeonfish and parrotfish) and the sea urchin *Echinometra lucunter* forage in the reef areas. Hay *et al.* found that herbivory was greatest in the reef-slope area. Fish had the greatest effect on algal loss; sea urchins have very little effect. In a series of transplant studies, reef-flat algae had higher growth rates on the reef-slope than on reef-flats, but suffered higher predation on the slopes. Thus, it appears that grazing by herbivorous fish limits the distribution of several algal species to a less optimal habitat on the reef-flat. In this study it was concluded that reef-flat algae are competitively superior (because of high growth rates), and in the absence of grazers outcompete corallines on the reef-slope.

This study contrasts with ones carried out in New England by Lubchenco (1980), who found that competition was the primary factor influencing the lower limit of *Fucus* spp. In *Fucus*-dominated shores the zone below the mid-zone is dominated by *Chondrus crispus* and *Fucus* is virtually absent from this zone. When Lubchenco totally removed *Chondrus* plants, *Fucus* colonized and grew into mature reproductive plants. *Fucus* grew faster in the low zone than in the mid-zone where it occurs. In the absence of herbivores (mainly *Littorina littorea*) *Fucus* can reach 100 per cent cover. When grazers are present, *Fucus* establishes but it is less abundant. Thus grazers reduce the abundance of *Fucus* but do not control its shore-distribution. In a series of cage experiments, Lubchenco and Menge (1978) found that the lower limit of *Chondrus* was set by sea urchin grazing, while desiccation set its upper limit.

The above studies suggest that on temperate shores sea urchins may be important in setting the lower limits of algal distribution; however, because urchins spend less time foraging in the higher intertidal zone they have little effect there. Sea urchins have no major effects on algae on some tropical shores (Hay *et al.* 1983; Walker 1987). At all latitudes, molluscan grazers are smaller, slower, and less efficient than urchins or fish. However, their effects may be enchanced by the slower growth rates of algae in the mid- to upper zones. Thus, we can predict that herbivorous gastropods exert their greatest effect where algae are in a sub-optimal habitat; they may be ineffective in zones where algal growth is high. The effectiveness of fish should only be limited by the amount of time they can spend foraging in the intertidal zone; as they are faster than molluscs, their effect should be greater at all levels on the shore.

3. Effect of algal productivity on herbivores

Algal productivity rates are important in determining the overall effects of herbivory. Algal growth rates vary seasonally at all latitudes, but in Oregon they are higher in the low zone and at less stressful periods of year (Castenholz 1961; Cubit 1984). Shores that experience seasonal upwelling of nutrient-rich water are likely to be more productive. Upwelling does not vary along a latitudinal gradient — some temperate and some tropical shores have seasonal upwelling, others do not. Changes in algal productivity have wider-ranging effects on the relative importance of herbivory and on the distribution of plant and animals on rocky shores. Recent studies on temperate shores have illustrated the importance of algal growth rates in relation to herbivory.

In a twist to the more usual approach, Underwood and Jernakoff (1981) showed how algal productivity can limit the distribution of a herbivore. They found that the limpet *Cellana tramoserica* was excluded from the lower intertidal by rapid growth rates of foliose algae.

Cubit (1984) showed that variation in seasonal abundance of algae on Oregon shores was due to a shift in the balance between the rate of algal production and loss to herbivores. Higher rates of algal production in winter more than compensated for increased herbivory over the period. Consequently, algal abundance was then greatest. In summer, despite a reduction in herbivory, rates of production were lower and did not offset the loss to herbivores, hence algal abundance decreased. Castenholz (1961) found similar results for diatom populations.

Some South African shores are exceptional in that the large limpet *Patella cochlear* dominates a narrow band in the low intertidal (in contrast to the usual pattern of an algal-dominated lower shore). *Patella* can reach densities of up to 1600 m^{-2} (Branch 1975). Foliose algae prevail above and below the limpet band; these algae soon dominate if *Patella* is removed and, once established, prevent subsequent recruitment of *Patella*. How does *Patella cochlear* maintain itself at such exceptional densities? In a neat study, Branch (1981) discovered that each limpet maintains a fringing garden of fine red algae. Exceptionally fast-growing, these algae have the high productivity essential to persistence of the limpets.

The above studies suggest that algal productivity may be a structuring force on temperate shores. But is there any evidence that tidal height and algal growth rates similarly affect interactions between tropical herbivores and algae? On Heron Island, Stephenson and Searles (1960) found that the herbivorous snail *Nerita* could not control the abundance of the green algae *Blidingia*, which developed into a thick unstable mat in the presence of *Nerita* alone; one-third of snails introduced into cages died during the experiment. However, grazing by fish and snails reduced *Blidingia* abundance and prevented the formation of thick unstable crusts.

Data obtained for growth rates of algae on artificial substrates in the Red Sea indicate a standing crop of about 25 g dry wt m^{-2} (Walker 1987). This is relatively low compared with temperate algal standing crops of the order of 1–2 kg dry wt m^{-2}. However, Walker found that production rates for Red Sea algae range from 1 g to 3 g dry wt m^{-2}; these rates are comparable with algal productivity in temperate regions. These values indicate that algal turnover time is 10 days. Clearly, biomass of algae on Red Sea reefs does not reflect their rate of production. The Red Sea supports high densities of herbivorous fish (surgeonfish and parrotfish) and Walker (1987) suggests that herbivory is responsible for the low standing crop. If this is so, then even a rapid algal growth rate is not sufficient to compensate for herbivory. We do not have growth rates for the Pacific coast of Panama, but foliose algal productivity is not high enough to escape fish grazing. Some algae could escape mollusc grazing alone, as suggested by the study of Stephenson and Searles (1960).

Many tropical studies refer to dramatic change in algal biomass between dry and wet seasons (Banaimoon 1988; Hodgkiss 1984; Murthy *et al.* 1989; Silva *et al.* 1987). For instance, Murthy *et al.* (1989) reported a fivefold increase in biomass of *Gracilaria*, *Gelidium*, and *Hypnea* between the dry and wet seasons. There is also seasonal upward movement of algae into the barnacle zone during the tropical wet season (e.g., Lawson 1966; Hodgkiss 1984; D. M. John, pers. comm.). During the dry season these algae persist lower on the shore. Physical factors are assumed to cause these changes on tropical shores, studies on possible roles of herbivores being absent. One plausible explanation for these phenomena is an increased effect of herbivory in the dry season and a decrease during the wet season. However, Menge (pers. comm.) observed that algal distributions extended upshore during the dry season in Panama. He suggested that this is probably due to increased nutrient supply during seasonal upwelling.

In conclusion, algal growth rates can be important on shores at all latitudes; both high and low productivity can limit herbivore distributions. There are, however, some latitudinal differences: high productivity in the tropics may be counteracted by high grazing pressures from herbivorous fish; this does not seem to be a feature of temperate shores. Seasonal changes in temperate algal abundance and distribution may result from changes in algal growth rates, even though herbivory remains the same or increases slightly. This may also be a feature of tropical shores, but currently there is not enough information to support the hypothesis. In addition to influencing algal growth rates, physical conditions can limit the foraging times of many herbivores, in turn affecting the outcome of herbivory.

5. Limitations to foraging on tropical and temperate shores

Life on a tropical shore is stressful. Organisms have to cope with high air and sea temperatures. Temperatures fluctuate less in tropical than temperate latitudes; even small changes can, nonetheless, impose high stress on tropical species. Moreover, Moore (1973) found that tropical marine species generally have lower temperature tolerance than temperate species. Such stresses potentially limit the foraging activities of tropical molluscs. How constrained are tropical gastropods compared with temperate counterparts? We do not have data on actual time spent foraging on tropical shores but we can make some latitudinal comparisons based on knowledge of foraging behaviour. Most cold temperate herbivorous gastropods feed only at high tide (Cubit 1984), hiding in crevices or under algal canopy during low tides. Grazers are less active during harsh conditions: *Littorina littorea* is inactive for much of the winter in New England (Menge, pers. comm.).

Browsing activities of tropical snails and limpets are confined by tidal cycle and weather; *Nerita* only forage on ebbing and flowing tides and gastropods remain in crevices during low tides (Garrity and Levings 1981; Levings and Garrity 1983, 1984; Ortega 1986). Overall, temperate herbivorous molluscs may well spend as much time avoiding desiccation and cold as do tropical species in avoiding heat and desiccation. Herbivorous molluscs may therefore have similar ecological roles at all latitudes. It would be interesting to compare the roles of herbivorous fishes between different tropical shores.

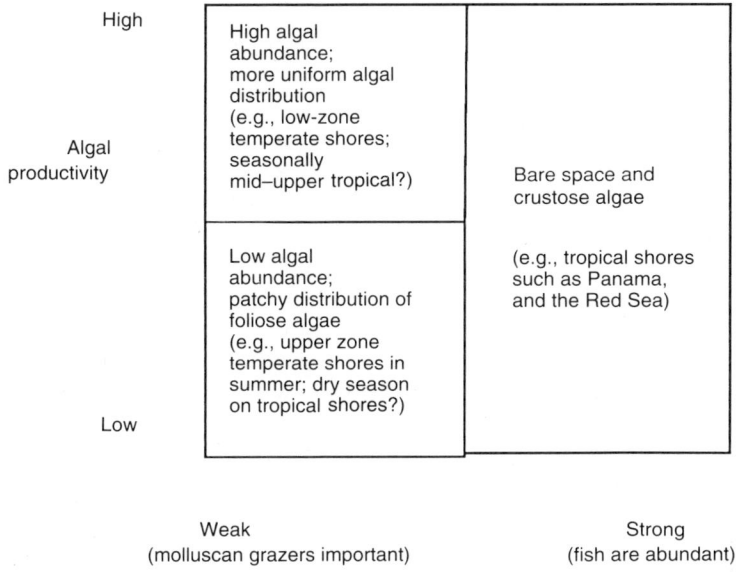

Fig. 5.1. A model of algal composition and distribution in relation to herbivore type (mollusc or fish) and algal productivity. This model predicts outcomes of interactions between herbivore type and algal productivity. When fish are more important (right rectangle), shores have a uniform appearance (bare space or crustose algae), competition between foliose algae is predicted to be low, and herbivory is important at all tidal heights. When slow-moving molluscan grazers predominate, their effectiveness depends on algal growth rates, and this in turn affects the relative importance of competition and physical stress.

A general model of herbivore impact

The results of studies mentioned above can be combined into a generalized model to predict the distribution of algal types in relation to herbivore size and efficiency (Fig. 5.1).

When herbivorous fish are abundant algal growth rates will not compensate for herbivory — shores will appear bare or dominated by resistant crustose forms. Invertebrate grazers have little impact on algal abundance (e.g., Hay *et al.* 1983; Menge *et al.* 1986a; Walker 1987), regardless of shore latitude. If slower-moving grazers predominate, the greatest effects occur where algal growth is too slow to compensate for herbivory (e.g., upper shore especially during the tropical dry season). Grazer impact decreases as algal growth rates increase (e.g., lower down shore and during wet season: Castenholz 1961; Lubchenco and Menge 1978; Lubchenco 1980; Underwood 1980; Cubit 1984).

How do such predictions translate into differences between tropical and temperate shores? Assuming that fish forage frequently the intertidal on tropical shores, these should have a less patchy appearance than temperate shores. This comparison and also the patchy nature of temperate shore algae has been noted by a number of researchers (e.g., Menge and Lubchenco 1981; Underwood 1981; Gaines and Lubchenco 1982; Menge et al. 1985, 1986a, b). Seasonality should be less pronounced, any increase in algal biomass being quickly eaten, and herbivores should limit upper distributions of algae more frequently than on temperate shores. Competition among foliose algae should be less important. On temperate rocky shores where molluscs dominate, seasonal effects in algal abundance should be more pronounced as algal productivity exceeds grazing (a relationship noted by Underwood 1980; Underwood 1981; Underwood and Jernakoff 1981, 1984; Cubit 1984; Branch 1985, 1986) and other factors (e.g., competitive and physical) should be relatively more important in setting distributional limits. This is because at low primary productivity spatial and temporal plant escapes will be important and tend to result in patchy algal distribution. At high growth rates, algae will be more evenly distributed as productivity exceeds herbivory, and algae begin to dominate spatially. These predictions apply regardless of latitude. Tropical shores may be located mostly in the right rectangle of Fig. 5.1, but their mid- to upper zones could oscillate back and forth between the upper left and lower left rectangles during wet and dry seasons respectively. For example, Menge et al. (1986a) found that removal of slow-moving grazers resulted in seasonal fluctuations in cover of foliose algae and a *Ralfsia* crust and small fish and crabs had no effect on seasonal algal fluctuations.

A second prediction of the model relates to herbivore size and rate of resource renewal. A fast-growing alga will be able to support a larger-sized grazer than will slow-growing algae. Large grazers should be more prevalent lower on the shore, and small grazers higher up. On many shores this pattern applies, although other factors such as desiccation might also be responsible. However, large limpets such as *Patella cochlear* are found low in the intertidal and their persistence depends on high algal productivity.

Herbivory and the evolution of plant defences

Recent theoretical and experimental work has emphasized the effect of herbivores on evolution of plant defences in terrestrial and marine systems (Feeny 1976; Rhoades and Cates 1976; Coley et al. 1985; Hay and Fenical 1988, and Chapter 14 this volume). Most theoretical studies have addressed terrestrial interactions, but could easily be adapted to those operating among marine species. For instance, several authors (e.g., Coley *et al*. 1985) have suggested that slow-growing plants should devote more resources to defence. Note that while research on plant defenses has stressed the evolution of growth forms and chemical defenses, morphological defences (such as calcification) are also important (Lubchenco and Gaines 1981; Padilla 1985).

In the marine intertidal, a crustose algal growth form and strong herbivory are positively correlated as crustose algae tend to be grazer-resistant (e.g., Paine and Vadas 1969; Estes and Palmasino 1974; Mann 1982; Lubchenco and Cubit 1980; Levings and Garrity 1983, 1984; Jara and Moreno 1984; Lubchenco and Gaines 1981; Gaines and Lubchenco 1982). Many ephemeral and annual algae are known to have heteromorphic life histories, existing as either crustose or upright morphs. Lubchenco and Cubit (1980) demonstrated that in New England and Oregon certain algae persist as crusts in the presence of grazers but upright morphs develop when grazers are removed. Lubchenco and Cubit suggest that a heteromorphic life history develops in response to herbivory, although resistance to desiccation may also be important. Similarly Jara and Moreno (1984) showed that the crustose form of *Iridaea boryana* is prevalent in the presence of limpets but is replaced by the canopy-forming morph when grazers are absent.

Not all crustose algae are equally resistant to grazers. Levings and Garrity (1984) found a negative relationship between algal crusts and the limpet *Siphonaria gigas*. Earlier, Levings and Garrity (1983) showed how crustose algae are inhibited by the snail *Nerita funiculata* that graze down to bare rock, although the same algae are more resistant to *Nerita scabricosta*.

The role of chemical compounds as a defence against marine herbivory has received considerable attention in recent times (e.g., Paul and Hay 1986; Hay and Fenical 1988; Hay and Fenical, Chapter 14 this volume). Most of the research has been on tropical algae, particularly on subtidal algae and herbivorous fish.

Van Alstyne (1988) showed that attacks by molluscan grazers induce production of secondary chemical deterrents by *Fucus* in Washington State. Are chemical defences important for intertidal tropical algae? Given the predominance of crustose and calcified forms, it is possible that tropical algae rely mainly on morphological defences. Note that these morphs are also more resistant to desiccation (Lubchenco and Cubit 1980). Coley *et al*. (1985) have discussed environmental constraints through which defence mechanisms can evolve under particular circumstances. It may be that morphological defences predominate in tropical areas because of resistance to desiccation. An alternative suggestion is that such defences are most effective against fish, the dominant tropical herbivore. If calcification is primarily a defence against grazers then when fish are rare the foliose algae should be more resistant regardless of desiccation stress. The results of Hay *et al.* (1983) support this interpretation.

Community-level effects of plant–animal interactions

It is not just a grazer and its algal food that are affected by herbivory. The outcome of plant–animal interactions can have major effects on community structure. For instance, the nature and density of herbivores affect the outcomes of herbivory; these effects will cascade through whole communities, leading off in different trajectories, and producing assorted patterns on intertidal shores. While this is true for all communities, a comparison of temperature and tropical intertidal communities illustrates conditions under which particular interactions may be either pivotal to community structure or be swamped by other effects.

The levels of response within a community vary with individual herbivore species. Levings and Garrity (1983) found very different community responses to the foraging behaviour of two species of the tropical snail *Nerita*. *Nerita scabricosta* lives in the upper zone but forages behind the retreating tide and above the incoming tide, making frequent feeding excursions to the low intertidal. *Nerita funiculata* lives in crevices in the mid-shore and also forages during flowing and ebbing tides. Removal of *N. funiculata* led to an increase in algal cover close to crevices, but not to other changes in the community. Removal of *N. scabricosta* led to much larger changes in the community over many intertidal zones: the abundance of blue-green algal crusts increased; barnacle densities increased, indicating that bulldozing was important; two sympatric species of littorines increased. These results show the sensitivity of communities to the presence of single species. Density effects are also possible. Underwood *et al.* (1983) described an interesting case where the survival of barnacles depended on intermediate densities of limpets.

Results from Panama indicate that interactions between different herbivores produce a variety of effects on this community (Lubchenco and Menge 1981; Lubchenco *et al.* 1984; Menge *et al.* 1986*a*, *b*). In Panama, no one consumer group had a dominant effect on overall community structure; however, particular prey groups could be strongly affected by either slow-moving herbivores, or by fish and crabs. For example, the effect of slow-moving herbivores on foliose algae was as great as the effect of all consumer groups combined. A similar outcome was found for the effect of small fish and crabs on colonial animals. Menge and Lubchenco concluded that the effect of consumer groups on sessile prey is interdependent; they all interact to maintain the dominance of crustose algae. For example, predatory fish and gastropods reduce the abundance of sessile animals, and thereby inhibit recruitment of algae that settle on these animals.

An interesting set of interactions between consumer groups and prey was reported by Stephenson and Searles (1960) from experiments conducted at Heron Island. In this community, fish appeared to have a more 'keystone' role than in the Panama community. Stephenson and Searles concluded that herbivorous fish, feeding in the intertidal to avoid predation, were primarily responsible for the barer appearance of shores on Heron Island Australia, than mainland shores. A film of the blue-green alga *Calothrix*, and a chiton (*Acanthozostera gemmata*) dominate the mid-shore of Heron Island, while *Blidingia minima* (Chlorophyta) prevailed lower on the shore. Using cage experiments, Stephenson and Searles discovered that, in the absence of all grazers, *Calothrix* developed into a thick mat. When only chitons were present, a mat (not quite so thick) of *Calothrix* also flourished. *Acanthozostera* cannot browse thick mats of *Calothrix* and depends on fish to prune the alga to a sufficiently low level on which it can then feed. Thus, in this community, chitons depend primarily on the grazing activities of fish; in the absence of fish, disturbance may account for their persistence. Dominance by *Calothrix* is also maintained by herbivores (primarily fish and other grazers) as, in the absence of chitons, foliose algae developed but were later outcompeted by *Calothrix*.

The interplay between fish, chitons, and algae on Heron Island, is similar to that reported by Dethier and Duggins (1984) for chitons and limpets on the Washington coast of North America. In this study, limpets, which feed on microalgae, depend on chitons to graze macroalgae, thus clearing space for spores to settle. Dethier and Duggins (1984) found that the chiton–limpet interaction was a major factor structuring the Washington community. It is interesting that the same interactions did not assume great importance in Alaska because there algal growth rates were lower (Dethier and Duggins 1988).

It is interesting to speculate whether interactions, similar to those on Heron Island, exist between fish and invertebrate herbivores on other tropical shores. Does the importance of these interactions vary and why? And what effects do invertebrates have on fish, as in many cases diets appear to overlap?

The examples serve to illustrate how interactions between plants and herbivores reverberate through communities. They also highlight how the nature of the interactions are the same on temperate and tropical shores, but differences in the relative importance of interactions may be greater between two shores at the same latitude than between a temperate and tropical shore. For example, an interaction of similar importance occurs between chitons and fish on Heron Island as between chitons and limpets in Washington State. Its importance varies, however, over a small latitudinal gradient between two temperate shores (from Washington to Alaska).

1. Spatial and temporal heterogeneity

Spatial and temporal refuges may allow coexistence by providing refuges from predation (Menge 1976, 1978a, b; Lubchenco 1978, 1980; Lubchenco and Menge 1978; Lubchenco and Cubit 1980; Menge and Lubchenco 1981; Menge et al. 1985). Refuges may also provide escapes from environmental harshness (Cubit 1984; Levings and Garrity 1983, 1984; Garrity 1984; Branch 1986; Ortega 1986).

On temperate shores, algal canopy acts as a refuge for many species (Menge 1976, 1978; Lubchenco 1978, 1980; D'Antonio 1985); the absence of foliose algae from the tropics means that this refuge is generally unavailable to tropical species. Physical refuges such as crevices and holes may perhaps be more important in tropical shores. In a comparison of prey refuges in relation to consumer pressure, Menge and Lubchenco (1981) compared refuges from tropical and temperate shores. They determined that rates of predation and herbivory are higher on Panamanian shores, and that crevices are essential for the persistence of foliose algae. Moreover, temporal and size refuges, so common on temperate shores (Lubchenco 1978, 1980; Lubchenco and Cubit 1980; Menge and Lubchenco 1981), are less important in Panama because consumer pressure is intense and constant. Sessile animals and foliose algae are practically confined to crevices and holes, suggesting that competition for refuge-space is potentially important. However, Menge and Lubchenco concluded that this is not the case; even in crevices, consumer pressure is strong, and bare space is abundant.

Consumer pressure is not solely responsible for the importance of spatial refuges. Escapes from harsh physical factors may interact with the need to avoid predation. Which of these two is the dominant factor may shift between shores at similar latitudes. Levings and Garrity (1983) showed that both predation and stress were influencing the importance of refuges in Panama. In the absence of crevices, *Nerita* suffered high mortality during periods of tidal emersion but it was also preyed on more heavily by carnivorous fish. However, Ortega (1986) reported that predation rates by carnivorous fish are similar on Costa Rican and on many temperate shores, but are lower than in Panama. Despite this, herbivorous snails exhibit the same behaviour on Costa Rican shores as on Panamanian shores. Ortega interpreted this as evidence for stress being the dominant factor controlling refuge use in Costa Rica.

As spatial refuges appear to be important in tropical shores, it follows that refuge availability will in turn influence the importance of species that use them. Stephenson and Searles (1960) found that there were few refuges available to *Nerita* on the mid-upper shore on Heron Island and that, as a result, the snail had very little effect on algal abundance in these zones. However, as the number of crevices increased further down-shore, so did the importance of *Nerita*. In conclusion, three-dimensional refuges appear to be the main ones available to tropical invertebrate herbivores. This contrasts with the spatial and temporal refuges so frequently used by temperate species. Consequently, crevices and holes should become increasingly more important with decreasing latitude.

2. *Indirect effects*

Recent theoretical work has emphasized the potential importance of indirect interactions (e.g., Holt 1977; Abrams 1984), and this has been further highlighted by experimental studies (e.g., Brown *et al.* 1986; Dungan 1986; Menge *et al.* 1986a; Sutherland 1990). Menge *et al.* (1986a) described a suite of indirect effects weaving through the intertidal community in Panama. For example, all consumers indirectly benefit grazer-resistant crusts by removing sessile animals and foliose algae. By eliminating foliose algae, herbivores also prevent recruitment of the oyster *Chama* (which preferentially recruits to moist algal turfs), but enhance barnacle recruitment. While these interactions are important because they can potentially lead to dominance by different forms on the shores, they are not major factors in structuring this community. Sessile animals are quickly consumed by predators. This in turn prevents recruitment of species which settle on to the shells of bivales and barnacles, using them as a refuge from predation and environmental stress. Yet the indirect effect of consumers on crustose algae is a consistent feature of this community. Menge *et al.* (1986a) suggest that these many complex interactions on this shore may be temporary phenomena, related to reductions in consumer pressure. However, given the constant and intense level of consumer attack, it seems unlikely that many of these indirect effects ever become important.

We do not yet know enough about indirect effects to make any firm statement as to their importance or occurrence, except that they are phenomena of all communities, regardless of latitude or type (terrestrial or marine). Clearly, this is a fertile field for further attention.

Non-trophic interactions

Most research on plant–animal interactions has stressed the trophic relationships of these two groups. Yet many plant–animal interactions are non-trophic, i.e. they involve a relationship other than feeding. Moreover, non-trophic interactions can be as important to ecology and evolution of the organisms concerned as are the more traditionally viewed trophic relations. What are the important non-trophic interactions between plants and animals on rocky shores? Do they differ between latitudes?

Many intertidal animals act as settlement substrates for plants (and vice versa). For instance, on both temperate and tropical shores algae are known to settle on barnacles and bivalves (Menge 1976; Hawkins 1981; Underwood *et al.* 1983; Menge *et al.* 1986). These interactions can be important in maintaining diversity on some shores. For instance, Dayton (1973) noted that on the north-west coast of America, *Postelsia palmaeformis* frequently settles on mussels, the dominant competitor for space on these shores. In Panama, Menge *et al.* (1986a) demonstrated that in the absence of grazers, the alga *Jania* (initially dominant on primary space) was subsequently outcompeted. *Jania* then persisted as a co-dominant in the community by becoming an epibiont on barnacles. Lee and Ambrose (1989) have similarly shown that adopting an epibiotic lifestyle can result in some species avoiding competitive exclusion.

Recent studies have shown how similar interactions are important in the successional sequence on many shores. On the west coast of America, barnacles are an intermediate successional stage following disturbance, subsequently providing settlement substrates for both algae and mussels (Sousa 1979; Farrell 1987, 1991; Brosnan, personal observations).

While sessile animals may prevent local extinction of some algae, epibionts for their part can have dramatic effects on the host survival. Epibionts have been shown to increase the rates of dislodgement of mussels (Witman and Suchanek 1984; Brosnan, unpublished data), barnacles (Hawkins 1981), and algal hosts (Lubchenco 1986; D'Antonio 1985). Algae will also overgrow and smother barnacles (Farrell 1987; Brosnan, personal observations). However, epiphytes can also enhance the survival of the host, particularly during extremes of heat and cold. I have shown that, during cold weather, mussels with epiphytes survived freezing conditions, whereas those without epiphytes suffered high mortality (Brosnan, unpublished data).

In many temperate communities, the algal canopy provides a spatial refuge for predators. For example, Menge (1976, 1978) showed how fucoid canopy increased the foraging efficiency of *Thais* on New England shores. In New South Wales, *Gelidium* growing on barnacles protects them from predation by whelks; however, because whelks hide out in the canopy of *Gelidium*, nearby barnacles without a covering of algae are heavily preyed on (Underwood *et al.* 1983). Other recent studies have shown that epibionts reduce predation on their hosts (e.g., Bloom 1975; Vance 1978). However, algal canopy can also have detrimental effects: for example, whiplash inhibits larval and algal sporeling settlement (Menge 1976, 1978*b*; Hawkins 1981).

Although little is known about non-trophic relationships on tropical coasts, some patterns might be predicted from results of the above studies. Firstly, given the intense herbivory on some tropical shores such as that of Panama, algal epibionts are probably few and unimportant. In addition, barnacles are sufficiently rare that benefit resulting from the removal of epibionts by grazers may not be a major factor. These predictions may not be true for other tropical shores, e.g. the Guinea coast of Africa where bivalves, barnacles, and algae are more abundant, or Costa Rica where predation is lower.

The effect of algal canopy is probably unimportant in the tropics, although algal crusts provide refuges for some species (Stephenson and Searles 1960; Hixon, pers. comm.); little is known of the ecological importance of this relationship.

Disturbance and succession

Disturbance has been well studied on temperate shores (e.g., Dayton 1971; Sousa 1979, 1985; Paine and Levin 1981; Sousa *et al.* 1981). Storm-generated disturbances have major effects on intertidal communities on the Pacific coast of North America by increasing diversity and preventing competitive exclusion by mussels (Dayton 1971; Sousa 19779; Paine and Levin 1981).

Almost nothing is known about the role of disturbance on tropical shores. Bare patches are occasionally created on Panamanian shores by 'objects' floating in the sea (Cubit, reported in Sousa 1985). Stephenson and Searles (1960) reported an interesting seasonal and fairly large-scale disturbance on some shores; female turtles, coming ashore to lay their eggs, gouge large tracts of intertidal rock as they drag themselves across the shore, creating wide areas of bare space. Extended low tides caused by climatic conditions can also cause high mortalities on some tropical shores (Hay *et al.* 1983; Hixon, pers. comm.). These are unpredictable events.

Ice scour is a major cause of disturbance on New England shores (Menge 1976; Wethey 1985), occurring in 35–40 per cent of winters (Wethey 1985). This coastline is dominated by early maturing, fast-growing species, in contrast to the larger and more slow-growing forms on the west coast of America. In a computer simulation model, Wethey (1985) concluded that longer lived, late maturing species would go extinct in the disturbance regimes of New England. The effects of ice scour in New England resemble disturbances caused by heat and desiccation on some tropical shores. For example, seasonal local extinctions of algae have been reported from shores in the tropics (e.g., Lawson 1966; Hodgkiss 1984; Banaimoon 1988; Murthy *et al.* 1989). Although the frequency of disturbance is higher on tropical shores than that reported for New England, it might also be predicted that short-lived, early-maturing species should dominate tropical shores; this interpretation is supported by work of Hodgkiss (1984) and Banaimoon (1988). Heat stress does not seem to be an important source of disturbance in Panama. A possible explanation for this is that the Pacific Coast of Panama has a higher tidal range (3–6 m) than any of the shores where high heat-stress mortality has been reported (usually < 1 m). Tides in Panama are predictable, and there is no appreciable drop in sea-level during the dry season. In contrast, a dry-season drop in sea-level, coupled with calm seas, may increase emersion times for intertidal species on other shores.

The intermediate disturbance hypothesis (Connell 1978) predicts that highest diversity will occur at intermediate levels of disturbance. High rates of disturbance will maintain communities in an early successional stage, dominated by species with high recruitment rates, fast growth rates, and early maturation. The mid- to upper-zones of some tropical shores are often dominated by ephemeral algae, such as *Enteromorpha* and *Ulva* (Hodgkiss 1984; Banaimoon 1988). High disturbance rate is one possible explanation for this phenomenon.

In one of the first experimental studies of intertidal succession in the tropics, Lawson (1966) cleared plots on a rocky shore in Ghana (West Africa) and monitored succession on both plots and controls for nine years. The overall pattern of succession followed the same as that reported for temperate shores (Castenholz 1961; Sousa 1979, 1985; Sousa *et al.* 1981; Paine and Levin 1981; Brosnan, personal observations). Diatoms appeared first, and within weeks were replaced by green algae (*Enteromorpha* and *Ulva*). At this stage herbivorous molluscs arrived, barnacles became common, and thereafter Lithothamnia appeared in the mid- to upper-zone quadrats. Crustose algae gradually began to invade the upper quadrats, *Gigartina* was a late successional species though never dominant. Lawson (1966) noted that recolonization took about six years, proceeding more quickly in lower-shore quadrats.

In summary, although succession may follow similar patterns on most shores, we know little about the causes, frequency, and intensity of disturbances on tropical shores, nor about their effects on succession.

A global approach to plant–animal interactions

Throughout this paper I have emphasized that ecological processes do not vary with latitude. Rather it is the relative importance of interactions that changes. These shifts in importance produce the observed diversity of patterns. If we can understand the conditions under which one or another process becomes dominant, then this will allow predictive approach to plant–animal interactions. For example, Underwood and Fairweather (1986) provided a stimulating discussion of how patterns and processes affecting the same species can vary across shores.

Recent models by Menge and Sutherland (1976, 1987) predict which factors will be most important in communities under different conditions. They consider how environmental harshness affects the relative importance of competition and predation at different trophic levels, and how trophic complexity will be affected by stress. They assume that mobile consumers are more constrained than sessile species by environmental harshness. Consequently, trophic complexity should increase as harshness decreases. Specific predictions include:

(1) in physically benign environments, trophic complexity will be high, and consumer pressure will be the main structuring force in the community;
(2) the effects of predation will increase as trophic level decreases;
(3) trophically intermediate species will be regulated by competition and predation;
(4) competition will occur among consumers that escape predation.

This model has been tested in temperate and tropical latitudes (Menge and Sutherland 1976, 1987; Menge *et al.* 1986*b*; Menge and Farrell 1989). Results are summarized in Table 5.2 and Fig. 5.2. In a comparison between the East and West coasts of America, Menge and Sutherland (1976) concluded that in the harshest environment tested (exposed New England shores) predation rates were low and competition between basal species was the dominant ecological factor. By contrast, predation rates were higher on protected New England shores, and competition among basal species was sporadic; herbivory

Table 5.2. Tests of the Menge and Sutherland (1976, 1987) model of the relative importance of ecological factors along an environmental harshness gradient (compiled from Menge and Sutherland 1976, 1987; Menge et al., 1986a, b; Menge and Farrell 1989).

Site	Harshness	Description	Relative importance of ecological factors
Temperate Exposed New England East Coast USA	Seasonally harsh and variable	Space is limiting. There is one predator, *Thais*, which is rare and has little effect. Mussels and barnacles are the dominant space occupiers.	Competition is more important and is strong between mussels and barnacles.
Temperate Protected New England	Less harsh	Bare space is more abundant. Predation, by *Thais*, is intense.	Predation is more important. Occasional prey-escapes (by mussels) lead to competition for space.
Temperate West Coast USA	Less harsh and less variable than East Coast	More diverse than the East Coast. A secondary carnivore, *Pisaster*, is present. *Thais* is secondarily important.	Predation dominates. Competition at lower trophic levels is sporadic.
Tropical Pacific Coast Panama	Constant physical conditions prevail and fairly benign	Fish, crabs, and molluscan consumers are abundant. Sessile animals and foliose algae are rare. Grazer-resistant, crustose algae predominate.	Predation is most important. Crustose algae compete.

is an important factor controlling algal distribution on these shores (Lubchenco 1978, 1980). Physical conditions in Panama are more benign and less variable than either of the coasts of North America (Menge et al. 1986b). As predicted, consumer pressure is the driving force in this community; the effects of consumers are more important

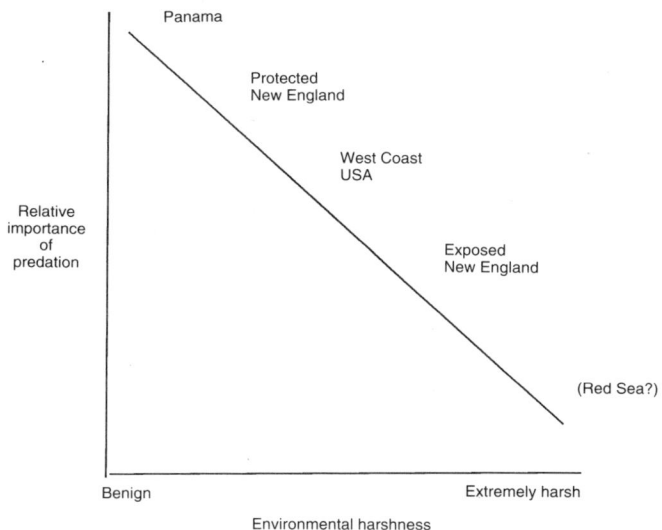

Fig. 5.2. A geographical comparison of the relative importance of consumer pressure with increasing environmental stress, based on Menge and Sutherland (1987). According to this model, the impact of consumers should increase as environmental conditions become more benign. Tests of the model support this prediction. Panama, the most benign shore in this comparison, has the highest level of predation (Menge et al. 1986 a, b). This model is not based on geographical areas, and for instance, although it has not been tested, the hot, dry season on some tropical shores (e.g., Red Sea and South Indo-China Sea) could place them on the far right of this figure, implying that physical factors are dominant and herbivory is unimportant.

on the lower trophic levels (e.g., there are no foliose algae); competition occurs between species that escape predation (e.g., crustose algae). However Menge et al. did not find that trophically intermediate species (herbivorous gastropods) were influenced by predation: the role of competition was not studied. Ortega (1985) found that in Costa Rica, interspecific competition between limpets only became important during the dry season. On shores where seasonality is strong (e.g., the Red Sea), the relative importance of consumers and competition may shift, and shores may move along the gradient from harsh (the dry season), when physical conditions are the overriding factor, to less harsh (wet season), when herbivory increases in importance.

How does this model relate to tropical–temperate comparisons of plant interactions? Specifically the model predicts that as harshness increases herbivory becomes less important. This may be true for some shores during the dry season where temperatures are high (e.g.,

Hodgkiss 1984; Jones *et al.* 1987; Murthy *et al.* 1989). On these shores physical factors are the major structuring forces, and the model predicts that plant–animal trophic relationships should be relatively unimportant, and that sessile species should dominate. This means that some tropical shores could be ecologically closer to exposed New England shores than to a protected New England shore, where herbivory is an important plant–animal interaction.

Recruitment

One needs only to look at the devastation wreaked on coral reefs from sporadic increases in *Acanthaster planci* to appreciate just how important is recruitment. While variations have long been acknowledged, only in recent years have there been any experimental manipulations of recruitment and theoretical studies (Gaines and Roughgarden 1985; 1987; Menge and Sutherland 1987; Possingham and Roughgarden 1990; Menge 1991) on its implications for species interactions and community structure (e.g., Denley and Underwood 1979; Gaines *et al.* 1985; Gaines and Roughgarden 1985, 1987; Menge *et al.* 1987; Underwood and Fairweather 1986). Little can be said about recruitment rates on most temperate shores, much less on tropical shores. Nevertheless, Menge and Sutherland (1987) reported low rates in Panama, orders of magnitude less than in New England. Sutherland (1990) studied barnacle recruitment in Costa Rica, and concluded that recruitment to Costa Rican shores is low. He also noted that Panama has a lower recruitment rate than Costa Rica. How might variable rates of recruitment affect community structure? In their model of community dynamics, Menge and Sutherland (1987) predict that low recruitment will reduce the relative importance of competition between species. This prediction is borne out by the above results from Panama and New England (Menge 1991). It has been reported that in Costa Rica competition between barnacles only becomes important during sporadic increases in recruitment (Sutherland and Ortega 1986; Sutherland 1990). Menge and Sutherland (1987) argued that when recruitment is low predation will also be less important. This does not mean that individual acts of predation or herbivory will be less important (Menge, pers. comm.) — they may in fact be relatively more important if the density of prey is low.

One assumption, that needs testing, of the Menge–Sutherland model is that recruitment is correlated at all trophic levels. It seems reasonable to assume that planktonic larvae and sporelings are similarly affected by currents and local shore topography. However, fish are the main consumers in Panama and these do not recruit into the intertidal zone. Factors affecting fish recruitment can depend on availability of suitable subtidal habitats, density of conspecifics, of competitors, and of predators (Booth 1991). These do not necessarily correlate with any intertidal phenomena. If fish recruitment varies independently of intertidal phenomena, then low recruitment rates may lead to increase in competition for space among sessile species; predation and herbivory on lower trophic levels will then become more important (as more carnivorous and herbivorous invertebrates escape predation). The relative importance of non-trophic interactions will also increase. If, however, fish recruitment increases and there is no change in recruitment of intertidal species, competition at lower trophic levels will decrease and non-trophic interactions will become less important.

Menge and Sutherland (1987) acknowledge that their model is difficult to test, but testing is feasible (see Menge and Sutherland 1987; Menge and Farrell 1989; for testing protocols). Quantifying a number of the variables may be difficult. Very few marine ecologists have the luxury, or the life-span to work on rocky shores on a global scale. Quantifications of wave-exposure are usually subjective and based on the relative experience of the researcher. A more objective measure of wave-exposure is desirable, but thus far more difficult to achieve. Some predictions of the model have not been met and others not yet tested (see Menge and Sutherland 1987; Menge and Farrell 1989; for full discussion). Nonetheless, the model has considerable predictive value as a tool for studying key issues in ecology, including plant–animal interactions.

One of the aims of this review is to highlight unanswered questions, and areas for future research. It is clear that little is known about tropical rocky shores, and that there is an urgent need for experimental studies on marine plant–animal interactions in the tropics.

Acknowledgements

Thanks are due to Steven Courtney for stimulating discussions, arguments, and translating from the Irish. Bruce Menge, Annette Olson, and Dave Booth provided me with information, scientific discussion, and critique.

References

Abrams, P. A. (1984). Shell fighting and competition between two hermit crab species in Panama. *Oecologia* (Berlin), **51**, 84–90.

Banaimoon, S. A. (1988). The marine algal flora of Khalf and adjacent regions, Hadramout, P. D. R. Yemen. *Botanica Marina*, **31**, 215–21.

Bloom, S. A. (1975). The mobile escape response of a sessile prey: a sponge scallop mutualism. *Journal of Experimental Marine Biology and Ecology*, **17**, 311–21.

Booth, D. (1991). The effects of sampling frequency on estimates of recruitment of the domino damselfish *Dascyllus albisella*. *Journal of Experimental Marine Biology and Ecology*, **45** (2) 149-50.

Branch, G.M. (1975). Ecology of *Patella* species from the Cape Peninsula, S. Africa. IV. Desiccation. *Marine Biology*, **32**, 179-200.

Branch, G.M. (1981). The biology of limpets: physical factors, energy flow and ecological interactions. *Oceanography and Marine Biology Annual Review*, **21**, 195-282.

Branch, G.M. (1986). Limpets: their role in littoral and sublittoral community dynamics. In *The ecology of rocky coasts*, (ed. P.G. Moore and R. Seed), pp. 97-116. Columbia University Press.

Brown, J.H., Davidson, D.W., Munger, J.C., and Inouye R.S., (1986). Experimental community ecology: the desert granivore system. In *Community ecology*, (ed. J. Diamond and T.J. Case), pp. 41-62. Harper and Row, New York.

Castenholz, R.W. (1961). The effect of grazing on marine littoral diatom populations. *Ecology*, **42**, 783-94.

Coley, P.D., Bryand, J.P., and Chapin III, F. Stuart (1985). Resource availability and plant antiherbivore defense. *Science*, **230**, 895-9.

Connell, J.H. (1978). Diversity in tropical rainforests and coral reefs. *Science*, **199**, 1302-10.

Cubit, J.D. (1984). Herbivory and the seasonal abundance of algae on a high intertidal rocky shore. *Ecology*, **65**, 1904-17.

D'Antonio, C. (1985). Epiphytes in the rocky intertidal alga *Rhodomela larix* (Turner) 1. Agardhi: negative effects on the host, and food for herbivores. *Journal of Experimental Marine Biology and Ecology*, **86**, 197-218.

Dayton, P.K. (1971). Competition, disturbance and community organization: the provision and subsequent utilization of space in a rocky intertidal community. *Ecological Monographs*, **41**, 351-89.

Dayton, P.K. (1973). Dispersion, dispersal, and persistence of the annual intertidal alga *Postelsia palmaeformis* Ruprecht. *Ecology*, **54**, 433-8.

Dayton, P.K. (1975). Experimental evaluation of ecological dominance in a rocky intertidal algal community. *Ecological Monographs*, **49**, 1075-91.

Denley, E.J. and Underwood, A.J. (1979). Experiments on factors influencing settlement, survival and growth of two species of barnacle in New South Wales. *Journal of Experimental Marine Biology and Ecology*, **36**, 269-93.

Dethier, M.N. and Duggins, D.O. (1984). An indirect 'commensalism' between marine herbivores and the importance of competitive hierarchies. *American Naturalist*, **124**, 205-19.

Dethier, M.N. and Duggins, D.O. (1988). Variation in strong interactions in the intertidal zone along a geographical gradient. *Marine Ecology Progress Series*, **50**, 97-105.

Duffy, J. Emmet (1990). Amphipods on seaweeds: patterns or pests. *Oecologia* (Berlin), **83**, 267-76.

Dungan, M. L. (1986). Three-way interactions: barnacles, limpets and algae in a Sonoran desert rocky intertidal zone. *American Naturalist*, **127**, 292-316.

Estes, J. A. and Palmisano, J. F. (1974). Sea otters: their role in structuring nearshore communities. *Science*, **185**, 1058-60.

Farrell, T. M. (1987). Succession and stability in two rocky intertidal communities on the Central Oregon Coast. Unpublished Ph.D. Thesis. Department of Zoology, Oregon State University, Corvallis, USA.

Farrell, T. M. (1991). Models and mechanisms of succession: an example from a rocky intertidal community. *Ecological Monographs*, **61**(1), 95-113.

Feeny, P. (1976). Plant apparency and chemical defenses. *Recent Advances in Phytochemistry*, **10**, 1-40.

Gaines, S. D. and Lubchenco, J. (1982). A unified approach to marine plant–herbivore interactions. II. Biogeography. *Annual Review of Ecology and Systematics*, **13**, 111-38.

Gaines, S. D. and Roughgarden, J. (1985). Larval settlement rate: a leading determinant of structure in an ecological community of the marine intertidal zone. *Proceedings of the National Academy of Sciences*, **82**, 3707-11.

Gaines, S. D. and Roughgarden, J. (1987). Fish in offshore kelp forests affect recruitment to intertidal barnacle populations. *Science*, **235**, 479-81.

Gaines, S. D., Brown, S., and Roughgarden, J. (1985). Spatial variation in larval concentrations as a cause of spatial variation in settlement for the barnacle *Balanus glandula*. *Oecologia* (Berlin), **67**, 267-72.

Garrity, S. D. (1984). Some adaptations of gastropods to physical stress on a tropical rocky shore. *Ecology*, **65**, 559-74.

Garrity, S. D. and Levings, S. C. (1981). A predator–prey interaction between two physically and biologically constrained tropical rocky shore gastropods: direct, indirect and community effects. *Ecological Monographs*, **51**, 267-86.

Glynn, P. W. (1972). Observations on the ecology of the Caribbean and Pacific coasts of Panama. *Bulletin of the Biological Society of Washington*, **2**, 15-30.

Gauld, P. J. and Buchanan, J. B. (1959). The principal features of the rock shore fauna in Ghana. *Oikos*, **10**, 121-32.

Hawkins, S. J. (1981). The influence of season and barnacles on the algal recolonization of *Patella vulgata* exclusion areas. *Journal of the Marine Biological Association of the UK*, **61**, 1-15.

Hawkins, S. J. and Hartnoll, R. G. (1983). Grazing of intertidal algae by marine invertebrates. *Oceanography and Marine Biology Annual Review*, **21**, 195-282.

Hay, M. E. and Fenical, W. (1988). Marine plant–herbivore interactions: the ecology of chemical defenses. *Annual Review of Ecology and Systematics*, **19**, 111-45.

Hay, M. E., Colburn, T., and Downing, D. (1983). Spatial and temporal patterns in herbivory on a Caribbean fringing reef: the effects on plant distribution. *Oecologia* (Berlin), **58**, 299-308.

Hixon, Mark. A. (1986). Fish predation and local prey diversity. In *contemporary studies on fish feeding*, (ed. Charles A. Simenstad and Gregor M. Cailliet), pp. 235-58. Dr. W. Junk Publishers, Dordrecht, Netherlands.

Hodgkiss, I.J. (1984). Seasonal patterns of intertidal algal distribution in Hong Kong. *Asian Marine Biology*, **1**, 49-57.

Holt, R.D. (1977). Predation, apparent competition, and the structure of prey communities. *Theoretical Population Biology*, **12**, 197-229.

Jara, H.J. and Moreno, C.A. (1984). Herbivory and structure in a midlittoral rocky community: a case in southern Chile. *Ecology*, **65**, 28-38.

John, D.M. and Pople, W. (1973). The fish grazing of rocky shore algae in the Gulf of Guinea. *Journal of Experimental Marine Biology and Ecology*, **11**, 81-90.

Jones, D.A., Ghamrawy, M., and Wahbeh, M. (1987). Littoral and shallow subtidal environments. In *Red Sea (key environments)*, (ed. A.J. Edwards and S.M. Head), pp. 169-93. Pergamon Press.

Lawson, G.W. (1957). Seasonal variation in intertidal zonation on the coasts of Ghana in relation to tidal factors. *Journal of Ecology*, **45**, 381-860.

Lawson, G.W. (1966). The littoral ecology of West Africa. *Oceanography and Marine Biology Annual Review*, **4**, 405-48.

Lee, H. and Ambrose, W.G. (1989). Life after competitive exclusion: an alternative strategy for a competitive inferior. *Oikos*, **56**, 424-30.

Levings, S.C. and Garrity, S.D. (1983). Diel and tidal movement of two co-occurring neritid snails; differences in grazing patterns on a tropical rocky shore. *Journal of Experimental Marine Biology and Ecology*, **67**, 261-78.

Levings, S.C. and Garrity, S.D. (1984). Grazing patterns in *Siphonaria gigas* (Mollusca, Pulmonata) on the rocky Pacific coast of Panama. *Oecologia* (Berlin), **64**, 152-9.

Longhurst, A.R. and Pauly, D. (1987). *Ecology of tropical oceans*. Academic Press, London.

Lubchenco, J. (1978). Plant species diversity in a marine intertidal community: importance of herbivore food preference, and algal competitive abilities. *American Naturalist*, **112**, 23-39.

Lubchenco, J. (1980). Algal zonation in the New England rocky intertidal community: an experimental analysis. *Ecology*, **61**, 333-44.

Lubchenco, J. (1986). Relative importance of competition and predation: early colonization by seaweeds in New England. In *Community ecology*, (ed. J. Diamond and T.J. Case), pp. 537-555. Harper and Row, New York.

Lubchenco, J. and Cubit, J. (1980). Heteromorphic life histories of certain marine algae as adaptations to variations in herbivory. *Ecology*, **61**, 676-87.

Lubchenco, J. and Gaines, S. D. (1981). A unified approach to marine plant–herbivore interactions. I. populations and communities. *Annual Review of Ecology and Systematics*, **12**, 405–37.

Lubchenco, J. and Menge, B. A. (1978). Community development and persistence in a low rocky intertidal zone. *Ecological Monographs*, **48**, 67–94.

Lubchenco, J., Menge, B. A., Garrity, S. D., Lubchenco, P. J., Ashkenas, L. R., and Gaines, S. D. (1984). Structure, persistence, and role of consumers in a tropical rocky intertidal community (Taboguilla Island, Bay of Panama). *Journal of Experimental Marine Biology and Ecology*, **78**, 23–73.

Mann, K. H. (1982). Kelp, sea urchins, and predators: a review of strong interactions in rocky subtidal systems of eastern Canada 1970–1988. *Netherlands Journal of Sea Research*, **16**, 414–23.

Menge, B. A. (1976). Organization of the New England rocky intertidal community: role of predation, competition, and environmental heterogeneity. *Ecological Monographs*, **46**, 355–93.

Menge B. A. (1978a). Predation intensity in a rocky intertidal community: relation between predator foraging activity and environmental harshness. *Oecologia* (Berlin), **34**, 1–16.

Menge, B. A. (1978b). Predation intensity in a rocky intertidal community: effect of algal cover, wave action and desiccation on predator feeding rates. *Oecologia* (Berlin), **34**, 17–35.

Menge, B. A. (1982). Reply to a comment by Edwards, Conover, and Sutter. *Ecology*, **63**, 1180–4.

Menge B. A. (1991). Relative importance of recruitment and other causes of variation in rocky intertidal community structure. *Journal of Experimental Marine Biology and Ecology*, **145**(1), 69–100.

Menge, B. A. and Farrell, T. (1989). Community structure and interaction webs in shallow marine hard-bottom communities: tests of an environmental stress model. *Advances in Ecological Research*, **19**, 189–262.

Menge, B. A. and Lubchenco, J. (1981). Community organization in temperate and tropical rocky intertidal habitats: prey refuges in relation to consumer pressure gradients. *Ecological Monographs*, **51**, 429–50.

Menge, B. A. and Sutherland, J. P. (1976). Species diversity gradients: synthesis of the roles of predation, competition, and temporal heterogeneity. *American Naturalist*, **110**, 351–69.

Menge, B. A. and Sutherland, J. P. (1987). Community regulation: variation in disturbance, competition and predation in relation to environmental stress and recruitment. *American Naturalist*, **130**, 730–57.

Menge, B. A., Ashkenas, L. R., and Matson, A. (1983). Use of artificial holes in studying community development in cryptic marine habitats in a tropical rocky intertidal region. *Marine Biology*, **77**, 129–42.

Menge, B. A., Lubchenco, J., Ashkenas, L. R., and Ramsey, F. (1986a). Experimental separation of effects of consumers on sessile prey in the low zone of a rocky shore in the Bay of Panama: direct and indirect consequences of food web complexity. *Journal of Experimental Marine Biology and Ecology*, **100**, 225-69.

Menge, B. A., Lubchenco, J., Gaines, S. D., and Ashkenas, L. R. (1986*b*). A test of the Menge-Sutherland model of community organization in a tropical rocky intertidal food web. *Oecologia* (Berlin), **71**, 75-89.

Moore, H. B. (1973). Aspects of stress in the tropical marine environment. *Advances in Marine Biology*, **10**, 217-69.

Murthy, M. S., Ramakrishna, T., Rao, Y. N., and Ghose, D. K. (1989). Ecological studies on some agarophytes from Veraval Coast (India) I. Effects of aerial conditions on the biomass dynamics. *Botanica Marina*, **32**, 515-20.

Ortega, S. (1985). Competitive interactions among tropical intertidal limpets. *Journal of Experimental Marin Biology and Ecology*, **90**, 11-25.

Ortega, S. (1986). Fish predation on gastropods on the pacific coast of Costa Rica. *Journal of Experimental Marine Biology and Ecology*, **97**, 181-91.

Padilla, D. K. (1985). Structural resistance of algae to herbivores: a biomechanical approach. *Marine Biology*, **90**, 103-9.

Paine, R. T. and Levin, S. (1981). Intertidal landscapes: disturbance and the dynamics of pattern. *Ecological Monographs*, **51**, 145-178.

Paine, R. T. and Vadas, R. L. (1969). Effects of grazing by sea urchins, *Strongylocentrotus* species on benthic algal populations. *Limnology and Oceanography*, **14**, 710-17.

Paul, V. J. and Hay, M. E. (1986). Seaweed susceptibility to herbivory: chemical and morphological correlates. *Marine Ecology Progress Series*, **33**, 255-64.

Possingham, H. P. and Roughgarden, J. (1990). Spatial population dynamics of a marine organism with a complex life cycle. *Ecology*, **71** (3), 973-85.

Rao, K. S. and Sundaram, K. S. (1974). Ecology of intertidal molluscs of gulf of Mannar and Palk Bay. *Indian National Science Academy Bulletin*, **47**, 462-74.

Rao, M. U. (1974). Ecological observations on some intertidal algae of Mandapam coast. *Indian National Science Academy Bulletin*, **47**, 298-307.

Rhoades, D. F. and Cates, R. G. (1976). Towards a general theory of plant antiherbivore chemistry. *Recent Advances in Phytochemistry*, **10**, 1-40.

Sousa, W. P. (1979). Disturbance in marine intertidal boulder fields: the nonequilibrium maintenance of species diversity. *Ecology*, **60**, 1225-39.

Sousa, W. P. (1985). Disturbance and patch dynamics on rocky intertidal shores. In *The ecology of natural disturbance and patch dynamics*, (ed. S. T. A. Pickett and P. S. White), pp. 101-24. Academic Press, London.

Sousa, W. P., Schoeter, S. C., and Gaines, S. D. (1981). Latitudinal variation in intertidal algal community structure: the influence of grazing and vegetative propagation. *Oecologia* (Berlin), **48**, 297-307.

Stephenson, W. and Searles, R. B. (1960). Experimental studies on the ecology of intertidal environments at Heron Island, Australia. *Australian Journal of Marine and Freshwater Research*, **11**, 241–67.

Stephenson, T. A. and Stephenson, A. (1972). *Life between tidemarks on rocky shores*. W. H. Freeman, San Francisco.

Sutherland, J. P. (1990). Recruitment regulates demographic variation in a tropical intertidal barnacle. *Ecology*, **71**, 955–72.

Sutherland, J. P. and Ortega, S. (1986). Competition conditional on recruitment and temporary escape from predators on a tropical rocky shore. *Journal of Experimental Marine Biology and Ecology*, **95**, 155–166.

Underwood, A. J. (1979). The ecology of intertidal gastropods. *Advances in Marine Biology*, **16**, 111–210.

Underwood, A. J. (1980). The effects of grazing by gastropods and physical factors on the upper limits of distribution of intertidal macroalgae. *Oecologia* (Berlin), **46**, 201–13.

Underwood, A. J. (1981). Structure of a rocky intertidal community in New South Wales: patterns of vertical distribution and seasonal changes. *Journal of Experimental Marine Biology and Ecology*, **51**, 57–85.

Underwood, A. J. (1984). Vertical and seasonal patterns in competition for microalgae between intertidal gastropods. *Oecologia* (Berlin), **64**, 211–22.

Underwood, A. J. and Fairweather, P. G. (1986). Intertidal communities: do they have different ecologies or different ecologists? *Proceedings of the Ecological Society of Australia*, **14**, 7–16.

Underwood, A. J. and Fairweather, P. G. (1989). Supply-side ecology and benthic marine assemblages. *Trends in Ecology and Evolution*, **4**, 16–19.

Underwood, A. J. and Jernakoff, P. (1981). Effects of interactions between algae and grazing gastropods on the structure of a low-shore intertidal algal community. *Oecologia* (Berlin), **48**, 221–33.

Underwood, A. J. and Jernakoff, P. (1984). The effects of tidal height, wave exposure, seasonality and rock-pools on grazing and the distribution of intertidal macroalgae in New South Wales. *Journal of Experimental Marine Biology and Ecology*, **75**, 71–96.

Underwood, A. J., Denley, E. J. and Moran, M. J. (1983). Experimental analyses of the structure and dynamics of mid-shore rocky intertidal communities in New South Wales. *Oecologia* (Berlin), **56**, 202–19.

Van Alstyne, K. L. (1988). Herbivore grazing increases polyphenolic defences in the intertidal brown alga *Fucus distichus*. *Ecology*, **63**(3), 655–63.

Vance, R. R. (1978). A mutualistic interaction between a sessile marine clam and its epibionts. *Ecology*, **59**, 679–85.

Walker, D. I. (1987). Benthic algae. In *Red Sea (Key environments)*, (ed. A. J. Edwards and S. J. Head), pp. 152–168. Pergamon Oxford.

Wethey, D. S. (1985). Catastrophe, extinction, and species diversity: a rocky intertidal example. *Ecology*, **66**, 445–56.

Witman, J. D. and Suchanek, T. H. (1984). Mussels in flow: drag and dislodgement by epizoans. *Marine Ecology Progress Series*, **16**, 259–68.

6. Grazing on sediment shores

KARSTEN REISE

Wadden Sea Institute Sylt of Biologische Anstalt Helgoland, List, Germany

Abstract

In intertidal mud and sand, diatoms are the dominant microalgae, with about 10^6 cells cm^{-2} and an annual net production of roughly 100 g C m^{-2}. One-third may be consumed by the zoobenthos. Large deposit feeders ingest microalgae together with sediment particles. Smaller ones scrape algal cells from sand grains or capture mobile diatoms. These are either ingested intact or are punctured and sucked out. Bulk measurements on primary production and on consumption provide little insight into plant–animal interactions in shallow sediments. Analyses should differentiate between type of algae and mode of feeding. Large epipelic diatoms are occasionally subject to overgrazing, while the small epipsammic diatoms are utilized to a lesser extent. At moderate density some grazers stimulate algal growth. Gardening is achieved by excretion of ammonia and secretion of mucus. The extracellular mucopolysaccharides produced by diatoms increase the surface stability of sediments and enhance the accretion of organic deposits. This is beneficial to deposit feeders. Heavy grazing on mud-flats reduces diatom density, and the habitat may become more sandy because of interference with the accretion of mud. In sand, bioturbating macrofauna depresses microalgal growth by frequent burial. Large burrowers occasionally anchor strings of green macroalgae, which subsequently grow into coherent algal mats, changing the entire habitat profoundly.

Introduction

Vast areas of tidal flats and long stretches of sandy shores are virtually free from vegetative cover. Only here and there, seagrasses abound in patches, red algae (*Gracilaria*) are cultured, and green algae

Plant–Animal Interactions in the Marine Benthos, (ed. D. M. John, S. J. Hawkins, and J. H. Price), Systematics Association Special Volume No. 46, pp. 133-45. Clarendon Press, Oxford, 1992. © The Systematics Association 1992.

accumulate on sheltered flats. The dominant plants on sediment shores are of microscopic size. These unicellular algae follow the pattern of the mobility of the sediments. From the small size of these plants it follows that all grazers are larger, and most of them considerably so. This is in contrast to the rocky shore where the phytobenthos tends to exceed the zoobenthos in size.

On sediment shores, animals are more conspicuous than plants. They have attracted the greatest scientific curiosity, while very little is known about the small algae between the grains of sand and upon the mud surface. Even less is known about the interactions between the benthic microalgae and their grazers. Consequently, this article is mostly a review of the unknown and contains suggestions for further research.

Benthic microalgae

Microalgae are the dominant plants on sediment shores and these are dominated by diatoms (Fig. 6.1). Other important primary producers are flagellates and Cyanobacteria. The composition of the microphytobenthos of a typical sand-flat in the temperate zone is detailed in Table

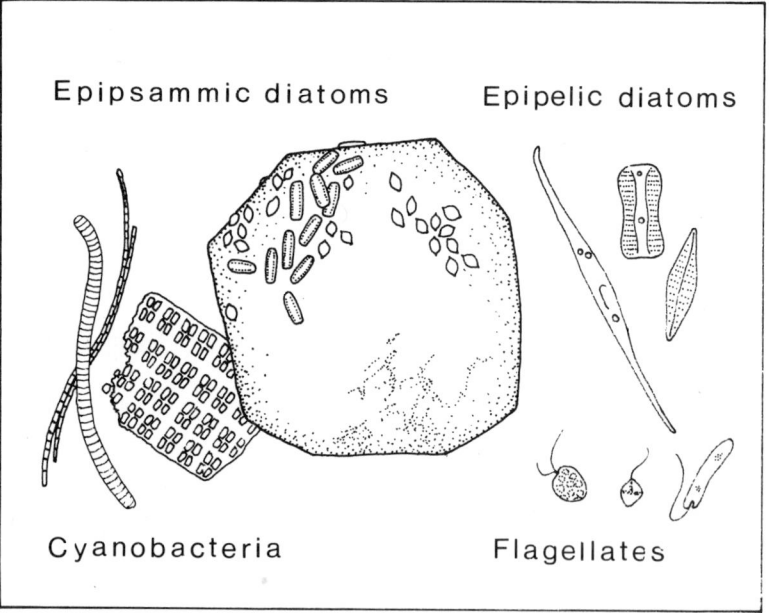

Fig. 6.1. Schematic presentation of forms in the microphytobenthos of sediment shores.

Table 6.1. Benthic microalgae of an *Arenicola* sand-flat at mid-tide level during summer near the Island of Sylt, North Sea (data compiled from Asmus 1984; Asmus and Asmus 1985; Werner 1990)

Benthic microalgae	size of cells, length in μm	cells cm^{-2}	Biomass μg C cm^{-2}
Cyanobacteria	4 to 8	0.3 to 4 $\times$ 10^6	2
Flagellates	20 to 50	1 to 4 $\times$ 10^3	< 1
Epipsammic diatoms	3 to 20	3 to 6 $\times$ 10^6	8
Epipelic diatoms	10 to 100	4 to 21 $\times$ 10^3	4

6.1. Diatoms range in size from three to more than 100 μm in length, with a cell weight of 5–5000 pg C. The smaller diatoms (up to 20 μm or 200 pg C cell^{-1}) are mostly ovoid in shape, attached to sand grains, and have been termed epipsammic diatoms (Round 1971). They outnumber all other microalgae and contribute to most of the algal biomass. The larger diatoms are mobile, easily resuspended from the sediment, and most of them are elongated in shape (epipelic diatoms).

On sand-flats, some diatoms move to the surface at daytime low tides and move downwards when the tide comes in (Aleem 1950; Palmer and Round 1967; Joint *et al.* 1982). Massive growths of diatoms have been observed to develop seasonally (spring: Asmus 1982; summer: Admiraal and Peletier 1980; autumn: Schwinghamer 1983). On mud in particular they may form coherent mats on the surface, often as a monospecific assemblage (Admiraal 1984). The extracellular mucus produced by these diatoms is conspicuous and may amount to 20 per cent of all microalgal carbon in the sediment (Grant *et al.* 1986).

The photosynthetic flagellates (dinoflagellates and euglenophytes) are in the size range of 5–50 μm, but there may be numerous individuals that are smaller than this. They are also highly mobile.

The photosynthetic blue-green algae or cyanobacteria form sheet-like colonies on sandy flats. Some are attached to sand grains, while others occur freely in the interstices of sand. In the upper intertidal zone, filamentous colonies of Cyanobacteria may form mats which consolidate the sediment and dominate over diatoms.

The annual production of benthic microalgae is usually in the range of 20–200 g C m^{-2}, with an average of 130 (Colijn and de Jonge 1984; Plante-Cuny 1984; Varela and Penas 1985). As is to be expected from the light regime, benthic primary production is high in the upper intertidal, low in the lower intertidal, and almost absent in the subtidal zone (Cadée and Hegemann 1977). On tropical flats, primary production of up to 300 g C m^{-2} yr^{-1} has been found. Partitioning of primary production between groups or species of the benthic microalgae has not been accomplished. Primary production of marsh grasses, mangroves, kelp beds, or coral reefs is often five to ten times higher (see Mann 1982). However, microalgae are more easily digested by marine invertebrates and thus constitute an important food source in the coastal ecosystem.

Grazers on benthic microalgae

Two types of grazing may be distinguished on sediment shores: selective capture of algal cells, and bulk feeding of sediment particles which includes microalgae to a varying degree. Benthic microalgae are embedded in a complex sediment structure. Therefore, only members of the microfauna, meiofauna, and small macrofauna are capable of preying selectively on algal cells. These grazers move through the interstitial system or upon the surface and capture mobile flagellates and diatoms, or else browse the epigrowth on sand grains.

Important selective grazers are ciliates which may occur in numbers of up to 4000 individuals cm^{-2}. According to Fenchel (1968, 1969), several species feed exclusively on flagellates and diatoms or on both. Diatom species are selected according to size. Averaged over several sediment shores, approximately half of all ciliates are algivores.

All major groups of meiofauna, (e.g., nematodes, copepods, and platyhelminths) include algivores (Tietjen and Lee 1977; Hicks and Coull 1983; Jensen 1987; Reise 1988). Among 177 species of platyhelminths on sediment shores around the island of Sylt, Germany, 27 per cent are diatom feeders. These dominate on mud-flats, while carnivores prevail in sand. Algal cells are ingested whole but some nematodes puncture diatoms to suck them out. Although mixed diets were usually recognized in gut analysis studies, preferences for certain prey species are common. Admiraal *et al.* (1983) demonstrate that a nematode species discriminates between two congeneric diatoms and preys more on one than the other. Diatoms are the most frequently recorded algae in meiofaunal guts, while flagellates and, particularly, Cyanobacteria are rarely noticed.

Small members of the macrofauna also seem to be capable of selective feeding on benthic microalgae. The snail *Hydrobia* is known to ingest individual mobile diatoms (Fenchel and Kofoed 1976), to scrape microalgae from the surface of sand grains (Lopez and Kofoed 1980: 'epipsammic browsing'), and to swallow entire sediment particles from which associated microflora is digested (Lopez and Levinton 1978). Diatoms are the principal food of *Hydrobia ulvae*, while bacteria may account for up to 10 per cent of consumption (Jensen and Siegismund

1980). In experiments, diatoms are preferred over other food (López-Figuerda and Niell 1987).

With increasing size of grazers, the potential to ingest individual algal cells and to scrape them from sand grains is lost, and deposit-feeding is the only means of obtaining benthic microalgae. Differential use of algal species may then depend on differential food absorption. Not all deposit feeders on the sediment shore utilize microalgae. For example, the hemichordate *Ptychodera bahamensis* digested bacteria, while the chlorophyll content in the sediment remained unchanged (Dobbs and Guckert 1988). With increasing size, the dependence on microalgal food probably decreases because other micro-organisms and detrital particles dominate in the sediment. The ecology of deposit-feeding animals in marine sediments is reviewed by Lopez and Levinton (1987).

A dependence on microalgal food may be indirectly inferred from the occurrence of dense populations of animals on intertidal flats and their absence on subtidal bottoms where light is insufficient for primary producers. This is particularly the case for gastropods like *Littorina* and *Ilyanassa* on temperate flats, or potamidid and cerithoid snails on tropical flats. The same applies to some tellinid bivalves which pipette their food from the sediment surface. On tropical flats, several deposit-feeding ocypodid crabs (*Ocypode, Uca, Scopimera, Mictyris, Macrophthalmus*) probably depend on a certain amount of microalgal food. Also some fish are microalgal feeders on sedimentary shores, like the mullet *Mugil* or the mudskipper *Boleophthalmus*.

Effects of grazers on benthic microalgae

Kuipers *et al.* (1981) suggested that most of the organic carbon produced on a tidal flat is utilized within the 'small food web', i.e. grazers of the micro- and meiofauna (see Fig. 6.2). Investigations by Admiraal *et al.* (1983), Montagna (1984), and Alongi (1988) do not support this hypothesis. Only a small fraction of microalgal biomass production seems to be consumed by the meiofauna. No decrease of standing stock was observed which could be attributed to grazing. However, selective feeding may change the dominance pattern within the microalgal assemblage. On a mud-flat, the abundant nematode *Eudiplogaster pararmatus* preferred the diatom *Navicula salinarum* over *N. pygmaea*. The latter developed into a monospecific algal mat and the former became rare (Admiraal *et al.* 1983).

The number of diatom species on a particular tidal flat may be more than 60 (Asmus 1984) and the number of micro- and meiofaunal species feeding on them is probably even higher. Considering the often

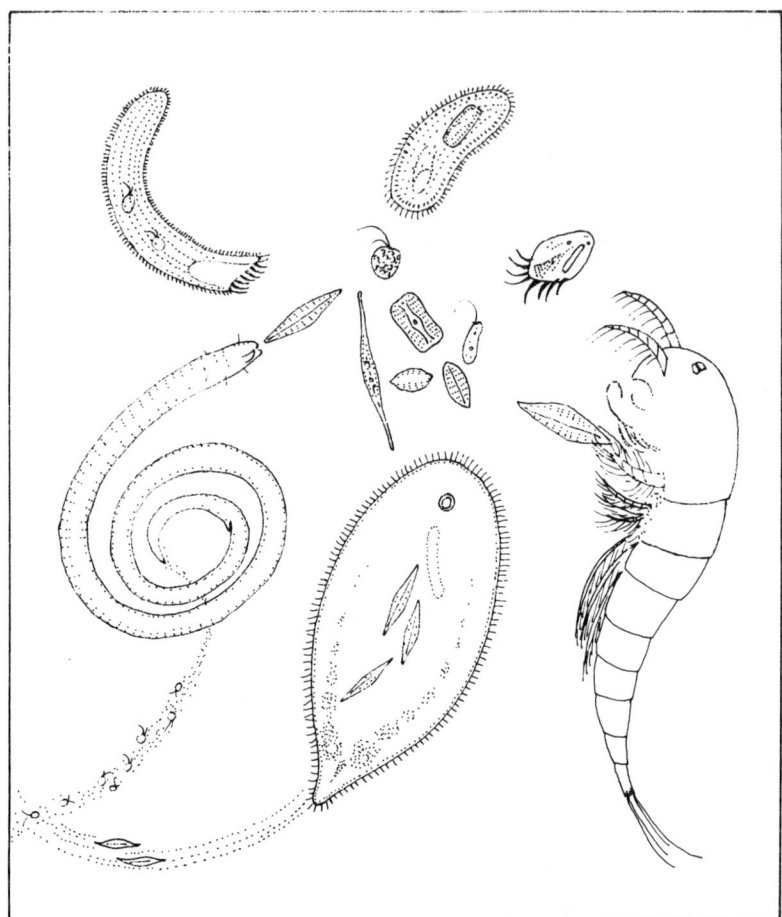

Fig. 6.2. The paradox of the 'small food web': the total amount of microflora is not depressed by the many algivores of the micro- and meiofauna. Do selective feeding and anti-grazer adaptations, gardening, and interalgal competition provide the answer?

very specific requirements in terms of quality and density of food, as well as competition between microalgae, it may indeed be unlikely that a single variable takes over. As a consequence, no conspicuous effects of meiofaunal algivores on total biomass may be observed.

Moderate exploitation may keep microalgae in an optimal growth phase, as was suggested for grazing on bacteria (Fenchel 1972; Tenore and Rice 1980). Some nematodes (Riemann and Schrage 1978; Warwick 1981) and platyhelminths (Klauser 1986) secrete mucus which enhances growth of bacteria and microalgae. Altogether, these microflora–meiofauna interactions generate an ecological complexity where bulk measurements provide no insight, correlation studies show spurious links, and research for causal factors is virtually endless (the 'tropical rainforest syndrome').

The situation changes when the effects of grazing by the macrofauna are considered. Asmus and Asmus (1985) investigated primary and secondary production on three tidal flats, and conclude that grazers (mainly hydrobiid snails) prevent the accumulation of diatom biomass. Roughly one-third of microalgal production may be consumed by the macrozoobenthos. Several field experiments with altered densities of gastropod grazers demonstrate conspicuous effects on the microbenthic algae. On a mud-flat, McClatchie et al. (1982) excluded and included the pulmonate snail *Amphibola crenata* with fences, and observed increasing and decreasing diatom numbers, respectively. Similar results were obtained with the littorinid snail *Bembicium auratum* by following changes in chlorophyll levels in mangrove mud (Branch and Branch 1980).

At high densities, the gastropod *Ilyanassa obsoleta* depressed diatom growth (Pace et al. 1979), while at moderate grazing-pressure growth was stimulated (Connor et al. 1982). The diatom-feeding snail *Hydrobia ulvae* showed evidence of intraspecific competition at natural density, and caused a marked decline in diatom numbers (Morrisey 1987, 1988). A stimulation effect on microalgal growth at low densities was not observed in this case, but it was demonstrated in laboratory experiments by Fenchel and Kofoed (1976). *Corophium arenarium* is even more efficient in digesting sediment microalgae than *Hydrobia* (Morrisey 1988). *Corophium volutator* was estimated to consume 27 per cent of the net epipelic microalgal production on an intertidal flat (Hawkins 1985), while in another study the same species turned out to be primarily a bacteria-feeder (Murdoch et al. 1986). The subsurface deposit-feeding polychaete *Leitoscoloplos fragilis* depends on benthic diatoms for food. In its presence, diatom production is increased, mediated by a high concentration of pore-water ammonia (Bianchi and Rice 1988).

These examples show that natural densities of macrofaunal grazers may often exert a strong effect on the benthic microalgae of sediment shores. Not only do grazers on the sediment surface affect microalgae but bioturbating subsurface feeders also have an influence. Branch and Pringle (1987) excluded the thalassinidean prawn *Callianassa kraussi*, and, in the absence of the sediment turnover caused by the prawns, benthic microalgae accumulated on the surface. This result is probably applicable to other bioturbating macrofauna.

Epilogue

In general, macrofaunal grazers may be less-efficient utilizers of microalgal biomass than micro- and meiofaunal grazers. This may in part explain why effects on the microphytobenthos by the larger algivores and deposit-feeders are more evident than those caused by algivores belonging to the small fauna. It is clear that macrofaunal grazers are capable of reducing the overall quantity of benthic microalgae on sediment shores. However, how this grazing affects the different groups of photosynthetic micro-organisms, i.e. Cyanobacteria, flagellates, and epipelic and epipsammic diatoms, is poorly known. *Hydrobia ulvae* may consume a large amount of Cyanobacteria (*Merismopedia*) but many cells seem to be still intact in the faecal pellets (Flothmann 1990).

Among diatoms, most of the grazing pressure is directed towards the epipelon while the small-sized epipsammon is less affected. Where competition occurs among benthic microalgae, grazing is expected to change the dominance pattern and species composition. McClatchie *et al.* (1982) showed that grazer exclusion decreased the number of diatom species, while heavy grazing increased the species number. No studies have been carried out on anti-grazing strategies of the microphytobenthos, although these are to be expected on intertidal flats with abundant grazers. Massive microalgal blooms occur at various seasons of the year and involve different species. To what extent these are triggered by a release from grazing pressure or are finished by an assault of grazers is not known.

Most of the evidence of macrofaunal grazing on sedimentary shores is from studies on gastropods. The abundant ocypodid crabs on tropical shores, and other taxa which include algivores, have received less attention. Juveniles of many deposit feeders are expected to feed selectively on the benthic microflora. In the omnivore gastropod *Ilyanassa obsoleta*, the juveniles are obligatory algivores (Brenchley 1987).

A potentially important subject is the indirect effects caused by grazing on microbenthic algae. Extracellular mucopolysaccharides produced by mobile epipelic diatoms reduce the erodibility of the mud surface and trap organic matter (Grant *et al.* 1986; Paterson 1989). Mud accretion by epipelic diatoms was recorded by Coles (1979). Accretion could be reversed by introducing grazers, the snail *Hydrobia ulvae* and the amphipod *Corophium volutator*. In a sandy area the removal of *Corophium* resulted in a dramatic explosion of diatom numbers and a subsequent accumulation of mud. Linke (1939) observed that the bumpy surface topography of a mud-flat was caused by patches of *Hydrobia ulvae*. The snails removed the diatom mat and exposed the mud

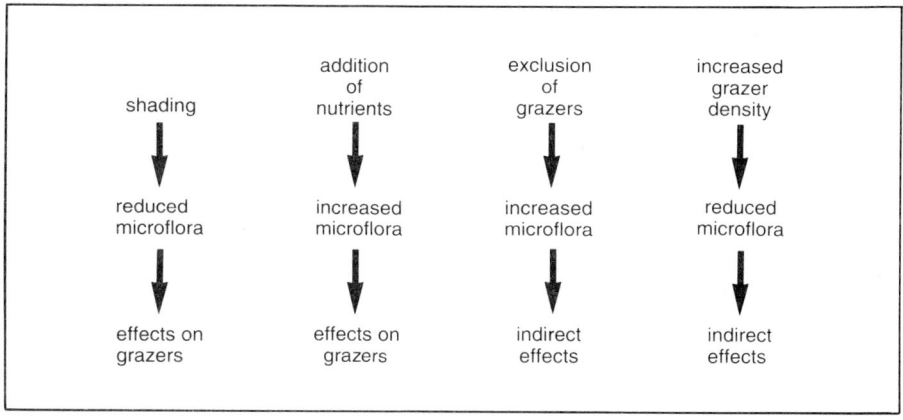

Fig. 6.3. Proposed set of field experiments to elucidate the role of grazing on benthic microalgae on sediment shores.

surface to erosion so resulting in the formation of pits. These interactions will affect many other species on the mud-flat.

Field experiments are a powerful tool to elucidate the role of grazing on sediment shores. A scheme is proposed in Fig. 6.3. Shading of the sediment surface decreases the growth of microalgae, reduces the number of grazers, and also reveals various indirect effects (Siebert 1989). Addition of inorganic nutrients to the sediment causes microalgal blooms (Flothmann 1990; Werner 1990). Provided that these algae represent suitable food, grazers may show specific responses. In experiments with altered grazer densities, changes in sediment properties and other indirect effects caused by microalgae should be taken into account. These experiments should also include bioturbators which seem to prevent the development of microalgal mats and sheets of mucilage produced by them. On the other hand, infauna may supply benthic microalgae with ammonia.

Finally, grazers may play a role in the development of green algal mats. Sporelings start their growth in the size range of benthic microalgae, and thus are expected to be subject to the same grazing pressure. Where these green algae grow into coherent mats, they strongly modify the entire habitat and its biota (Reise 1983). Strings of green algae are often anchored in the sediment by burrowing or tube-building macrofauna.

References

Admiraal, W. (1984). The ecology of estuarine sediment-inhabiting diatoms. *Progress in Phycological Research*, **3**, 269-322.

Admiraal, W, and Peletier, H. (1980). Influence of seasonal variations of temperature and light on the growth rate of cultures and natural populations of intertidal diatoms. *Marine Ecology Progress Series*, **2**, 35-43.

Admiraal, W., Bouwman, L.A., Hoekstra, L., and Romeyn, K. (1983). Qualitative and quantitative interactions between microphytobenthos and herbivorous meiofauna on a brackish intertidal mud flat. *Internationale Revue der gesamten Hydrobiologie*, **68**, 175-91.

Aleem, A.A. (1950). The diatom community inhabiting the mudflats at Whitstable, Kent. *New Phytologist*, **49**, 174-88.

Alongi, D.M. (1988). Microbial-meiofaunal interrelationships in some tropical intertidal sediments. *Journal of Marine Research*, **46**, 349-65.

Asmus, R. (1982). Field measurements on seasonal variation of the activity of primary producers on a sandy tidal flat in the northern Wadden Sea. *Netherlands Journal of Sea Research*, **16**, 389-402.

Asmus, R. (1984). Benthische und pelagische Primärproduktion und Nährsalzbilanz — eine Freilanduntersuchung im Watt der Nordsee. *Berichte des Instituts für Meereskunde, Kiel*, **131**, 1-148.

Asmus, H. and Asmus, R. (1985). The importance of grazing food chain for energy flow and production in three intertidal sand bottom communities of the northern Wadden Sea. *Helgoländer Meeresuntersuchungen*, **39**, 273-301.

Bianchi, T.S. and Rice, D.L. (1988). Feeding ecology of *Leitoscoloplos fragilis*. II. Effects of worm density on benthic diatom production. *Marine Biology*, **99**, 123-31.

Branch, G.M. and Branch, M.L. (1980). Competition in *Bembicium auratum* (Gastropoda) and its effect on microalgal standing stock in mangrove muds. *Oecologia* (Berlin), **46**, 106-14.

Branch, G.M. and Pringle, A. (1987). The impact of the sand prawn *Callianassa kraussi* Stebbing on sediment turnover and on bacteria, meiofauna, and benthic microflora. *Journal of Experimental Marine Biology and Ecology*, **107**, 219-35.

Brenchley, G.A. (1987). Herbivory in juvenile *Ilyanassa obsoleta* (Neogastropoda). *The Veliger*, **30**, 167-72.

Cadée, G.C. and Hegemann, J. (1977). Distribution of primary production of the benthic microflora and accumulation of organic matter on a tidal flat area, Balgzand, Dutch Wadden Sea. *Netherlands Journal of Sea Research*, **11**, 24-41.

Coles, S.M. (1979). Benthic microalgal populations in intertidal sediments and their role as precursors of salt marsh development. In *Ecological processes in coastal environments*, (ed. R. L. Jeffries and A. J. Davy), pp. 25-42. Blackwell, Oxford.

Colijn, F. and Jonge, V. N. de (1984). Primary production of microphytobenthos in the Ems-Dollard estuary. *Marine Ecology Progress Series*, **14**, 185-96.

Connor, M. S., Teal, J. M., and Valida, I. (1982). The effect of feeding by mud snails, *Ilanassa obsoleta* (bay), on the structure and metabolism of a laboratory benthic algal community. *Journal of Experimental Marine Biology and Ecology*, **65**, 29-45.

Dobbs, F. C. and Guckert, J. B. (1988). Microbial food resources of the macrofaunal deposit feeder *Ptychodera bahamensis* (Hemichordata: Enteropneusta). *Marine Ecology Progress Series*, **45**, 127-36.

Fenchel, T. (1968). The ecology of marine microbenthos. II. The food of marine benthic ciliates. *Ophelia*, **5**, 73-121.

Fenchel, T. (1969). The ecology of marine microbenthos. IV. Structure and function of the benthic ecosystem, its chemical and physical factors and the microfauna communities with special reference to the ciliated protozoa. *Ophelia*, **6**, 1-182.

Fenchel, T. (1972). Aspects of decomposer food chains in marine benthos. *Verhandlungen der Deutschen Zoologischen Gesellschaft*, **1972**, 14-23.

Fenchel, T. and Kofoed, L. H. (1976). Evidence for exploitative interspecific competition in mudsnails (Hydrobiidae). *Oikos*, **27**, 367-76.

Flothmann, S. (1990). Auswirkungen von Eutrophierungsexperimenten auf Makro- und Meiozoobenthos. Unpublished Diplomarbeit. Institut für Meereskunde, Kiel.

Grant, J., Bathmann, U. K., and Mills, E. L. (1986). The interaction between benthic diatom films and sediment transport. *Estuarine, Coastal and Shelf Science*, **13**, 225-238.

Hawkins, C. M. (1985). Population carbon budgets and the importance of the amphipod *Corophium volutator* in the carbon transfer on a Cumberland Basin mudflat, upper Bay of Fundy, Canada. *Netherlands Journal of Sea Research*, **19**, 165-76.

Hicks, G. R. F. and Coull, B. C. (1983). The ecology of marine meiobenthic harpacticoid copepods. *Oceanography and Marine Biology Annual Review*, **21**, 67-175.

Jensen, K. T. and Siegismund, H. R. (1980). The importance of diatoms and bacteria in the diet of *Hydrobia* species. *Ophelia*, Suppl., **17**, 193-9.

Jensen, P. (1987). Feeding ecology of free-living aquatic nematodes. *Marine Ecology Progress Series*, **35**, 187-96.

Joint, I. R., Gee, J. M., and Warwick, R. M. (1982). Determination of fine-scale vertical distribution of microbes and meiofauna in an intertidal sediment. *Marine Biology*, **72**, 157-64.

Klauser, M. D. (1986). Mucous secretions of the acoel turbellarian *Convoluta sp.* Orsted: an ecological and functional approach. *Journal of Experimental Marine Biology and Ecology*, **97**, 123-33.

Kuipers, B. R., Wilde, P. A. W. J. de, and Creutzberg, F. (1981). Energy flow in a tidal flat ecosystem. *Marine Ecology Progress Series*, **5**, 215-21.

Linke, O. (1939). Die Biota des Jadebusens. *Helgoländer Wissenschaftliche Meeresuntersuchungen*, **1**, 201-348.

Lopez, G. R. and Kofoed, L. H. (1980). Epipsammic browsing and deposit-feeding in mudsnails (Hydrobiidae). *Journal of Marine Research*, **38**, 585-99.

Lopez, G. R. and Levinton, J. S. (1978). Availability of micro-organisms attached to sediment particles as food for *Hydrobia ventriosa* Montagu (Gastropoda: Prosobranchia). *Oecologia* (Berlin), **32**, 263-75.

Lopez, G. R. and Levinton, J. S. (1987). Ecology of deposit-feeding animals in marine sediments. *The Quarterly Review of Biology*, **62**, 235-60.

López-Figuerda, F. and Niell, F. X. (1987). Feeding behaviour of *Hydrobia ulvae* (Pennant) in microcosms. *Journal of Experimental Marine Biology and Ecology*, **114**, 153-67.

Mann, K. H. (1982). *Ecology of coastal waters, a systems approach.* Blackwell, Oxford.

McClatchie, S., Juniper, S. K., and Knox, G. A. (1982). Structure of a mudflat diatom community in the Avon-Heathcote Estuary, New Zealand. *New Zealand Journal of Marine and Freshwater Research*, **16**, 299-309.

Montagna, P. A. (1984) In situ measurement of meiobenthic grazing rates on sediment bacteria and edaphic diatoms. *Marine Ecology Progress Series*, **18**, 119-30.

Morrisey, D. J. (1987). Effect of population density and presence of a potential competitor on the growth rate of the mud snail *Hydrobia ulvae* (Pennant). *Journal of Experimental Marine Biology and Ecology*, **108**, 275-95.

Morrisey, D. J. (1988). Differences in effects of grazing by deposit-feeders *Hydrobia ulvae* (Pennant) (Gastropoda: Prosobranchia) and *Corophium arenarium* Crawford (Amphipoda) on sediment microalgal populations. *Journal of Experimental Marine Biology and Ecology*, **118**, 33-53.

Murdoch, M. H., Bärlocher, F., and Laltoo, M. (1986). Population dynamics and nutrition of *Corophium volutator* (Pallas) in the Cumberland Basin (Bay of Fundy). *Journal of Experimental Marine Biology and Ecology*, **103**, 235-49.

Pace, M. C., Shimmel, S., and Darley, W. M. (1979). The effect of grazing by a gastropod, *Nassarius obsoletus*, on the benthic microbial community of a salt marsh mudflat. *Estuarine and Coastal Science*, **9**, 121-34.

Palmer, J. D. and Round, F. E. (1967). Persistent, vertical migration rhythms in benthic microflora. VI. The tidal and diurnal nature of the rhythm in the diatom *Hantzschia virgata. Biological Bulletin*, **132**, 44-55.

Paterson, D. M. (1989). Short-term changes in the erodibility of intertidal

cohesive sediments related to the migratory behaviour of epipelic diatoms. *Limnology and Oceanography*, **34**, 223-34.
Plante-Cuny, M.R. (1984). Le microphytobenthos et son rôle à l'échelon primaire dans le milieu marin. *Oceanis*, **10**, 417-27.
Reise, K. (1983). Sewage, green algal mats anchored by lugworms, and the effect on Turbellaria and small Polychaeta. *Helgoländer Meeresuntersuchungen*, **36**, 151-62.
Reise, K. (1988). Plathelminth diversity in littoral sediments around the island of Sylt in the North Sea. *Progress in Zoology*, **36**, 469-80.
Riemann, F. and Schrage, M. (1978). The mucus-trap hypothesis on feeding of aquatic nematodes and implications for biodegradation and sediment texture. *Oecologia* (Berlin), **34**, 75-88.
Round, F.E. (1971). Benthic marine diatoms. *Oceanography and Marine Biology Annual Review*, **9**, 83-139.
Schwinghamer, P. (1983). Generating ecological hypotheses from biomass spectra using causal analysis: a benthic example. *Marine Ecology Progress Series*, **13**, 151-66.
Siebert, I. (1989). Veränderungen der Lebensbedingungen im Nordseewattboden durch experimentelle Beschattung. Unpublished Diplomarbeit. Universität Göttingen.
Tenore, K.R. and Rice, D.L. (1980). A review of tropic factors affecting secondary production of deposit-feeding. In *Marine benthic dynamics*, (ed. K.R. Tenore and B.C. Coull), pp. 325-40. University of South Carolina Press, Columbia, S.C.
Tietjen, J.H. and Lee, J.J. (1977). Feeding behaviour of marine nematodes. In *Ecology of marine benthos*, (ed. B.C. Coull), pp. 21-35. University of South Carolina Press, Columbia, S.C.
Varela, M. and Penas, E. (1985). Primary production of benthic microalgae in an intertidal sand flat of the Ria de Arosa, NW Spain. *Marine Ecology Progress Series*, **25**, 111-19.
Warwick, R.M. (1981). Survival strategies of meiofauna. In *Feeding and survival strategies of estuarine organisms*, (ed. N.V. Jones and W.J. Wolff), pp. 39-52. Plenum Press, New York.
Werner, I. (1990). Auswirkungen von Eutrophierungsexperimenten auf das Microphytobenthos im Wattenmeer. Unpublished Diplomarbeit. Institut für Meereskunde, Kiel.

7. A perspective on plant–animal interactions in seagrasses: physical and biological determinants influencing plant and animal abundance

ROBERT J. ORTH

School of Marine Science, Virginia Institute of Marine Science, College of William and Mary, Gloucester Point, Virginia, USA

Abstract

Quantitative studies on benthic faunal communities associated with seagrass beds have shown these ecosystems to contain a significantly higher diversity and density of fauna when compared with adjacent, unvegetated areas. Hypotheses suggested for this abundance and diversity include habitat complexity, refuge from predation, increased food supplies, hydrodynamic effects on larval supply, and stable substrates. Manipulative laboratory and field experiments have shown that this high diversity may result from complex, and highly interrelated plant–plant, plant–animal, plant–physical environment, and animal–animal interactions. In particular, predator–prey studies have shown that foraging success is reduced by macrophyte complexity, and that the relationship between habitat complexity and predation success is non-linear. Important components of predator–prey interactions are the relative sizes as well as the behavioural repertoire of both predator and prey. Current debate centres around the role of structural complexity, initial settlement densities of recruiting individuals, the physical location of a seagrass meadow in relation to larval supply or source of adults or juveniles, and the ability of individuals to move within and between seagrass areas once settled, in determining population abundances in a given seagrass area. Since seagrasses modify hydrodynamic processes, they may play an important role in determining larval supply rates and larval settlement.

Plant–Animal Interactions in the Marine Benthos, (ed. D. M. John, S. J. Hawkins, and J. H. Price), Systematics Association Special Volume No. 46, pp. 147–164. Clarendon Press, Oxford, 1992. © The Systematics Association 1992.

Recent interest has focused on understanding interactions that influence plant dynamics, in particular the seagrass–epiphyte–nutrient–grazer interactions. Grazers have been shown to enhance the productivity of seagrasses by removing epiphytes, but the interrelationships are complex, involving both direct and indirect effects that may change seasonally.

Introduction

Seagrass beds are dynamic, biologically productive systems, which are present in many shallow-water coastal habitats of the world (Den Hartog 1970). A distinguishing characteristic of seagrass beds is the significantly greater abundance of animals and number of species as compared with adjacent unvegetated areas (Orth et al. 1984a). This contrast provides a forum for tests of ecological theory that have evolved from studies of terrestrial, freshwater, and marine rocky intertidal communities.

Seagrasses, of which there are approximately 50 species worldwide (Den Hartog 1970), may occur as continuous stands of single or mixed species assemblages covering hundreds of square kilometres, such as those found in the US Gulf Coast and western Australia's Shark Bay, or as narrow, fringing beds, only metres or tens of metres wide, present in many coastal estuaries world-wide. Seagrass beds may persist in a mosaic of various-sized vegetated patches surrounded by unvegetated sand or mud, superimposing a high degree of habitat heterogeneity on a particular area. In addition, a particular species can exhibit a continuum of leaf widths and lengths, normally related to substrate type, sediment nutrient concentration, and water depth. Different morphological traits among species (Den Hartog 1970) in mixed species stands offer additional above-ground microsites for faunal colonization. The root–rhizome component of seagrasses provides further belowground habitat complexity in vegetated patches. The configuration and extent of the root–rhizome system, which varies from millimetres to tens of centimetres below the sediment–water interface, is species specific (Den Hartog 1970).

The hydrodynamics of a vegetated site are modified to varying degrees (e.g., reduction of current flow and baffling of waves) depending on the species (Ginsburg and Lowenstam 1958; Ward et al. 1984; Fonseca and Fisher 1986). Coupled with the stabilizing effects of roots and rhizomes (Orth 1977), the sedimentary characteristics of a site vegetated with seagrasses is usually significantly different (greater percentage of fine-grained sediments, in particular silts and clays, and organic carbon) than an adjacent, unvegetated site (Orth 1977; Kenworthy et al. 1982).

The above characteristics noted for seagrasses (plant canopy, meadow formation, the root–rhizome system, nutrient interactions between the water column and sediments, hydrodynamic and sedimentary effects) all interact to underpin the seagrass faunal community, where complex plant–animal, animal–animal, and plant–epiphyte–animal–nutrient interactions occur. The goal of this review is to provide a brief overview of some of the major factors regulating faunal abundances in seagrass meadows. Relevant laboratory and field studies are cited that have been aimed at testing various hypothesis generated from earlier studies conducted to explain the increase in faunal density and diversity. More detailed reviews can be found in Orth *et al.* (1984*b*), Larkum *et al.* (1989), and Heck and Crowder (1990).

The seagrass faunal community

The fact that seagrass meadows harbour a diverse and dense faunal assemblage was first recognized in the early 1900s, when the Danish scientist Peterson described the richness of the seagrass fauna in European waters (Peterson 1918). This richness was highlighted by the dramatic changes in associated fauna, as well as sediments, that occurred subsequent to the pandemic decline of *Zostera marina* in the 1930s (Rasmussen 1973; Christiansen *et al.* 1981).

Faunal research in the 1950s and 1960s centred around European and Japanese studies describing communities or associations in these habitats (Kikuchi and Peres 1977). However, little experimental work was conducted to elucidate mechanisms accounting for these associations, the differences between these associations, and any interactions among the faunal components.

In the 1970s and 1980s, field manipulation experiments using techniques, such as caging and tethering and laboratory microcosm and mesocosm experiments, have been used to elucidate mechanisms contributing to increased abundances of animals in seagrass beds. Much of this work has revolved around such benefits to fauna as increased food supplies (Peterson *et al.* 1984), refuge from predation (Heck and Thoman 1981; Heck and Wilson 1987; Summerson and Peterson 1984), habitat complexity (Gotceitas and Colgan 1989; Heck and Crowder 1990), sediment stability (Orth 1977), and hydrodynamic effects on larval recruitment dynamics (Ekman 1983, 1987). The results of these studies, highlighted in this review, have provided insightful data on the roles of various factors in determining population and community dynamics in seagrass beds, as well as setting the stage for second- and third-generation hypothesis testing.

Plant–animal interactions

1. The roles of refuge and habitat complexity

Much of the early mechanistic work centred on the refuge value of seagrasses by examining the relationship of some concentration of a vegetation parameter, e.g., shoot density, shoot biomass, or surface area, to survivorship of prey (Nelson 1979; Heck and Thoman 1981; Stoner 1980, 1982; Savino and Stein 1982; Gotceitas and Colgan 1987). These studies have shown that:

(1) foraging success of predators is reduced by macrophyte complexity, whether measured as density of shoots or biomass;
(2) there is a non-linear relationship between habitat complexity and predation success, i.e. a threshold level of complexity was necessary before foraging success was significantly reduced.

Nelson (1979) first described this step function (see Fig. 7.1) but the data from Nelson and other studies supporting this hypothesis (Heck and Thoman 1981; Stoner 1980, 1982; Savino and Stein 1982; Gotceitas and Colgan 1987) were limited to only a few levels of macrophyte density. Gotceitas and Colgan (1989) confirmed the 'threshold' effect by incorporating eight levels of complexity into the experimental design. Moreover, they predicted that this 'threshold' level could shift in either direction depending on vegetational characteristics, such as plant morphology, or characteristics of the predator, such as body size.

Although this 'threshold' theory has been applied to epifaunal organisms and above-ground vegetation, this effect may also apply to

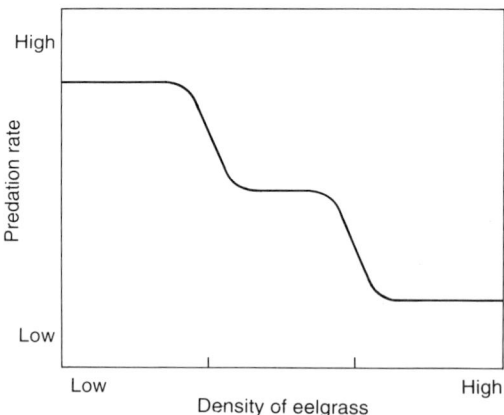

Fig. 7.1. Conceptual model of the relationship between predation intensity and density of the eelgrass *Zostera marina* (from Nelson 1979).

infaunal populations inhabiting seagrass meadows and receiving protection from the root–rhizome mat (Orth 1977; Blundon and Kennedy 1982; Peterson 1982). This 'threshold' value should similarly be related to the seagrass species because of the differences in complexity and depth penetration of the root–rhizome layer. Prey behaviour would also influence this relationship, e.g., shallow-dwelling animals (if they lived at or above the root–rhizome layer) would be predicted to be more vulnerable to predation than deeper-dwelling species (Virnstein 1979; Peterson 1982; Peterson and Quammen 1982).

The relative sizes of predator and prey, as well as their behavioural repertoires, can ultimately determine the effect of vegetative structure on predator–prey interactions (Main 1985, 1987; Ryer 1988). Ryer (1988) showed that foraging success by small pipe-fish on amphipods was unaffected by vegetative structure, whereas larger fish decreased in foraging efficiency with increased habitat complexity. Amphipod vulnerability increased with increased body size of the amphipod. Main (1987) found that the shrimp *Tozeuma carolinense*, which clings to grass blades, moved around the grass blade in the presence of a fish predator, *Lagodon rhomboides*, to avoid predation. Main found that the combination of active predator-avoidance behaviours and the presence of a structurally complex, opaque substrate (seagrass) resulted in a visual barrier between predator and prey. Main and Ryer stressed the importance of measuring prey characteristics (e.g., size, coloration, motion, morphology) as well as predator foraging characteristics. (e.g., ambush vs stalking) in developing a clearer understanding of the mechanistic interactions between predator and prey in seagrass habitats.

2. Habitat selection processes

Prey abundances will not only depend on predation rate, but also on the physical complexity of the habitat. In the absence of predation, prey densities would be expected to increase with increased physical complexity (shoot density, degree of branching, etc.) (Heck and Orth 1981). However, prey abundances will be a balance between refuge from predation and the availability of increased space. Prey would be expected to optimize energetic considerations in selecting certain types of structured habitats, both in the presence and absence of predation.

The relative importance of active and passive habitat selection by seagrass fauna is currently under debate. Several critical questions are being addressed:

1. Is there active selection for some physical attribute of the seagrass, e.g., degree of branching, leaf width?

2. Are abundances related to passive depositional processes and subsequent post-settlement mortality factors?
3. Is selection active based on some chemical cue derived from the seagrass and over what scales would this occur?

Heck and Orth (1980) first proposed a model that related faunal abundance in seagrass beds to plant surface-area, with abundance increasing with increasing plant surface per unit bottom area to a threshold beyond which it declines because of physical and environmental constraints imposed by very dense vegetation. This model did not explore species-specific selection for some plant attribute. Bell and Westoby (1986a, 1986b) proposed an 'indiscriminate settlement' or 'settle and stay' hypothesis. They suggested that larvae of fish and decapods are distributed patchily when competent to settle, and do not discriminate among beds when they settle regardless of physical complexity. Once settled, individuals do not leave beds but redistribute within the bed to select microsites favouring survival. Subsequent work (Bell *et al.* 1987, 1988; Sogard 1989) amended this 'settle and stay' hypothesis. They suggested that the actual physical location of a bed in relation to larval supply or source of adults or juveniles is significant and overrides the physical complexity of the seagrass component on broad scales.

Sogard's (1989) and Virnstein and Curran's (1986) data do not support aspects of the 'settle and stay' hypothesis. Their work showed rapid colonization of artificial substrates by both juvenile and adult fishes and decapods and smaller, less-mobile invertebrates at varying distances from existing vegetation. They suggested that benefits of searching and finding more 'optimal' habitats may outweigh the risks of migrating across bare sand.

Virnstein and Curran (1986) proposed a 'nearest refuge' hypothesis, suggesting that animals emigrate from vegetated habitats for a variety of reasons (escape predators, avoid competitors, search for food, mate selection), and that, once dispersed over bare sand, seek out the nearest refuge; this emigration may be nocturnal. With dawn, animals must settle in vegetated habitats or they face higher predation risk in unvegetated areas.

These risks may vary both within and between regions depending on the suite of predators in a particular area. Predation intensity has been shown to increase with decreasing latitude for terrestrial and other marine systems (Bertness *et al.* 1981; Menge and Lubchenco 1981). Heck and Wilson (1987) conducted tethering experiments in seagrass beds along a latitudinal gradient. They found an increase in predation rates with decreasing latitude in the vegetated habitats but not in the unvegetated habitats. Their data suggest that, regardless of latitude, unvegetated areas are areas of extremely high vulnerability, while in vegetated areas protection from predation is dependent on geographical location — and thus the suite of predators in that location — as well as time.

3. Hydrodynamics and larval supply

The burgeoning interest in the larval supply question (Connell 1985; Roughgarden *et al.* 1988) has sparked much interest in the role of hydrodynamics in both mesoscale and microscale dispersal of larvae. Butman suggested that, since swimming speeds of many invertebrate larvae are much less than current speeds and sinking rates, and larval responses to small-scale flows are in the range of similar-sized passive particles, passive deposition may determine the large-scale distribution of larvae (Butman 1986, 1987, 1989; Butman *et al.* 1988). Only at scales of centimetres, or possibly millimetres, does swimming play a role in the selection process. Since seagrasses modify hydrodynamic processes (Fonseca *et al.* 1982; Ekman 1983, 1987; Fonseca and Fisher 1986), they may play an important role in determining larval supply rates and larval settlement. Within a bed of seagrass current speed is substantially reduced (Peterson *et al.* 1984; Fonseca and Fisher 1986) and so enhancing passive particle, thus larval deposition. If current speed within the bed is below the swimming speed of a larva, active microsite selection is possible, either for a particular sediment (e.g., grain size or sediment organic content) or vegetation (e.g., leaf width or degree of branching) characteristic (Fig. 7.2). Concomitant with this flow-reduction is the potential alteration of the larval supply. As larvae enter or encounter a bed, they may passively settle out or set on the first blades they encounter, reducing the supply of larvae toward the centre or inner edge of the bed. This produces a so-called 'settlement shadow' (Fig. 7.2) (Roughgarden *et al.* 1988).

A re-examination of data collected by Marsh (1970) along a depth gradient in a densely vegetated seagrass (*Zostera marina*, eelgrass) bed suggests hydrodynamic and, possible, shadow effects (Fig. 7.3). Monthly quantitative data were collected for the eelgrass epifauna. Data for species showing distinct peaks in abundance were re-examined. For those species that produce planktonic larvae and recruit to new habitats from the offshore plankton (the polychaetes *Polydora ligni* and *Sabella microphthalma*, the cirriped *Balanus improvisus*, and the ascidian *Molgula manhattensis*), densities were much higher in the offshore, deeper station. For species that brood and release young with no larval stage (the amphipods *Ampithoe longimana* and *Cymadusa compta*, the isopod *Erichsonella attenuata*, and the gastropod *Crepidula convexa*), the above pattern was not evident, with abundances higher at inshore, shallower

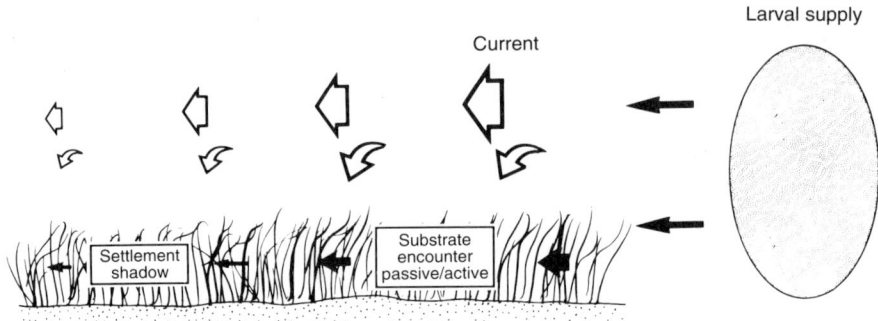

Fig. 7.2. Conceptualization of the interactions in seagrasses involving hydrodynamic processes and larval supply. A parcel of water entering a seagrass bed with a supply of larvae is affected by its interaction with the seagrass canopy. As water velocity decreases, larvae can passively descend into the seagrass bed as current velocities are reduced, and allowing for either passive or active selection of microsites within the canopy or sediment. In addition as larvae either pass through the bed or the numbers of larvae in the patch are reduced, the number of settlers in sections of the bed furthest from the source is reduced, producing a 'settlement shadow'.

stations. Because these latter species recruit from within the bed, other factors such as predation may have produced the observed patterns.

4. Chemical cues

Chemical cues have been shown to be important in inducing settlement of infaunal invertebrates (Crisp 1974, see Chapter 17 this volume). However, the scales at which these cues operate have been questioned (Butman 1987). Hydrodynamic processes discussed above may control large-scale settlement-abundance patterns, but microscale patterns may

---

Fig. 7.3. Abundance data (numbers g^{-1} of grass (solid line) and 0.25 m^{-2} (dashed line) of sea-bottom) for epifaunal species with two different recruitment strategies: species that recruit from an offshore, planktonic source (*Polydora ligni*, *Sabella microphthalma*, *Balanus improvisus*, and *Molgula manhattensis*) and those which brood their young (*Ampithoe longimana*, *Cymadusa compta*, *Erichsonella attenuata*, and *Crepidula convexa*). Data are from three stations (A, B, and C) linearly arranged from inshore to offshore along a depth gradient. Horizontal axis is distance from shore while depth is in metres at mean low water. The top figure shows the mean length of eelgrass at each station location for the period when the eelgrass reaches maximum standing stock; see text for details (data from Marsh (1970).

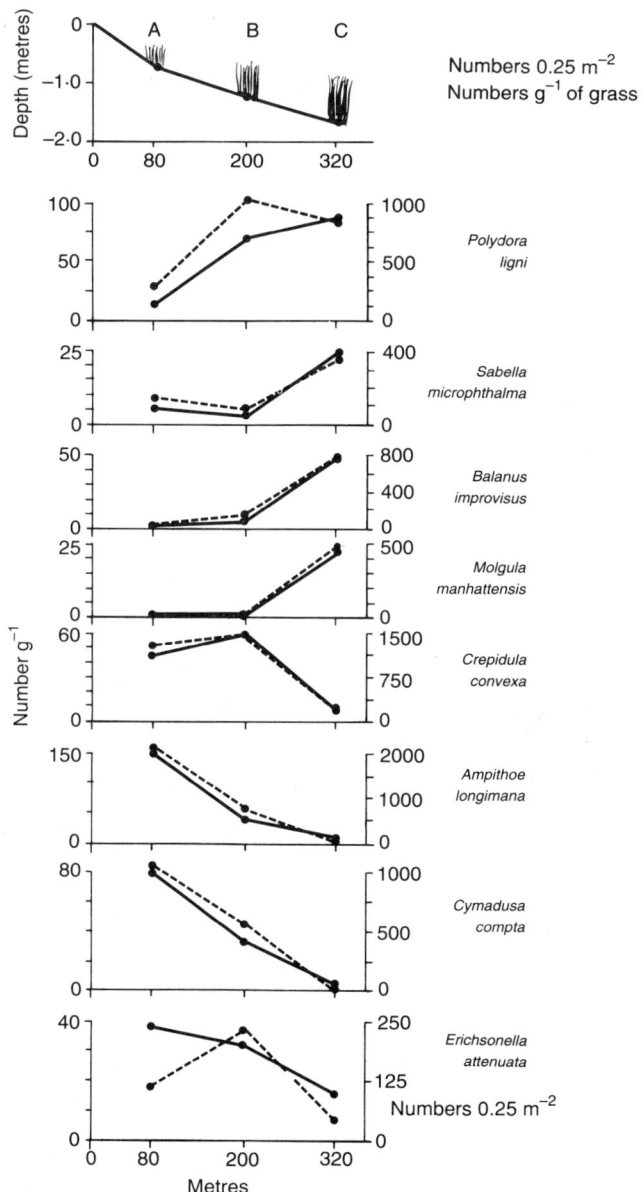

be determined by chemical attractants that can induce settlement, or repellants that retard this process.

Seagrasses contain chemicals which may serve as inducers for larvae to settle. Some of our preliminary research with the post-larvae of the portunid crab, *Callinectes sapidus*, in experimental mesocosms, revealed that settlement was greater in treatments with live seagrass (*Zostera marina*) than in those with artificial seagrass, sand, mud, and live oysters and oyster shell, suggesting the possibility of a chemical attractant (unpublished data). Although there is a dearth of information in seagrasses, the area of chemical ecology as it applies to faunal relationships in seagrass meadows promises to be a fruitful area of research.

Plant–animal–epiphyte interactions

Much of the research conducted as discussed above relates to the outome of animal interactions as mediated by the physical structure of the plant. Recently, much interest has been generated in ecosystem interactions: predator–herbivore–plant interactions or nutrient–plant–predator–prey interactions (Heck and Crowder 1990).

Epiphytes or periphyton (often collectively defined to include micro- and macroalgae, as well as the organic and inorganic debris and associated micro-organisms (protozoa, bacteria, microfauna) found in the mucus-like layer coating seagrass blades) are an important component of the seagrass community. They contribute substantially to the overall productivity of the system, estimated to be from 20 to 60 per cent of the total production (Penhale and Smith 1977; Mazzella and Alberte 1986).

Epiphytes can have both beneficial and deleterious effects on the productivity of the seagrass. Beneficial effects include reduction of desiccation of the host in very shallow or intertidal areas (Penhale and Smith 1977; Richardson 1980), as well as protection from ultraviolet radiation (Trocine *et al.* 1981). Deleterious effects include shading and competition for light (Borum and Wium-Anderson 1980; Bulthuis and Woelkerling 1983; Sand-Jensen and Borum 1983), and interference with uptake of carbon and phosphorus (Sand-Jensen 1977; Johnstone 1979).

Epiphyte biomass increases with increasing nutrient concentrations and has been a proposed mechanism for the decline of seagrasses in highly eutrophic areas (Orth and Moore 1983; Silberstein *et al.* 1986). Twilley *et al.* (1985) experimentally increased water-column nutrient concentrations in ponds containing species of brackish-water submersed macrophytes and demonstrated that both plankton and epiphytes contributed to light attenuation, especially in the high dose treatment. Their data suggested that macrophyte growth under the combined stresses of water-column turbidity and epiphytic growth would be impossible to maintain.

Epiphytes are an abundant food for a variety of herbivores such as amphipods, isopods, and snails (Orth and van Montfrans 1984; van Montfrans *et al.* 1984). This has spawned much interest in questions relating to the extent to which grazers control epiphytic production and biomass, the influence that their removal has on macrophyte production, the indirect effects of nutrient enrichment and grazing on macrophyte production, the seasonal effects on these interactions, and the importance of such interactions to the long-term macrophyte stability.

Grazers significantly reduce epiphytic biomass on seagrass leaves when compared with ungrazed leaves (for reviews, see Orth and van Montfrans 1984; van Montfrans *et al.* 1984; also Howard and Short 1986). Epiphytic biomass reduction results in enhanced macrophyte production in both temperate and tropical systems (Hootsmans and Vermaat 1985; Howard and Short 1986). In particular, Howard and Short (1986) found significant increases in the growth and weight of the above-ground vegetation of the seagrass *Halodule wrightii* as well as the number of shoots and below-ground biomass in treatments with grazers compared to treatments without grazers after four months.

Many studies have only measured the direct effects of grazer abundance and worked with one level of treatment: presence or absence of grazers and their effect on the periphyton. Several studies examined indirect effects on consequences for macrophyte production (Hootsmans and Vermaat 1985; Howard and Short 1986). Under natural conditions, the macrophyte responds to a range of light, nutrient, temperature, and grazer abundance over the course of a growing season. There have been very few studies examining the interactive effects of these parameters (Neckles 1990).

Neckles (1990) tested the simultaneous effects of nutrient enrichment and epiphytic grazers on epiphytic biomass and macrophyte growth in microcosms of the seagrass *Zostera marina* (eelgrass) over four seasons. Neckles also extended a carbon-based simulation model (Wetzel and Neckles 1986) to test the potential long-term effects of nutrient enrichment and epiphytic grazing on macrophyte community stability. Her microcosm experiments showed seasonal differences in response by eelgrass and attached epiphytes. Although grazing was generally more important than nutrient supply in controlling epiphyte biomass, the indirect effects on macrophyte growth depended on the relative magnitude of each factor and the physiological demands of the macrophyte. For example, under low grazer densities typical of early summer in Chesapeake Bay habitats, both grazer removal and nutrient enrichment reduced the macrophyte production, whereas under high grazer densities of late summer, enrichment reduced the production only when grazers were absent. There were no indirect effects of either factor on macrophyte production during the autumn or spring, regardless of intermediate effects on epiphytes on epiphytic biomass.

Model simulations indicated that either nutrient enrichment or a loss of grazers alone could reduce macrophyte production but would not result in long-term instability. Nutrient enrichment in the absence of grazers could result in significant changes in macrophyte community dynamics. These interactions were further influenced by levels of light reaching the macrophyte canopy.

Results from the above experiments suggest that grazing on epiphytes may be important in the maintenance of growth, productivity, and depth-distribution of seagrasses, particularly in light- and nutrient-stressed environments. However, Neckles' (1990) work suggests that caution should be urged in making generalizations from individual interactions from only one season, as indirect effects may change seasonally.

Summary

Seagrasses provide the foundation for the complex ecological interactions that result in the large number of species and abundances reported for these systems. A variety of mechanisms appear to be operating that result in these enhanced abundances, including habitat complexity, increased food supplies, prey *refugia*, stable substrates, and the influence on larval supply rates by altering hydrodynamic processes. However, these relationships are clearly interrelated and non-linear in their effects, e.g., predator–prey interactions. They are dependent on both the characteristics of the prey and predator (e.g., body size) as well the plant (e.g., plant morphology).

Recruitment dynamics appears to be an increasingly critical area in seagrass ecology. Future research should be directed toward understanding spatial and temporal patterns in larval supply rates, pre- and post-settlement mortality factors in setting initial faunal densities, factors affecting faunal movement within and between beds, hydrodynamic effects on larval supply rates, and the influence of seagrasses via hydrodynamic processes on larval supply and depletion rates.

Elucidation of ecosystem-level interactions are crucial because of the above-noted complexities and their proposed role in influencing the distribution and abundance of the plant system. In particular, the relative roles of epiphytes, herbivorous grazers, and nutrient supplies need to be examined most critically in light of the increasing anthropogenic stresses placed on seagrasses in estuarine and coastal areas.

Acknowledgements

I greatly appreciate the timely and insightful reviews by Mark Luckenbach, Jacques van Montfrans, Hilary Neckles, and Eric Garnick. The National Sea Grant College Program of the National Oceanic and Atmospheric Administration, US Department of Commerce, Grant No. NA86AA-D-SG042 to the Virginia Graduate Marine Science Consortium and the Virginia Sea Grant College Program, and the Allied–Signal Foundation, provided financial support for my seagrass research that helped in the synthesis of much of the relevant information and the generation of new ideas and concepts. I am indebted to them for their backing. Contribution No. 1641 from the Virginia Institute of Marine Science, School of Marine Science.

References

Bell, J. D. and Westoby, M. (1986a). Variation in seagrass height and density over a wide spatial scale: effects on fish and decapods. *Journal of Experimental Marine Biology and Ecology*, **104**, 275–95.

Bell, J. D. and Westoby, M. (1986b). Importance of local changes in leaf height and density to fish and decapods associated with seagrasses. *Journal of Experimental Marine Biology and Ecology*, **104**, 249–74.

Bell, J. E., Westoby, M., and Steffe, A. S. (1987). Fish larvae settling in seagrass: do they discriminate between beds of different leaf density? *Journal of Experimental Marine Biology and Ecology*, **111**, 133–44.

Bell, J. E., Steffe, A. S., and Westoby, M. (1988). Location of seagrass beds in estuaries; effects on associated fish and decapods. *Journal of Experimental Marine Biology and Ecology*, **122**, 127–46.

Bertness, M. D., Garrity, S. D., and Levings, S. C. (1981). Predation pressure and gastropod foraging: a tropical–temperate comparison. *Evolution*, **35**, 995–1077.

Blundon, J. A. and Kennedy, V. S. (1982). Refuges for infaunal bivalves from blue crab, *Callinectes sapidus* (Rathbun), predation in Chesapeake Bay. *Journal of Experimental Marine Biology and Ecology*, **65**, 67–81.

Borum, J. and Wium-Andersen, S. (1980). Biomass and production of epiphytes on eelgrass (*Zostera marina* L.) in the Oresund, Denmark. *Ophelia*, suppl., **1**, 57–64.

Bulthuis, D. and Woelkerling, W. J. (1983). Biomass accumulation and shading effects of epiphytes on leaves of the seagrass, *Heterozostera tasmanica*, in Victoria, Australia. *Aquatic Botany*, **16**, 137–48.

Butman, C. A. (1986). Larval settlement of soft-sediment invertebrates: some predictions based on an analysis of near-bottom velocity profiles. In *Marine interfaces ecohydrodynamics*, (ed. J.C. Nihoul), Elsevier Oceanography Series, No. 42, pp. 487–513. Elsevier, Amsterdam.

Butman, C. A. (1987). Larval settlement of soft-sediment invertebrates: the spatial scales of pattern explained by active habitat selection and the merging role of hydrodynamical processes. *Oceanography and Marine Biological Review*, **25**, 113–65.

Butman, C. A. (1989). Sediment-trap experiments on the importance of hydrodynamical processes in distributing settling invertebrate larvae in near-bottom waters. *Journal of Experimental Marine Biology and Ecology*, **134**, 37–77.

Butman, C. A., Grassle, J. P., and Webb, C. M. (1988). Substrate choices made by marine larvae in still water and in a flume flow. *Nature*, **333**, 771–3.

Christiansen, C., Christoffersen, H., Dalsgaard, J., and Nornberg, P. (1981). Coastal and near-shore changes correlated with die-back in eelgrass (*Zostera marina* L.). *Aquatic Botany* **28**, 163–173.

Connell, J. H. (1985). The consequences of variation in initial settlement vs. post-settlement mortality in rocky intertidal communities. *Journal of Experimental Marine Biology and Ecology*, **93**, 11–45.

Crisp, D. J. (1974). Factors influencing the settlement of marine invertebrate larvae. In *Chemoreception in marine organisms*, (ed. P. T. Grant and A. M. Mackie), pp. 177–265. Academic Press, New York.

Hartog, C. Den (1970). *The seagrasses of the world. Verhandelingen der Koninklijke Nederlandsche Akademie van Wetenschappen. Afdeeling Natuurkunde.* Reeks II **59**, pp. 1–275.

Eckman, J. E. (1983). Hydrodynamic processes affecting benthic recruitment. *Limnology and Oceanography*, **28**, 241–57.

Eckman, J. E. (1987). The role of hydrodynamics in recruitment, growth, and survival of *Argopecten irradians* (L) and *Anomia simplex* (D'Orbigny, within eelgrass meadows. *Journal of Experimental Marine Biology and Ecology*, **106**, 165–91.

Fonseca, M. S. and Fisher, J. S. (1986). A comparison of canopy friction and sediment movement between four species of seagrass with reference to their ecology and restoration. *Marine Ecology Progress Series*, **29**, 15–22.

Fonseca, M. S., Zieman, J. C., Thayer, G. W., and Fisher, J. S. (1982). Influence of the seagrass, *Zostera marina*, on current flow. *Estuarine, Coastal and Shelf Science*, **15**, 351–64.

Ginsburg, R. N. and Lowenstam, H. A. (1958). The influence of marine bottom communities on the depositional environment of sediments. *Journal of Geology*, **66**, 310–18.

Gotceitas, V. and Colgan, P. (1987). Selection between densities of artificial vegetation by young bluegills avoiding predation. *Transactions of the American Fisheries Society*, **116**, 40-49.

Gotceitas, V. and Colgan, P. (1989). Predator foraging success and habitat complexity: quantitative test of the threshold hypothesis. *Oecologia* (Berlin), **80**, 158-66.

Heck, K. L. Jr. and Crowder, L. B. (1990). Habitat structure and predator-prey interactions. In *Habitat complexity: the physical arrangement of objects in space*, (ed. S. Bell, E. McCoy, and H. Mushinsky), pp. 281-299. Chapman and Hall, New York.

Heck, K. L. Jr. and Orth, R. J. (1980). Seagrass habitats: the roles of habitat complexity, competition and predation in structuring associated fish and motile macroinvertebrate assemblages. In *Estuarine Perspectives*, (ed. V. S. Kennedy), pp. 449-64. Academic Press, New York.

Heck, K. L. Jr. and Thoman, T. A. (1981). Experiments on predator-prey interactions in vegetated aquatic habitats. *Journal of Experimental Marine Biology and Ecology*, **53**, 125-34.

Heck, K. L. Jr. and Wilson, K. A. (1987). Predation rates on decapod crustaceans in latitudinally separated seagrass communities: a study of spatial and variation using tethering techniques. *Journal of Experimental Marine Biology and Ecology*, **107**, 87-91.

Hootsmans, M. J. M. and Vermaat, J. E. (1985). The effect of periphyton-grazing by three epifaunal species on the growth of *Zostera marina* L. under experimental conditions. *Aquatic Botany*, **22**, 83-8.

Howard, R. K. and Short, F. T. (1986). Seagrass growth and survivorship under the influence of epiphyte grazers. *Aquatic Botany*, **24**, 287-302.

Johnstone, I. M. (1979). Papua New Guinea seagrasses and aspects of the biology and growth of *Enhalus acoroides* (L.f.) Royle. *Aquatic Botany*, **7**, 197-208.

Kenworthy, W. J., Zieman, J. C., and Thayer, G. W. (1982). Evidence for the influence of seagrasses on the benthic nitrogen cycle in a coastal plain estuary near Beaufort, North Carolina (U.S.A). *Oecologia* (Berlin), **54**, 152-8.

Kikuchi, T. and Peres, J. M. (1977). Consumer ecology of seagrass beds. In *Seagrass ecosystems: a scientific perspective*, (ed. C. P. McRoy and C. Helfferich), pp. 147-94. Marcel Dekker, New York.

Larkum, A. W. D., McComb, A. J., and Shepherd, S. A. (ed.) (1989). *Biology of seagrasses*. Elsevier, Amsterdam.

Main, K. L. (1985). The influence of prey identity and size on selection of prey by two marine fishes. *Journal of Experimental Marine Biology and Ecology*, **98**, 145-52.

Main, K. L. (1987). Predator avoidance in seagrass meadows: prey behavior, microhabitat selection and cryptic coloration. *Ecology*, **68**, 170-80.

Marsh, G. A. (1970). A seasonal study of *Zostera* epibiota in the York River,

Virginia. Unpublished Ph.D. Dissertation. College of William and Mary, Williamsburg, Virginia.

Mazzela, L. and Alberte, R. S. (1986). Light adaptation and the role of autotrophic epiphytes in primary production of the temperate seagrass, *Zostera marina* L. *Journal of Experimental Biology and Ecology* **100**, 165–80.

Menge, B. A. and Lubchenco, J. (1981). Community organization in temperate and tropical rocky intertidal habitats: prey refuges in relation to consumer pressure gradients. *Ecological Monographs*, **51**, 429–50.

Neckles, H. A. (1990). Relative effects of nutrient enrichment and grazing on epiphyton–macrophyte (*Zostera marina* L.) dynamics. Unpublished Ph.D. Dissertation. College of William and Mary, Williamsburg, Virginia.

Nelson, W. G. (1979). Experimental studies of selective predation on amphipods: consequences for amphipod distribution and abundance. *Journal of Experimental Marine Biology and Ecology*, **38**, 224–45.

Orth, R. J. (1977). The importance of sediment stability in seagrass communities. In *Ecology of marine benthos*, (ed. B. C. Coull), pp. 281–300. University of South Carolina Press, Columbia, S.C.

Orth, R. J. and Van Montfrans, J. (1984). Epiphyte-seagrass relationships with an emphasis on the role of micrograzing: a review. *Aquatic Botany*, **18**, 43–69.

Orth, R. J. and Moore, K. A. (1983). Chesapeake Bay: an unprecedented decline in submerged aquatic vegetation. *Science*, **222**, 51–3.

Orth, R. J., Heck, K. L. Jr., and van Montfrans, J. (1984a). Faunal communities in seagrass beds: a review of the influence of plant structure and prey characteristics on predator–prey relationships. *Estuaries*, **7**, 339–50.

Orth, R. J., Heck, K. L. Jr., and Weinstein, M. P. (ed.)(1984b). Faunal relationships in seagrass and marsh ecosystems. *Estuaries*, **7**, 273–470.

Penhale, P. A. and Smith, Jr. W. O. (1977). Excretion of dissolved organic carbon by eelgrass (*Zostera marina*) and its epiphytes. *Limnology and Oceanography*, **22**, 400–7.

Peterson, C. G. L. (1918). The sea bottom and its production of fish foods; a survey of the work done in connection with valuation of Danish waters from 1883–1917. *Report of the Danish Biological Station*, **25**, 1–62.

Peterson, C. H. (1982). Clam predation by whelks (*Busycon* spp.): experimental tests of the importance of prey size, prey density, and seagrass cover. *Marine Biology*, **66**, 159–70.

Peterson, C. H. and Quammen, M. L. (1982). Siphon nipping: it importance to small fishes and its impact on growth of the bivalve *Prototheca stanubea* (Conrad). *Journal of Experimental Marine Biology and Ecology*, **63**, 249–68.

Peterson, C. H., Summerson, H. C., and Duncan, P. B. (1984). The influence of seagrass cover on population structure and individual growth rate of a suspension-feeding bivalve, *Mercenaria mercenaria*. *Journal of Marine Research*, **42**, 123–38.

Rasmussen, E. (1973). Systematics and ecology of the Isefjord marine fauna (Denmark). *Ophelia*, **11**, 1–495.

Richardson, F. D. (1980). Ecology of *Ruppia maritima* L. in New Hampshire (U.S.A.) tidal marshes. *Rhodora*, **82**, 403–39.

Roughgarden, J., Gaines, S., and Possingham, S. (1988). Recruitment dynamics in complex life cycles. *Science*, **241**, 1460–6.

Ryer, C. H. (1988). Pipefish foraging: effects of fish size, prey size and altered habitat complexity. *Marine Ecology Progress Series*, **48**, 37–45.

Sand-Jensen, K. (1977). Effect of epiphytes on eelgrass photosynthesis. *Aquatic Botany*, **3**, 55–63.

Sand-Jensen, K. and Borum, J. (1983). Regulation of growth of eelgrass (*Zostera marina* L.) in Danish waters. *Marine Technology Society Journal*, **17**, 15–21.

Savino, J. and Stein, R. A. (1982). Predator–prey interaction between largemouth bass and bluegills as influenced by simulated, submerged vegetation. *Transactions of American Fisheries Society*, **111**, 255–66.

Silberstein, K., Chiffings, A. W., and McComb, A. J.(1986). The loss of seagrass in Cockburn Sound, Western Australia. III. The effect of epiphytes on productivity of *Posidonia australis* Hook. f. *Aquatic Botany*, **24**, 355–71.

Sogard, S. M. (1989). Colonization of artificial seagrass by fishes and decapod crustaceans: importance of proximity to natural eelgrass. *Journal of Experimental Marine Biology and Ecology*, **133**, 15–37.

Stoner, A. W. (1980). Perception and choice of substratum by epifaunal amphipods associated with seagrasses. *Marine Ecology Progress Series*, **3**, 105–11.

Stoner, A. W. (1982). The influence of benthic macrophytes on the foraging behavior of pinfish, *Lagodon rhomboides* (Linnaeus). *Journal of Experimental Marine Biology and Ecology*, **58**, 271–84.

Summerson, H. C. and Peterson, C. H. (1984). Role of predation in organizing benthic communities of a temperate-zone seagrass bed. *Marine Ecology Progress Series*, **15**, 63–77.

Trocine, R. P., Rice, J. D., and Wells, G. N. (1981). Inhibition of seagrass photosynthesis by ultraviolet B radiation. *Plant Physiology*, **68**, 74–81.

Twilley, R. R., Kemp, W. M., Staver, K. W., Stevenson, J. C., and Boynton, W. R. (1985). Nutrient enrichment of estuarine submersed vascular plant communities. I. Algal growth and effects on production of plants and associated communities. *Marine Ecology Progress Series*, **23**, 179–91.

Van Montfrans, J., Wetzel, R. L., and Orth, R. J. (1984). Epiphyte-grazer relationships in seagrass meadows: consequences for seagrass growth and production. *Estuaries*, **7**, 289–309.

Virnstein, R.W. (1979). Predation on estuarine infauna: response patterns of component species. *Estuaries*, **2**, 69–86.

Virnstein, R.W. and Curran, M.C. (1986). Colonization of artificial seagrass versus time and distance from source. *Marine Ecology Progress Series*, **29**, 279–88.

Ward, L.G., Boynton, W.R., and Kemp, W.M. (1984). The influence of waves and seagrass communities and suspended particulates in an estuarine embayment. *Marine Geology*, **59**, 85–103.

Wetzel, R.L. and Neckles, H.A. (1986). A model of *Zostera marina* L. photosynthesis and growth: simulated effects of selected physical-chemical variables and biological interactions. *Aquatic Botany*, **26**, 307–23.

8. Plant–animal trophic relationships in the *Posidonia oceanica* ecosystem of the Mediterranean Sea: a review

L. MAZZELLA, M.C. BUIA, M.C. GAMBI,
M. LORENTI, G.F. RUSSO, M.B. SCIPIONE,
and V. ZUPO
Laboratorio di Ecologia del Benthos, Stazione Zoologica 'Anton Dohrn' di Napoli, Ischia, Napoli, Italy

Abstract

In the Mediterranean Sea, beds of the endemic phanerogam *Posidonia oceanica*, characterized by both high subsurface and high photosynthetic biomass, form one of the most productive systems of the basin. The biomass values can exceed those of other phanerogams and the plant shows distinct partitioning, often mainly directed into the lignified rhizomes, which can account for up to 90 per cent of total biomass. Leaf-borne algal flora can at times contribute up to 30 per cent of overall production. Part of the energy produced by the plant is exported to adjacent systems through leaf detritus, but the major part is stored by the plant and recycled within the system. Because of some intrinsic features of *P. oceanica* living tissue few direct herbivores, apparently relatively unselective or generalist feeders (such as echinoids, isopods or the fish *Sarpa salpa*), have been found. Yet 'herbivores' and 'herbivores–deposit feeders' constitute the dominant feeding categories among the animal communities of the leaf stratum. These very rich and diversified communities are mainly represented by Mollusca (e.g., the gastropods *Gibbula umbilicaris* and *G. ardens*), Amphipoda, Isopoda, Decapoda, and Polychaeta, which feed preferentially on microalgae and on macroalgae of the upright layer of the epiphytic community. The grazing pathway, therefore, also seems to play a fundamental role in the consumption of

Plant–Animal Interactions in the Marine Benthos, (ed. D.M. John, S.J. Hawkins, and J.H. Price), Systematics Association Special Volume No. 46, pp. 165–187. Clarendon Press, Oxford, 1992. © The Systematics Association, 1992.

the system's production, over and above the detrital pathway. Leaf detritus, in fact, despite its importance as a food source, assumes a controversial meaning from the nutritional point of view. From this overview, the most striking feature of the *P. oceanica* ecosystem is the conservative role of the plant itself, primarily forming a multidimensional habitat for organisms directly participating in the trophic dynamics. Maintenance of the necessary complex vertical architecture causes the plant to invest the major part of its resources, while dynamic interactions occur between different components of the ecosystems.

Introduction

In the Mediterranean Sea, *Posidonia oceanica* (L.) Delile beds exert a multifunctional role within coastal systems, comparable with that of other seagrasses in temperate and tropical areas. Yet this plant is endemic to the Mediterranean basin and its peculiar functions influence the adjacent benthic biotopes in different ways (Boudouresque et al. 1984, 1989).

The great importance of *P. oceanica* beds arises from their vast extent (beds typically cover several square kilometres within subsurface depth ranges of 1 to about 40 metres), from their high productivity and, finally, from their stability, being a 'climax' of successional processes (Den Hartog 1977). The crucial element in this ecosystem is the plant itself: with its phenological features, growth dynamics, and biomass partitioning, it forms the physical and trophic framework for the associated algal and animal communities.

Features of *Posidonia oceanica* ecosystem

1. Bed structure and plant biomass and production

Since *P. oceanica* colonizes large areas of sea-bottom over relatively wide depth ranges, the physiognomy of the 'prairies' so formed shows different aspects (e.g., continuous coverage; patchy distribution) varying with the geomorphology of the bottom. Moreover, within these depths, beds show plant density decreasing from 1200 shoots m^{-2} in shallow stands (1–5 m depths) to fewer than 100 shoots m^{-2} at depths around 30 m. The physiognomy and density of beds, which also constitute substrate for settlement, food availability, and shelter, greatly influence the composition and structure of the associated communities (Chimenz et al. 1989).

Another important parameter, partly related to density, is the high standing crop of both subsurface (rhizomes and roots) and erect (leaves) components, a balance which nevertheless varies remarkably at different depths of the stands (1617 g dry wt m^{-2} at 5 m; 316 at 22 m; Mazzella and Buia, unpublished data). These values exceed those of other phanerogams, including the Australian *Posidonia* species (Ott 1980; Romero 1985; Hillman *et al.* 1989; Buia *et al.* 1991). A striking feature is the distinct partitioning of the biomass, often mainly directed into the lignified rhizomes, which can account for up to 90 per cent of total biomass (Pirc 1983; Francour 1985). An opposite trend is seen in production rates, where leaves can account for more than 90 per cent (Wittmann 1984). This is due to different turnover rates of the two components: higher growth rates and an evident and consistent seasonality for leaf tissue, compared with very low rates for rhizomes, whose major roles are storage and stability of the system. These aspects have been addressed by several authors, who have considered seasonal storage and translocation of constituents, including structural carbohydrates, free amino acids and starch (Pirc and Wollenweber 1988; Lawrence *et al.* 1989). All these compounds seem to support growth, mainly in leaves of deep plants where irradiance could be a limiting factor for photosynthesis (Pirc 1986; Buia *et al.* 1991). Notwithstanding the high energy content shown by Mediterranean seagrasses, in particular by *P. oceanica*, synthesized metabolites are mainly structural carbohydrates apparently unavailable to direct herbivory (Pirc 1989; Lawrence *et al.* 1989). Another important feature is the high C/N ratios in both dead leaves and leaf detritus, demonstrating thereby a high resistance to mechanical factors (water-movement, fragmentation by animals) and to microbial decomposition (Pirc and Wollenweber 1988). Velimirov (1986), moreover, demonstrated that the contribution of leaves to the dissolved organic carbon (DOC) in the system was only 0·61 per cent, while the high overall values found in the beds seemed to be partly attributable to the rhizomes/roots system, which can act as a sink. In *P. oceanica*, the synthesis of secondary metabolites, such as polyphenolic compounds, is relevant although its significance as a deterrent to herbivores is debatable (Cariello and Zanetti 1979).

2. *Microbial and algal epiphytic communities*

Studies on epiphytic communities of *P. oceanica* have been performed at both leaf and rhizome layer levels. For the leaf stratum, studies have dealt with spatial and temporal distribution, density, and the role of both microbial and macroalgal components, while, for rhizomes, only the macroflora has been investigated.

Along the leaf axis, a predictable succession and zonation of epiphytes is identifiable according to the blade age-gradient, regardless of season and depth. In young leaf tissues, or those near the meristematic area, at an age of about one week, bacterial and diatom coverage is dominant; in the median parts of leaves, at ages of up to about 100 days, an encrusting layer of brown and red macroalgae is observed; towards the apical portions, which can attain an age of 300 days, an upright macroalgal layer is found, in addition to the earlier assemblages (Wittmann et al. 1981a; Novak 1984; Buia et al. 1989). In this colonization sequence, environmental parameters act on composition, mainly of the last successional stage, and on settlement time, resulting in diverse community structures at different depths and seasons (Battiato et al. 1982; Casola et al. 1987). The varied algal composition and structure represent a wide spectrum of food sources located according to different spatial and temporal scales.

The bacterial population can reach densities of 10^5 cells mm^{-2} of leaf area; both leaf surface and encrusting macroalgae are colonized (Novak 1984). The diatom components, almost exclusively the order Pennales, show different composition and colonization patterns according to leaf age, position along the shoot (e.g., leaf bases vs apices), and ambient exposure to light and water movement (Mazzella 1983). The greatest species-richness and abundance (up to 10^3 cells mm^{-2}) are reached in the intermediate leaf part (Novak 1984). Diatoms with different growth forms (based on the rate of movement, methods of adhesion to the substratum, and life histories) can be distinguished on *P. oceanica* leaves (Mazzella, unpublished data). These growth forms have ecological implications for nutrient and light availability, migration and colonization, and grazing susceptibility, according to Hudon and Legendre (1987). Beyond specific colonization patterns of both bacteria and diatoms, competitive phenomena between these two epiphytic groups are also important determinants of community structure, although both are far from being completely understood.

For the macrophytic community inhabiting *Posidonia* beds, two assemblages can be distinguished, one occurring in the leaf layer and a second in the rhizome layer.

The algal flora of *P. oceanica* leaves is mainly composed of ephemeral species dictated by substrate and environment. The highest cover is consistently constituted by encrusting members of the family Corallinaceae (*Pneophyllum* and *Fosliella* spp.) and by the brown alga *Myrionema orbicularis*. These algae form a layer on which filamentous erect algae successively settle, mainly towards the leaf apices. This upright layer, unlike the encrusting base, shows species composition and abundance varying with depth and season (Battiato et al. 1982; Mazzella et al. 1989). Macroflora composition, species abundance, life-history, zonation pattern along the leaf axis, and spatial temporal variability could all have implications for variable grazer activity, as shown for other situations (Lubchenco and Cubit 1980).

The algal flora of the rhizomes, although involving a high number of species, shows no particularly characteristic taxon and is similar to that found in any environment characterized by reduced light conditions (Panayotidis 1980; Boudouresque et al. 1981).

Among macroalgae, morpho-functional categories have been proposed in relation to the different patterns of energy balance and grazing susceptibility (Littler and Littler 1980; Steneck and Watling 1982). For *Posidonia* epiphytes, seven main morpho-functional categories, based on the Steneck and Watling (1982) classification, have been identified (Mazzella and Buia, unpublished data). For both rhizomes and leaves, the dominant group in terms of species-number is of small filamentous forms, characterized by a simple morphology and structure, the highest diversity being in the shallowest stands. However, because of the small size and short life history of these epiphytes, and possibly also grazing pressure, the group typically shows lower coverage and biomass than among other groups, such as the soft and calcareous encrusting forms in some seasons quantitatively the dominants.

3. Animal communities

No fully integrated picture of the fauna associated with the *Posidonia* ecosystem is so far available. A major obstacle has been the different sampling methods used, as a result of which the respective infaunal assemblages of the 'matte', leaves, and epiphytic community are often viewed as different stratocoenoses having only gap relationships (Kerneis 1960; Bianchi et al. 1989). A whole complex of structural and functional interrelationships actually occurs between these different layers (Lorenti and Scipione 1990; Gambi et al., 1991a). Microfauna and meiofauna of *Posidonia* beds are poorly known, although their importance, both in terms of biomass and as trophic links between epiphytic organisms and macrofaunal predators has been recognized (Chessa et al. 1982; Novak 1989).

The infaunal communities of both living and dead *Posidonia* 'matte' are determined by the mosaic nature of the substrate, supporting species with different ecological requirements (Harmelin 1964; Willsie 1983). From a trophic viewpoint this assemblage (dominated by polychaetes as to both species richness and biomass) is represented mostly by surface and subsurface deposit feeders and by omnivorous forms. Predation by epibenthic species is also likely to have a role in structuring the infauna, at least influencing its vertical distribution within the structure of the meadow (Willsie 1987).

Rhizome epifauna is exposed to a less stressful and variable environment than that of the leaf stratum (Pansini and Pronzato 1985;

Gambi *et al.* 1989), while the multiplicity of microhabitats again supports forms with different ecological requirements. It is noteworthy that detrital litter, mainly the product of dead *Posidonia* leaves deposited at this level, allows the development of a rich fauna (Wittmann *et al.* 1981*b*). Echinoderms, an important component of the system in both biomass and energy flow (Zupi and Fresi 1984), are very abundant within the rhizome layer; holothurians play a central role in reworking prairie surface sediment.

The sessile epifauna of rhizomes also seems to be less structured than in the leaf stratum: the latter dominated by hydroids and bryozoans. These taxocoenoses show clear zonation and succession in growth form related to depth and to position along the leaf axis (Boero 1981; Fresi *et al.* 1982; Balduzzi *et al.* 1983; Casola *et al.* 1987). Animal epiphytes parallel algal epiphytes in forming very complex 'canopy' architecture.

Community structure within vagile fauna of the leaf stratum has received increasing attention following studies by Ledoyer (1962, 1966). Distributional patterns have been studied both for the whole community (Scipione *et al.* 1983; Gambi *et al.* 1991; Chimenz *et al.* 1989; Mazzella *et al.* 1989) and for individual taxocoenoses (polychaetes: Gambi *et al.* 1989; molluscs: Idato *et al.* 1983; amphipods: Scipione and Fresi 1984; isopods: Lorenti and Fresi 1983; decapods: Falciai 1985–86). Generally, patterns have been related to the influence of environmental factors such as water motion and temperature. Trophic spectra of vagile leaf macrofauna, collected by direct sampling (e.g., hand-towed net by scuba diving), give a clear idea of the occurrence of herbivory. A study by Gambi *et al.* (1991) on the principal taxa reveals the numerical dominance of plant feeders (over 71 per cent of total population), mainly represented by micro- and mesograzers (Hay *et al.* 1987), among which different specializations are likely to occur. Herbivory–deposit feeding, as described in Scipione (1989), based on plant epiphytes and associated particulate organic matter, is a widespread mode. Depth-related zonation and seasonal variability of trophic groups occur. An example is the true herbivore group whose strong representation in shallow stands seems to be influenced by patterns of algal epiphytes, while the lower abundance of carnivores (some polychaetes and decapods) seems linked to the more stable and diverse habitats found in deeper stands.

Posidonia beds host a diversified fish population, over 60 per cent composed of resident species (Bell and Harmelin-Vivien 1982), *Labridae* being the overall dominant family. Almost all prairie fish species are considered to be carnivores, mainly on crustaceans (Bell and Harmelin-Vivien 1983; Khoury 1984). However, the herbivore sparid *Sarpa salpa*, a schooling bream, can be an important component of total fish biomass, at least in sheltered shallow meadows (Velimirov 1984; Verlaque 1985; Francour 1990). Species widely distributed in coastal biotopes (e.g., *Mullus surmuletus*) show optima in *Posidonia* beds (Arculeo et al. 1989).

Biotic interrelations

1. Gut content analysis

A basic approach to the definition of food webs in *Posidonia* systems is gut content analysis of macrofaunal samples obtained by indirect methods, principally bottom trawling (Chessa *et al.* 1982; Zupi and Fresi 1984, 1985). This method yields a picture quite different from that obtained by feeding-guilds classification of leaf macrofauna related to body-size range, ethology, and taxonomic spectrum of the animals collected (mostly large echinoderms, decapods, and molluscs). A study carried out on 91 macrozoobenthic species, among whose gut contents 21 food items were distinguished, showed the central role played in the overall structure of the food web by 'dead *Posidonia*', 'living *Posidonia*', 'macroalgae', and 'diatoms' (Zupo, unpublished data). A cluster analysis of these data indicated that plant and animal prey occur unevenly in the gut contents.

Choice of food is dependent on type (plant or animal) and size, permitting distinction between macrophagous and microphagous feeders (Zupo 1990). Microherbivores and detritivores appear to be the main links in energy transfer within the trophic web. *Posidonia* tissues and macroalgae are fed upon by echinoderms (*Paracentrotus lividus* and *Holothuria tubulosa*), decapods (Majiidae) and amphipods (*Dexamine spinosa*). *Posidonia* detritus, diatoms, and micro-organisms occur mostly in molluscs (*Rissoa ventricosa*, *Tricholia speciosa*), amphipods (*Lysianassa pilicornis*), polychaetes (Serpulidae) and the echinoderm *Echinaster sepositus*. Carnivores are mainly represented among fishes and by a number of molluscs, polychaetes, and decapods.

These results emphasize the primary importance of *Posidonia* tissues and associated periphyton in supporting the food webs of the system.

2. Direct consumption of living P. oceanica

In this section we consider the removal of relevant plant portions by animals during feeding, both for inherent nutritive qualities and as a substrate for other food sources. Notwithstanding the high biomass and energetic values, some intrinsic features (high proportion of structural carbohydrates, high C/N ratio, occurrence of phenolic compounds) make *P. oceanica* living tissue unattractive to many direct grazers. Existing studies do not clearly establish the mode of direct utilization of the living tissue as nutritive matter. Assimilation of viable *Posidonia* seems in general to be negligible. The sparid fish *Sarpa salpa* seems to assimilate about 90 per cent of the soluble sugars and free amino-acid fractions of *Posidonia* leaf organic matter (Velimirov 1984), based on comparisons between fresh food and faecal pellets. Assimilation of non-soluble organic matter is calculated by the same author to be at least 10 per cent. Feeding behaviour of large *S. salpa* has considerable effect on *Posidonia* and its epiphyte biomass; intensive browsing by this school-forming fish can lead to the cropping of extensive prairie areas (Laborel-Deguen and Laborel 1977; Velimirov 1984).

Living *Posidonia* enters the diet of various echinoderms. The sea-urchin *Paracentrotus lividus* is classically viewed as the major consumer of *Posidonia* leaves because of the dramatic reduction that its grazing can induce in leaf standing crop (Kirkman and Young 1981; Verlaque and Nedelec 1983). *Paracentrotus lividus* behaves fundamentally as an opportunistic feeder able widely to adapt to availability of food resources (Zupi and Fresi 1984). Somewhat conflicting views exist on the proportion of living tissue in the diet of this echinoid. According to Ott and Maurer (1977) and Traer (1980), consumption of unshed leaves mostly affects the epiphytized brown tips, so that *P. lividus* is defined by Ott (1981) as a 'pseudograzer', feeding on plant portions which are actually debris. On the other hand, Nedelec and Verlaque (1984) state that green leaf fragments are constantly dominant in the urchin gut contents, the proportion of *Posidonia* leaves in the diet seeming to increase with age. Consumption rates by the largest individuals can reach values as high as $1 \cdot 13$ g dry wt day^{-1}, in spring (Nedelec and Verlaque 1984); the average estimated for individuals — of a test diameter between 4·9–7·3 cm — is $0 \cdot 51$ g dry wt day^{-1} (Nedelec *et al.* 1981). Absorption efficiency is reported by Traer (1984) to be lower for non-epiphytized than for epiphytized leaves, thus confirming preference by *P. lividus* for algal epigrowth.

Living *Posidonia* is thought to have a less important role in the diet of echinoids (other than *P. lividus*), although opportunism seems to be a consistent trait in their feeding behaviour (Zupi and Fresi 1984). *Psammechinus microtuberculatus*, a common species in *Posidonia* meadows, feeds mainly on shed leaves and attached epiphytes; occurrence of pseudograzing behaviour is also hypothesized for this urchin, based on its known nocturnal migrations along the blade axis (Paul *et al.* 1984).

As shown by gut content analysis in decapod crustaceans, green *Posidonia* seems to be part of the diet of various omnivorous species

(Zupi and Fresi 1985; Chessa *et al.* 1989*a*, *b*). Two crab species of the family Majiidae (*Pisa muscosa* and *P. nodipes*) are reported to cut and ingest up to 2 mm pieces of leaves (Vadon 1981). Some idoteid isopods typical of *P. oceanica* systems (e.g., *Idotea hectica*) are able rapidly to ingest large living leaf portions. *Idotea baltica basteri* is commonly found in *Posidonia* leaf litter, on which it feeds (Wittmann *et al.* 1981*b*) although in laboratory experiments the adults and juveniles were shown actively to consume living *Posidonia* tissue (Lorenti and Fresi 1983). It is worth noting that idoteid isopods are recorded as among the major seagrass grazers in various extra-Mediterranean systems (Klumpp *et al.* 1989).

On the whole, available information suggests that consumption of *P. oceanica* is restricted to a few, relatively unselective, species able to deal with large food masses in which nutritive resources are mixed. Relationships of a biomechanical nature allow such species, of comparatively large size and provided with robust mouth pieces, to overcome *P. oceanica* tissue toughness.

Lastly, it must be kept in mind that nutritional values of living *P. oceanica* vary with age and season of the year (Pirc and Wollenweber 1988). Juvenile leaves and meristem, which have lower proportions of structural carbohydrate and polyphenols and lower C/N ratios, appear attractive to potential grazers (Lorenti and Fresi 1983), although strategies developed by the plant in nature make access to them difficult to consumers (Velimirov 1987).

3. Epiphyte–herbivore interactions

In several seagrass systems, algal epiphytes are found to be more attractive food for herbivores than the leaves themselves (Kitting *et al.* 1984; Orth and Van Montfrans 1984; Van Montfrans *et al.* 1984), because of high biomass of epiphytes and several features that may increase the epiphyte trophic value (e.g., more energy allocated in production processes than in stabilizing storage capability) as well as their ability to survive grazing pressure (e.g., heteromorphic and short life histories) (Littler and Littler 1980; Lubchenco and Cubit 1980; Littler *et al.* 1983). Epiphytic algae can contribute up to 40 per cent to the total biomass of seagrass leaf systems as in *P. oceanica* (Mazzella and Ott 1984), although the complex epiphyte-grazer relationships, in terms of energy transfer and community structuring, are not well understood.

Molluscs are the animal group to which the main research effort has been directed in recent years. In laboratory experiments, two species of *Gibbula* (Gastropoda) showed different preferences for algal periphyton based on size (small vs large diatoms), shape (rounded vs elongated diatoms), and toughness (thin filamentous vs calcareous algae). When offered epiphytized *P. oceanica* leaves as a food source,

Gibbula ardens showed preference for early successional stages of colonization (such as bacteria and diatoms) but seldom for encrusting red algae; *G. umbilicaris*, by contrast, showed preference for erect macroalgae and seldom for soft encrusting species (Mazzella and Russo 1989). These feeding habits were in accord with the species arrangement along the leaves and with the spatio-temporal distribution in the bed (Russo *et al.* 1984). For neither species was ingestion of living and dead *Posidonia* observed, corroborating other authors' findings (Peduzzi 1987). For other mollusc species, micro- and macroalgae seem to be preferred to the plant leaves. In particular, application of the functional group approach proposed by Steneck and Watling (1982) showed that prosobranchs, with radulae of low excavation capability ('brushers' and 'scrapers'), comprise almost exclusively the mollusc group present on the *P. oceanica* leaves. This trophic convergence allows the removal of the periphyton, probably exercising a sort of control against overgrowth and at the same time avoiding damage to the leaf substratum and therefore to the plants. It is easy to speculate on coevolutionary processes as the basis of this adaptation of the mollusc community (Russo 1986), also supported by analysis of its degree of 'habitat fidelity', which is very high, mainly in shallow beds (Russo 1989). Another important aspect of the role of prosobranch grazers could be their low absorption efficiency compared with their high feeding rates. In fact, it has been calculated that more than 70 per cent of the ingested organic carbon from macro-epiphytes is again lost by *G. umbilicaris* during defaecation; with regard to this, faecal pellet remineralization seems to be important in the overall carbon cycle (Peduzzi 1987).

For other macrofaunal components of the leaf stratum community, feeding guild analyses give results in favour of indirect herbivory. Herbivores and herbivores–deposit feeders are the most represented categories in amphipods, isopods, and polychaetes. However, distribution of these categories seems to be related to availability of food sources along depth transects and at different seasons. Among amphipod species, different trophic habits are present; some prefer epiphytic micro- and macroalgae to leaf tissue, while others feed preferentially on leaf detritus or debris. Scraping edible material by using antennae or mouth parts has been observed on *P. oceanica* as on other seagrasses (Zimmermann *et al.* 1979; Howard 1982; Scipione 1989). Microalgae seem to be assimilated preferentially over macrophytes; SEM observations of faecal pellets of *Dexamine spinosa* showed almost exclusively the presence of slow-moving diatoms (Scipione and Mazzella, in press). Some species, by contrast, showed evidence of being capable of digesting residual plant material (Zimmermann *et al.* 1979). The high abundance of amphipods (as herbivores) within vagile fauna communities attests to an important role at the level of *P. oceanica* leaf stratum.

Among polychaetes, micro- and mesoherbivore feeding guilds can be distinguished on the basis of plant items consumed (micro- and macroflora, respectively). Microherbivores are mainly represented by some species of syllids (Exogoninae); these are distributed mostly among intermediate and deep *Posidonia* stands, with low seasonal variability (Gambi et al. 1989). Mesoherbivores are represented by Nereididae (and in particular by *Platynereis dumerilii*) distributed mainly in shallow stands with a maximum in summer, probably because of seasonally higher epiphyte development (Gambi et al. 1989, 1991). In other phytal systems, however, this species is observed to feed especially on algal detritus (Bedford and Moore 1985). Elsewhere in seagrass beds (e.g., *Zostera marina*), few polychaetes seem, even occasionally, to consume seagrass detritus (Mangum et al. 1968; Tenore 1977).

Among the periphyton consumers, those feeding on colonial and sedentary animals (FCSA), such as sponges, hydroids, and bryozoans are mainly represented by polychaetes (polynoids and syllids), and molluscs (mainly opisthobranchs). Animal grazers are particularly abundant in deeper beds (Russo 1989; Gambi et al. 1991).

On the whole, available information suggests strong relationships between plants and epiphytes, as between epiphytes and grazers, probably as a result of coevolutionary processes. Herbivory is certainly the main trophic adaptation in the most important groups of vagile fauna (polychaetes, molluscs, and amphipods) but its role in the control of epiphytic overgrowth requires more accurate assessment. It is easy to speculate on the one hand about importance of the removal by grazing of microphytic and macrophytic upright layers, while on the other hand constant presence and high abundance of prostrate and encrusting macroalgae can be seen as coevolutionary adaptative protection against intense grazing.

4. Trophic role of P. oceanica-*derived detritus*

Within the *Posidonia* ecosystem, detritus available to animal consumers is roughly divisible into three:

(1) litter of plant material in various stages of fragmentation;
(2) particles suspended above and within the canopy;
(3) detritus trapped in the epiphytic cover.

Litter is essentially formed by decaying *P. oceanica* leaves, with minor proportions of rhizome, sheath, and seaweed remains. Velimirov et al. (1981) adopt the term 'debris' for this material, including in it dead leaf tips still on the shoot. Great quantities of shed leaves are driven out of stands (by wave-action) to form 'wrackbeds', both submerged and washed ashore, yet a remarkable amount is retained within the beds, in proportions increasing with depth (Romero et al., 1991). Studies based on gut content analysis in macroinvertebrates and fishes show that *P. oceanica* debris is the basis of trophic webs in the prairie (Chessa et al. 1982; Zupi and Fresi 1984, 1985). The nutritional value of *Posidonia* debris for detritivores is actually probably of minor importance, as usual for vascular plants (Tenore 1981). Soluble sugars become rapidly depleted from senescent leaf tissue (Velimirov et al. 1981), while the C/N ratio of dead leaves and rhizomes has high values (55 to 77) indicating resistance to decomposition (Pirc and Wollenweber 1988). Velimirov (1987) infers that *P. oceanica* debris is unlikely by itself effectively to support consumer nutrition. The same conclusion is reached by Dauby and Mosora (1988) based on the $^{13}C/^{12}C$ ratio analysis of some detritivore species from the Gulf of Calvi (Corsica). In the absence of data on direct assimilation, microbial epigrowth on debris is regarded as the major nutritional source for macro-consumers. Bacteria on dead *Posidonia* leaves can reach densities as high as 4×10^4 cells mm^{-2} (Velimirov et al. 1981), while fungal mycelia can also raise the proteinaceous nitrogen content of a detrital mass (Cuomo et al. 1982). Microbial activity and nitrogen content increase with reduction of particle size (Velimirov et al. 1981). In this regard, fragmentation of debris by macrofauna associated with the litter plays a major role (Wittmann et al. 1981*b*).

The contribution of *Posidonia*-originated detritus to suspended particulate organic matter (POM) is unclear. According to Velimirov (1987), chemical and morphological features of suspended particles in the prairie water column differ markedly from such features in the seagrasses, indicating a different origin. This could be reflected in suspension feeder nutrition. Dauby and Mosora (1988) and Dauby (1989) point out the dissimilarity between the $^{13}C/^{12}C$ ratio found for *Posidonia* tissue and the ratio shown by a number of sestonophagous species from *Posidonia*-dominated coastal systems. Clearly, however, without experiments on detritus assimilation or analyses of a significant number of consumer species, conclusions cannot be drawn. Among macrofaunal components, holothurians represent a fundamental link in detritus recycling, mainly in view of their role in reworking sediments enriched with detrital particles (Verlaque 1981; Coulon and Jangoux 1991).

Conclusions

From this overview, the most striking feature of the *P. oceanica* ecosystem appears to be the 'conservative' role of the plant itself with respect to the other components. Its high energy content and biomass are only minimally available for direct trophic interactions, although they become of major importance for stabilizing the system. This apparent paradox is explained by the primary role of *P. oceanica* as a multidimensional habitat for organisms directly participating in the trophic dynamics of the system. This role is displayed at various spatio-temporal levels; ground biomass (rhizomes and stems) provides a multiplicity of cryptic habitats and settling surfaces, while fronds (with the higher turnover and growth rates) form a dynamic environment for all epibiota. As a result we receive the impression of a complex infrastructure for the maintenance of which the plant invests the major part of its resources. Within this variegated ecosystem, highly dynamic interaction occurs between different components. Adaptive mechanisms affecting plant–animal trophic relationships derive from other factors mediated by plant infrastructure. This is the case with predation pressures on smaller herbivores and detritivores showing complex behavioural-patterns such as diel migration or cryptic habits (Lorenti and Scipione 1990; Gambi *et al.* 1991). The central role in this respect is played by mesograzers (*sensu* Hay *et al.* 1987) both as controllers of algal periphyton and as vehicles for energy transfer from epiphytes to higher consumer levels (Chessa *et al.* 1982; Khoury 1984; Harmelin-Vivien and Francour, 1991).

As a synthetic overview, a conceptual scheme of the vertical structure of the *Posidonia oceanica* ecosystem is presented to provide a simplified image of biotic relationships, taking the *Posidonia* as a reference (Fig. 8.1). A broad zonation is recognizable within the plant components and, accordingly, within animal taxocoenoses and functional guilds. Interactions among different compartments reflect gradients in plant and animal overall biomass and diversity, as well as in habitat complexity. Biotic processes such as microhabitat shifts (e.g., migrations) and predation by species active at several levels of the canopy and/or 'matte' promote dynamic biomass and energy exchanges within the system. Spatio-temporal variations linked to seasonal phenomena (environmental variability, plant phenology, animal life cycles) further complicate biotic interactions in the *Posidonia* ecosystem.

Acknowledgements

This work was partly supported by the European Community Commission (Environmental Programme; N.EV4VI 0139-B).

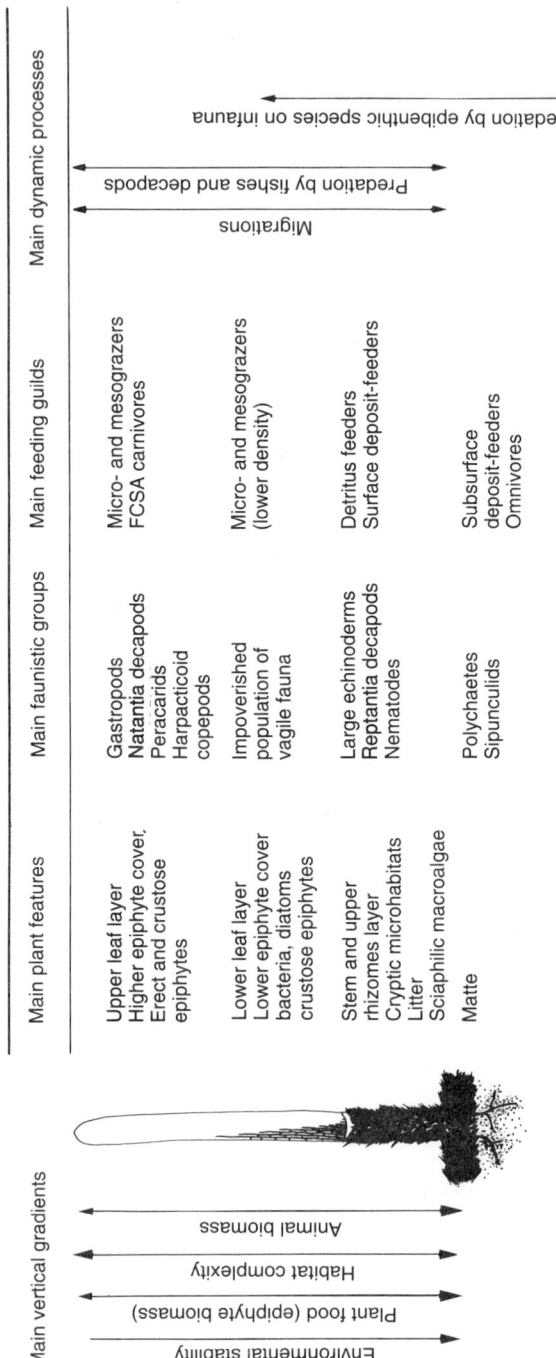

Fig. 8.1. Conceptual scheme of *Posidonia oceanica* ecosystem vertical structure.

References

Arculeo, M., Pipitone, C., and Riggio, S. (1989). Aspetti del regime alimentare di *Mullus surmuletus* L. (Pisces, Mullidae) nel Golfo di Palermo. *Oebalia*, **15**, 76-7.

Balduzzi, A., Barbieri, M., and Gobetto, F. (1983). Distribution des Bryozoaires Gymnolemes en deux herbiers de Posidonies italiens. Analyse de correspondences. *Rapports et procès-verbaux des réunions de la Commission Internationale pour l'Exploration Scientifique de la Mer Méditerranée, Monaco*, **28** (3), 137-8.

Battiato, A., Cinelli, F., Cormaci, M., Furnari, G., and Mazzella, L. (1982). Studio preliminare della macroflora epifita della *Posidonia oceanica* (L.) Delile di una prateria di Ischia (Golfo di Napoli) (Potamogetonaceae, Helobiae). *Naturalista Siciliano*, Serie IV, **6**, 15-27.

Bedford, A.P. and Moore, P.G. (1985). Macrofaunal involvement in the sublittoral decay of kelp debris: the Polychaete *Platynereis dumerilii* (Audouin and Milne-Edwards) (Annelida: Polychaeta). *Estuarine, Coastal and Shelf Science*, **20** (2), 117-34.

Bell, J.D. and Harmelin-Vivien, M.L. (1982). Fish fauna of French Mediterranean *Posidonia oceanica* seagrass meadows. 1. Community structure. *Téthys*, **10**, 1-14.

Bell, J.D. and Harmelin-Vivien, M.L. (1983). Fish fauna of French Mediterranean *Posidonia oceanica* seagrass meadows. II Feeding habits. *Téthys*, **12**, 1-14.

Bianchi, C.N., Bedulli, D., Morri, C., and Occhipinti-Ambrogi, A. (1989). L'herbier de Posidonies: écosystème ou carrefour écoéthologique? In *II International workshop on* Posidonia *beds*, (ed. C.F. Boudouresque, A. Meinesz, E. Fresi, and V. Gravez), Vol. 2, pp. 257-72. GIS Posidonie, Marseilles, France.

Boero, F. (1981). Systematics and ecology of the Hydroid population on two *Posidonia oceanica* meadows. *P.S.Z.N.I.: Marine Ecology*, **2**, 181-97.

Boudouresque, C.F., Cinelli, F., Mazzella, L., and Richard, M. (1981). Algal undergrowth of *Posidonia oceanica* beds in the Gulf of Naples: floristic study. *Rapports et procès-verbaux des réunions de la Commission Internationale pour l'Exploration Scientifique de la Mer Méditerranée, Monaco*, **27** (2), 195-6.

Boudouresque, C.F., Jeudy de Grissac, A., and Olivier, A. (ed.) (1984). *I International workshop on* Posidonia *beds*. GIS Posidonie, Marseilles, France.

Boudouresque, C.F., Meinesz, A., Fresi, E., and Gravez, V. (ed.) (1989). *II International workshop on* Posidonia *beds*. GIS Posidonie, Marseilles, France.

Buia, M.C., Cormaci, M., Furnari, G., and Mazzella, L. (1989). *Posidonia oceanica* off Capo Passaro (Sicily, Italy): leaf phenology and leaf algal epiphytic community. In *II International workshop on* Posidonia *beds*, (ed. C. F. Boudouresque, A. Meinesz, E. Fresi, and V. Gravez), Vol. 2, pp. 127–43. GIS Posidonie, Marseilles, France.

Buia, M. C., Zupo, V., and Mazzella, L. (1991). Primary production and growth in *Posidonia oceanica*. *P.S.Z.N.I.: Marine Ecology*, in press.

Cariello, L. and Zanetti, L. (1979). Effect of *Posidonia oceanica* extracts on the growth of *Staphylococcus aureus*. *Botanica Marina*, **22**, 269–85.

Casola, E., Scardi, M., Mazzella, L., and Fresi, E. (1987). Structure of the epiphytic community of *Posidonia oceanica* leaves in a shallow meadow. *P.S.Z.N.I.: Marine Ecology*, **8**, 285–96.

Chessa, L. A., Fresi, E., and Soggiu, L. (1982). Primi dati sulla rete trofica dei consumatori in una prateria di *Posidonia oceanica* (L.) Delile. *Bollettino dei Musei ed degli Istituti Biologici dell'Università di Genova*, **50** (suppl.), 156–61.

Chessa, L. A., Scardi, M., Fresi, E., and Russu, P. (1989a). Consumers in *Posidonia oceanica* beds: 1. *Processa edulis* (Risso), (Decapoda, Caridea). In *II International workshop on* Posidonia *beds*, (ed. C. F. Boudouresque, A. Meinesz, E. Fresi, and V. Gravez), Vol. 2, pp. 243–50. GIS Posidonie, Marseilles, France.

Chessa, L. A., Scardi, M., Fresi, E., and Saba, S. (1989b). Consumers in *Posidonia oceanica* beds: 2. *Galathea squamifera* Leach (Decapoda, Anomura). In *II International workshop on* Posidonia *beds*, (ed. C. F. Boudouresque, A. Meinesz, E. Fresi, and V. Gravez), Vol. 2, pp. 251–6. GIS Posidonie, Marseilles, France.

Chimenz, C. *et al.* (1989). Studies on animal populations of the leaves and rhizomes of *Posidonia oceanica* (L.) Delile on the rocky bottom of Torvaldaliga. In *II International workshop on* Posidonia *beds*, (ed. C. F. Boudouresque, A. Meinesz, E. Fresi, and V. Gravez), pp. 145–56. GIS Posidonie, Marseilles, France.

Coulon, P. and Jangoux, M. (1991). Rate and rhythm of feeding of the holothuroid *Holothuria tubulosa* in the seagrass beds of Ischia island (Bay of Naples, Italy). *P.S.Z.N.I.: Marine Ecology*. (In press).

Cuomo, V., Vanzanella, F., Fresi, E., Mazzella, L., and Scipione, M. B. (1982). Micoflora delle fanerogame dell'isola d'Ischia: *Posidonia oceanica* (L.) Delile e *Cymodocea nodosa* (Ucria) Aschers. *Bollettino del Museo e dell'Istituto di Biologia dell'Università di Genova*, **50**, 162–6.

Den Hartog, C. (1977). Structure, function, and classification in seagrass communities. In *Seagrass ecosystems. A scientific perspective*, (ed. C. P. McRoy and C. Helfferich), pp. 89–121. Marcel Dekker, New York.

Dauby, P. (1989). The stable carbon isotope ratios in benthic food webs of the Gulf of Calvi, Corsica. *Continental Shelf Research*, **9**, 181–5.

Dauby, P. and Mosora, F. (1988). Analyse à l'aide des isotopes stables du carbone des sources de nutrition des détritivores et suspensivores benthiques. *Bulletin de la Société Royal de Liège*, **57** (4–5), 241–8.

Falciai, L. (1985–86). Fauna vagile di una prateria a *Posidonia oceanica*: i Crostacei Decapodi. *Quaderni dell'Istituto di Idrobiologia e Acquacoltura Brunelli*, **5–6**, 75–84.

Francour, P. (1985). Root and rhizome biomass of *Posidonia oceanica* bed. *Rapports et procès-verbaux des réunions de la Commission Internationale pour l'Exploration Scientifique de la Mer Méditerranée, Monaco*, **29** (5), 183–5.

Francour, P. (1990). Dinamique de l'ecosysteme à *Posidonia oceanica* dans le Parc National de Port-Cros. Analyse des compartiments matte, letière, fauna vagile, échinodermes et poissons. Unpublished Thesis. University of Paris.

Fresi, E., Chimenz, C., and Marchio, G. (1982). Zonazione di Briozoi ed Idroidi epifiti in una prateria di *Posidonia oceanica* (L.) Delile. *Naturalista Siciliano*, **67**, 499–508.

Gambi, M.C., Giangrande, A., Chessa, L.A., Manconi, R., and Scardi, M. (1989). Distribution and ecology of Polychaetes in the foliar stratum of a *Posidonia oceanica* bed in the bay of Porto Conte (N.W. Sardinia). In *II International workshop on* Posidonia *beds*, (ed. C.F. Boudouresque, A. Meinesz, E. Fresi, and V. Gravez), Vol. 2, pp. 175–88. GIS Posidonie, Marseilles, France.

Gambi, M.C., Lorenti, M., Russo, G.F., Scipione, M.B., and Zupo, V. (1991). Depth and seasonal distribution of some groups of the vagile fauna of *Posidonia oceanica* leaf stratum: structural and trophic analyses *P.S.Z.N.I.: Marine Ecology*. (In press).

Harmelin, J.G. (1964). Étude de l'endofaune des 'mattes' d'herbiers de *Posidonia oceanica* Delile. *Recueil des Travaux de la Station Marine d'Endoume*, **35** (51), 43–106.

Harmelin-Vivien, M.L. and Francour, P. (1991). Trawling or visual censuses? Methodological bias in the assessment of fish populations in seagrass beds. *P.S.Z.N.I.: Marine Ecology*. (In press).

Hay, M.E., Duffy, J.E., Pfister, C.A., and Fenical, W. (1987). Chemical defenses against different marine herbivores: are amphipods insect equivalents? *Ecology*, **68**, 1567–80.

Hillman, K., Walker, D.I., Larkum, A.W.D., and McComb, A.J. (1989). Productivity and nutrient limitation. In *Biology of seagrasses*, (ed. A.W.D. Larkum, A.J. McComb, and S.A. Shepherd), Vol. 2, pp. 635–68. Elsevier, Amsterdam.

Howard, R.K. (1982). Impact of feeding activities of epibenthic Amphipods on surface-fouling of eelgrass leaves. *Aquatic Botany*, **14**, 91–7.

Hudon, C. and Legendre, P. (1987). The ecological implications of growth forms in epibenthic diatoms. *Journal of Phycology*, **23**, 434–41.

Idato, E., Fresi, E., and Russo, G. F. (1983). Zonazione verticale della fauna vagile di strato foliare in una prateria di *Posidonia oceanica* Delile: 1. Molluschi. *Bollettino Malacologico*, **19** (5), 109–20.

Kerneis, A. (1960). Contribution a l'étude faunistique et ecologique des herbiers de Posidonies de la region de Banyuls. *Vie Milieu*, 11, 145–87.

Khoury, C. (1984). Ethologies alimentaires de quelques espèces de poissons de l'herbier de Posidonies du Parc National de Port- Cros. In *I International workshop on* Posidonia oceanica *beds*, (ed. C. F. Boudouresque, A. Jeudy de Grissac, and J. Olivier), Vol. 1, pp. 335–47. GIS Posidonie, Marseilles, France.

Kirkman, H. and Young, P. C. (1981). Measurements of health and echinoderm grazing on *Posidonia oceanica* (L.) Delile. *Aquatic Botany*, **10**, 329–38.

Kitting, C. L., Fry, B., and Morgan, M. D. (1984). Detection of inconspicuous epiphytic algae supporting food webs in seagrass meadows. *Oecologia* (Berlin), **62**, 145–9.

Klumpp, D. W., Howard, R. K., and Pollard, D. A. (1989). Trophodynamics and nutritional ecology of seagrass communities. In *Biology of seagrasses*, (ed. A. W. D. Larkum, A. J. McComb, and S. A. Shepherd), Vol. **2**, pp. 394–457. Elsevier, Amsterdam.

Laborel-Deguen, F. and Laborel, J. (1977). Broutage de Posidonie à la plage du Sud. *Travaux scientifiques du Parc national de Port-Cros*, **3**, 213–4.

Lawrence, J. M., Boudouresque, C. F., and Maggiore, F. (1989). Proximate constituents, biomass, and energy in *Posidonia oceanica* (Potamogetonaceae). *P.S.Z.N.I.: Marine Ecology*, **10**, 263–70.

Ledoyer, M. (1962). Étude de la faune vagile des herbiers superficiels de Zosteracées et de quelques biotopes d'Algues littorales. *Recueil des travaux de la station marine d'Endoume*, **39** (25), 117–235.

Ledoyer, M. (1966). Ecologie de la faune vagile des biotopes méditerranéens accessibles en scaphandre autonome. 2. Données analytiques sur les herbiers de Phanérogames. *Recueil des travaux de la station marine d'Endoume*, **57** (41), 135–64.

Littler, M. M. and Littler, D. S. (1980). The evolution of thallus form and survival strategies in benthic marine macroalgae: field and laboratory tests of a functional form model. *American Naturalist*, **116**, 25–44.

Littler, M. M., Littler, D. S., and Taylor, P. R. (1983). Evolutionary strategies in a tropical barrier reef system: functional form groups of marine macroalgae. *Journal of Phycology*, **19**, 229–37.

Lorenti, M. and Fresi, E. (1983). Vertical zonation of vagile fauna from the foliar stratum of a *Posidonia oceanica* bed. Isopoda. *Rapports et procès-verbaux des réunions de la Commission Internationale pour l'Exploration Scientifique de la Mer Méditerranée, Monaco*, **28** (3), 143–5.

Lorenti, M. and Scipione, M. B. (1990). Relationships between trophic structure and diel migrations of Isopods and Amphipods in a *Posidonia oceanica* bed of the Island of Ischia. *Rapports et procès-verbaux des réunions de la Commission Internationale pour l'Exploration Scientifique de la Mer Méditerranée, Monaco*, **2** (1), p. 17.

Lubchenco, J. and Cubit, J. (1980). Heteromorphic life histories of certain marine algae as adaptations to variations in herbivory. *Ecology*, **61**, 676–87.

Mangum, C. P., Santos, S. L., and Rhodes, W. R. (1968). Distribution and feeding on the Onuychid Polychaete *Diopatra cuprea*. *Marine Biology*. **2**, 33–40.

Mazzella, L. (1983). Studies on the epiphytic diatoms of *Posidonia oceanica* (L.) Delile leaves. *Rapports et procès-verbaux des réunions de la Commission Internationale pour l'Exploration Scientifique de la Mer Méditerranée, Monaco*, **28** (3) 123–4.

Mazzella, L. and Ott, J. (1984). Seasonal changes in some features of *Posidonia oceanica* (L.) Delile leaves and epiphytes at different depths. In *I International workshop on* Posidonia oceanica *beds*, (ed. C. F. Boudouresque, A. Jeudy de Grissac, and J. Olivier), Vol. 1, pp. 119–27. GIS Posidonie, Marseilles, France.

Mazzella, L. and Russo, G. F. (1989). Grazing effect of two *Gibbula* species (Mollusca, Archaeogastropoda) on the epiphytic community of *Posidonia oceanica* leaves. *Aquatic Botany*, **35**, 357–73.

Mazzella, L., Scipione, M. B., and Buia, M. C. (1989). Spatio-temporal distribution of algal and animal communities in a *Posidonia oceanica* (L.) Delile meadow. *P.S.Z.N.I.: Marine Ecology*, **10**, 107–31.

Nedelec, H. and Verlaque, M. (1984). Alimentation de l'oursin *Paracentrotus lividus* (Lamarck) dans un herbier à *Posidonia oceanica* (L.) Delile en Corse. In *I International workshop on* Posidonia oceanica *beds*, (ed. C. F. Boudouresque, A. Jeudy de Grissac, and J. Olivier), Vol. 1, pp. 349–64. GIS Posidonie, Marseilles, France.

Nedelec, H., Verlaque, M., and Diapoulis, A. (1981). Preliminary data on *Posidonia* consumption by *Paracentrotus lividus* in Corsica (France). *Rapports et procès-verbaux des réunions de la Commission Internationale pour l'Exploration Scientifique de la Méditerranée, Monaco*, **27**(2), 203–4.

Novak, R. (1984). A study in ultra-ecology: microorganisms on the seagrass *Posidonia oceanica* (L.) Delile. *P.S.Z.N.I.: Marine Ecology*, **10**, 107–31.

Novak, R. (1989). Ecology of nematodes in the Mediterranean seagrass *Posidonia oceanica* (L.) Delile. 1. General part and faunistics of the Nematode community. *P.S.Z.N.I.: Marine Ecology*, **10**, 335–63.

Orth, R. J. and Van Montfrans, J. (1984). Epiphyte–seagrass relationships with an emphasis on the role of micrograzing: a review. *Aquatic Botany*, **18**, 43–69.

Ott, J. A. (1980). Growth and production in *Posidonia oceanica* (L.) Delile. *P.S.Z.N.I.: Marine Ecology*, **1**, 145–87.

Ott, J. A. (1981). Adaptive strategies at the ecosystem level: examples from two benthic marine systems. *P.S.Z.N.I.: Marine Ecology*, **2**, 113–58.

Ott, J. A. and Maurer, L. (1977). Strategies of energy transfer from marine macrophytes to consumer levels: the *Posidonia oceanica* example. In *Biology of benthic organisms*, (ed. B. F. Keegan, P. O. Ceidigh, and P. J. S. Boaden), pp. 493–502. Pergamon Press, Oxford.

Panayotidis, P. (1980). Contribution à l'étude qualitative et quantitative de l'association Posidonietum oceanicae Funk 1927. Unpublished Thesis. University of Aix-Marseilles II, France.

Pansini, M. and Pronzato, R. (1985). Distribution and ecology of epiphytic *Porifera* in two *Posidonia oceanica* (L.) Delile meadows of the Ligurian and Tyrrhenian Sea. *P.S.Z.N.I.: Marine Ecology*, **6**, 1–11.

Paul, O., Verlaque, M., and Boudouresque, C. F. (1984). Étude du contenu digestif de l'oursin regulier *Psammechinus microtuberculatus* dans l'herbier a *Posidonia oceanica* de la baie de Port-Cros (Var, France). In *I International workshop on* Posidonia oceanica *beds*, (ed. C. F. Boudouresque, A. Jeudy de Grissac, and J. Olivier), Vol. 1, pp. 365–71. GIS Posidonie, Marseilles, France.

Peduzzi, P. (1987). Dietary preferences and carbon absorption by two grazing Gastropods, *Gibbula umbilicaris* (Linné) and *Jujubinus striatus* (Linné). *P.S.Z.N.I.: Marine Ecology*, **8**, 359–70.

Pirc. H. (1983). Belowground biomass of *Posidonia oceanica* (L.) Delile and its importance to the growth dynamics. In *Proceedings International Symposium on Aquatic Macrophytes*, (ed. C. den Hartog), pp. 177–81. Faculteit der Wiskunde en Natuurwetenschappen, Katholieke Universiteit, Nijmigen.

Pirc, H. (1986). Seasonal aspects of photosynthesis in *Posidonia oceanica*: influence of depth, temperature and light intensity. *Aquatic Botany*, **26**, 203–12.

Pirc. H. (1989). Seasonal changes in soluble carbohydrates, starch, and energy content in Mediterranean seagrasses. *P.S.Z.N.I.: Marine Ecology*, **10**, 97–105.

Pirc, H. and Wollenweber, B. (1988). Seasonal changes in nitrogen, free amino acids, and C/N ratio in Mediterranean seagrasses. *P.S.Z.N.I.: Marine Ecology*, **9**, 167–79.

Romero, J. (1985). Estudio ecologico de las fanerogamas marinas de la costa catalana: produccion primaria de *Posidonia oceanica* (L.) Delile en las Islas Medes. Unpublished Thesis. University of Barcelona.

Romero, J., Pergent, G., Pergent-Martini, C., Mateo, M. A., and Regnier, C. (1991). The detritic compartment in a *Posidonia oceanica*

meadow: litter features, decomposition rates and mineral stocks. (In press).
Russo, G. F. (1986). Evoluzione ed adattamenti trofici nei Prosobranchi: spunti per una analisi funzionale del popolamento Malacologico di una prateria a *Posidonia oceanica* dell'isola d'Ischia. *Nova Thalassia*, **8**, 643-4.
Russo, G. F. (1989). La scelta dei descrittori morfo-funzionali nell'analisi dei sistemi bentonici: un approccio con la componente malacologica di una prateria a *Posidonia oceanica*. *Oebalia*, **15**, 213-28.
Russo, G. F., Fresi, E., Vinci, D., and Chessa, L. A. (1984). Mollusk syntaxon of foliar stratum along a depth gradient in a *Posidonia oceanica* (L.) Delile meadow: seasonal variability. In *I International workshop on* Posidonia oceanica *beds*, (ed. C. F. Boudouresque, A. Jeudy de Grissac, and J. Olivier), Vol. 1, pp. 311-18. GIS Posidonie, Marseilles, France.
Scipione, M. B. (1989). Comportamento trofico dei Crostacei Anfipodi in alcuni sistemi bentonici costieri. *Oebalia*, **15**, 249-60.
Scipione, M. B. and Fresi, E. (1984). Distribution of Amphipod Crustaceans in *Posidonia oceanica* (L.) Delile foliar stratum. In *I International workshop on* Posidonia oceanica *beds*, (ed. C. F. Boudouresque, A. Jeudy de Grissac, and J. Olivier), Vol. 1, pp. 319-29. GIS Posidonie, Marseilles, France.
Scipione, M. B, Fresi, E., and Wittmann, K. J. (1983). The vagile fauna of *Posidonia oceanica* (L.) Delile foliar stratum: a community approach. *Rapports et procès-verbaux des réunions de la Commission Internationale pour l'Exploration Scientifique de la Mer Méditerranée, Monaco*, **28** (3), 141-2.
Scipione, M. B. and Mazzella, L. Epiphytic diatoms in the diet of Crustacean Amphipods of *Posidonia oceanica* leaf stratum. *Oebalia*. (In press).
Steneck, R. S. and Watling, L. (1982). Feeding capabilities and limitation of herbivorous molluscs: a functional group approach. *Marine Biology*, **68**, 299-319.
Tenore, K. R. (1977). Growth of the polychaete, *Capitella capitata*, cultured on different levels of detritus derived from various sources. *Limnology and Oceanography*, **22**, 936-41.
Tenore, K. R. (1981). Organic nitrogen and caloric content of detritus. *Estuarine, Coastal and Shelf Science*, **12**, 39-47.
Traer, K. (1980). The consumption of *Posidonia oceanica* Delile by Echinoids at the Isle of Ischia. In *Echinoderms: present and past*, (ed. M. Jangoux), pp. 241-4. Balkema, Rotterdam.
Traer, K. (1984). Ernahrung und energetik regularer Seeigel in Bestanden des mediterranen Seegrases *Posidonia oceanica*. Unpublished Thesis. University of Vienna.
Vadon, C. (1981). Les Brachyoures des herbiers de Posidonies de la région de Villefranche-sur-mer: biologie, écologie et variations quantitatives des populations. Unpublished Thesis. University of Paris.

Van Montfrans, J., Wetzel, R. L., and Orth, R. J. (1984). Epiphyte-grazers relationships in seagrass meadows: consequences for seagrass growth and production. *Estuaries*, **7**, 289-309.

Velimirov, B. (1984). Grazing of *Sarpa salpa* L. on *Posidonia oceanica* and utilization of soluble compounds. In *I International workshop on* Posidonia oceanica *beds*, (ed. C. F. Boudouresque, A. Jeudy de Grissac, and J. Olivier), Vol. 1, pp. 381-7. GIS Posidonie, Marseilles, France.

Velimirov, B. (1986). DOC dynamics in a Mediterranean seagrass system. *Marine Ecology Progress Series*, **28**, 21-41.

Velimirov, B. (1987). Organic matter derived from a seagrass meadow: origin, properties, and quality of particles. *P.S.Z.N.I.: Marine Ecology*, **8**, 143-73.

Velimirov, B., Ott, J. A., and Novak, R. (1981). Microorganisms on macrophyte debris: biodegradation and its implication in the food web. *Kieler Meeresforschungen Sonderheft*, **5**, 333-44.

Verlaque, M. (1981). Preliminary data on some *Posidonia* feeders. *Rapports et procès-verbaux des réunions de la Commission Internationale pour l'Exploration Scientifique de la Mer Méditerranée, Monaco*, **27** (2), 201-2.

Verlaque, M. (1985). Note preliminaire sur le comportement alimentaire de *Sarpa salpa* (L.) (Sparidae) en Méditerranée. *Rapports et procès-verbaux des réunions de la Commission Internationale pour l'Exploration Scientifique de la Mer Méditerranée, Monaco*, **29** (5), 193-4.

Verlaque, M. and Nedelec, H. (1983). Note préliminaire sur les relations biotiques *Paracentrotus lividus* (LMK.) et herbier de Posidonies. *Rapports et procès-verbaux des réunions de la Commission Internationale pour l'Exploration Scientifique de la Mer Méditerranée, Monaco*, **28** (3), 157-8.

Willsie, A. (1983). Zonation de la macrofaune endogée de la matte d'herbier de *Posidonia oceanica* (L.) Delile. *Rapports et procès-verbaux des réunions de la Commission Internationale pour l'Exploration Scientifique de la Mer Méditerranée, Monaco*, **28** (3), 165-6.

Willsie, A. (1987). Structure et fonctionnement de la macrofaune associée à la matte morte et d'herbier vivant de *Posidonia oceanica* (L.) Delile: influence des facteurs abiotiques et biotiques. Unpublished Thesis. University of Aix- Marseilles II.

Wittmann, K. (1984). Temporal and morphological variations of growth in a natural stand of *Posidonia oceanica* (L.) Delile. *P.S.Z.N.I.: Marine Ecology*, **5**, 301-16.

Wittmann, K., Mazzella, L., and Fresi, E. (1981*a*). Age specific patterns of leaf growth: their determination and importance for epiphytic colonization in *Posidonia oceanica* (L.) Delile. *Rapports et procès-verbaux des réunions de la Commission Internationale pour l'Exploration Scientifique de la Mer Méditerranée, Monaco*, **27** (2), 189-91.

Wittmann, K., Scipione, MB., and Fresi, E. (1981*b*). Some laboratory experiments on the activity of the macrofauna in the fragmentation of detrital leaves of *Posidonia oceanica* (L.) Delile. *Rapports et procès-verbaux des réunions de la Commission Internationale pour l'Exploration Scientifique de la Mer Méditerranée, Monaco*, **27** (2), 205–6.

Zimmerman, R., Gibson, R., and Harrinton, J. (1979). Herbivory and detritivory among Gammaridean Amphipods from a Florida seagrass community. *Marine Biology*, **54**, 41–7.

Zupi, V. and Fresi, E. (1984). A study on the food web of the *Posidonia oceanica* ecosystem: analysis of the gut contents of Echinoderms. In *I International workshop on* Posidonia oceanica *beds*, (ed. C. F. Boudouresque, A. Jeudy de Grissac, and J. Olivier), Vol. 1, pp. 373–9. GIS Posidonie, Marseilles, France.

Zupi, V. and Fresi, E. (1985). A study on the food web of the *Posidonia oceanica* (L.) Delile ecosystems: analysis of the gut contents of Decapod Crustaceans. *Rapports et procès-verbaux des réunions de la Commission Internationale pour l'exploration Scientifique de la Mer Méditerranée, Monaco*, **29** (5), 189–92.

Zupo, V. (1990). The food web of *Posidonia oceanica* beds around the island of Ischia (Gulf of Naples — Italy): a new trophic index. *Rapports et procès-verbaux des réunions de la Commission Internationale pour l'exploration scientifique de la Mer Méditerranée, Monaco*, **32** (1), p. 16.

9. Interactions between macrofaunal epiphytes and their host algae

GRAY A. WILLIAMS* and R. SEED[†]

* Swire Marine Laboratory, University of Hong Kong, Cape d'Aguilar, Hong Kong
[†] School of Ocean Sciences, University of Wales Bangor, Menai Bridge, Gwynedd, Wales, UK

Abstract

Marine algae provide a suitable habitat for a wide range of mobile and sessile macrofauna. Many of these organisms are permanent members of the epiphyton, whereas others utilize the host in a more transient fashion. Different types of algae support different suites of epiphytes according to the resources which the former provide. These include food and shelter as well as surfaces for attachment. Although some epiphytes are beneficial, others may impose a high selective cost on their host algae. Consequently the dynamic interactions between algae and their epiphytes are often extremely complex and not easy to categorize. In general, algae attempt to maximize associations with beneficial epiphytes and minimize encounters with those which impose excessive cost. Such complex interactions and the often close interdependence of epiphytes and host suggest a considerable degree of coevolution, though at present there is little conclusive evidence for this. This intriguing yet surprisingly neglected field of coastal ecology is clearly one which merits further detailed study.

Introduction

Coastal macroalgae, in temperate waters, often host a wide range of animal species from virtually all the main phyla. Many species are permanent members of the epiphyton; others are more transient; some are sessile and others mobile. The sessile fauna occurs in solitary and

Plant-Animal Interactions in the Marine Benthos, (ed. D. M. John, S. J. Hawkins, and J. H. Price), Systematics Association Special Volume No. 46, pp. 189–211. Clarendon Press, Oxford, 1992. © The Systematics Association 1992.

colonial forms. Colonial forms are generally better spatial competitors (Seed and O'Connor 1981) and tend to predominate on many low-intertidal, subtidal or floating weeds (e.g., Ryland 1974; Boaden et al. 1975; Fletcher and Day 1983). Algae situated in the more desiccating conditions of the high intertidal, by contrast, are more typically dominated by mobile taxa that can migrate into the humid recesses of the weed beds during periods of aerial exposure (e.g., Moore 1977; Dunstone et al. 1979). Many sessile taxa occur on only a limited number of algal species and some are even restricted to particular regions within colonized plants.

Mobile taxa are often subdivided into macrofaunal, meiofaunal, and microfaunal components. They generally colonize a wider range of algae than the sessile taxa, although their relative abundance is often related to the size and morphology of the algae colonized (Gunnill 1982; Preston and Moore 1988). Edgar and Moore (1986) recognize four categories:

(1) primary phytal species;
(2) cryptofaunal species;
(3) sediment-associated species;
(4) algal borers.

Primary phytal species can occur on non-algal substrata but achieve their maximum abundance on macrophytic algae. Cryptofaunal species utilize algal holdfasts and turf-forming algae and generally exhibit little habitat specificity. Algal borers and sediment-associated species are generally less important although the latter are commonly encountered in kelp holdfasts or among rigid algae which trap sediments. This review is largely restricted to macrofaunal species in categories (1) and (2).

The range of macroalgal species colonized is equally diverse and includes crustose corallines, filamentous forms, foliose fucoids, and the large laminarians. These macroalgae provide an array of resources for exploitation by epiphytes. Hayward (1980) has characterized these as:

(1) surface area for attachment;
(2) shelter, either as a permanent or temporary habitat;
(3) sediment traps;
(4) food, directly in the case of algal grazers, indirectly in the case of microalgal browsers and detritivores (see also Beckley and McLachlan 1980).

Sessile faunas may also benefit nutritionally by consuming algal exudates (de Burgh and Fankboner 1979; Oswald and Seed 1986).

Different classes of algae do not, however, offer these resources to the same extent. Brown algae, by virtue of their overall dominance on many rocky shores, are generally favourable. In particular their large size, broad fronds, and often massive holdfasts provide suitable habitats for diverse assemblages of both sessile and mobile organisms (e.g., Boaden et al. 1975; Cancino and Santelices 1981; Moore 1985). Green algae typically accommodate fewer sessile species but may hold significant assemblages of mobile taxa (e.g., Preston and Moore 1988, 1989). Red algae usually support rather less diverse faunas (e.g., D'Antonio 1985; Cancino et al. 1987; Grahame and Hanna 1989) although some species may be more extensively used (e.g., Anadon 1988), especially where brown algae are scarce.

Given their widespread distribution, the faunas associated with macroalgae until recently have been surprisingly neglected. Early accounts were largely restricted to distributional studies and only within the past decade or so has the emphasis shifted to an understanding of community organization and dynamics (e.g., Seed and O'Connor 1981; Hicks 1985; Seed 1985; Cancino 1986; Edgar and Moore 1986, and references therein) as well as of the complex interactions that can exist between these faunas and their algal hosts (e.g., Oswald and Seed 1986). It is also becoming increasingly evident that, despite their generally low biomass, the secondary production of these faunas provides a major trophic pathway between the primary production of benthic macrophytes and higher-level consumers including fish (Edgar and Moore 1986).

This review examines the epiphytes associated with coastal macroalgae, focusing in particular on the dynamic interactions that exist between these faunas and the plants on which they live. The costs and benefits of these interactions to both host and epiphyte, and the possible evolutionary implications of these relations, will also be briefly considered. While we accept that 'epiphyte' is often restricted to epibiotic plants, we use this term to refer to any macrofaunal species that are found on algae (the host) either permanently or temporarily (but see Wahl 1989 for alternative terminology of marine epibioses).

Epiphyte–host interactions

1. Costs to the host algae

Many epiphyte–plant relations are deleterious to the success of the host. The precise adverse effect suffered is not always simply due to tissue removal, but can often be a sheer weighing down of the plant by the epiphytes, leading to reduced photosynthesis, growth, and hence decreased reproductive success.

Many mobile epiphytes are, however, herbivores that actually consume the host tissue. In many cases this is not deleterious to the host. *Littorina obtusata* is a macroalgal grazer that lives predominantly on *Ascophyllum nodosum*. On shores in Great Britain there is physical damage to the alga, but this does not appear to affect the success of individual plants or the overall *Ascophyllum* population. Populations of *L. obtusata* maintained on the more foliose low-shore alga *Fucus serratus*, by contrast, rapidly decimate the fronds reducing the plant to stipe and mid-rib (Williams 1987). Other epiphytes appear to be prudent grazers, and it is only when the epiphyte population rapidly expands that gross effects on the host occur. A 'huge' population explosion of *Lacuna vincta* on *Laminaria saccharina* in New Hampshire resulted in perforations (3–10 mm in diameter) of the algal blades. At some sites only residual holdfasts and stipes remained (Fralick *et al.* 1974). A similar population explosion of *L. vincta* on *Fucus edentatus* (up to 280 m^{-2}) resulted in the snail removing 79 per cent of net production during the summer months (Thomas and Page 1983; Fig. 9.1). Grazing by epiphytic limpets can also cause perforations in laminarian blades; *Patina pellucida* may weaken kelp holdfasts, thus resulting in increased algal detachment rates during storms (Kain and Svendsen 1969). Hay *et al.* (1987) suggest that grazing by amphipods and polychaetes causes damage to 1–20 per cent of the blade area of *Dictyota*.

Damage to the host is often initiated when a preferred food source (e.g., microalgae on the algal fronds) becomes scarce. This has been demonstrated for various amphipods (Hay *et al.* 1987) and isopods grazing on *Fucus vesiculosus* (Kangas *et al.* 1982; Shacklock and Doyle 1983; Salemma 1987). Indeed, grazing by *Idotea* is thought to be a major

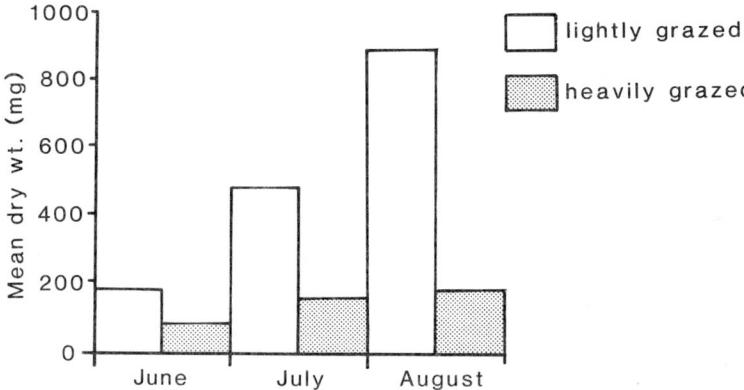

Fig. 9.1. Reduction in biomass of *Fucus edentatus* due to grazing by *Lacuna vincta* (after Thomas and Page 1983).

factor accounting for the decrease of *F. vesiculosus* in the Baltic Sea. A rise in salinity and nutrients in the late 1970s and early 1980s caused a bloom of microalgae on Baltic *F. vesiculosus*. This bloom in turn supported a population increase in *Idotea* species, as the juveniles feed on this resource. Adults, however, feed on the *Fucus* plant itself and the increased numbers exceeded the carrying capacity of the host, rapidly depleting the *Fucus* population in the Baltic (Kangas *et al.* 1982). It appears that there is a maximum density of herbivores that an alga can support without large-scale deleterious grazing effects.

Many herbivores crop only limited amounts of their host tissues. *Idotea* prefers old, eroding thalli of *Fucus vesiculosus* (Salemma 1987). *Hyale*, an amphipod which lives on the high-intertidal algae *Fucus spiralis* and *Pelvetia canaliculata*, grazes in such a way as to minimize physical damage to the host (Moore 1986). Many species eat only non-meristematic tissue or non-reproductive tissue, concentrating their grazing on vegetative tissues, such as the lateral frills of *Laminaria* in the case of *Lacuna* (Johnson and Mann 1986). Other species are restricted to specific parts of colonized plants. Many epiphytic limpet species (Table 9.1) have a precise location on the host plant which can be associated with reducing the effect of limpet-induced damage on the host's meristematic and reproductive tissue (Table 9.1 Choat and Black 1979; Muñoz and Santelices 1989).

Extensive colonization by encrusting invertebrates can also decrease plant success. Under particularly favourable conditions, epiphytic growth may cover over 75 per cent of the frond surface and effectively double the weight of the plant. Encrusting bryozoans can significantly reduce photosynthetic activity of *Fucus serratus* (Table 9.2). Photosynthesis in frond segments heavily colonized by *Flustrellidra* was less than 5 per cent of unencrusted controls. *Alcyonidium*, another fleshy bryozoan, also significantly depressed photosynthesis, while the effect of the thinner colonies of *Electra pilosa* was less pronounced. Oswald (1986) calculated that the potential reduction in net photosynthesis attributable to these bryozoans could be as much as 40 per cent, given the typical levels of encrustation that can often prevail on *F. serratus*. Fouling by *Membranipora* can impair the growth rate of *Macrocystis pyrifera* and the increased weight caused by particularly heavy encrustations may cause some plants to sink (Wing and Clendenning 1971; Woollacott and North 1971). Cancino *et al.* (1987), however, found that while *Membranipora* significantly reduced the photosynthetic rate of *Gelidium rex*, the growth rate was apparently unaffected.

Encrusting bryozoans may have other deleterious effects on their hosts. Calcareous species can reduce the flexibility of algal fronds, thus resulting in increased frond losses particularly in turbulent conditions

Table 9.1. A review of epiphytic limpet–algal relations (after Muñoz and Santelices 1989)

	Patina on *Laminaria*	*Notoacmaea* on *Egregia*	*Patella* on *Ecklonia*	*Scurria* on *Lessonia*
Position on host	Frond and holdfast (in cavities)	Stipe and rachis (in cavities)	Stipes	Below first dichotomy (in cavities)
Motility	High Migrate between stipe and holdfast	Low Move from old to new scars	High Along stipe	High Forage on high tide
Limpet-induced damage to meristematic or reproductive tissue	No/No	No/No	No/No	No/No
Limpet mortality due to algal loss	High (seasonal shedding)	High (water movement)	High (storm damage)	High (storm damage)
Fate of limpet after algal loss	Attached to holdfast	Death—desiccation on cast weed	Death—desiccation on cast weed	c. 20 per cent on stipe; death if remain attached

Table 9.2. The effect of encrusting bryozoans on photosynthetic activity in *Fucus serratus* (after Oswald et al. 1984)

	Photosynthetic activity (DPM mg^{-1})[a]				
Encrusting species	Mean	± SD	t^b	P	percent[c]
Unencrusted control	652.5	241.7	—	—	—
Electra pilosa	349.5	147.0	2.1	NS	53.4
Alcyonidium hirsutum	107.4	59.9	4.3	*	16.4
Flustrellidra hispida	24.9	12.1	5.2	*	3.8

[a] Disintegrations of ^{14}C min^{-1} mg dry wt^{-1}
[b] Student's *t*-test comparing the effect of each bryozoan with the unencrusted control
[c] Reduction in photosynthetic activity as a percent of value for control plants
* P < 0.05; NS = not significant.

(e.g., Dixon et al. 1981). Fouled blades are often damaged more readily by fish which forage on the attached invertebrates (Bernstein and Jung 1979). *Membranipora* can inhibit sporulation in *Laminaria hyperborea* (Kain 1975). The loading effects of sessile epiphytes and algae are sometimes sufficient to weigh down *Ascophyllum* plants in the water column and increase detachment rates during storms (Menge 1975). Several algal species have been shown to be smothered by microalgal growths (Sousa 1979; Steneck 1982, 1983) and fouling by mussels and more importantly epiphytic algae, may induce increased axis breakage and decreased growth rates and reproduction in *Rhodomela larix* (D'Antonio 1985).

Heavy loading of both sessile and mobile epiphytes can thus have negative effects on host algae. These can be lethal (e.g., intense herbivory, smothering, or detachment) or sublethal if growth rate and hence reproductive output are impaired. The ultimate effect on reproduction will have a selective influence, and the development of strategies by the algae to accommodate trade-offs between epiphyte loading and reproduction is an area worthy of further investigation.

2. Benefits to the host algae

The effects of epiphytes on the host plant need not always be detrimental, indeed some may even be beneficial. Mobile epiphytes achieve this either directly by pruning host tissue or indirectly by removing harmful epiphytes. Sessile species can benefit their hosts via nutrient-exchange mechanisms and by providing protection from grazing and desiccation.

Prudent herbivory can often be of positive benefit to the plant. Grazing of peripheral tissue decreases frond surface area and results in reduced drag in wave-exposed environments (Johnson and Mann

1986). Many species remove senescing tissue from their host algae (Johnson and Mann 1986; Salemma 1987) and the feeding action of many grazers facilitates the release and consequent dispersal of the host's gametes (Moore 1986; Buschmann and Santelices 1987). Some algal spores are apparently resistant to digestion by herbivores and are excreted in faecal pellets (e.g., amphipods, Santelices and Ugarte 1987; Santelices, Chapter 16 this volume).

Feeding by microphagous epiphytes can reduce the abundance of other surface epiphytes. Such grazers release the host algae from the full costs of supporting fouling epiphytes, thereby increasing photosynthesis rates and decreasing detachment rates. Limpets epiphytic on coralline algae are vital to survival of the alga. Removal of *Acmaea* resulted in the death of *Clathromorphum* due to epiphyte smothering (Steneck 1982, 1983). It has also been suggested that by releasing their hosts from the negative effects of harmful epiphytes, mesograzers can affect population densities of macroalgae, and ultimately, therefore, community structure (Brawley and Adey 1981a, b). This has a positive feedback to the grazers, since a denser algal stand probably provides the grazers with greater protection from predators. Crypsis from predators is a major benefit that many epiphytes obtain from algae (Hay *et al*, 1987; Hay and Fenical 1988). The application of amphipod 'cleaning' has been realized in aquaculture systems, where increased plant growth is attained by the removal of microalgal foulers (Shacklock and Doyle 1983; Brawley and Fei 1987). While the action of many microphagous epiphytes may increase the fitness of the host plant (Norton and Benson 1983), this has rarely been tested rigorously. D'Antonio (1985), however, has shown that by removing fouling epiphytes the grazing actions of amphipods and snails increase growth and reproduction in *Rhodomela*. Predation on sessile epiphytes by nudibranchs and pycnogonids may also benefit the host. Removal, either total or partial, of bryozoan zooids can make the epiphyton community patchy (Todd and Havenhand 1989) and affect the success of the host alga.

The effects of sessile epiphytes are also not always detrimental to the plant. Some algae readily absorb ammonia (e.g., Fujita and Goldman 1985; Lobban *et al*. 1985), and as this is the main nitrogenous excretory product of many marine invertebrates, it seems likely that epiphyte metabolites could enhance algal growth rates, especially when nitrogen is limiting. In laboratory experiments, encrusting bryozoans reduced the susceptibility of *Fucus serratus* to extremes of temperature at low tide and enabled faster recovery from desiccation and freezing of the algal thallus (Oswald 1986; Oswald and Seed 1986). Fig. 9.2 illustrates the photosynthetic ability of encrusted and non-encrusted *F. serratus* fronds following different periods of exposure to extreme low tempera-

Interactions between macrofaunal epiphytes and their host algae 197

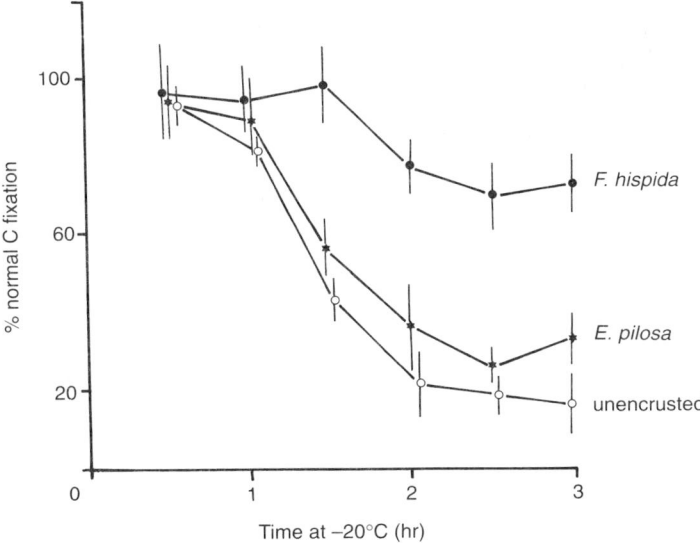

Fig. 9.2. Levels of carbon fixation in non-encrusted segments of *Fucus serratus* and in segments with epiphytic bryozoans (*Electra pilosa*, *Flustrellidra hispida*) after periods of exposure to −20°C. Fixation expressed as percentage of values for unfrozen controls (after Oswald 1986).

ture. The benefits of encrustation in protecting temperate-water algae from desiccation and temperature extremes may, therefore, be quite substantial and the potential damage to algal beds that might be caused by severe frosts or hot spells could well provide a strong selective force.

Epiphyte–host strategies — maximizing benefits and minimizing costs

If epiphytes can exert such severe costs and/or benefits on their host algae, it follows that plants should exhibit adaptations either to increase contact with beneficial species or decrease contact with high-cost species. This will lead to two main strategies, either plant-epiphyte mutualism or an 'arms race' (Thompson 1989). Similarly, epiphytes should exhibit behavioural responses to enable them to locate their host and, in the case of mobile species, to re-locate the host plant if displaced. Sessile species should select settlement sites which would maximize their subsequent success.

1. The host algae — The role of secondary chemicals and other antifouling strategies

Macroalgae, like other attached marine organisms, must either defend their exposed surfaces or tolerate varying levels of fouling and grazing damage (Hay and Fenical 1988; Wahl 1989). Antifouling and grazing defences, especially those involving the production of toxic chemicals are, however, energetically costly; tolerating fouling or having 'cleaning' epiphytes would, therefore, release energy resources that could then be allocated to growth and reproduction. Much attention has recently been paid to the role of secondary plant compounds in reducing the effects of grazing by herbivorous epiphytes. A confusing array of 'biologically' active compounds have been isolated (see review by Hay and Fenical 1988). Much of the early work was of a correlative nature and only recently has the role of secondary metabolites been adequately tested in field situations. Van Alstyne (1988) has demonstrated that grazing by the temporary epiphytes *Littorina sitkana* and *L. scutulata* inhibits growth of *Fucus distichus*. Grazing increased polyphenol exudation from the algae and this made the 'damaged' plants repellent to snails.

Laboratory evidence for secondary plant compounds as grazing deterrents is convincing in its quantity alone; areas of the plant avoided by grazers are those which are high in polyphenols (e.g. Johnson and Mann 1986; Norton and Benson 1983). This field has been well reviewed (Hay *et al.* 1987; Hay and Fenical 1988; Hay and Fenical, chapter 14 this volume). The species that are most susceptible to the chemical defences of plants tend to be very mobile, generalist, grazers and not the smaller, permanent, epiphytic grazers (Hay *et al.* 1988). For example, herbivorous fish were deterred by secondary compounds in *Dictyota*, whereas the small epiphytic amphipod *Ampithoe* was not (Hay *et al.* 1987).

Hay and his coworkers suggest that this epiphyte-host specialism accounts for the tolerance of these herbivores to plant defences which affect more generalist (non-epiphyte) grazers. This host-specific relationship is common in many sessile species and also in a number of mobile species. Given a choice of algae, the gastropods *Littorina obtusata* and *L. mariae* actively select their preferred host alga (Watson and Norton 1987). When displaced in the field, both winkle species show orientated movement towards the zones occupied by these algae (Williams 1987). The ecologies of these two species are intimately linked with their respective host algae. *Littorina mariae* browses epiphytes from *F. serratus* and by doing so may be of benefit to its host (Williams 1990). *Littorina obtusata* is a macroalgal grazer and is the only grazer

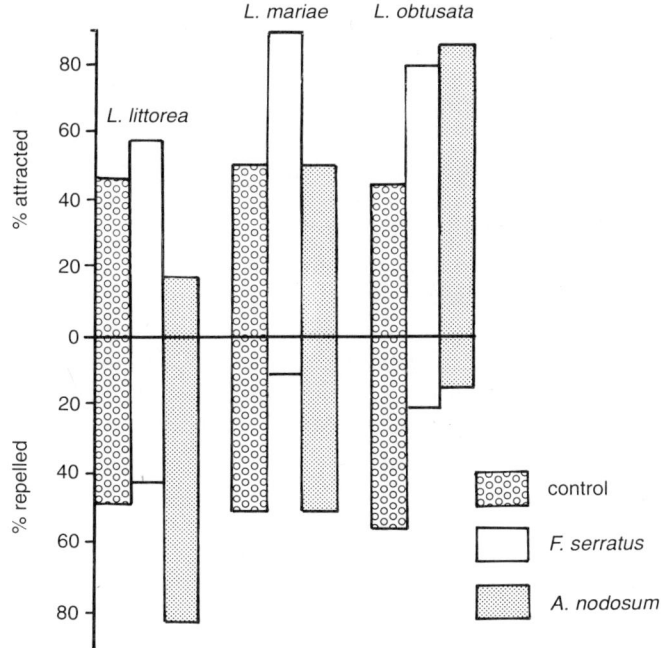

Fig. 9.3. Attraction or repulsion of littorinids to extracts of fucoid algae as compared with a control of filtered seawater (after Norton and Manley 1990).

effectively to utilize *Ascophyllum nodosum*. *Ascophyllum* has few epiphytes (Round 1984) and sheds its epidermis regularly (Filion-Myklebust and Norton 1981). *Ascophyllum* also produces noxious secondary chemicals which may repel other winkle species but attract *L. obtusata* (Fig. 9.3 Norton and Manley 1990; Norton *et al.* 1990). Grazing by *L. obtusata* does not appear to be deleterious to the alga which forms large stands on sheltered shores. *L. obtusata* may, however, graze microalgae and the removal of epiphytes by *L. obtusata* decreases the detachment rates of *Ascophyllum* during storms on shores in the USA (Menge 1975).

Algal exudates are also often used as cues by sessile species in order to ensure that their larvae locate an appropriate host. Species of macroalgae vary in their attractiveness to prospecting larvae. Some larvae are relatively eurytopic, others highly specific in their settlement preferences, although the precise choice can vary from one locality to another. Inter-population differences, for example, have been reported for *Spirorbis spirorbis (borealis)* in which larvae exhibit a marked preference for the algal species supporting their parents (MacKay and Doyle 1978). Laboratory experiments have demonstrated that certain intertidal bryozoans (e.g., *Flustrellidra hispida*, *Alcyonidium hirsutum*) settle preferentially on precisely those algae whereon they normally occur on the shore (e.g., Oswald and Seed 1986). Such habitat selection seems to be related to the chemical nature of the settlement surface, since inert panels can be made more attractive when covered with a film of specific algal extracts.

Larvae are selective not only with respect to different algal species but also discriminate between different regions within individual plants. The fronds of *F. serratus* have recognizable concave and convex surfaces and most sessile species are both larger and more abundant on the concave surface (Boaden *et al.* 1976; Wood and Seed 1980). For some species this reflects larval preferences at settlement (Oswald and Seed 1986), but established colonies may also obtain some protection or benefit from local food-bearing eddies on this surface. Many epiphyte species settle on to and may even orientate their subsequent growth towards the younger, less heavily-fouled regions of the host plant. This is clearly adaptive for sessile species, since it utilizes areas of the plant which are least likely to bear potential competitors; on rapidly growing kelp fronds with an age-dependent mortality, it also maximizes the survival time of the colonizing organisms (Cancino 1986). The mechanisms by which larvae identify the age gradients involved is uncertain, although bacterial films (e.g., Mihn *et al.* 1981; Brancato and Woollacott 1982), physiological gradients (e.g., Dieckman 1980; Oswald *et al.* 1984), or even gradients in the production of antimicrobial compounds (e.g., Al-Ogily and Knight-Jones 1977), could all provide the necessary cues.

Algal polyphenols and tannins are known to have a deleterious effect on settlement of bacteria (e.g., Hornsey and Hide 1976; Al-Ogily 1985) and may also be toxic, even at very low concentrations, to certain invertebrates (e.g., Geiselman and McConnell 1981). By inhibiting microbial films, these antibacterial substances influence larval settlement patterns and may also serve to suppress the growth of established colonies. Antibiotic production is not uniform along the fronds and also varies seasonally. Production can also vary with different stages of plant growth; thus the vulnerable upper part of the 'button' stage of *Himanthalia* has antibacterial chemicals which prevent epiphyte settlement (Al-Ogily and Knight-Jones 1977; Kitching 1987).

Frond shedding, especially as plant branches die back after fruiting (e.g., Knight and Parke 1950; Wood 1983), and epidermis sloughing (e.g., Filion-Mykelbust and Norton 1981; Moss 1982; Russell and Veltkamp 1982), may also be viewed as an antifouling strategy. Similarly, accelerated plant growth which produces new tissue at a faster rate than the local rate of fouling could provide a further channel of escape from excessive epiphyte loads. Plants could thus maintain a relatively clean, photosynthetically-active zone in the proximity of the meristematic tissue, while epiphyte cover increases elsewhere along the age gradient of the frond.

2. The epiphytes — the accommodation of host variability

The abundance and diversity of epiphytes is primarily a reflection of the complexity of the host algae as a habitat. Each algal species will provide a different number of micro-environments for epiphytes to exploit (Beckley and McLachlan 1980; Gunnill 1982; Salemma 1987; Preston and Moore 1988). Within these environments the epiphytes may gain protection from physical (especially desiccation, wave effects) and biological (predation, interspecific competition) extremes and also gain food resources (from the host itself, other epiphytes, detritus, plankton). Sediment accumulation is especially important for the meiofaunal component of the epiphyton (Hicks 1985). The host obviously benefits the mobile epiphytes by providing a substratum on which they can live, feed, and lay egg masses. Settlement on macroalgal fronds effectively elevates sessile organisms above the rock surface and thus results in a generally more favourable hydrodynamic regime. Improved water flow over their surfaces ensures a better supply of oxygen and food as well as more efficient removal of wastes.

Many marine macroalgae exude substantial amounts of dissolved organic carbon (DOC). Fouling can significantly increase the rate of exudation and there is now compelling evidence that such exudates provide an important nutritive source for many encrusting species. It is estimated that *Flustrellidra* takes up the equivalent of approximately 20 per cent of the total carbon fixed by *Fucus serratus* and can potentially gain some 17 per cent of its metabolic requirements from this source (Oswald 1986), a factor which presumably contributes to the competitive superiority of this species within the *F. serratus* epiphyton. Moreover, the uptake of DOC by *Flustrellidra* continues even when *Fucus* plants are emersed for periods of up to 12 hours (Fig. 9.4). DOC exuded by the bull kelp *Nereocystis luetkeana* is similarly utilized by *Membranipora* (de Burgh and Fankboner 1979).

Perhaps the major disadvantage of living on macroalgae is the relatively unstable and ephemeral nature of the substratum. Frond tissue, especially in rapidly growing kelps, may be short-lived, while frond-shedding in fucoids and other long-lived algae may be unpredictable. Cancino (1986) found that the main cause of mortality of the bryozoan *Celleporella hyalina* on the kelp *Laminaria saccharina* was

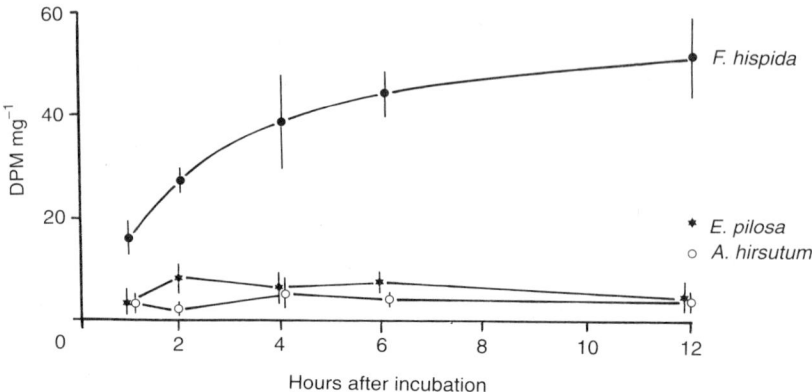

Fig. 9.4. Uptake of DOC from emersed *Fucus serratus* by encrusting bryozoans (*Alcyonidium hirsutum*, *Electra pilosa*, *Flustrellidra hispida*). Uptake expressed as radioisotope (disintegrations ^{14}C min^{-1} mg dry wt^{-1}) remaining within the colony (after Oswald 1986).

disintegration of the kelp surface. On *Fucus serratus*, 17 per cent of the total mortality of *Alcyonidium hirsutum* was attributable to frond loss (Wood 1983), while the number of spirorbids dying from this cause increased from the older basal regions to the younger distal regions, where it accounted for over 80 per cent of the total mortality (Seed and O'Connor 1981).

Mobile species often avoid this limitation of algae as a substratum by utilizing the resource in a transient fashion. Algae fulfil an important role as a nursery ground for many mobile species (Beckley and McLachlan 1980; Preston and Moore 1988). For some, this simply represents a change in host algae between adult and juvenile stages (e.g., *Idotea* migrates from *Cladophora* to *Fucus*; Salemma 1987). In others, it represents a whole new substrate type (e.g., *Dynamene*, Holdich 1968; *Nucella*, Crothers 1985). Algae can also act as primary settlement sites; *Mytilus edulis* (size range 1–1·5 mm) is a temporary epiphyte using algae as a primary attachment site before migrating to the final attachment site on adult mussel beds (Bayne 1964; Seed 1969). While they remain on the alga, some of these temporary epiphytes can have an important effect, *Nucella*, for example, eats spirorbids.

Given this temporal variation in algal usage, it is clear that many epiphytic faunas are highly dynamic and exhibit marked seasonal changes. As many species of epiphyte are obligate, the temporal patterns of the epiphyte and host species are often closely linked. While sessile species can only vary their position on the plant by growth orientation and settlement patterns, mobile species do not suffer from such restrictions. Consequently, mobile species can exploit different parts of the alga, often moving from the frond to the stipe during periods of frond shedding (e.g., *Patina*, although many epiphytic limpets suffer heavy mortality during frond shedding, Table 9.1). Some mobile species have life cycles which are closely synchronized with temporal variations in the life history of the host alga. The annual life patterns of *Lacuna pallidula* and *Littorina mariae*, for example, are both closely synchronized with the growth and subsequent die-back of *F. serratus* (Smith 1973; Grahame 1985; Williams 1987). Epiphytic limpets also show specific life-history adjustments involving reproductive and behavioural modifications in order best to exploit their algal hosts (see Muñoz and Santelices 1988 and Table 9.1).

Many species, especially amphipods, gastropods and nudibranchs, lay benthic egg masses on the young growing fronds of algae, thus ensuring a reliable resource for the next generation. Other mobile species (and most sessile species) have a planktonic larval stage. These dispersal stages must be able to locate the host algae, often using secondary chemicals as cues. Species which lay benthic egg capsules on algae are often dispersed via 'rafting' when pieces of detached alga float to another shore. This may be a better method of long-distance dispersal as compared with an unpredictable planktonic phase (Johannesson 1988). The dispersal capacity of sessile species, however, is especially important for competitively inferior fugitive species dependent on reaching isolated and/or temporary habitat patches. Some species such as the bryozoans *Electra pilosa* and *Membranipora membranacea* produce relatively long-lived (several weeks) planktotrophic larvae, but the general pattern for most sessile epiphytes is of relatively short-lived (several hours) lecithotrophic larvae (Seed 1985).

Seasonal patterns of larval settlement and plant growth effectively enable the two competitively dominant bryozoans, *Alcyonidium* and *Flustrellidra*, within the *Fucus serratus* community to coexist (Seed and O'Connor 1981). Larval selectivity and resource regeneration through plant growth have been identified as the two most important factors influencing community organization in sessile algal epiphyton (Seed and O'Connor 1981). Mobile species are not limited by such constraints, although some species might be limited to the canopy of a single plant for their entire life cycle. The importance of new plant growth for fugitive sessile species is emphasized by the predominance of such species on the ephemeral blades of many kelps. Whereas fucoids are comparatively long-lived, some kelps often regenerate their blades several times each year and survival times of frond tissue can be as short as two months (Cancino 1986). Consequently these algae are effectively colonized only by rapidly growing species with early and extended reproduction. Longer-lived, seasonally reproducing, species do occur on kelps but usually only on the perennial holdfasts and stipes.

Conclusions

The relationship between epiphytes and their host algae is extremely complex and dynamic. Despite the diversity of the epiphyton and host algae many of the processes affecting these species-interactions are similar. In general, the host alga would be expected to attempt to maximize the association with beneficial epiphytes and minimize encounters with epiphytes that incur a cost. The strategies of the epiphytes and their host are therefore different. The epiphyte needs to locate and remain on the host and generally benefits from the association. The host, however, has three strategies: it can tolerate its epiphytic load, defend itself again epiphytic fouling, or positively encourage beneficial epiphytes. Such close interdependence and interactions suggest a degree of coevolution between epiphytes and their host algae (specific coevolution *sensu* Thompson 1989). Although there is little direct evidence for this (but see Steneck 1982, 1983), the complex relationships between these groups of species do support such conclusions (Bernstein and Jung 1979; Hay *et al.* 1987; Hay and Fenical 1988; Muñoz and Santelices 1989).

In some partnerships there is no conflict of interest, and a mutualistic relationship exists between the epiphytes and the host (Brawley and Adey 1981a; Steneck 1982; D'Antonio 1985; see Brönmark 1989 for a review of interactions in freshwater systems). When there is a conflict of interest, however, plants opt to defend themselves while the epiphytes try to overcome these defences; an 'arms race' of escalating selection for defences and strategies that overcome these defences thus develops (see Steneck 1983). In both these scenarios the epiphyte will try to accommodate seasonal variation in the host algae, either through growth and movement patterns or via synchronization of life histories.

The life histories of epiphytes and their hosts are therefore very intimate (Bernstein and Jung 1979). As algal fronds are often ephemeral, natural selection ought to operate such that species will tend to avoid settling on those algae, or on areas within individual plants, which have a life expectancy that is shorter than the life cycle of the colonizing species (Cancino 1986). Moreover, ephemeral algae and those parts of plants with relatively short life expectancies (e.g., kelp fronds) will be favourable only to fugitive species with short life cycles and will, therefore, tend to support rather less diverse faunas than longer-lived algae (e.g., fucoids) or perennial structures such as kelp stipes and holdfasts (Seed and O'Connor 1981).

In cases where the host appears to tolerate the presence of epiphytes, there are often examples where the relationship between host benefit and cost is unbalanced and the epiphyte has a negative impact on the host. Such events are unpredictable, and therefore, although having a stochastic short-term effect, will have little selective impact on the host algae. Over evolutionary time the host often appears to tolerate epiphytes rather than to adopt costly defences. Defences are most often effective against generalist epiphytes and specialized species have either overcome these defences, are participating in an 'arms race' or are of mutualistic benefit to the host. The host algae are therefore defended against a wide range of epiphytes but are able to be utilized by species which are host specific. These species may provide valuable insights into the dynamics and exact costs and benefits of host–epiphyte relations.

Many algal–epiphyte interactions, however, cannot easily be classified as either beneficial or detrimental for either the colonizing organism or the host plant, since there are inevitable costs and benefits for each partner. It seems likely that, with time, these plant-animal associations have become increasingly complex and that varying degrees of interdependence have developed. Future studies on algal epiphytes should, therefore, look for and not neglect the importance of such intimate interrelationships. Moreover, the dynamics of algal populations themselves must be significantly altered by the very heavy levels of epiphytic growth frequently encountered on many macroalgae from temperate waters and this may prove to be an important factor in community-level dynamics.

Acknowledgements

One of us (GAW) acknowledges a conference grant from Hong Kong University to aid attendance at the Symposium.

References

Al-Ogily, S. M. (1985). Further experiments on larval behaviour of the tubiculous polychaete *Spirorbis inornatus* L'Hardy & Quiévreux. *Journal of Experimental Marine Biology and Ecology*, **86**, 285–95.

Al-Ogily, S. M. and Knight-Jones, E. W. (1977). Antifouling role of antibiotics by marine algae and bryozoans. *Nature*, **265**, 728–9.

Anadon, N. (1988). Ciclo annual de la epifauna sesil de *Gelidium* spp en la zona de Cabo Peñas (Asturias, Norte de España). *Revista de Biologia de la Universidad de Oviedo*, **6**, 67–82.

Bayne, B. L. (1964). Primary and secondary settlement in *Mytilus edulis* (Mollusca). *Journal of Animal Ecology*, **33**, 513–23.

Beckley, L. E. and McLachlan, A. (1980). Studies on the littoral seaweed epifauna of St Croix Island. 2. Composition and summer standing stock. *South African Journal of Zoology*, **15**, 170–6.

Bernstein, B. B. and Jung, N. (1979). Selective pressures and coevolution in a kelp canopy community in southern California. *Ecological Monographs*, **49**, 335–55.

Boaden, P. J. S., O'Connor, R. J., and Seed, R. (1975). The composition and zonation of a *Fucus serratus* community in Strangford Lough, Co. Down. *Journal of Experimental Marine Biology and Ecology*, **17**, 111–36.

Boaden, P. J. S., O'Connor, R. J., and Seed, R. (1976). The fauna of a *Fucus serratus* L. community: ecological isolation in sponges and tunicates. *Journal of Experimental Marine Biology and Ecology*, **21**, 249–67.

Brancato, M. S. and Woollacott, R. M. (1982). Effect of microbial films on settlement of bryozoan larvae (*Bugula simplex*, *B. stolonifera* and *B. turrita*). *Marine Biology*, **71**, 51–6.

Brawley, S. H. and Adey, W. H. (1981*a*). The effect of micrograzers on algal community structure in a coral reef microcosm. *Marine Biology*, **61**, 167–77.

Brawley, S. H. and Adey, W. H. (1981*b*). Micrograzers may affect macroalgal density. *Nature*, **292**, p. 177.

Brawley, S. H. and Fei, X. G. (1987). Studies of mesoherbivory in aquaria and in an unbarricaded mariculture farm on the Chinese coast. *Journal of Phycology*, **23**, 614–23.

Brönmark, C. (1989). Interactions between epiphytes, macrophytes and freshwater snails: a review. *Journal of Molluscan Studies*, **55**, 299–311.

Burgh, M. E. de and Fankboner, P. V. (1979). A nutritional association between the bullkelp *Nereocystis luetkeana* and its epizooic bryozoan *Membranipora membranacea*. *Oikos*, **31**, 69–72.

Buschmann, A. and Santelices, B. (1987). Micrograzers and spore release in *Iridaea laminarioides* Bory (Rhodophyta: Gigartinales). *Journal of Experimental Marine Biology and Ecology*, **108**, 171–9.

Cancino, J. M. (1986). Marine macroalgae as a substratum for sessile invertebrates: a study of *Cellepora hyalina* (Bryozoa) on fronds of *Laminaria saccharina* (Phaeophyta). In *Simposio Internacional. Usos y funciones ecológicas de las algas marinas bentónicas*, (ed. B. Santelices), Monografias Biologicas, Vol. 4, pp. 279–308.

Cancino, J. M. and Santelices, B. (1981). The ecological importance of kelp-like holdfasts as a habitat of invertebrates in Central Chile. II. Factors affecting community organization. In *International seaweed symposium*, (ed. T. Levring), **10**, pp. 241–6. Walter de Gruyter, New York.

Cancino, J. M., Muñoz, J., Muñoz, M., and Orellana, M. C. (1987). Effects of the bryozoan *Membranipora tuberculata* (Bosc) on the photosynthesis and growth of *Gelidium rex* Santelices et Abbott. *Journal of Experimental Marine Biology and Ecology*, **113**, 105-12.

Choat, J. H. and Black, R. (1979). Life histories of limpets and the limpet-laminarian relationship. *Journal of Experimental Marine Biology and Ecology*, **41**, 25-50.

Crothers, J. H, (1985). Dog-whelks: an introduction to the biology of *Nucella lapillus* (L.). *Field Studies*, **6**, 291-360.

D'Antonio, C. (1985). Epiphytes on the rocky intertidal red alga *Rhodomela larix* (Turner): negative effects on the host and food for herbivores? *Journal of Experimental Marine Biology and Ecology*, **86**, 197-218.

Dieckman, G. S. (1980). Aspects of the ecology of *Laminaria pallida* (Grev) J. Ag. of the Cape Peninsula, South Africa. I. Seasonal growth. *Marine Botany*, **23**, 579-85.

Dixon, J., Schroeter, S. C., and Kastendiek, J. (1981). Effects of the encrusting bryozoan *Membranipora membranacea* on the loss of blades and fronds by the giant kelp *Macrocystis pyrifera* (Laminariales). *Journal of Phycology*, **17**, 341-5.

Dunstone, M. A., O'Connor, R. J., and Seed, R. (1979). The epifaunal communities of *Pelvetia canaliculata* and *Fucus spiralis*. *Holarctic Ecology*, **2**, 6-11.

Edgar, G. J. and Moore, P. G. (1986). Macroalgae as habitats for motile macrofauna. In *Simposio Internacional. Usos y funciones ecológicas de las algas marinas bentónicas*, (ed. B. Santelices), *Monografias Biologicas*, Vol. 4, pp. 255-77.

Filion-Mykelbust, C. and Norton, T. A. (1981). Epidermis shedding in the brown seaweed *Ascophyllum nodosum* (L.) Le Jolis and its ecological significance. *Marine Biology Letters*, **2**, 45-51.

Fletcher, W. J. and Day, R. W. (1983). The distribution of epifauna on *Ecklonia radiata* (C. Agardh) J. Agardh and the effect of disturbance. *Journal of Experimental Marine Biology and Ecology*, **71**, 205-20.

Fralick, R. A., Turgeon, K. W., and Mathieson, A. C. (1974). Destruction of kelp populations by *Lacuna vincta* (Montagu). *Nautilus*, **88**, 112-14.

Fujita, R. M. and Goldman, J. C. (1985). Nutrient flux and growth of the red alga *Gracilaria tikvahiae* McLachlan (Rhodophyta). *Botanica Marina*, **28**, 265-8.

Geiselman, J. A. and McConnell, O. J. (1981). Polyphenols in brown algae *Fucus vesiculosus* and *Ascophyllum nodosum*: chemical defences against the marine herbivorous snail, *Littorina littorea*. *Journal of Chemical Ecology*, **7**, 1115-33.

Grahame, J. (1985). The population biology of two species of *Lacuna* (Chinkshells) at Robin Hood's Bay. In *The ecology of rocky coasts: essays presented to J. R. Lewis, D.Sc.*, (ed. P. G. Moore and R. Seed), pp. 136-42. Hodder and Stoughton, London.

Grahame, J. and Hanna, F. S. (1989). Factors affecting the distribution of the epiphytic fauna of *Corallina officinalis* (L.) on an exposed rocky shore. *Ophelia*, **30**, 113–29.

Gunnill, F. C. (1982). Effects of plant size and distribution on the numbers of invertebrate species and individuals inhabiting the brown alga *Pelvetia fastigiata*. *Marine Biology*, **69**, 263–80.

Hay, M. E. and Fenical, W. (1988). Marine plant–herbivore interactions: the ecology of chemical defense. *Annual Review of Ecology and Systematics*, **19**, 111–45.

Hay, M. E., Duffy, E. J., Pfister, C. A., and Fenical, W. (1987). Chemical defense against different marine herbivores: are amphipods insect equivalents? *Ecology*, **68**, 1567–80.

Hay, M. E., Renaud, P. E., and Fenical, W. (1988). Large mobile versus small sedentary herbivores and their resistance to seaweed chemical defenses. *Oecologia* (Berlin), **75**, 246–52.

Hayward, P. J. (1980). Invertebrate epiphytes of coastal marine algae. In *The shore environment*, Vol. 2: *Ecosystems*, Systematics Association Special Volume No. 17, (ed. J. H. Price, D. E. G. Irvine, and W. F. Farnham), pp. 761–87. Academic Press, London and New York.

Hicks, G. R. F. (1985). Meiofauna associated with rocky shore algae. In *The ecology of rocky coasts: essays presented to J. R. Lewis, D.Sc.*, (ed. P. G. Moore and R. Seed), pp. 36–56. Hodder and Stoughton, London.

Holdich, D. M. (1968). Reproduction, growth and biomass of *Dynamene bidentata* (Crustacea: Isopoda). *Proceedings of the Zoological Society of London*, **156**, 137–63.

Hornsey, I. S. and Hide, D. (1974). The production of antimicrobial compounds by British marine algae. I. Antibiotic-producing marine algae. *British Phycological Journal*, **9**, 353–61.

Johannesson, K. (1988). The paradox of Rockall: why is a brooding gastropod (*Littorina saxatilis*) more widespread than one having a planktonic larval dispersal stage (*L. littorea*)? *Marine Biology*, **99**, 507–13.

Johnson, C. R. and Mann, K. H. (1986). The crustose coralline alga *Phymatolithon* Foslie inhibits the overgrowth of seaweeds without relying on herbivores. *Journal of Experimental Marine Biology and Ecology*, **96**, 127–46.

Kain, J. M. (1975). The biology of *Laminaria hyperborea*. VII. Reproduction of the sporophyte. *Journal of the Marine Biological Association of the United Kingdom*, **55**, 567–82.

Kain, J. M. and Svendsen, P. (1969). A note on the behaviour of *Patina pellucida* in Britain and Norway. *Sarsia*, **38**, 25–30.

Kangas, P., Autio, H., Hallfors, G., Luther, H., Niemi, A., and Salemma, H. (1982). A general model for the decline of *Fucus vesiculosus* at Tvärminne, south coast of Finland in 1977–81. *Acta Botanica Fennica*, **118**, 1–27.

Kitching, J. A. (1987). The flora and fauna associated with *Himanthalia elongata* (L.) S. F. Gray in relation to water current and wave action in the Lough Hyne Marine Nature Reserve. *Estuarine, Coastal and Shelf Science*, **25**, 663-76.

Knight, M. and Parke, M. (1950). A biological study of *Fucus vesiculosus* (L.) and *F. serratus* (L.). *Journal of the Marine Biological Association of the United Kingdom*, **29**, 439-514.

Lobban, C. S., Harrison, P. J., and Duncan, M. J. (1985). *The physiological ecology of seaweeds*. Cambridge University Press, London, 242 pp.

MacKay, T. F. C. and Doyle, R. W. (1978). An ecological genetic analysis of the settlement behaviour of a marine polychaete. I. Probability of settlement and gregarious behaviour. *Heredity*, **40**, 1-12.

Menge, J. L. (1975). Effect of herbivores on community structure of the New England rocky intertidal region: distribution, abundance and diversity of algae. Unpublished Ph.D. Thesis, University of Harvard. 164 pp.

Mihn, J. W., Banta, W. C., and Loeb, G. I. (1981). Effects of absorbed organic and primary fouling films on bryozoan settlement. *Journal of Experimental Marine Biology and Ecology*, **54**, 167-79.

Moore, P. G. (1977). Organisation in simple communities: observations on the natural history of *Hyale nilssoni* (Amphipoda) in high littoral seaweeds. In *Biology of benthic organisms*, (ed. B. F. Keegan, P. O. Ceidigh, and P. J. S Boaden), pp. 443-51. Pergamon Press, Oxford.

Moore, P. G. (1985). Levels of heterogeneity and the amphipod fauna of kelp holdfasts. In *The ecology of rocky coasts: essays presented to J. R. Lewis, D. Sc.*, (ed. P. G. Moore and R. Seed), pp. 274-89. Hodder and Stoughton, London.

Moore, P. G. (1986). Seaweed-associated animal communities in the Firth of Clyde, with special reference to the population biology of the amphipod *Hyale nilssoni* (Rathke). *Proceedings of the Royal Society of Edinburgh*, **90B**, 271-86.

Moss, B. (1982). The control of epiphytes by *Halidrys siliquosa* (L.) Lyngb. (Phaeophyta, Cystoseiraceae). *Phycologia*, **21**, 185-191.

Muñoz, M. and Santelices, B. (1989). Determination of the distribution and abundance of the limpet *Scurria scurra* on the stipes of the kelp *Lessonia nigrescens* in Central Chile. *Marine Ecology Progress Series*, **54**, 277-85.

Norton, T. A. and Benson, M. R. (1983). Ecological interactions between the brown seaweed *Sargassum muticum* and its associated fauna. *Marine Biology*, **75**, 169-77.

Norton, T. A. and Manley, N. L. (1990). The characteristics of algae in relation to their vulnerability to grazing snails. In *Behavioural mechanisms of food selection*, NATO ASI Series. Vol. G20, (ed. R. N. Hughes), pp. 461-78. Springer-Verlag, Berlin and Heidelberg.

Norton, T.A., Hawkins, S.J., Manley, N.L., Williams, G.A., and Watson, D.C. (1990). Scraping a living: a review of littorinid grazing. In *Progress in littorinid and muricid biology*, (ed. K. Johannesson, D. Raffaelli, and C.J. Hannaford-Ellis), *Hydrobiologia*, Vol. 193, pp. 117-38. Kluwer Academic Publishers, Belgium.

Oswald, R.C. (1986). The epifaunal community of *Fucus serratus* (L.): ecology and physiology of association. Unpublished Ph.D. Thesis, University of Wales. 282 pp.

Oswald, R.C. and Seed, R. (1986). Organisation and seasonal progression within the epifaunal communities of coastal macroalgae. *Cahiers de Biologie Marine*, **27**, 29-40.

Oswald, R.C., Telford, N., Seed, R. and Happey-Wood, C.M. (1984). The effect of encrusting bryozoans on the photosynthetic activity of *Fucus serratus* L. *Estuarine, Coastal and Shelf Science*, **19**, 697-702.

Preston, A. and Moore, P.G. (1988). The flora and fauna associated with *Cladophora albida* (Huds.) Kütz. from rockpools on Great Cumbrae Island, Scotland. *Ophelia*, **29**, 169-86.

Preston, A. and Moore, P.G. (1989). Seasonal cycles of abundance of the flora and fauna associated with *Cladophora albida* (Huds.) Kütz. in rockpools. *Journal of Natural History*, **23**, 983-1002.

Round, F.E. (1984). *The ecology of the algae*. Cambridge University Press. 653 pp.

Russell, G. and Veltkamp, C.J. (1984). Epiphyte survival on skin-shedding macrophytes. *Marine Ecology Progress Series*, **18**, 149-53.

Ryland, J.S. (1974). Observations on some epibionts of gulf-weed *Sargassum natans* (L.) Meyer. *Journal of Experimental Marine Biology and Ecology*, **14**, 17-25.

Salemma, H. (1987). Herbivory and microhabitat preferences of *Idotea* spp. (Isopoda) in the Northern Baltic Sea. *Ophelia*, **27**, 1-15.

Santelices, B. and Ugarte, R. (1987). Algal life-history strategies and resistance to digestion. *Marine Ecology Progress Series*, **35**, 267-75.

Seed, R. (1969). The ecology of *Mytilus edulis* L. (Lamellibranchiata) on exposed rocky shores. I. Breeding and settlement. *Oecologia*, **3**, 277-316.

Seed, R. (1985). Ecological patterns in the epifaunal communities of coastal macroalgae. In *The ecology of rocky coasts: essays presented to J.R. Lewis, D.Sc.*, (ed. P.G. Moore and R. Seed), pp. 22-35. Hodder and Stoughton, London.

Seed, R and O'Connor, R.J. (1981). Community organization in marine algal epifaunas. *Annual Review of Ecology and Systematics*, **12**, 49-74.

Shacklock, P.F. and Doyle, R.W. (1983). Control of epiphytes in seaweed culture using grazers. *Aquaculture*, **31**, 141-51.

Smith, D.A. (1973). The population biology of *Lacuna pallidula* (da Costa) and *Lacuna vincta* (Montagu) in North East England. *Journal of the Marine*

Biological Association of the United Kingdom, **53**, 493–520.

Sousa, W. P. (1979). Experimental investigations of disturbance and ecological succession in a rocky intertidal algal community. *Ecological Monographs*, **49**, 227–54.

Steneck, R. S. (1982). A limpet–coralline alga association: adaptations and defences between a selective herbivore and its prey. *Ecology*, **63**, 507–22.

Steneck, R. S. (1983). Escalating herbivory and resulting adaptive trends in calcareous crusts. *Paleobiology*, **9**, 44–61.

Thomas, M. L. H. and Page, F. H. (1983). Grazing by the gastropod, *Lacuna vincta*, in the lower intertidal area at Musquash Head, New Brunswick, Canada. *Journal of the Marine Biological Association of the United Kingdom*, **63**, 725–36.

Thompson, J. N. (1989). Concepts of coevolution. *Trends in Ecology and Evolution*, **4**, 179–83.

Todd, C. D. and Havenhand, J. N. (1989). Nudibranch–bryozoan associations: the quantification of ingestion and some observations on partial predation among Doridoidea. *Journal of Molluscan Studies*, **55** 245–59.

Van Alstyne, K. L. (1988). Herbivore grazing increases polyphenolic defenses in the intertidal brown alga *Fucus distichus*. *Ecology*, **69**, 655–63.

Wahl, M. (1989). Marine epibiosis. I. Fouling and antifouling: some basic aspects. *Marine Ecology Progress Series*, **58**, 175–89.

Watson, D. C. and Norton, T. A. (1987). The habitat and feeding preferences of *Littorina obtusata* (L.) and *L. mariae* Sacchi et Rastelli. *Journal of Experimental Marine Biology and Ecology*, **112**, 61–72.

Williams, G. A. (1987). Niche partitioning in *Littorina obtusata* and *L. mariae*. Unpublishd Ph.D. Thesis, University of Bristol. 265 pp.

Williams, G. A. (1990). *Littorina mariae*: a factor structuring low shore communities? In *Progress in littorinid and muricid biology*, (ed. K. Johannesson, D. Raffaelli, and C. J. Hannaford-Ellis), *Hydrobiologia*, Vol. 193, pp. 139–46. Kluwer Academic Publishers, Belgium.

Wing, B. L. and Clendenning, K. A. (1971). Kelp surfaces and associated invertebrates. *Beihefte zür Nova Hedwigia*, **32**, 319–41.

Wood, V. (1983). A population study of the major sedentary faunal associates of *Fucus serratus* (L.). Unpublished Ph.D. Thesis, University of Wales. 237 pp.

Wood, V. and Seed, R. (1980). The effects of shore level on the epifaunal communities associated with *Fucus serratus* (L.) in the Menai Strait, North Wales. *Cahiers de Biologie Marine*, **21**, 135–54.

Woollacott, R. M. and North, W. J. (1971). Bryozoans of California and North Mexico kelp beds. *Beihefte zür Nova Hedwigia*, **32**, 455–79.

10. A trophodynamic model of fish production on a windward reef tract

N.V.C. POLUNIN* and D.W. KLUMPP[†]

Department of Marine Sciences and Coastal Management, University of Newcastle upon Tyne, Newcastle upon Tyne, England, UK, [†]Australian Institute of Marine Science, Townsville MC, Queensland, Australia.

Abstract

The biological basis of fish-biomass production on coral reefs has not been explored, but in this paper a trophodynamic model is constructed for the first time. Previously published data on benthic primary productivity, rates of macrograzing and planktivory, and abundances of mesofauna of a windward reef-tract at Davies Reef, Great Barrier Reef, Australia are synthesized to provide a view of the dominant sources and immediate fates of organic carbon in this system. These fluxes are postulated to support fish production primarily through eight different types of food-chain (five algal-based and three plankton-based) which vary in numbers of links and the food-chain efficiencies postulated for them. Fish production is estimated by making extra assumptions about the level of meso-grazing and calculating detritus production by difference. Using the model, the potential fishery yield of macro-grazers over the whole reef is $4\,t\,km^{-2}$, invertebrate-feeders $7-16\,t\,km^{-2}$, piscivores $0-2\,t\,km^{-2}$ and of planktivores $3\,t\,km^{-2}$. Maximum predicted yield is $14-23\,t\,km^{-2}$ for a fishery targeted at macrograzing, invertebrate-feeding, and planktivorous fishes, which in carbon terms is 0·5–0·8 per cent of gross primary productivity or 1·1–1·9 per cent of net primary productivity (NPP). If the predicted fish production were realized, this coral reef would have as high a fish yield in relation to its primary productivity as other productive marine ecosystems. Detritus-based food-chains account for some 70 per cent of the NPP, 45 per cent of the production of invertebrate-feeding fishes and 32 per cent of the piscivorous fish yield; they therefore warrant greater attention than they have been previously accorded.

Introduction

High photosynthetic rates (Sargent and Austin 1954) and fish biomass values (Goldman and Talbot 1976) have been known from coral reefs for many years, and some substantial fishery yields have been reported from tropical reef areas (Munro and Williams 1985). Although there is evidence that coral-reef fisheries are often under-exploited (Smith 1978; Wright and Richards 1985), there are indications also that the biological communities involved may be prone to over-exploitation (Munro and Williams 1985). The response of these fisheries to intensive exploitation and other types of perturbation can scarcely be predicted, but community models, such as those based on carbon flux, have the potential to do this, as well as giving estimates of potential yield.

Coral reefs are unusual among aquatic ecosystems for the readiness with which the large organisms can be observed and important processes measured. In many respects there has been little attempt to make full use of such advantages. An obvious place to increase understanding of the dynamics of coral reefs is with grazing food-chains, but even here progress has been limited (Hatcher 1983). This review will draw in particular on measurements of benthic primary productivity and grazing on a particular reef, and use these data along with general inferences on food-chains and transfer efficiencies to construct a trophodynamic model. The model will be used to predict potential fishery yields and should provide a basis for understanding some aspects of coral reef community dynamics and the response of these to human perturbations, of which fishing is one.

1. Primary productivity and herbivory

In spite of their prominence and gross productivity hard corals cannot support substantial fish production and this is not merely a result of their limited coverage. The reason is that photosynthesis of the zooxanthellae is compensated for by respiration of the symbiotic unit as a whole (e.g., Gladfelter and Kinsey 1985). In fact, the main net primary producers of most coral reefs are benthic microalgae, including mats of filamentous forms and crustose corallines. Fishes have a remarkable ability to feed intensively on the whole range of algal types and are typically the dominant macrograzers (Steneck 1988). Sea urchins can be very substantial algal consumers also, although in the Caribbean this may be an artefact of over-fishing (Hay 1984). On the Great Barrier Reef, by contrast, sea urchins do not appear to be important grazers (Klumpp and Pulfrich 1989).

It is important to examine rates of algal processing by grazers and their assimilation efficiencies, because these will ultimately determine levels of secondary productivity. Several studies have estimated net photosynthesis and herbivory in particular reef habitats (Montgomery 1980; Chartock 1983; Carpenter 1986; Polunin 1988; Klumpp and Polunin 1989), but very few have attempted to quantify these fluxes on larger scales (Hatcher 1981). Research on the Great Barrier Reef of Australia provided an opportunity to estimate accurately both algal photosynthesis across a range of reef zones and associated ingestion by the different macrograzers involved.

2. Potential food-chains

Each heterotroph is supported by food from many sources, and reef trophodynamics should therefore be concerned with the intricate web of predator–prey interactions. A great deal of the carbon flux, however, will be along a limited number of pathways. One sea urchin population is reported to consume 20–100 per cent of the net primary productivity (NPP) on a reef (Hawkins and Lewis 1982; Carpenter 1986). A small damsel-fish ingested 71 per cent of NPP in its outer reef-flat territories in winter (Polunin 1988) and a greater percentage in summer (Polunin and Brothers 1989). Hatcher (1983) has expressed the opinion that detrital pathways of reefs may be governed by a small number of generalist consumer species. The number of links in such food-chains will influence how efficiently carbon is passed from primary producers to top consumers in a sequence.

A simple way in which NPP is used is the conversion of algal biomass into macrograzer tissue which is then ingested by piscivores. Where the herbivore is small, two piscivores might be involved in an entire sequence as the initial piscivore could be relatively small (Fig. 10.1a). Adult fish grow slowly, however, and allocation to reproductive tissue could be large in populations dominated by older individuals. A consequence of this is that the immediate consumers of a large proportion of herbivore secondary production (gametes) may not be piscivores, but carnivores which can pick at small food-items (Polunin and Brothers 1989). These consumers (e.g., plankton-feeding caesionids in the case of pelagic-spawning fishes) will only then be available to piscivores, with the NPP being dissipated through three or four links at most (Fig. 10.1b). The relative significance of these two food-chains will depend on several factors, but a herbivore converting its food into body-growth inefficiently and allocating intake more to gametes should support the longer type of food-chain (Fig. 10.1b). The shorter will pertain far more to young fishes which are fast-growing and not reproducing.

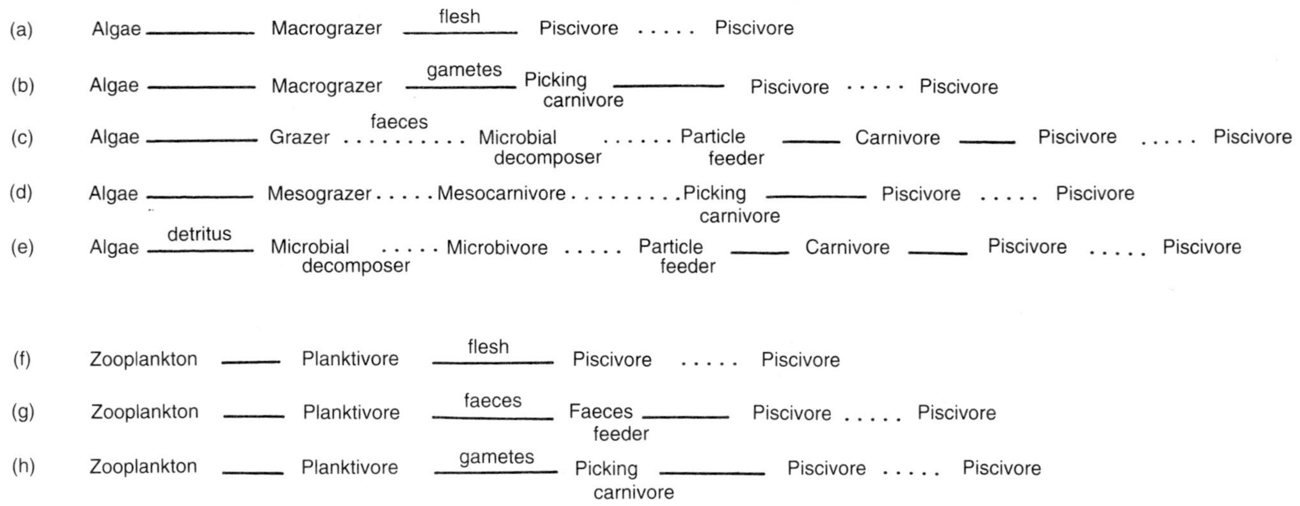

(a) Algae ——— Macrograzer ——flesh—— Piscivore ······ Piscivore

(b) Algae ——— Macrograzer —gametes— Picking ——— Piscivore ······ Piscivore
                                    carnivore

(c) Algae ·····faeces····· Grazer ·····Microbial····· Particle ——— Carnivore ——— Piscivore ······ Piscivore
                          decomposer   feeder

(d) Algae ·····Mesograzer·····Mesocarnivore·········Picking ——— Piscivore ······ Piscivore
                                                   carnivore

(e) Algae ·····Microbial····· Microbivore ······ Particle ——— Carnivore ——— Piscivore ······ Piscivore
        detritus decomposer   feeder

(f) Zooplankton ——— Planktivore —flesh— Piscivore ······ Piscivore

(g) Zooplankton ——— Planktivore —faeces— Faeces ——— Piscivore ······ Piscivore
                                         feeder

(h) Zooplankton ——— Planktivore —gametes— Picking ——— Piscivore ······ Piscivore
                                          carnivore

Fig. 10.1. Food-chains potentially supporting fish production on a coral reef, based on benthic algal productivity (a–e) and oceanic zooplankton (f–h). The solid lines are 'obligatory' trophic links in the food-chains, while the dashed lines are 'optional' links.

Dietary absorption efficiency of marine herbivores can be high, particularly where microalgae are concerned (Valiela 1984), but, if grazing is intense, much detritus will be generated in the form of faeces. Defaecated material will be subject to microbial degradation and ingestion by particle-feeders such as corals or brittle-stars (Polunin 1988) in a sequence which could plausibly dissipate NPP at most through four to six links (Fig. 10.1*c*). The importance of this pathway relative to others will depend upon the levels of grazer ingestion and dietary absorption. Mesograzers such as copepods and polychaetes may be important algal consumers (Klumpp *et al.* 1988). Because of their size they will be susceptible to predation by benthic picking carnivores such as small labrid fishes. A food-chain so constituted might have a maximum of five steps (Fig. 10.1*d*).

The fate of the material unaccounted for by the grazers is less easily characterized. Substantial pools of carbon in the form of dissolved organic carbon (DOC) and particulate organic carbon (POC) exist in the water over coral reefs. This organic carbon can follow several routes:

(1) uptake by microbial decomposers;
(2) ingestion by particle-feeders;
(3) incorporation into sediments;
(4) export (Hatcher 1983).

A possible food-chain for detrital carbon taken up by reef-associated organisms indicates that four to six links could be involved (Fig. 10.1*e*).

Aside from benthic photosynthesis, the other main source of food is from the plankton. Many fishes feed intensely upon zooplankton (Hamner *et al.* 1988) and these are evidently preyed upon directly by piscivores, indicating a relatively simple food-chain (Fig. 10.1*f*). In addition, faecal loss from planktivores may be taken up by other fishes, as indicated by Robertson (1982), and thus provides a subsidiary, albeit longer, mode of uptake by the reef community (Fig. 10.1*g*). If planktivore reproductive effort is high, then separate account may need to be taken of the gametes, as for macrograzers (Fig. 10.1*h*).

Sites, methods, and sources of data

1. Davies Reef and the algal community

Davies is an intensively studied lagoonal platform reef, located some 50 km offshore in the mid-shelf region of the Great Barrier Reef (18·8° S, 147·6° E). The area has been described elsewhere (e.g., Daniel *et al.* 1985, Bradbury *et al.* 1986). The trophodynamic model is constructed for a 410 m long transect of a windward tract of the outer rim. For convenience this is taken to be 1 m wide and all subsequent calculations concerning the tract refer to its 410 m² area. Information on sand coverage of the substratum is that presented by Klumpp and McKinnon (1989, table VI). Where necessary we have used rugosity data of Klumpp and McKinnon (1989) to relate figures to actual planar reef area.

Data on algal community carbon flux on hard substrata come from extensive measurements using automated *in situ* respirometry (see Klumpp *et al.* 1987). Community respiration of hard substrata, derived from respirometry measurement of night-time oxygen uptake by natural and artificial rock surfaces, is contributed to by all organisms colonizing experimental plates or natural rock surfaces. Primary productivity data available for June–September inclusive are taken here to be indicative of 'winter' (six months), while those for October–January are regarded as approximating 'summer' conditions. Estimated carbon flux for the sand habitat is regarded as similar to that reported by Kinsey (1978, table 3) for One Tree Island in early summer.

2. Trophodynamics of grazers

Ingestion by fishes has been estimated by variations on a gut-filling/feeding-activity technique (Polunin 1988) combined with observations of fishes foraging in known areas of hard reef substratum. Most of the data have been gathered by Klumpp and Polunin (1989, 1990; Polunin and Klumpp, unpublished data), the only exception being grazer activity at the reef-crest which is taken to be largely that of parrotfishes at a level similar to that of Zone 4 of Klumpp and Polunin (1990). Klumpp and Pulfrich (1989) have assessed the impact of large herbivorous invertebrates. Benthic mesograzers have also been enumerated: locality 3 inside-territory data of Klumpp *et al.* (1988, figs 3–4, table 5) are indicative of zones B and C of the present chapter; outside-territory data from the same locality are indicative of mesofaunal abundances of zone A here; outside-territory data from locality 3 apply to zones D and E here, and Zone F of this chapter is taken to be an average of inside and outside-territory data from locality 4.

Ingestion rates and dietary absorption efficiencies have been measured or estimated for certain damselfish, blennies, parrotfish, surgeonfish and molluscs by us or coworkers (see Polunin 1988; Klumpp and Polunin 1989, 1990; Klumpp and Pulfrich 1989; Polunin and Klumpp, unpublished data). Ingestion by mesofauna has been predicted on the basis of size and a dietary content of 36 per cent organic carbon (Klumpp *et al.* 1988, table 7). Only limited data are available on standard respiration, body-growth, and excretion in a damselfish

(Polunin and Koike 1987; Polunin and Klumpp 1989; Polunin and Brothers 1989), but data on growth of certain adult parrotfishes, surgeonfishes, blennies, and other damselfishes are available (Polunin and Klumpp, unpublished data).

Results

1. Primary productivity

Several different habitats were recognized along the 410 m tract on the basis of the prominent grazers and main types of algal community present (Fig. 10.2). Zones A–C are at the windward edge of the reef and are susceptible to high wave energy, but do not emerge at low tide. Zone E is a particularly extensive, but quite uniform, habitat-type which is subject to current, turbulence and extreme shallowing, or emergence, at low tide. Zone D is similar to E, but with blennies relatively uncommon and a weaker territorial dominant fish than in B and C. The mean ratio of gross primary production (GPP) to community respiration (R) for the four zones of Klumpp and McKinnon (1989) in winter and summer was 1.95:1 ($\pm$1SD 0.10). The GPP:R ratio did not vary significantly between winter and summer for this reef-flat, but there were differences between zones (Fig. 10.2). The greatest habitat-specific GPP and net primary productivity (NPP) per unit planar area of reef occur at the reef crest (Zone A) and on the outer reef-flat (zones B–D), where preside certain territorial fishes including species of damselfish and surgeonfish (Table 10.1). The areally-

Table 10.1. Annual gross primary productivity (GPP, g C m^{-2} d^{-1}), winter and summer net primary productivity (NPP, g C m^{-2} d^{-1}) and percentage of NPP macrograzed in winter and summer in sandy habitat and zones A–F of the Davies Reef reef-flat transect.

		Winter		Summer	
Zone	GPP	NPP	Per cent Macrograzed	NPP	Per cent Macrograzed
A	4.33	1.81	27.3	2.43	52.9
B	5.32	2.47	62.4	3.06	100
C	5.32	2.47	40.0	3.06	68.0
D	4.18	1.74	47.0	2.6	58.9
E	2.84	1.13	21.5	1.56	27.4
F	3.41	1.33	11.3	1.70	13.1
Sand	0.86	0.48	?	0.48	?

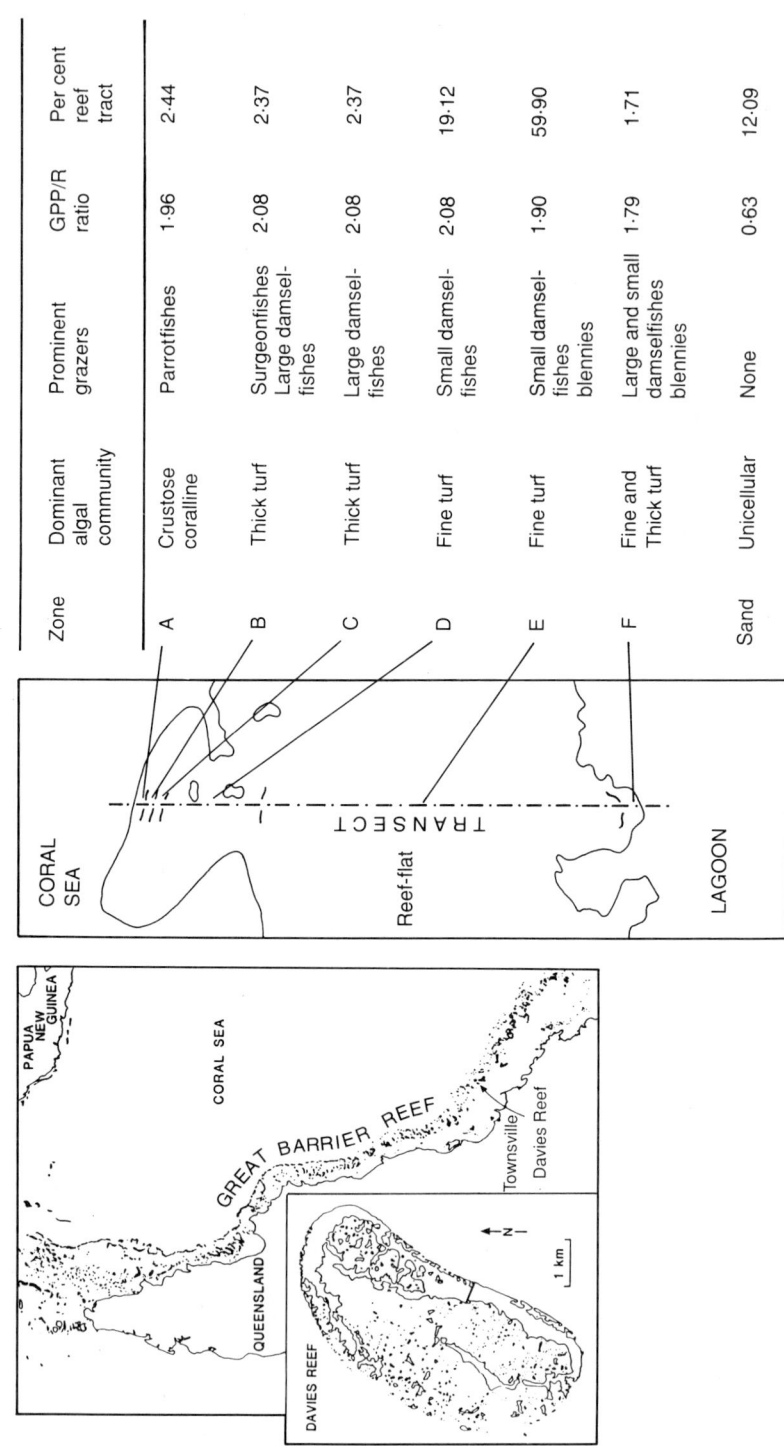

Fig. 10.2. Location of Davies Reef on the Great Barrier Reef, showing the extent of zones A–F on the reef-flat transect and the algal types, prominent grazers, ratio of gross primary productivity (GPP) to algal-community respiration (R), and the percentage of the reef-flat transect covered by each zone/habitat at Davies Reef.

weighted mean GPP for the reef tract is estimated to be $3 \cdot 02$ g C $m^{-2} d^{-1}$ (Table 10.1), producing approximately 450 kg C yr^{-1} in total. On the same basis the mean NPP is $1 \cdot 13$ g C m^{-2} d^{-1} in winter and 1.61 g C $m^{-2} d^{-1}$ in summer (Table 10.1); this produces approximately 200 kg C yr^{-1} for the whole reef-tract.

2. Algal processing by grazers and planktivory

The productive zones of the crest and outer reef-flat (zones A–D) are those with the most intense feeding (52–100 per cent of NPP in summer) by large herbivores. Most of this is by fishes, although these consumers decline somewhat in importance from front (>99 per cent of macro-grazing in summer) to back (zones E–F, 80–82 per cent) of the reef (Table 10.2). Of the areally weighted NPP for the whole reef (Table 10.1, Fig. 10.2), some 60 per cent in summer and 75 per cent in winter are not macrograzed. The seasonal difference occurs because these consumers increase their feeding into the summer more rapidly than the algae increase the NPP as the temperature rises. Some 650 g C yr^{-1} are taken up by macrograzers over the entire reef-tract, leaving 135 kg C yr^{-1}.

Mesofaunal ingestion per unit area of reef (Table 10.3) is predicted to be greatest in zones A, B and C (330–376 g C m^{-2} yr^{-1} and least in zone F (143 g C m^{-2} yr^{-1}). On an areally-weighted basis, mesofaunal consumption as a whole would be equivalent to approximately 95 kg C yr^{-1}. For the purposes of this review it is assumed that 75 per cent of this ingestion is herbivory, amounting to some 70 kg C yr^{-1}. Strict herbivores might in fact constitute a smaller proportion of the total mesofauna, but it is considered that in practice it will prove very difficult to differentiate between 'herbivores' and 'detritivores' among these animals, while in energetic terms the distinction might not prove too important in any case.

Available data for adult small damselfish (*Plectroglyphidodon lacrymatus*) and large damselfishes (*Stegastes apicalis*) and parrotfishes (*Scarus frenatus*, *S. sordidus*) indicate gross growth efficiencies of much less than 1 per cent (Polunin and Brothers 1989; Polunin and Klumpp, unpublished data). A figure of 1 per cent is therefore thought to be reasonable for the conventional grazing food-chain efficiency for whole populations of such macrograzers. In addition, it is considered that reproductive effort could be substantial, accounting for maybe a quarter of organic-carbon intake. These estimates provide a basis for allocating the macrograzer intake to food-chain types 'a' and 'b' (Figs. 10.1a–b and 10.4).

Hamner *et al.* (1988) have made a detailed study of the zooplankton-feeding fish community at the windward edge of Davies

Table 10.2. Percentage of NPP in different zones of the Davies Reef reef flat ingested by various macrograzer groups in winter/summer

Zone	Molluscs	Parrotfishes	Damselfish	Damselfish	Blennies	Surgeonfishes	All
A	0.3/0.3	27.0/52.6	0/0	0/0	0/0	0/0	27.3/52.9
B	0.4/0.4	4.4/5.6	9.6/14.9	0.9/0.9	0/0	46.4/78.2	61.7/100
C	0.4/0.4	14.0/29.0	25.0/38.0	0/0	0/0	0/0	39.4/67.4
D	4.5/4.5	28.2/48.9	0/0	2.2/3.3	1.4/2.2	0/0	36.3/58.9
E	5.3/5.3	1.5/2.4	0/0	8.2/8.5	6.5/11.2	0/0	21.5/27.4
F	2.4/2.4	0.5/0.8	0/0	5.4/4.3	3.0/5.6	0/0	11.3/13.1

Table 10.3. Total mesofaunal ingestion (g C m^{-1} yr^{-1}) predicted on the basis of mean body-sizes and abundances of the taxa involved, conversion of surface areas to planar areas of reef, and a dietary organic carbon content of 36 per cent. No estimate is made of seasonality of ingestion

Zone	Amphipods	Copepods	Other crustaceans	Molluscs	Polychaetes	All
A	57.1	17.9	57.2	87.8	110.4	330.4
B	64.5	42.1	48.7	80.2	140.9	376.4
C	64.5	42.1	48.7	80.2	140.9	376.4
D	79.0	12.0	28.1	93.0	40.7	252.8
E	79.0	12.0	28.1	93.0	40.7	252.8
F	26.0	40.3	23.4	22.5	30.8	143.0

Reef and indicated possible organic inputs to the reef-front of 21.5 g dry wt m^{-1} d^{-1} by this route. If this matter is 50 per cent carbon, then the zooplanktivorous-fish input to the reef-tract as a whole is approximately 4 kg C yr^{-1}. If zooplanktivorous fishes are relatively fast-growing, then 10 per cent of their intake might be allocated to body growth. Reproductive effort might be another 15 per cent, making an overall production efficiency (body plus reproductive growth in relation to ingestion) comparable with that suggested for the grazing fishes.

3. Production of detritus

Dietary organic-carbon absorption efficiencies of herbivorous reef-fishes can be high, even in parrot-fishes which account for approximately half of the macrograzing at Davies Reef (Table 10.1). It is suggested that an indicative figure for the most important macrograzers is 45 per cent. On this basis some 35 kg C yr^{-1} will be produced as faeces by macrograzers on the whole reef-tract. If absorption efficiency of mesograzers is similar to that of macrograzers in the same community, then these consumers could produce approximately 38 kg C yr^{-1} of faeces. This faecal material will be available to organisms such as those of a type 'c' food-chain (Fig. 10.1c). Because planktivores are feeding upon animal material, an absorption efficiency of perhaps 85 per cent of carbon seems feasible. Planktivores may therefore produce faeces at approximately 0.6 kg C yr^{-1} on the reef tract, material which is likely to be taken up by specialist planktivores and herbivores (Polunin, unpublished data).

After allowing for macrograzing and mesograzing, some 65 kg C yr^{-1} of the NPP (Fig. 10.3) remain unaccounted for. It is proposed that this material is mostly in the form of detritus, which may be used in food-chains of the type depicted in Fig. 10.1e.

4. Model of fish production

Fig. 10.1 can be used as a basis for predicting fish production on the Davies Reef flat if the inputs to the various food-chains are known together with the food-chain (gross ecological) efficiencies supported by each link. Although some processes, particularly algal photosynthesis and macrograzing, have been quite extensively studied, the greatest uncertainties derive from poor knowledge of food-chain transfer.

Organic-carbon inputs to the postulated food-chains, and food-chain efficiencies assumed to influence their transfer through successive consumers in the present trophodynamic model, are summarized in Fig. 10.4. For mesocarnivores, planktivores and piscivores, a food-

Model of fish production on a windward reef tract

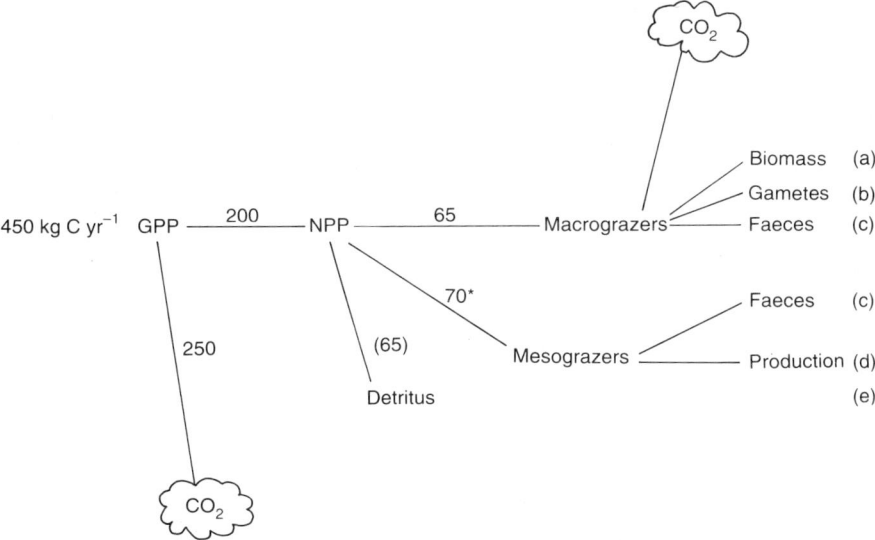

Fig. 10.3. Organic-carbon fluxes derived from benthic algal photosynthesis on the Davies reef-flat transect (410 m × 1 m) estimated from direct measurement, primarily from body-size predictions *, or by difference ( ). The food-chain types to which these fluxes contribute in the model are indicated by the letters a–e (Fig. 10.1a–e).

chain efficiency of 10 per cent is assumed to prevail (see Slobodkin 1961). This is also supposed for most invertebrate-feeders, except for feeding on macrograzer gametes, where the consumption (ecotrophic) efficiency may be halved relative to other prey. The basis of this assumption is that gametes being small, transparent, often produced on an ebb tide and at dawn or dusk, are more likely to escape predation than larger mobile benthic prey such as crustaceans. These consumers are thus taken to have a food-chain efficiency of 5 per cent Mesograzers may channel 8 per cent of the intake of Fig. 10.3 into secondary production, if their net production efficiency is 18 per cent (Bayne and Newell 1983, table 1); their food-chain efficiency might be 7 per cent if most of their production is efficiently utilized by predators. Food-chain efficiencies of particle-feeders may be 12 per cent, slightly higher than the small grazers because of absorption and net production efficiencies being slightly higher than herbivores, averaging perhaps 54 per cent and 27 per cent, respectively (Bayne and Newell 1983, table 1). Finally, microbial decomposers may have relatively high food-chain efficiencies of 25 per cent, because it is considered that production efficiencies of these organisms may be large (Pomeroy and Wiebe 1988, table 1).

On this basis, fish yield is estimated for the reef-tract and for the whole reef, assuming a carbon content of 40 per cent of dry weight and water content 70 per cent of fresh weight. The estimated production depends on the food-chain characteristics, and for this purpose the minimum estimate is derived by including all possible links and the maximum by excluding the 'optional' links (dotted lines in Figs. 10.1 and 10.4). The yields obtained also rely upon ecological tactics of the fishery (Table 10.4). By following a 'piscivore strategy', which would target only fish-feeding species, a small yield of $0-2 \, t \, km^{-2}$ might be

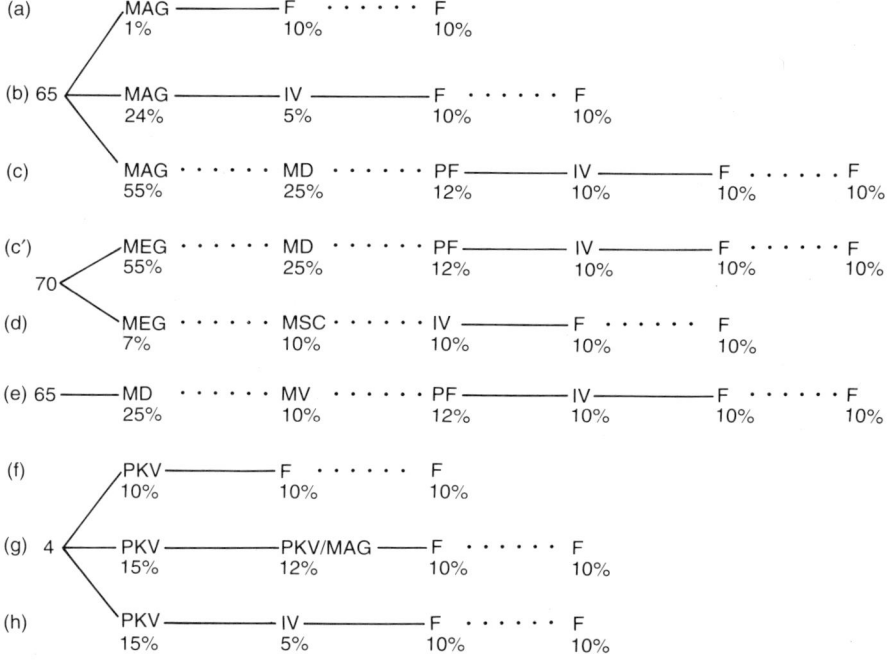

Fig. 10.4. Trophodynamic model of fish production at Davies Reef, including the estimated carbon inputs (benthic fluxes from Fig. 10.3) to the various food-chains (lower-case letters on left of diagram refer to the food-chains in Fig. 10.1) and food-chain efficiencies postulated for all the links involved (see text for details). These data are used to estimate fish production (Table 10.4) by incorporating all trophic links (maximum value) or omitting the optional links indicated by dotted lines (minimum value). The upper-case letters in the food-chains symbolize the trophic links of Fig. 10.1: MAG, macrograzer; F, piscivore; IV, invertebrate-feeder (picking carnivore or carnivore); MD, microbial-decomposer; PF, particle-feeder; MEG, mesograzer; MSC, mesocarnivore; MV, microbivore; PKV, planktivore.

Table 10.4. Fishery yields predicted for the reef-flat transect, and whole, of Davies Reef, based on Fig. 10.4: fish production is estimated by ecological group of fishes targeted in the fishery. MAG, macrograzer; IV, invertebrate-feeder; F, piscivore; PKV, planktivore. Column A is calculated from organic carbon yield converted to dry total weight by multiplying by 2.5 (dry weight 40 per cent) and fresh weight by a factor of 3.33 (water content of 30 per cent). Column B is derived from A by multiplying by 0.33 (reef flat as proportion of total reef area)

Fishery resource targeted	Yield from reef tract (t km^{-2} tract)	Yield from whole of Davies Reef (t km^{-2} reef)
	A	B
MAG	13.9	4
IV	22.3–48.4	7–16
F	0.5–6.7	0–2
PKV	8.8	3
MAG + IV	36.2–62.3	11–20
MAG + IV + PKV	45.0–71.1	14–23

obtained from all of the food-chains of the model (Fig. 10.4a–h). This targeting would, however, preclude heavy exploitation at lower trophic levels simultaneously because these earlier links in food-chains support piscivore production. A low yield of 4 t km^{-2} at most might be obtainable by targeting grazing fishes on their own (Fig. 10.4a). It is worth noting that although macrograzers enter into food-chain types 'b' and 'c' as well, the macrograzer output to the food-chains in those cases is in the form of gametes and faeces, respectively, which are not directly exploitable by fishing. A maximal yield is conceivable from a fishery aimed at macrograzers, invertebrate-feeders (Fig. 10.4b–e and h) and planktivores (Fig. 10.4f); the total yield in this case might be 14–23 t km^{-2}.

Discussion

Small changes in some parameters have substantial effects on the secondary production figures calculated, but the basis of the model as outlined in Table 10.1 and Figs 10.2–4 is explicit and can be varied as new data become available. Key parameters include not only food-chain efficiencies, but also such estimates as the level of mesograzing. In spite of their importance, zones D and E, which constitute 80 per cent of the present reef-flat, have been little-studied. In contrast, fore-reef habitats have been given a great deal of attention. Increased knowledge of the central and back reef-flat is likely therefore to considerably improve understanding of the trophodynamics of fish production on this reef as a whole.

The fish yields derived from our trophodynamic model of Davies reef-flat reach the upper range of harvests actually reported for coral-reef fisheries. A well-studied and intensely-exploited reef in the Philippines has been reported to produce $25 \, t \, km^{-2} \, yr^{-1}$ (Bellwood 1988), but yields elsewhere in that country and in the southern Pacific are reported in places to be greater than $40 \, t \, km^{-2} \, yr^{-1}$ (cf. Munro and Williams 1985). Galzin (1987) suggested on the basis of three dominant fish species, but with several assumptions, that the fish yield of a Polynesian fringing-reef might be $23 \, t \, km^{-2} \, yr^{-1}$. The maximum yield predicted for Davies Reef on the basis of the present model is $14-23 \, t \, km^{-2} \, yr^{-1}$ (Table 10.3). This is $0.5-0.8$ per cent of the GPP and $1.1-1.9$ per cent of the NPP, percentages which if realized would mean that, contrary to a previous assessment (Marten and Polovina 1982), coral reefs may have a high fishery yield in relation to their primary productivity.

The issue as to whether loosely reef-associated species should be included does not arise in this model because assumptions are made about how much organic carbon flows through various food-chains, but no assumptions are made about where consumption takes place. Some of the piscivore production would undoubtedly be in the form of 'pelagic' species (e.g., *Scomberomorus*), while much of the detrital carbon especially can be expected to support secondary production away from the reef itself. Apart from planktivory, there is no obvious way in which fishes would have access to organic carbon sources derived outside of the reef itself. There are, for example, no extensive shallows colonized by seagrasses at Davies Reef, although elsewhere these are known to support some reef fish production. Water column primary productivity is likely to be higher in the lagoon than outside, but the level involved is much smaller than that for the reef-flat (Furnas *et al.* 1990). Water-column productivity is the basis for the planktivore food intake incorporated in the model.

The data for Davies Reef indicate large differences in the trophodynamic basis of fish production across the reef-flat. On the reef-flat overall, approximately one-third of the NPP is taken up by macrograzers, one-third may be utilized by mesograzers, and the remaining one-third may be detritus (Fig. 10.3). More than half of the grazed material seems likely to be recycled as faeces. Some 70 per cent of the NPP is probably used ultimately in the form of detritus, and the single best explanation for this lies in the low proportion of nitrogen and phosphorus to organic carbon in reef-algal production (Polunin 1988). The emphasis on detritus will be greater in central and back areas of the reef-flat because much less of the NPP is macrograzed there than in the fore reef-zones. There may be many reasons for the low level of macrograzing in these areas, e.g. shallowness and wave-exposure, lack of hiding places, and low algal biomass density. Conversely, these conditions appear to be more favourable to macrograzing in the fore-reef zones. Klumpp and Polunin (1990) show that there is marked habitat partitioning between parrotfish species feeding on the outer reef-flat. This probably reflects significant differences in their ability to forage in the face of water movement and territorial defence under slightly different conditions. At the windward edge of the reef, the NPP is high and a very large proportion of it is processed by macrograzers (Tables 10.1 and 10.2), but overall the yield available from grazing fishes specifically may be small (Table 10.4). The most productive component of the fish community would be the invertebrate-feeders, 45 per cent of the yield of which is based upon detrital food-chains (Fig. 10.4c, c' and e), while 32 per cent of the piscivore yield will be detritus-based (Fig. 10.4c, c', e, and g).

It is worth noting that because many herbivorous reef fishes forage most actively in the afternoon (Polunin and Klumpp 1989; Klumpp and Polunin, unpublished data), the production of fish-derived detritus will be greatest later in the day. This production should be most easily measured in the fore-reef zones, but because of the predominant current at Davies Reef, both this and any other detrital material will tend to be transported back towards the lagoon. The extent of such transport will depend on the density of the material and intensity of water movements that will tend to resuspend it, but the magnitude of this flux has not been estimated (Riddle et al. 1990) even though it might easily be quantified.

If the proposed food-chains are correct, fish production relative to primary productivity should be more efficient (cf. Fig. 10.1a–b) in the fore-reef where the NPP is greater than in the central and back-reef areas, longer detrital food-chains predominate and dispersal over a larger area seems likely. These latter zones are nevertheless very important in determining the overall pattern of organic carbon flux (Fig. 10.2, Table 10.1). Russ (1984) has noted the relative abundance of certain herbivorous fishes on the reef-crest and the lower densities on the reef-flats of mid-shelf reefs.

Much macrograzing is likely to be inefficiently passed up the food-chain because a large allocation to reproduction may mean the insertion of an extra trophic link in the food-chain (Polunin and

Brothers 1989) and perhaps significant export from the reef. In addition, it is clear that targeting piscivorous fish will greatly reduce the potential yield, yet this is evidently what happens where fishing relies heavily on hand-lining (e.g., Rohan and Church 1979).

The present prediction of maximum yield applies to one site which has been relatively well-studied. The trophodynamics of coral reefs differ in patterns of primary productivity (Klumpp and McKinnon, unpublished data) and in their fish communities (e.g., Williams and Hatcher 1983) even within the Great Barrier Reef system. It is believed that the present model may be applicable to other coral reefs.

Acknowledgements

Thanks go to the Royal Society of London and the British Government Overseas Development Administration for grants to NVCP under their Overseas Field Research Scheme and Fish Management Science Programme, respectively. Many individuals have helped with data collection upon which this model is based and have been acknowledged, in previous papers. Alistair Robertson provided useful comments on the draft paper. Australian Institute of Marine Science contribution No. 541.

References

Bayne, B. L. and Newell, R. C. (1983). Physiological energetics of marine molluscs. In *The Mollusca*, (ed. A. S. M. Saleuddin and K. M. Wilbur), Vol. 4, pp. 407–515. Academic Press, New York and London.

Bellwood, D. R. (1988). Seasonal changes in the size and composition of the fish yield from reefs around Apo Island, central Philippines, with notes on methods of yield estimation. *Journal of Fish Biology*, **32**, 881–93.

Bradbury, R. H., Loya, Y., Reichelt, R. E., and Williams, W. T. (1986). Patterns in the structural typology of benthic communities on two coral reefs of the central Great Barrier Reef. *Coral Reefs*, **4**, 161–7.

Carpenter, R. C. (1986). Partitioning herbivory and its effects on coral reef algal communities. *Ecological Monographs*, **56**, 345–63.

Chartock, M. A. (1983). The role of *Acanthurus guttatus* (Bloch and Schneider 1801) in cycling algal production to detritus. *Biotropica*, **15**, 117–21.

Daniel, P. A., Johnson, D. B., De Vantier, L. M., and Barnes, G. R. (1985). Davies Reef. In *Studies in the assessment of coral reef ecosystems*, (ed. R. H. Bradbury and R. E. Reichelt), Part 7, 25 pp. Australian Institute of Marine Science, Cape Ferguson, Queensland, Australia.

Furnas, M. J., Mitchell, A. W., Gilmartin, M., and Revelante, N. (1990).

Phytoplankton biomass and primary production in semi-enclosed reef lagoons of the central Great Barrier Reef, Australia. *Coral Reefs*, **9**, 1-10.

Galzin, R. (1987). Potential fisheries yield of a Moorea fringing reef (French Polynesia) by the analysis of three dominant fishes. *Atoll Research Bulletin*, **305**, 21 pp.

Gladfelter, E. H. and Kinsey, D. W. (1985). Metabolism, calcification and carbon production. In *Proceedings of the 5th International Coral Reef Congress*, 4, pp. 505-40. EPHE, Tahiti.

Goldman, B. and Talbot, F. H. (1976). Aspects of the ecology of coral reef fishes. In *Biology and geology of coral reefs*, (ed. O. A. Jones and R. Endean), Vol. IV, *Biology* 2, pp. 125-154. Academic Press, New York and London.

Hamner, W. M., Jones, M. S., Carleton, J. H., Hauri, I. R., and Williams, D. M. (1988). Zooplankton, planktivorous fish, and water currents on a windward reef face: Great Barrier reef, Australia. *Bulletin of Marine Science*, **42**, 459-79.

Hatcher, B. G. (1981). The interactions between grazing organisms and the epilithic algal community of a coral reef: a quantitative assessment. In *Proceedings of the Fourth International Coral Reef Symposium*, 2, pp. 515-24.

Hatcher, B. G. (1983). The role of detritus in the metabolism and secondary production of coral reef ecosystems. In *Proceedings of the Great Barrier Reef Inaugural Conference*, (ed. J. T. Baker *et al.*), pp. 317-25. James Cook University, Townsville, Queensland, Australia.

Hawkins, C. M. and Lewis, J. B. (1982). Ecological energetics of the tropical sea urchin *Diadema antillarum* Philippi in Barbados, West Indies. *Estuarine, Coastal and Shelf Science*, **15**, 645-69.

Hay, M. E. (1984). Patterns of fish and urchin grazing on Caribbean coral reefs: are previous results typical? *Ecology*, **65**, 446-54.

Kinsey, D. W. (1978). Productivity and calcification estimates using slack-water periods and field enclosures. In *Coral reefs: research methods*, (ed. D. R. Stoddart and R. E. Johannes), pp. 439-69. UNESCO, Paris.

Klumpp, D. W. and McKinnon, D. (1989). Temporal and spatial pattern in primary productivity of a coral-reef epilithic algal community. *Journal of Experimental Marine Biology and Ecology*, **131**, 1-22.

Klumpp, D. W. and Polunin, N. V. C. (1989). Partitioning among grazers of food resources within damselfish territories on a coral reef. *Journal of Experimental Marine Biology and Ecology*, **125**, 145-69.

Klumpp, D. W. and Polunin, N. V. C. (1990). Algal production, grazers and habitat partitioning on a coral reef: positive correlation between grazing rate and food availability. In *Trophic relationships in the marine environment*, (ed. M. Barnes and R. N. Gibson), Proceedings of the 24th Marine Biology Symposium, pp. 372-88. Aberdeen University Press.

Klumpp, D.W. and Pulfrich, A. (1989). Trophic significance of herbivorous macroinvertebrates on the central Great Barrier Reef. *Coral Reefs*, **8**, 135-44.

Klumpp, D.W., McKinnon, D., and Daniel, P. (1987). Damselfish territories: zones of high productivity on coral reefs. *Marine Ecology Progress Series*, **40**, 41-51.

Klumpp, D.W., McKinnon, D., and Mundy, C.N. (1988). Motile cryptofauna of a coral reef: abundance, distribution and trophic potential. *Marine Ecology Progress Series*, **45**, 95-108.

Marten, G.G. and Polovina, J.J. (1982). A comparative study of fish yields from various tropical ecosystems. In *Theory and management of tropical fisheries*, (ed. D. Pauly and G.I. Murphy), pp. 255-89. International Center for Living Aquatic Resources Management, Manila.

Montgomery, W.L. (1980). Comparative feeding ecology of two herbivorous damsel-fish (Pomacentridae: Teleostei) from the Gulf of California, Mexico. *Journal of Experimental Marine Biology and Ecology*, **47**, 9-24.

Munro, J.L. and Williams, D.M. (1985). Assessment and management of coral reef fisheries: biological, environmental and socio-economic aspects. In *Proceedings of the Fifth International Coral Reef Congress*, 4, pp. 544-81. EPHE, Tahiti.

Polunin, N.V.C. (1988). Efficient uptake of algal productivity by a single resident herbivorous fish on the reef. *Journal of Experimental Marine Biology and Ecology*, **123**, 61-76.

Polunin, N.V.C. and Brothers, E.B. (1989). Low efficiency of dietary carbon and nitrogen conversion to growth in an herbivorous coral-reef fish in the wild. *Journal of Fish Biology*, **35**, 869-79.

Polunin, N.V.C. and Klumpp, D.W. (1989). Ecological correlates of foraging periodicity in herbivorous reef fishes of the Coral Sea. *Journal of Experimental Marine Biology and Ecology*, **126**, 1-20.

Polunin, N.V.C. and Koike, I. (1987). Temporal focussing of nitrogen release by a periodically feeding herbivorous reef fish. *Journal of Experimental Marine Biology and Ecology*, **111**, 285-96.

Pomeroy, L.R. and Wiebe, W.J. (1988). Energetics of microbial food webs. *Hydrobiologia*, **159**, 7-18.

Riddle, M.J., Alongi, D.M., Dayton, P.K., Hansen, J.A., and Klumpp, D.W. (1990). Detrital pathways in a coral reef lagoon. I. Macrofaunal biomass and estimates of production. *Marine Biology*, **104**, 109-18.

Robertson, D.R. (1982). Fish faeces as fish food on a Pacific coral reef. *Marine Ecology Progress Series*, **7**, 253-65.

Rohan, G. and Church, A. (1979). *A review of the northern territory mackerel and reef fisheries*. Northern Territory (Australia) Department of Primary Industry Fishery Report, No. 3, 25 pp.

Russ, G. (1984). Distribution and abundance of herbivorous grazing fishes in the central Great Barrier Reef. II. Patterns of zonation of mid-shelf and outershelf reefs. *Marine Ecology Progress Series*, **20**, 35–44.

Sargent, M.C. and Austin, T.M. (1954). Biologic economy of coral reefs. Bikini and nearby atolls, Part 2 Oceanography (Biologic). *United States Geological Survey Professional Paper*, **260-E**, 293–300.

Slobodkin, L.B. (1961). *Growth and regulation of animal populations*. Holt, Reinhart and Winston. 184 pp.

Smith, S.V. (1978). Coral-reef area and contributions of reefs to processes and resources of the world's oceans. *Nature*, **273**, 225–6.

Steneck, R.S. (1988). Herbivory on coral reefs: a synthesis. In *Proceedings of the Sixth International Coral Reef Symposium*, **1**, 37–49. Townsville, Queensland, Australia.

Valiela, I. (1984). *Marine ecological processes*. Springer–Verlag, Berlin.

Williams, D.M. and Hatcher, A.I. (1983). Structure of fish communities on outer slopes of inshore, mid-shelf and outer shelf reefs on the Great Barrier Reef. *Marine Ecology Progress Series*, **10**, 239–50.

Wright, A. and Richards, A.H. (1985). A multispecies fishery associated with coral reefs in the Tigak Islands, Papua New Guinea. *Asian Marine Biology*, **2**, 69–84.

11. Mesoherbivores

SUSAN H. BRAWLEY

Department of Plant Biology and Pathology, University of Maine, Orono, USA

Abstract

Arthropod mesoherbivores affect the species composition and stability of marine communities. Many mesograzers attain densities of thousands of animals m^{-2}, but their nocturnal activity has hindered recognition of their abundance and importance. The difference between the potential and realized diets of mesoherbivores is discussed in the context of factors (e.g., predation, short life-spans) that may limit overgrazing of macroalgae.

Introduction

Mesoherbivores are small invertebrate grazers (Brawley and Fei 1987). As understood here, animals larger than an average copepod but smaller than c. 2·5 cm will be considered to be mesograzers. This definition includes not only many arthropods (e.g., small crabs, shrimp, amphipods, isopods, and dipteran larvae) but also small molluscan grazers and polychaetes. Some animals grow through a 'mesograzer' stage, although as adults they may be macrograzers (e.g., some crabs and molluscs). The effects of some molluscans on marine communities have been compared directly with those of crustacean mesoherbivores (Lubchenco 1983), but molluscan grazing is considered in detail elsewhere in this volume, and will not be dealt with here. At least nine families of polychaetes include herbivorous species (Fauchald and Jumars 1979). Brostoff (1985) has found that nereids are present on and consume most of the common tropical algae on a Hawaiian back-reef. However, the ecological roles of herbivorous polychaetes remain largely unknown. This review will consider the varied roles of arthropod mesoherbivores in marine communities, including effects upon succession, recruitment, species diversity, competition, and marine food chains. This list foreshadows the possibility that mesoherbivores may play a pivotal role in some marine communities. Why, then, have we only recently begun an intensive inquiry into their ecological roles?

The Los Angeles commuter problem: activity and mobility of mesoherbivores

'Nocturnal excursions along the seashore of the . . . Baltic gave rise to the present investigation. The light of a torch revealed a great many animals aggregating in the free water above the algal belts where in the daytime they are seen only occasionally. Gammarids (amphipods) were seen to be swimming along the water's edge, or crawling on the rocks . . .' (Jansson and Källander 1968, p. 24).

Mesograzers are very abundant; dipteran larvae, amphipods and isopods often number thousands of individuals m^{-2} in both temperate and tropical habitats (Table 11.1). However, the small size and, especially, the nocturnal habits of many mesoherbivores have hindered a general appreciation of both their numbers and mobility. Consider the problem of appreciating the high mobility of a couple who commute from a rose-decked home on a tree-lined suburban street to jobs in Los Angeles, 60 miles away, if we dropped in on them at midnight! The nocturnal habits of mesoherbivores have been described, however (e.g., crustaceans: Tattersall 1913; Blegvad 1922; Longley 1927; Nicholls 1931; Jansson and Källander 1968; Williams and Bynum 1972; Fincham 1974; Brenner *et al.* 1976; Montouchet 1979; Alldredge and King 1980; Heck and Orth 1980; Brawley and Adey 1981; Robles 1982; Edgar 1983*b*, Robertson and Klumpp 1983; Kitting *et al.* 1984; Brawley and Fei 1987; Dean and Connell 1987; Willows 1987; Buschmann 1990; see Ricketts *et al.* 1985 for similar references to other groups of mesoherbivores). Blegvad (1922) was one of the first investigators to suggest that nocturnal activity patterns in these organisms were selected because they decreased predation by visual predators (see pp. 249–50. Even tube-dwelling amphipods such as *Ampithoe* spp. are mobile at night (e.g., Williams and Bynum 1972; Brawley and Adey 1981; Edgar 1983*b*; Brawley and Fei 1987).

Part of the enhanced activity of mesoherbivores at night relates to mating (see references in Williams and Bynum 1972; Alldredge and King 1980), but many authors cite increased mobility in the context of nocturnal feeding (e.g., Nicholls 1931; Brenner *et al.* 1976; Robles 1982; Kitting *et al.* 1984; Brawley and Fei 1987; Willows 1987; Buschmann 1990). Buschmann (1990) found that the density of amphipods (*Hyale* spp.) increased significantly on a preferred food, the red macroalga *Iridaea laminarioides*, at night; he suggested that this seaweed might not be a good refuge from predators during the day (also see Edgar 1983b). Kitting *et al.* (1984) demonstrated that mesoherbivores, especially two species of shrimp, fed nocturnally on seagrass epiphytes, although they had previously been considered to be detrital feeders.

Some amphipods, such as caprellids, are active throughout a 24-hour period (Brawley and Fei 1987). Likewise, although *Hyale frequens* (a gammarid amphipod) were more active at night, Dean and Connell (1987) noted that they, also, fed during the daytime. The mobility of the tropical tube-dwelling amphipod *Ampithoe ramondi* increased in mesocosm experiments as food became limiting and amphipods were displaced from their tubes by other amphipods; under such circumstances, these animals were active during the day, as well as at night (Brawley and Adey 1981). Kennelly and Underwood (1985) found that only polychaetes, among potential mesoherbivores, were collected more commonly at night than during the day; however, these workers sampled only surfaces covered with encrusting algae, not areas covered with foliose and filamentous algae, because quantitative sampling with their benthic suction-sampler was impossible in spatially complex areas. In general, if mesoherbivores are in a protected habitat that is also a preferred food source, some feeding throughout a 24-hour period is likely, but many mesoherbivores appear to consume much of their food at night.

Mobilities of mesoherbivores can be quite high. Nichols (1931) observed the supralittoral isopod *Ligia oceanica* running down into the *Fucus* zone to feed at night during low tide; this isopod moves at speeds of 0.5 m s^{-1} (Manton 1952). Caprellids in the field-study by Brawley and Fei (1987) were so mobile that a pseudoreplicated design had to be used in this study of their effects upon epiphytes in the field. These temperate caprellids moved at speeds of 0.5 m min^{-1}, and gammarids in a rocky intertidal habitat swam 1 m min^{-1} (*Ampithoe lacertosa*) and 4 m min^{-1} (*Pontogeneia rostrata*), the latter a combination of swimming and displacement by water movement (Brawley and Fei, unpublished observations). Ledoyer (1969) observed that the gammarid amphipod *Cymadusa crassicornis* moved up stems of *Posidonia*, a seagrass of about 1 m length, at night; this has also been observed by Scipione and coworkers (pers. comm.) in the Bay of Naples and is related to the fact that the upper, older leaves are covered with a preferred food, epiphytic algae. Dye-marking and recapture studies (Brenner *et al.* 1976) may be helpful in future quantitative studies of the mobility of mesoherbivores, especially to assess turnover of individuals on plants (e.g., on *Sargassum*, Edgar 1983b; on *Ulva*, Buschmann 1990) where overall densities are stable on a 24-hour basis.

In summary, mesoherbivores are abundant intertidally and subtidally, in temperate and tropical habitats. Direct observations and measurements, as well as information from plankton tows, demonstrate that many species move over distances of several meters, mostly at night, on a daily basis. The mobilities of mesoherbivores compare favourably with the foraging ranges of sea urchins (e.g., $c.\ 1\ \mathrm{m}^2$, Carpenter 1984, 1986) and intertidal fishes such as blennies (Gibson 1967) and territorial fishes such as damselfish (see Lobel 1980 for general references); the assertion in Hay et al. (1987) that mesoherbivores are relatively sedentary is not consistent with the literature.

Effects of mesoherbivores upon marine communities

1. Succession and species composition

I now review the evidence that mesoherbivores have important effects upon some facets of community structure and then discuss their diets (pp. 246–9). It is through an evaluation of what they eat and the degree to which predators control potential vs realized diets that we will have a view of where to focus our future studies of the effects of mesoherbivores upon community structure.

Hruby and Norton (1979) found that algal propagules failed to penetrate existing turfs of ephemeral algae. This suggests that succession to a macroalgal community might be retarded if turfs of filamentous algae dominate space in a marine habitat. Since many mesoherbivores feed preferentially upon ephemeral algae, they may expose bare substrate, permitting successful settlement by macroalgal propagules. Several studies demonstrate that when mesoherbivores (e.g., dipteran larvae: Robles and Cubit 1981; amphipods: Brawley and Adey 1981; crabs: Sousa 1979) are experimentally excluded from bare substrate, early successional algal species become established and persist. Sousa (1979) found that succession from ephemeral green algae (e.g., *Ulva*) to a perennial red algal community was retarded when crabs (*Pachygrapsus crassipes*) were experimentally excluded from the low intertidal zone in California. Robles and Cubit's study (1981) in the high-intertidal zone of the California coast found that a thick diatom film persisted on substrata from which fly larvae were removed either by the insecticide Malathion or by hand. Seasonal blooms of foliaceous green algae are associated with high densities (Table 11.1) of dipteran larvae in areas from which other grazers (e.g., limpets) have been removed by storms on the California coast (Robles 1982). Brawley and

Table 11.1. Abundance of mesoherbivores in selected marine communities (mean number m^{-2})

Group	Animals m^{-2}	Locality	Reference
Fly larvae	12 000	High zone, rocky intertidal, USA	Robles and Cubit 1981
Fly larvae	8337	Mid-zone, rocky intertidal, UK	Colman 1939
Amphipods	6877	Rocky mid-zone, UK	Colman 1939
Amphipods	112–1488	Sandy/rocky subtidal (seasonal variation), temperate China	Brawley and Fei 1988
Amphipods	51–2102	Seagrass communities (geographical variation), USA and Canada	Nelson 1980
Amphipods	0–368 800	Various localities world-wide, members of Talitroidea	Wildish 1988
Amphipods	491–2682 1558–22 968	Outside damselfish territories Inside damselfish territories, reef, Australia	Klumpp *et al.* 1988
Isopods	4290	Mid-zone, rocky intertidal, UK	Colman 1939
Isopods	386–2478 696–3840	Outside damselfish territories Inside damselfish territories, reef, Australia	Klumpp *et al.* 1988
Isopods	*c.* 600–1600	Subtidal, Baltic Sea (decade-long variation within *Fucus vesiculosus* beds)	Kangas *et al.* 1982
Shrimp	0–54	Temperate seagrass beds (seasonal variation of *Palaemonetes pugio*), USA	Van Dolah 1978
Crabs	2.9	Low zone, rocky intertidal, USA	Sousa 1979

Adey (1981) found that 15–16 species of filamentous algae persisted, in a refuge tank connected to a coral-reef mesocosm, when amphipods were prevented from colonizing the tank. If amphipod emigration was permitted, most filamentous species disappeared, and the macroalga *Hypnea spinella* colonized and dominated the 'community'; subsequent removal of amphipods resulted in a return to an epiphyte-dominated community with greatly reduced macroalgal growth rates. Young *Hypnea* sporelings successfully colonized substrate along the mid-point of an amphipod tube, an area subject to less grazing. Mesoherbivores might control the species composition of reef communities in areas (e.g., algal ridges, reef crests) where wave turbulence and topography limit grazing by larger herbivores (Brawley and Adey 1981).

Mesoherbivores may commonly influence community structure by affecting the proportion of epiphytes and macrophytes that occur within a community. Epiphytes may have some beneficial effects upon macrophytes, for example, by retarding desiccation in intertidal algae and by protecting shallow-water plants from photo-inhibition (see discussion in Orth and van Montfrans 1984). Macroalgal growth is not always inhibited by epiphytes in short-term experiments (e.g., Brawley and Fei 1987; Duffy 1990), but there are many demonstrated negative effects of epiphytes, both on seagrasses and macroalgae. These include the loss of heavily epiphytized plants, probably during storms because of the extra drag created by the epiphytes (e.g., D'Antonio 1985), and decreased growth because of a reduction of photosynthetic capacity and nutrient availability by the opaque, diffusion barrier that epiphytes create (e.g., Caine 1980; Brawley and Adey 1981; Howard 1982; Sand-Jensen *et al.* 1985). Thus, those mesograzers that feed primarily upon epiphytes may prolong the longevity and fitness of macrophytes and such mesoherbivory probably occurs in nature. Duffy (1990) found that caprellid amphipods in tank experiments effectively reduced epiphytes on macrophytes at 1/17 the maximum density at which these animals were found at his field sites in North Carolina. Brawley and Fei (1987) tested the effect of placing caprellids on one group of macrophytes (vs controls vs insecticide-treated plants, all groups being subject to natural colonization by mesoherbivores in the field) in an unbarricaded, coastal mariculture farm in China that in many respects resembles the adjacent intertidal and shallow subtidal rocky coast. Plants in the caprellid-enrichment group had lower epiphyte/macrophyte biomasses at the end of the experiments.

Mesograzers may mediate competition between sessile invertebrates and algae in a fashion similar to the interactions described above for ephemeral and perennial algae. Grazing by the *Mithrax sculptus* (maximum size used was 2·5 cm carapace width) prevents the coral *Porites porites* from being smothered by algae (Coen 1988*a*). Control corals and treatments in which crabs were enclosed around corals had an algal cover of 10 per cent or less, whereas greater than 75 per cent of coral surface area was covered by algae such as *Dictyota* spp. within one month after crabs were excluded.

Whether mesoherbivores increase or decrease overall community productivity is an interesting question for future study. Epiphytic and ephemeral algae are frequently among the most productive algal species (references in Hackney *et al.* 1989).

2. Other effects on propagules

Succession is probably facilitated by mesoherbivory because this provides propagule-sized spaces through an inhibitory turf or film of 'early successional' species. There is also evidence of mesoherbivore-triggered release and dispersal of algal propagules. For example, Buschmann and Bravo (1990) demonstrated that *Hyale hirtipalma* dispersed carpospores of the red alga *Iridaea laminarioides* under an *Ulva* turf in the laboratory. Both in the field and in the laboratory, these investigators found carpospores stuck to amphipod legs. Additionally, *Hyale* spp. fed preferentially upon cystocarpic *Iridaea laminarioides*, and released more carpospores from the plants than were released in their absence (Buschmann and Santelices 1987). The carpospores that were released by amphipods were as capable of germination as were the controls. Carpospores that were ingested by amphipods sometimes survived digestion; those that did grew faster than controls, probably because they were deposited into a nutrient-rich microenvironment (faeces). Observations of plants in the field suggested that many cystocarps do not open on their own, and that amphipods may increase the fitness of the plants by aiding this process. The isopod *Idotea wosnesenskii* preferentially consumed cystocarpic *Iridaea cordata*, and Gaines (1985) suggested that this might reduce the plant's fitness; a re-examination of this relationship would be valuable in light of Buschmann and Santelices's (1987) study. Not all red algal cystocarps are fed upon preferentially, however; D'Antonio (1985) found that the amphipod *Ampithoe simulans* did not consume cystocarps of *Neorhodomela larix*.

Other mesograzer effects upon the reproductive capacity of plants and, therefore, on their fitness, may occur. For example, Brawley and Adey (1981) noted that amphipod-grazed *Hypnea* appeared more rapidly to become reproductive; the major effect of this grazing was to remove epiphytes, although tips of branches were also eaten, and this could affect concentrations of putative growth regulators. D'Antonio (1985) found a positive correlation between the amount of reproductive tissue and the amount of epiphyte-free surface of *N. larix*. The effect of mesograzers upon plant reproduction is an exciting area for additional study.

3. Overgrazing of macrophytes

There are several reports of destruction of macroalgal stands by mesograzers. One of these (see Tegner and Dayton 1987; Dayton and Tegner 1990) concerns the destruction of a *Macrocystis pyrifera* community by amphipods (*Ampithoe humernalis*) as part of an El Niño-related disturbance. Whether or not the densities and sizes of amphipods increased is unknown (Tegner, *pers. comm.*), because baseline data on the amphipod community are unavailable. The investigators hypothesized that predation pressure on the amphipods was reduced because associated populations of kelp fishes declined dramatically.

Kangas *et al.* (1982) observed a major reduction in the fucoid community of the Baltic Sea that was related to overgrazing by isopods. Both the total density of isopods (primarily *Idotea baltica*) and the number of adults on *F. vesiculosus* plants increased dramatically. This increase was related to a natural eutrophication event that increased the biomass of filamentous algae. Since smaller isopods feed almost exclusively on such filamentous algae, then migrate to *Fucus* plants as adults, the additional food was hypothesized by Kangas *et al.* (1982) to permit an increase in the isopod population, resulting in the destruction of the *Fucus* community. This community has now recovered; under 'normal' conditions, the isopods do not appear to pose a threat to these macrophytes (Salemaa 1987), although adults clearly graze on *Fucus*. The short life-spans of mesoherbivores (Salemaa 1979; references in Koch 1990), as well as food limitation and predation (see pp. 249–52), may effectively limit the potential of some species to overgraze macro-algal communities.

4. Detrital feeding

Some groups of mesoherbivores are detrital feeders, and this can have an important local effect upon nutrient availability to communities adjacent to the detritus. Many members of the amphipod superfamily Talitroidea (e.g., the beach hoppers) are detrital feeders, and an excellent recent review on them has been written by Wildish (1988). The materials that are consumed by these amphipods include the gamut of algal species and even driftwood; β-glucosidase and some cellulases have been identified in the guts of some *Orchestia* (Wildish 1988). Lopez *et al.* (1977) considered the effect of amphipods (*Orchestia*) on wrack in *Spartina* marshes, and they suggested that the effect upon remineralization of the detritus was due not to a reduction in particle size but to a mesograzer-related increase in microbial biomass. Robertson and Lucas (1983) studied the effect of the amphipod *Allorchestes compressa* in detrital turn-over in the surf zone of sandy Australian beaches adjacent to seagrass and kelp communities. They estimated that the amphipods turned over kelp biomass in the detritus twice a month during some seasons. The isopod *Ligia dilatata* on rocky shores of South Africa consumes annually up to 76 kg (wet wt) of drift kelp m^{-2} of shoreline (Koop and Field 1981), and they extrude 72 per cent of this biomass as faeces. Koop and Field suggested that this is an important food for the filter feeders in the adjacent sublittoral zone and that commercial collection of stranded kelp, by reducing food to these filter feeders, could adversely affect their commercially important predators, rock lobsters.

Algae and seagrasses often have chemical and mechanical defences (against herbivores) that can affect detrital feeders, too. Harrison (1982) found that *Eogammarus confervicolus* did not graze dead *Zostera* leaves if they were treated with water-soluble extracts of living leaves or with any of three phenolic acids, implying that seagrasses may be chemically protected from mesograzers by phenolic compounds. Similarly, grass shrimp (*Palaemonetes pugio*) feed on living epiphytes and detrital seagrass, but not on living seagrasses (Morgan 1980). Robertson and Lucas (1983) found that breaking the epidermis of living *Ecklonia* permitted *Allorchestes* to feed on the parenchyma of that kelp; since the epidermal cell layers were still not eaten, the epidermis appears to provide mechanical and/or chemical protection for living plants against some mesograzers (also see Gaines 1985). Some other algae that are chemically-defended against fish appear not to have chemical defences against mesoherbivores (Hay and Fenical, Chapter 14 this volume).

5. Other effects and summary

Not all effects of mesograzers result from their feeding activities. For example, Perry (1988) found that the boring isopod *Sphaeroma peruvianum* reduces root growth rate by 50 per cent and net root production by 62 per cent in the red mangrove *Rhizophora mangle* on the Pacific coast of Costa Rica.

As the studies reviewed in this section suggest, it is possible to enrich, maintain, or exclude mesoherbivores in some field experiments. However, many of these experiments have been very manpower-demanding because of the small size and high natural densities of animals. Another approach that some investigators have used to predict potential effects of mesoherbivores upon community structure is to examine their feeding capabilities in the laboratory.

The Andean plane crash problem: potential vs realized diets

1. Laboratory experiments

By definition, laboratory experiments can only provide good clues to the effects of mesoherbivores in natural communities. Whether the clues are to the potential or realized effects of a species will depend upon how an experiment is designed vis-à-vis the questions posed in Table 11.2, and it requires careful attention to the natural history of the community. For example, not only will abnormally high consumption (e.g., Sushchenya and Khmeleva 1967; Lukasheva 1971) be observed when animals larger than the natural size-frequency distribution are used in laboratory experiments, but larger animals often eat larger, tougher species (Kangas et al. 1982; but see Brawley and Fei 1987). Salemaa (1987) found that female *Idotea* consumed significantly more *Fucus vesiculosus* than males; mesoherbivore sex ratios are frequently unbalanced in nature (Wenner 1972, Salemaa 1979), and the natural ratio should be tested for clues about realized diets. Mesoherbivores have been starved before testing their feeding preferences in many studies, but starvation can increase food consumption (Lukasheva 1971). Feeding is, also, temperature-dependent (Morgan 1980; Robertson and Lucas 1983); thus, laboratory experiments on temperate species should be conducted over the range of seasonal temperatures by investigators wishing to estimate potential annual consumption rates. Lastly, the range of plant species offered to mesoherbivores and their condition (see Table 11.2, b, c) will affect an investigator's conclusions. This is the problem of the Andean plane

Table 11.2. Important considerations in interpretation of laboratory feeding experiments

---

A. How do the size-frequency, density, and sex-ratio of animals compare to those of the natural community?
B. Was a wide choice of algae (i.e., macroalgae, ephemerals, epiphytes) offered to the grazers?
C. How were foods presented and grazing observed (e.g., as whole plants or plugs of tissue, epiphyte-free or with epiphytes, singly or as a multispecific assemblage? Was it possible to observe the order in which foods were consumed)?
D. Were animals starved before the experiment?
E. What were the general laboratory conditions (e.g., temperature, light cycle, type of aquarium)?

crash victims (see Read 1974): humans are rarely cannibalistic, but may become so under tragic circumstances when their normal foods are unavailable. Robertson and Lucas (1983) demonstrated a shift in species preferences and in total food consumed during a study of the detrital-feeding gammarid *Allorchestes compta*, depending upon whether fresh or drift kelp was offered with three other foods (see their table IV). Some amphipods that feed on macroalgae are preferentially attracted to epiphyte-covered macrophytes (Norton and Benson 1983); this suggests that using epiphyte-free plants might increase macroalgal feeding. This possibility was demonstrated for both an amphipod (*Gammarus lawrencianus*) and an isopod (*Idotea baltica*) by Shacklock and Doyle (1983), who found that the amphipod, which consumed only one (*Palmaria palmata*) of four macrophytes tested, ate less of it when it was covered by epiphytes, which they preferred as food and consumed. Similarly, the isopod ate all of the macrophytes but, when they were covered with epiphytes, only *Palmaria palmata* was seriously damaged.

Another facet of the 'Andean plane crash problem' was demonstrated by Salemaa (1987) who found that *Idotea baltica*'s consumption of *Fucus* varied depending upon the part of the *Fucus* thallus offered to animals. Indeed, any *piece* of a plant may permit greater macroalgal feeding than would occur on intact plants since, as mentioned above, an intact epidermis offers mechanical protection. This suggests that plugs of macroalgal tissue should not be offered to mesoherbivores in tests of feeding preferences. However, mesoherbivores that do feed on intact macroalgae may permit macroalgal feeding by animals or species that would not be able to feed on intact plants (e.g., in a mixed assemblage of the isopod *Dynamene bidentata* and the amphipod *Hyale nilssoni*, isopod feeding on *Fucus* permitted amphipod feeding, Arrontes and Viejo, pers. comm.). Such facilitation demonstrates the complexity of the question, 'What do mesoherbivores eat?' Nicotri (1980) discussed the varying consumption rates obtained in experiments depending upon whether a single species or an array of species were offered to animals. Peterson and Renaud (1989) have discussed the statistical problems (i.e., lack of independence) inherent to presenting multiple species simultaneously to a grazer, but even 'clean' field plants are usually covered by some unicellular epiphytes; there are few, if any, monospecific stands in nature from a mesoherbivore's viewpoint. In multi-specific tests, observation of grazing activity over the full course of the experiment, is also important to interpretation. For example, *Pontogeneia rostrata* grazed the macrophyte *Gracilaria asiatica* on the last day of a week-long experiment, only after all of the *Enteromorpha* had been consumed (Brawley and Fei 1987).

To date, no experiments avoid all the design problems indicated above, but several studies (Zimmerman et al. 1979; Shacklock and Doyle 1983; Brawley and Fei 1987; Duffy 1990) offer good evidence for the differing potentials of various mesograzers to affect the structure of algal communities. As Duffy (1990) emphasized, the effects observed will depend upon the species composition of the mesograzer community; as indicated here, it will also depend very much upon other characteristics such as the size distribution of individuals within the mesoherbivore population.

Useful additions to laboratory protocols will include estimation of assimilation rates of different foods and tests of the fitness of a species feeding on different foods. Zimmerman et al. (1979) found that epiphytes were assimilated much better than macrophytes by the ampithoid *Cymadusa compta* (also see Brenner et al. 1976). Kitting et al. (1984) used stable carbon-isotope analysis to test assimilation of detritus v. epiphytes by shrimp in the field; this technique is potentially very useful in any situation where available foods have two significantly different isotopic signatures. Robertson and Lucas (1983) tested the fitness of amphipods feeding on different food sources over several months, determining their survival, growth, the proportion of ovigerous females, and their maturation rate. Such tests of fitness would seem to be especially useful in cases where foods are suspected of being chemically defended against amphipods (e.g., Harrison 1982) or other herbivore groups (e.g., Hay et al. 1987). It would also be useful to examine gut contents of amphipods fed on different foods in the laboratory to see how accurately these foods can be recognized. Likely diets of amphipods have been described by some investigators (e.g., Nelson 1979b; Robertson and Lucas 1983; D'Antonio 1985) from gut contents of field-collected animals; however, McBane and Croker (1983) found that a preferred food of *Hyale nilssoni*, the filamentous red alga *Polysiphonia lanosa*, was rarely identified in gut contents because of its rapid breakdown and assimilation.

2. Potential diets of different mesoherbivores

A summary of information from both laboratory and field studies of the potential diets of arthropod mesograzers is presented in Table 11.3. The preferred foods of mesoherbivores do not often correlate with the calorific value or nitrogen content of the foods (see Nicotri 1980; Coen 1988b). This may be because the calories in cell-wall polysaccharides are unavailable to many species (Nicotri 1980). Perhaps more importantly, mesoherbivores may eat foods of comparatively low nutritive value that lack mechanical or chemical grazing deterrents (Coen 1988b). The literature on self-selected diets in insects (e.g., Waldbauer et al. 1984; Cohen et al. 1987) suggests that there may be strong age-specific and species-specific effects upon preferred diets, even without the complications of plant defences against herbivores.

Table 11.3 demonstrates that there is some predictability for diets within different groups of mesoherbivores. For example, fly larvae (Diptera) and caprellid amphipods graze smaller algae, especially filamentous and unicellular algae; they also feed on ephemeral algae such as *Ulva* and *Enteromorpha*. Of course, not all amphipod and isopod species are herbivorous (Caine 1974, 1977, Barnes 1980). Some are predators, and even herbivorous species are often cannibalistic or predatory when food is limiting (Holmes 1901; Skutch 1926; McCain 1968; Brawley and Adey 1981). Likewise, many crabs and shrimp have omnivorous diets that include other mesoherbivores (e.g., Dean and Connell 1987, and see p. 249. Of the arthropod mesoherbivores examined thus far, isopods appear to have the greatest potential to damage living macroalgae, based on both field (Naylor 1955; Kangas *et al.* 1982; Gaines 1985) and laboratory (Nicotri 1977; Shacklock and Doyle 1983; Salemaa 1987) work. This may be a reflection, in part, of the larger size of adult idoteids compared to most amphipods. Nicotri (1977) found that *Ampithoe valida* and *Idotea baltica* of equivalent sizes grazed five of six macrophytes at similar rates. However, *Fucus* was one of *Idotea*'s preferred foods and it was not consumed by the amphipods.

Certain groups of isopods (e.g., Ligiidae) and gammarid amphipods (Talitridae and Hyalidae) are especially associated with feeding on beached or drift wrack, but some species (e.g., *Hyale* spp.) in each of these families feed predominantly on living seaweeds. *Ligia* spp. feed on wrack, but also on diatoms, ephemeral algae, and perennials such as *Fucus* (Table 11.3).

Several relatively well-studied amphipod families have very broad diets (Table 11.3). However, even within a single genus, the potential feeding capabilities of the various species show considerable divergence. For example, *Ampithoe ramondi* does not appear to have the potential to graze macroalgae (Brawley and Adey 1981), whereas other species (e.g., *Ampithoe mea*: Norton and Benson 1983: *A. lacertosa*: Brawley and Fei 1987; *A. longimana*: Hay *et al.* 1987; *A. marcuzii*: Duffy 1990) readily feed on macroalgae in laboratory studies, and their potential for overgrazing of macroalgal stands in nature is therefore demonstrated. Why is this overgrazing rarely observed in nature? Among several answers to this question is that of the designs by which macroalgal feeding has been demonstrated and interpreted. For example, although Brawley and Fei (1987) observed macroalgal feeding by both juvenile and adult *Ampithoe lacertosa*, they found that a common ephemeral species was preferred. Similarly Hay *et al.* (1987), although emphasizing macroalgal feeding of *Ampithoe longimana*, actually

Table 11.3. Potential diets of selected Arthropoda on drift (detrital seagrasses and algae, beached or drifting), perennials (e.g., *Fucus*, *Chondrus*, seagrasses), ephemerals (e.g., *Ulva*, *Chondria*), filamentous algae, and unicellular algae such as diatoms.

	Drift	Perennial	Ephemeral	Filamentous	Diatoms
Diptera			x	x	x
Amphipoda					
Caprellidea					
Caprellidae			x	x	x
Gammaridea					
Ampithoidae	x	x	x	x	x
Calliopiidae		x	x	x	x
Corophiidae			x	x	x
Eusiridae			x	x	x
Gammaridae	x	x	x	x	x
Hyalidae	x	x	x	x	x
Ischyroceridae		x	x	x	x
Talitridae	x			x	
Isopoda					
Idoteidae	x	x	x	x	x
Ligiidae	x	x	x	x	x
Decapoda					
Brachyura		x	x	x	x
Caridea	x			x	x

References: Diptera (Robles and Cubit 1981), Caprellidae (*Caprella*: Caine 1974, 1980; Norton and Benson 1983; Brawley and Fei 1987; Duffy 1990), Ampithoidae (*Ampithoe, Cymadusa*: Holmes 1901; Skutch 1926; Nicotri 1977; Zimmerman *et al.* 1979; Brawley and Adey 1981; Norton and Benson 1983; D'Antonio 1985; Brawley and Fei 1987; Hay *et al.* 1987), Calliopiidae (*Calliopius*: Hudon 1983; McGrouther 1983; Pederson and Capuzzo 1984), Corophiidae (*Grandidierella, Lembos*: Zimmerman *et al.* 1979; Shillaker and Moore 1987), Eusiridae (*Pontogeneia*: Brawley and Fei 1987), Gammaridae (*Melita, Gammarus*: Ravanko 1969; Zimmerman *et al.* 1979; Price and Hylleberg 1982; Harrison 1982; Shacklock and Doyle 1983), Hyalidae (*Allorchestes, Hyale*: McBane and Croker 1983; McGrouther 1983; Robertson and Lucas 1983; D'Antonio 1985; Buschmann and Santelices 1987; references in Wildish 1988; Buschmann 1990), Ischyroceridae (*Jassa*: Brawley and Fei 1987; Duffy 1990), Talitridae (*Talorchestia, Orchestia*, Brenner *et al.* 1976; references in Wildish 1988), Ligiidae (*Ligia*: Nicholls 1931; Koop and Field 1981; Willows 1987), Idoteidae (*Idotea*, Naylor 1955; Nicotri 1977; Kangas *et al.* 1982; Shacklock and Doyle 1983; Gaines 1985; Salemaa 1987), Brachyura (general references: Knudsen 1964; Wagner 1977; Morris *et al.* 1980; Majiidae: *Mithrax, Pugettia*: Leighton 1966; Coen 1988a, 1988b; Caine 1980; Grapsidae: *Pachygrapsus*, Robles 1982), Caridea (Palaemonidae: *Palaemonetes*: Welsh 1975; Morgan 1980; Warwick *et al.* 1982; Kitting *et al.* 1984) demonstrated (compare their figs 4 and 5) that little of the preferred macrophyte *Dictyota dichotoma* would likely be consumed in the field. This is because several common epiphytes were consumed at the same rate or at significantly greater rates than the *Dictyota* in their experiments and, in their region of North Carolina, Penhale (1977) estimated that 25 per cent of the above-ground biomass of seagrass beds consisted of epiphytic algae. Duffy (1990) demonstrated that *Ampithoe marcuzii* preferred *Sargassum* (plugs of tissue) compared with several other macrophytes (including *Enteromorpha*), but significant reductions in *Sargassum* biomass compared to controls were observed only when large animals at three times the highest observed field density were used in the laboratory experiments. The several cases where overgrazing of macroalgae has been observed in the field (see p. 242) include a case in which a release from predation was hypothesized to have allowed overgrazing by an ampithoid. This discussion of laboratory feeding experiments suggests how significant a release from predation might be if it results in increased recruitment and in higher numbers of large animals in the population. To what extent do predators limit the full expression of the potential diets of mesoherbivores?

The ecology and evolution of predator/mesoherbivore interactions

1. The cast of characters

Amphipods, isopods, polychaetes, small crabs, and other mesograzers are important components of the diets of many marine fishes (Gibson 1972; Hobson 1974; Nelson 1979*b*; Stoner 1979; Kelley 1987; MacDonald and Waiwood 1987), even of some largely herbivorous species (e.g., Lobel 1980). Birds are also predators of mesoherbivores, both in the rocky intertidal zone and on tidal flats (Bent 1921; Goss-Custard 1969, 1970; Baltz and Morejohn 1977; Schneider 1978; Gratto et al. 1984; Braune 1987). Especially when predation by fishes has been experimentally reduced or controlled, predation by carnivorous or omnivorous crabs and shrimps on small mesograzers has been demonstrated (Young et al. 1976; Virnstein 1977; Van Dolah 1978; Nelson 1979*a*). Even a jellyfish is reported by Holmes (1901) to capture swimming *Ampithoe longimana*.

In order to understand how predators might limit the ecological potential of mesoherbivores, we need to consider the ecology and evolution of predator–prey interactions. Most of the invertebrate predators of mesoherbivores are tactile predators. However, visual detection of prey is important to many fishes and some birds. Blegvad (1922) suggested that nocturnal vs diurnal activity patterns were evolved by mesoherbivores in response to predation pressure from visual predators; this hypothesis has been accepted by many subsequent workers (e.g., Jansson and Källander 1968). A possible evolutionary route to this behaviour is discussed by Hobson (1974), who studied the gut contents of 102 species of diurnal and nocturnal fishes on a Hawaiian reef. His data show that fish having a dietary volume with a significant (10 per cent or greater) crab component are split almost evenly between diurnal and nocturnal species, but fish with a quantitatively similar amphipod-based diet were all diurnal.

Alldredge and King (1980) found that amphipods and isopods avoided moonlit nights, returning to benthic refuges at moonrise and they suggested that this behaviour reduced predation. Jansson and Källander (1968) quantified the very low light intensities at which amphipods and isopods became more active. Mesoherbivores may reduce their overall susceptibility to predation by being relatively inactive by day (also see Morgan 1990; for general comments, Barnes 1980), but they are important prey to many generalized nocturnal predators (e.g., Hobson 1974). One omnivorous fish even feeds on algae by day but on mesoherbivores at night (Robertson and Klumpp 1983). Increased size (Van Dolah 1978, Nelson 1979a; Edgar 1983b; Main 1985), markings that make prey more visible (Clements and Livingston 1984), and movement (Main 1985; Russo 1987) make mesoherbivores more susceptible to some visual predators in laboratory experiments, consistent with predictions (Protasov 1970). It is not surprising that many mesoherbivores have cryptic coloration and body ornamentation (Gamble and Keeble 1900; Lee and Gilchrist 1972; Field 1974; Salemaa 1978; Wallerstein and Brusca 1982; Pederson and Capuzzo 1984). Wallerstein and Brusca (1982) suggested that characteristics of different species of isopods, such as maximum body length and size at reproductive maturity, may have evolved in response to a latitudinal gradient in the intensity of fish predation.

2. Experimental manipulations of predators in the field

It appears that predator-prey interactions may have had a significant effect upon mesoherbivore activity patterns, life history traits, and morphology. Clearly, mesoherbivores occupy a key, intermediate position in marine food chains. If predators are excluded in the field, do densities of mesoherbivores increase within the protected areas? The densities and/or sizes of mesoherbivores increased under cages in several experiments (Vince *et al.* 1976, Young *et al.* 1976; Van Dolah 1978; Kennelly 1983; Zeller 1988). Changes in vegetation under caged areas co-occurred with increased numbers of animals in Kennelly's (1983) and Zeller's (1988) experiments. As indicated earlier, some experiments of this nature did not result in increases in mesoherbivore populations; the major reason for this was an increase in decapod predators under the cages. However, it is clear that there are important natural population cycles in these short-lived organisms that are not damped by predation, and in some habitats, especially where benthic complexity is high (e.g., *Corallina* turf, Choat and Kingett 1982), seasonal fluctuations in mesoherbivore densities appear to be relatively independent of predator control. The converse experiment, enclosing predators under a cage to reduce mesoherbivore density, was successful in experiments by Nelson (1979b) and by Warwick et al. (1982); an *Enteromorpha* turf persisted under Warwick et al.'s cages, whereas it disappeared from adjacent areas of mud-flat (also see Brawley and Adey 1981).

A few field studies of mesoherbivores have been equivocal; for example, the imposed predation level was too low for levels of natural immigration in one study (Holmlund et al. 1990). Carpenter (1986) located mesoherbivore treatments off the reef in the lagoon to isolate them from predators and macroherbivores, but this may have retarded mesoherbivore colonization; the final densities of mesoherbivores in his treatments were low (cf. Lobel 1980; Klumpp et al. 1988; Zeller 1988). Additionally, sedimentation problems may have made Carpenter's colonization plates unsuitable as a mesoherbivore habitat. Nonetheless, Carpenter's (1986) assessment that the herbivorous sea urchins had a greater effect on the algal community of the reef, is probably true from the viewpoint of standing biomass (see p. 253). However, many Caribbean reefs recently experienced mass mortalities of the sea urchin *Diadema antillarum*, and some of these (e.g., in Jamaica) already had low fish populations because of overfishing. Macroalgae currently dominate the Jamaican reef; filamentous species are only common in damselfish territories (Hughes et al. 1987). The macroalgal community structure may well be maintained by high numbers of mesoherbivores and this possibility should be investigated (see pp. 238–41 this volume; Brawley and Adey 1981). Mesoherbivore densities are generally higher in damselfish territories than in other areas of the reef; this may not be true at present on the Jamaican reef. Since both damselfish and mesoherbivores eat filamentous algae, the normal interaction between the fish and their mesoherbivores deserves additional attention. Presumably the fish garden the algae and raise mesoherbivores (see especially Lobel 1980), with strong size-selective predation on the mesoherbivores to prevent overgrazing of the fish territories.

Progress in understanding predator control of mesoherbivore populations in the field will depend upon studies at the species-level of interactions between predators and prey, including the roles of habitat complexity and habitat selection. Plant-dwelling mesoherbivores are protected from some predators as the complexity of their habitat increases (Vince et al. 1976; Nelson 1979a; Stoner 1980; Coen et al. 1981; Coull and Wells 1983; Dean and Connell 1987); a threshold level of complexity is required, and 'complexity' appears to depend more strongly upon plant surface area than on morphology (e.g., Stoner 1980; Dean and Connell 1987). Active selection of habitat is affected by a plant's value as both refuge and food (e.g., Nicotri 1980), may change as the animal grows (Edgar 1983a), and is not yet well understood (Russo 1987). Hay et al. (1987, 1988) suggested that mesoherbivores that live and feed on algae that are chemically defended against herbivorous fishes might have higher fitness because they would avoid accidental ingestion. However, it is unknown as yet whether such a benefit would be overcome by predation from omnivorous and carnivorous fishes. Fish do have quite different feeding strategies, although this is one of the areas in which we most need additional data. For example, some fish pick their way through fronds, whereas other fish ambush prey (e.g., Coull and Wells 1983; Main 1985). Some fish overlook tube-dwelling amphipods inside their tubes (Nelson 1979a; Brawley and Adey 1981), whereas other species recognize these as a food source (Brawley and Adey 1981, Gunnill 1984). A very specialized amphipod (*Pseudoamphithoides incurvaria*) makes domiciles out of a chemically-defended alga; this appears to reduce its susceptibility to predation by the wrasse *Thalassoma bifasciatum* (Hay et al. 1990) but 20 per cent of the wrasse *Halichoeres bivittatus* examined by Lewis and Kensley (1982) on the reef had eaten 'podded' amphipods.

As Russo remarked (1987): 'The feeding behaviour of most marine fish predators is still unknown and therefore generalizations explaining fish predator–prey interactions in the marine environment cannot be made.' This clearly limits our ability to make predictions about the extent to which predation controls mesoherbivore populations. Some examples (Choat and Kingett 1982; Brawley and Fei 1987) exist in which mesoherbivores may exert an effect on the structure of their communities that is independent of, or in spite of, predation. In other cases (e.g., Vince et al. 1976), predators limit the size structure of the mesograzer population, which can obviously limit the potential of mesograzers to overgraze macrophytes.

Conclusions and recommendations

This review began by suggesting that mesoherbivores occupy a pivotal role in some marine communities. There is indeed evidence that mesoherbivores are important players in the first (propagule dispersal, facilitation of succession) and final (longevity of macrophytes and sessile invertebrates) acts, besides which, they keep most of the other cast (e.g., fishes, shore birds) well fed. An algal community is often defined more sharply by the presence of macroherbivores (e.g., fishes, urchins, some molluscs) and some molluscan mesoherbivores, than by arthropod mesoherbivores. Macrophytes that are eliminated from a community by grazers will usually be eliminated faster by macroherbivores than by mesoherbivores. The size of the animal often affects both consumption rates and the kinds of plants that are consumed. This does not diminish the importance of mesograzers in natural communities, because there is increasing evidence that many mesoherbivores keep macrophytes and some sessile invertebrates clean(er), which affects many aspects of community structure. This is a community role especially associated with mesoherbivores. However, macroalgal feeding can be substantial and significant, if additive in most cases. Indeed, it is too early for many generalizations about mesoherbivores, and generalizations even for the same species may turn awry in different ecosystems: mesoherbivores are at a pivotal point. The specific predators present, varying environmental conditions and the availability of appropriate foods for juveniles, among other factors, will affect the size-structure, overall density, and sex ratios of mesoherbivores. This variation will produce different effects, as the literature on isopods already suggests. Good natural history will help to interpret why there are differences between our experiments and expectations.

Understanding the natural history of these organisms is as much a part of experimental design as is good replication. Specifically, it would be helpful for investigators to report size-frequencies on a subsample of the natural mesoherbivore population selected for experimental work and to express the overall density as animals m^{-2} or as animals/plant weight along with a determination of plant density. Our present base of experimental and descriptive information on mesoherbivores is a treasure-trove of questions. For example, how often is food limiting for mesoherbivores in nature? Do mesoherbivore populations undergo strong fluctuations as successfully recruiting animals clean up macrophytes and become more evident to predators; or do they emigrate before being eaten? And what role do thousands m^{-2} of dipteran larvae play in the *Ascophyllum* zone (Table 11.1)?

Mesoherbivores may figure prominently in the outcome of some of the man-made environmental problems that we face. For example, eutrophication of coastal waters, which can increase epiphyte loads beyond the carrying capacity of macrophyte communities (Orth and van Montfrans 1984), can be offset by epiphyte-grazers (but only if the mesoherbivores have survived toxic pollutants). If this translates into large adults with different grazing capabilities from the normal-sized population, however, macrophyte overgrazing by mesoherbivores may be observed more commonly.

Acknowledgements

I thank the staff of the Vanderbilt Science Library for their assistance in locating older literature. Preparation of this manuscript was supported by National Science Foundation #8802065.

References

Alldredge, A. L. and King, J. M.(1980). Effects of moonlight on the vertical migration patterns of demersal zooplankton. *Journal of Experimental Marine Biology and Ecology*, **44**, 133-56.

Baltz, D. M. and Morejohn, G. V. (1977). Food habits and niche overlap of sea-birds wintering on Monterey Bay, California. *The Auk*, **94**, 526-43.

Barnes, R. D. (1980). *Invertebrate Zoology* (4th edn). W. B. Saunders Co., Philadelphia.

Bent, A. C. (1921). Life histories of North American gulls and terns. *United States National Museum Bulletin*, **113**, 1-345.

Blegvad, H. (1922). On the biology of some Danish gammarids and mysids. *Reports of the Danish Biological Station*, **28**, 1-103.

Braune, B. M. (1987). Seasonal aspects of the diet of Bonaparte's gulls (*Larus philadelphia*) in the Quoddy region, New Brunswick, Canada. *The Auk*, **104**, 167-72.

Brawley, S. H. and Adey, W. H. (1981). The effect of micrograzers on algal community structure in a coral reef microcosm. *Marine Biology*, **61**, 167-77.

Brawley, S. H. and Fei, X. G. (1987). Studies of mesoherbivory in aquaria and in an unbarricaded mariculture farm on the Chinese coast. *Journal of Phycology*, **23**, 614-23.

Brawley, S. H. and Fei, X. G. (1988). Ecological studies of *Gracilaria asiatica* and *Gracilaria lemaneiformis* in Zhan-shan Bay, Qingdao. *Chinese Journal of Oceanology and Limnology*, **6**, 22-34.

Brenner, D., Valiela, I., van Raalte, C. D., and Carpenter, E. J. (1976). Grazing by *Talorchestia longicornis* on an algal mat in a New England salt marsh. *Journal of Experimental Marine Biology and Ecology*, **22**, 161-9.

Brostoff, W. N. (1985). Seaweed grazing and attachment by the nereid polychaete *Platynereis dumerilii*. In *Proceedings of the Fifth International Coral Reef Congress*, 4, pp. 3–8. EPHE, Tahiti.

Buschmann, A. H. (1990). Intertidal macroalgae as refuge and food for amphipoda in central Chile. *Aquatic Botany*, 36, 237–45.

Buschmann, A. H. and Bravo, A. (1990). Intertidal amphipods as potential dispersal agents of carpospores of *Iridaea laminarioides* (Gigartinales, Rhodophyta). *Journal of Phycology*, 26, 417–20.

Buschmann, A. and Santelices, B. (1987). Micrograzers and spore release in *Iridaea laminarioides* Bory (Rhodophyta: Gigartinales). *Journal of Experimental Marine Biology and Ecology*, 108, 171–9.

Caine, E. A. (1974). Comparative functional morphology of feeding in three species of caprellids (Crustacea, Amphipoda) from the northwestern Florida Gulf coast. *Journal of Experimental Marine Biology and Ecology* 15, 81–96.

Caine, E. A. (1977). Feeding mechanisms and possible resource partitioning of the Caprellidae (Crustacea: Amphipoda) from Puget Sound, USA. *Marine Biology*, 42, 331–6.

Caine, E. A. (1980). Ecology of two littoral species of caprellid amphipods (Crustacea) from Washington, USA. *Marine Biology*, 56, 327–35.

Carpenter, R. C. (1984). Predator and population density control of homing behavior in the Caribbean echinoid *Diadema antillarum*. *Marine Biology*, 82, 101–8.

Carpenter, R. C. (1986). Partitioning herbivory and its effects on coral reef algal communities. *Ecological Monographs*, 56, 345–63.

Choat, J. H. and Kingett, P. D. (1982). The influence of fish predation on the abundance cycles of an algal turf invertebrate fauna. *Oecologia* (Berlin), 54, 88–95.

Clements, W. H. and Livingston, R. J. (1984). Prey selectivity of the fringed filefish *Monacanthus ciliatus* (Pisces: Monacanthidae): Role of prey accessibility. *Marine Ecology Progress Series*, 16, 291–5.

Coen, L. D. (1988a). Herbivory by crabs and the control of algal epibionts on Caribbean host corals. *Oecologia* (Berlin), 75, 198–203.

Coen, L. D. (1988b). Herbivory by Caribbean majid crabs: feeding ecology and plant susceptibility. *Journal of Experimental Marine Biology and Ecology*, 122, 257–76.

Coen, L. D., Heck K. L. Jr., and Abele, L. G. (1981). Experiments on competition and predation among shrimps of seagrass meadows. *Ecology*, 62, 1484–93.

Cohen, R. W., Heydon, S. L., Waldbauer, G. P., and Friedman, S. (1987). Nutrient self-selection by the omniverous cockroach *Supella longipalpa*. *Journal of Insect Physiology*, 33, 77–82.

Colman, J. (1939). On the faunas inhabiting intertidal seaweeds. *Journal of the*

Marine Biological Association of the United Kingdom, **24**, 129–83.

Coull, B.C. and Wells, J.B.J. (1983). Refuges from fish predation: experiments with phytal meiofauna from the New Zealand rocky intertidal. *Ecology*, **64**, 1599–1609.

D'Antonio, C. (1985). Epiphytes on the rocky intertidal red alga *Rhodomela larix* (Turner) C. Agardh: negative effects on the host and food for herbivores. *Journal of Experimental Marine Biology and Ecology*, **86**, 197–218.

Dayton, P.K. and Tegner, M.J. (1991). Bottoms beneath troubled waters: benthic impacts of the 1982–1984 El Niño in the temperate zone. In *Global ecological consequences of the 1982–1984 El Niño in the temperate zone*, (ed. P.W. Glynn), pp. 433–72. Elsevier, Amsterdam.

Dean, R.L. and Connell, J.H. (1987). Marine invertebrates in an algal succession. III. Mechanisms linking habitat complexity with diversity. *Journal of Experimental Marine Biology and Ecology*, **109**, 249–73.

Duffy, J.E. (1990). Amphipods on seaweeds: Partners or pests? *Oecologia* (Berlin), **83**, 267–76.

Edgar, G.J. (1983a). The ecology of south-east Tasmanian phytal animal communities. I. Spatial organization on a local scale. *Journal of Experimental Marine Biology and Ecology*, **70**, 129–57.

Edgar, G.J. (1983b). The ecology of south-east Tasmanian phytal animal communities. IV. Factors affecting the distribution of ampithoid amphipods among algae. *Journal of Experimental Marine Biology and Ecology*, **70**, 205–25.

Fauchald, K. and Jumars, P.A. (1979). The diet of worms: a study of polychaete feeding guilds. *Oceanography and Marine Biology Annual Review*, **17**, 193–284.

Field, L.H. (1974). A description and experimental analysis of Batesian mimicry between a marine gastropod and an amphipod. *Pacific Science*, **28**, 439–47.

Fincham, A.A. (1974). Periodic swimming behaviour of amphipods in Wellington Harbor. *New Zealand Journal of Marine and Freshwater Research*, **8**, 505–21.

Gaines, S.D. (1985). Herbivory and between-habitat diversity: the differential effectiveness of defenses in a marine plant. *Ecology*, **66**, 473–85.

Gamble, F.W. and Keeble, F.W. (1900). *Hippolyte varians*: a study in colour-change. *Quarterly Journal of Microscopical Science*, **43**, 589–698.

Gibson, R.N. (1967). Studies on the movements of littoral fish. *Journal of Animal Ecology*, **36**, 215–34.

Gibson, R.N. (1972). The vertical distribution and feeding relationships of intertidal fish on the Atlantic coast of France. *Journal of Animal Ecology*, **41**, 189–207.

Goss-Custard, J.D. (1969). The winter feeding ecology of the redshank *Tringa totanus*. *Ibis*, **111**, 338–56.

Goss-Custard, J. D. (1970). Feeding dispersion in some overwintering wading birds. In *Social behavior in birds and mammals*, (ed. J. H. Crook), pp. 3-35. Academic Press, London.

Gratto, G. W., Thomas, M. L. H., and Gratto, C. L. (1984). Some aspects of the foraging ecology of migrant juvenile sandpipers in the outer Bay of Fundy. *Canadian Journal of Zoology*, **62**, 1889-92.

Gunnill, F. C. (1984). Differing distributions of potentially competing amphipods, copepods and gastropods among specimens of the intertidal alga *Pelvetia fastigiata*. *Marine Biology*, **82**, 277-91.

Hackney, J. M., Carpenter, R. C., and Adey, W. H. (1989). Characteristic adaptations to grazing among algal turfs on a Caribbean coral reef. *Phycologia*, **28**, 109-19.

Harrison, P. G. (1982). Control of microbial growth and of amphipod grazing by water-soluble compounds from leaves of *Zostera marina*. *Marine Biology*, **67**, 225-30.

Hay, M. E., Duffy, J. E., Pfister, C. A., and Fenical, W. (1987). Chemical defense against different marine herbivores: are amphipods insect equivalents? *Ecology*, **68**, 1567-80.

Hay, M. E., Renaud, P. E., and Fenical, W. (1988). Large mobile versus small sedentary herbivores and their resistance to seaweed chemical defenses. *Oecologia* (Berlin), **75**, 246-52.

Hay, M. E., Duffy, J. E., and Fenical, W. (1990). Host-plant specialization decreases predation on a marine amphipod: an herbivore in plant's clothing. *Ecology*, **71**, 733-43.

Heck, K. L. Jr. and Orth, R. J. (1980). Structural components of eelgrass (*Zostera marina*) meadows in the lower Chesapeake Bay — Decapod Crustacea. *Estuaries*, **3**, 289-95.

Hobson, E. S. (1974). Feeding relationships of teleostean fishes on coral reefs in Kona, Hawaii. *Fishery Bulletin*, **72**, 915-1031.

Holmes, S. J. (1901). Observations on the habits and natural history of *Ampithoe longimana* Smith. *Biological Bulletin of the Marine Biological Laboratory, Woods Hole*, **2**, 165-93.

Holmlund, M. B., Peterson, C. H., and Hay, M. E. (1990). Does algal morphology affect amphipod susceptibility to fish predation. *Journal of Experimental Marine Biology and Ecology*, **139**, 65-83.

Howard, R. K. (1982). Impact of feeding activities of epibenthic amphipods on surface-fouling of eelgrass leaves. *Aquatic Botany*, **14**, 91-7.

Hruby, T. and Norton, T. (1979). Algal colonization on rocky shores in the Firth of Clyde. *Journal of Ecology*, **67**, 65-77.

Hudon, C. (1983). Selection of unicellular algae by the littoral amphipods *Gammarus oceanicus* and *Calliopius laeviusculus* (Crustacea). *Marine Biology (Berlin)*, **78**, 59-67.

Hughes, T. P., Reed, D. C., and Boyle, M. J. (1987). Herbivory on coral reefs: community structure following mass mortalities of sea urchins. *Journal of Experimental Biology and Ecology*, **113**, 39–59.

Jansson, B.-O. and Källander, C. (1968). On the diurnal activity of some littoral peracarid crustaceans in the Baltic Sea. *Journal of Experimental Marine Biology and Ecology*, **2**, 24–36.

Kangas, P., Autio, H., Hallfors, G., Luther, H., Niemi, A., and Salemaa, H. (1982). A general model of the decline of *Fucus vesiculosus* at Tvarminne, south coast of Finland in 1977–81. *Acta Botanica Fennica*, **118**, 1–27.

Kelley, D. F. (1987). Food of bass in U.K. waters. *Journal of the Marine Biological Association of the United Kingdom*, **67**, 275–86.

Kennelly, S.J. (1983). An experimental approach to the study of factors affecting algal colonization in a sublittoral kelp forest. *Journal of Experimental Marine Biology and Ecology*, **68**, 257–76.

Kennelly, S.J. and Underwood, A.J. (1985). Sampling of small invertebrates on natural hard substrata in a sublittoral kelp forest. *Journal of Experimental Marine Biology and Ecology*, **89**, 55–67.

Kitting, C.L., Fry, B., and Morgan, M.D. (1984). Detection of inconspicuous epiphytic algae supporting food webs in seagrass meadows. *Oecologia* (Berlin), **62**, 145–9.

Klumpp, D.W., McKinnon, A.D., and Mundy, C.N. (1988). Motile cryptofauna of a coral reef: abundance, distribution and trophic potential. *Marine Ecology Progress Series*, **45**, 95–108.

Knudsen, J.W. (1964). Observations of the reproductive cycles and ecology of the common Brachyura and crablike Anomura of Puget Sound, Washington. *Pacific Science*, **18**, 3–33.

Koch, H. (1990). Aspects of the population biology of *Traskorchestia traskiana* (Stimpson 1857) (Amphipoda, Talitridae) in the Pacific Northwest, U.S.A. *Crustaceana*, **59**, 35–52.

Koop, K. and Field, J.G. (1981). Energy transformation by the supralittoral isopod *Ligia dilatata* Brandt. *Journal of Experimental Marine Biology and Ecology*, **53**, 221–33.

Ledoyer, M (1969). Écologie de la faune vagile des biotopes Méditerranéens accessibles en scaphandre autonome. V. Étude des phénomènes nycthéméraux, les variations nycthémérales des populations animales dans les biotopes. *Téthys*, **1**, 291–308.

Lee, W.L. and Gilchrist, B.M. (1972). Pigmentation, color change and the ecology of the marine isopod *Idotea resecata* (Stimpson). *Journal of Experimental Marine Biology and Ecology*, **10**, 1–27.

Leighton, D.L. (1966). Studies of food preference in algivorous invertebrates of Southern California kelp beds. *Pacific Science*, **20**, 104–13.

Lewis, S.M. and Kensley, B. (1982). Notes on the ecology and behavior of *Pseudoamphithoides incurvaria* (Just) (Crustacea, Amphipoda,

Ampithoidae). *Journal of Natural History*, **16**, 267-74.
Lobel, P. S. (1980). Herbivory by damsel-fishes and their role in coral reef community ecology. *Bulletin of Marine Science*, **30**, 273-89.
Longley, W. H. (1927). Life on a coral reef: the fertility and mystery of the sea studied beneath the waters surrounding Dry Tortugas. *National Geographic Magazine*, **51**, 61-83.
Lopez, G. R., Kevinton, J. S., and Slobodkin, L. B. (1977). The effect of grazing by the detritivore *Orchestia grillus* on *Spartina* litter and its associated microbial community. *Oecologia* (Berlin), **30**, 111-27.
Lubchenco, J. (1983). *Littorina* and *Fucus*: effects of herbivores, substratum heterogeneity, and plant escapes during succession. *Ecology*, **64**, 1116-23.
Lukasheva, T. A. (1971). Quantity of food consumed as a function of body weight in *Idotea baltica* (Pallas). *Hydrobiological Journal*, **7**, 87-90.
MacDonald, J. S. and Waiwood, K. G. (1987). Feeding chronology and daily ration calculations for winter flounder (*Pseudopleuronectes americanus*), American plaice (*Hippoglossoides platessoides*), and ocean pout (*Macrozoarces americanus*) in Passamaquoddy Bay, New Brunswick. *Canadian Journal of Zoology*, **65**, 499-503.
Main, K. L. (1985). The influence of prey identity and size on selection of prey by two marine fishes. *Journal of Experimental Marine Biology and Ecology*, **88**, 145-52.
Manton, S. M. (1952). The evolution of arthropodan locomotory mechanisms. Part 2. *Journal of the Linnean Society, London (Zoology)*, **42**, 93-117.
McBane, C. D. and R. A. Croker (1983). Animal-algal relationships of the amphipod *Hyale nilssoni* (Rathke) in the rocky intertidal. *Journal of Crustacean Biology*, **3**, 592-601.
McCain, J. B. (1968). The Caprellidae (Crustacea: Amphipoda) of the western North Atlantic. *Bulletin of the United States National Museum*, **278**, 1-147.
McGrouther, M. A. (1983). Comparison of feeding mechanisms in two intertidal gammarideans, *Hyale rupicola* (Haswell) and *Paracalliope australis* (Haswell) (Crustacea: Amphipoda). *Australian Journal of Marine and Freshwater Research*, **34**, 717-26.
Montouchet, P. C. G. (1979). Sur la communauté des animaux vagile associés à Ubatuba, État de São Paulo, Brésil. *Studies of Neotropical Fauna*, **14**, 33-64.
Morgan, M. D. (1980). Grazing and predation of the grass shrimp *Palaemonetes pugio*. *Limnology and Oceanography*, **25**, 896-902.
Morgan, S. G. (1990). Impact of planktivorous fishes on dispersal, hatching, and morphology of estuarine crab larvae. *Ecology*, **71**, 1639-52.
Morris, R. H., Abbott, D. P., and Haderlie, E. C. (1980). *Intertidal invertebrates of California*. Stanford University Press, Stanford, California.

Naylor, E. (1955). The diet and feeding mechanism of *Idotea*. *Journal of the Marine Biological Association of the United Kingdom*, **34**, 347-55.

Nelson, W. G. (1979a). Experimental studies of selective predation on amphipods: consequences for amphipod distribution and abundance. *Journal of Experimental Marine Biology and Ecology*, **38**, 225-45.

Nelson, W. G. (1979b). An analysis of structural pattern in an eelgrass (*Zostera marina* L.) amphipod community. *Journal of Experimental Marine Biology and Ecology*, **39**, 231-64.

Nelson, W. G. (1980). A comparative study of amphipods in seagrasses from Florida to Nova Scotia. *Bulletin of Marine Science*, **30**, 80-9.

Nicholls, A. G. (1931). Studies on *Ligia oceanica*. Part II. The processes of feeding, digestion and absorption, with a description of the structure of the foregut. *Journal of the Marine Biological Association of the United Kingdom*, **17**, 675-707.

Nicotri, M. E. (1977). Impact of crustacean herbivores on cultured seaweed populations. *Aquaculture*, **12**, 127-36.

Nicotri, M. E. (1980). Factors involved in herbivore food preference. *Journal of Experimental Marine Biology and Ecology*, **42**, 13-26.

Norton, T. A. and Benson, M. R. (1983). Ecological interactions between the brown seaweed *Sargassum muticum* and its associated fauna. *Marine Biology*, **75**, 169-77.

Orth, R. J. and Montfrans, J. van (1984). Epiphyte-seagrass relationships with an emphasis on the role of micrograzing: a review. *Aquatic Botany*, **18**, 43-69.

Pederson, J. B. and Capuzzo, J. M. (1984). Energy budget of an omniverous rocky shore amphipod, *Calliopius laeviusculus* (Kröyer). *Journal of Experimental Marine Biology and Ecology*, **76**, 277-91.

Penhale, P. A. (1977). Macrophyte-epiphyte biomass and productivity in an eelgrass (*Zostera marina* L.) community. *Journal of Experimental Marine Biology and Ecology*, **26**, 211-24.

Perry, D. M. (1988). Effects of associated fauna on growth and productivity in the red mangrove. *Ecology*, **69**, 1064-75.

Peterson, C. H. and Renaud, P. E. (1989). Analysis of feeding preference experiments. *Oecologia* (Berlin), **80**, 82-6.

Price, L. H. and Hylleberg, J. (1982). Algal-faunal interactions in a mat of *Ulva fenestrata* in False Bay, Washington. *Ophelia*, **21**, 75-88.

Protasov, V. R. (1970). *Vision and near orientation of fish*, (trans. M. Raveh). Israel Program for Scientific Translations, Jerusalem.

Ravanko, O. (1969). Benthic algae as food for some invertebrates in the inner part of the Baltic. *Limnologica*, **7**, 203-5.

Read, P. P. (1974). *Alive*. Harper and Row, New York.

Ricketts, E. F., Calvin, J., Hedgpeth, J. W., and Phillips, D. W. (1985). *Between Pacific tides* (5th edn). Stanford University Press, Stanford, California.

Robertson, A. I. and Klumpp, D. W. (1983). Feeding habits of the southern Australian garfish *Hyporhamphus melanochir*: a diurnal herbivore and nocturnal carnivore. *Marine Ecology Progress Series*, **10**, 197–201.

Robertson, A. I. and Lucas, J. S. (1983). Food choice, feeding rates and the turnover of macrophyte biomass by a surf-zone inhabiting amphipod. *Journal of Experimental Marine Biology and Ecology*, **72**, 99–124.

Robles, C. D. (1982). Disturbance and predation in an assemblage of herbivorous Diptera and algae on rocky shores. *Oecologia* (Berlin), **54**, 23–31.

Robles, C. D. and Cubit, J. (1981). Influence of biotic factors in an upper intertidal community: dipteran larvae grazing on algae. *Ecology*, **62**, 1536–47.

Russo, A. R. (1987). Role of habitat complexity in mediating predation by the gray damselfish *Abudefduf sordidus* on epiphytal amphipods. *Marine Ecology Progress Series*, **36**, 101–5.

Salemaa, H. (1978). Geographical variability in the colour polymorphism of *Idotea baltica* (Isopoda) in the northern Baltic. *Heriditas*, **88**, 165–82.

Salemaa, H. (1979). Ecology of *Idotea* spp. (Isopoda) in the Northern Baltic. *Ophelia*, **18**, 133–50.

Salemaa, H. (1987). Herbivory and microhabitat preferences of *Idotea* spp. (Isopoda) in the northern Baltic Sea. *Ophelia*, **27**, 1–15.

Sand-Jensen, K., Revsbech, N. P., and Barker-Jörgensen, B. (1985). Microprofiles of oxygen in epiphyte communities on submerged macrophytes. *Marine Biology* (Berlin), **89**, 55–62.

Schneider, D. (1978). Equalization of prey numbers by migratory shorebirds. *Nature*, **271**, 353–4.

Shacklock, P. F. and Doyle, R. W. (1983). Control of epiphytes in seaweed cultures using grazers. *Aquaculture*, **31**, 141–51.

Shillaker, R. O. and Moore, P. G. (1978). Tube-building by the amphipods *Lembos websteri* Bate and *Corophium bonnellii* Milne Edwards. *Journal of Experimental Marine Biology and Ecology*, **33**, 169–85.

Skutch, A. F. (1926). On the habits and ecology of the tube-dwelling amphipod *Ampithoe rubricata* Montagu. *Ecology*, **7**, 481–502.

Sousa, W. P. (1979). Experimental investigations of disturbance and ecological succession in a rocky intertidal algal community. *Ecological Monographs*, **49**, 227–54.

Stoner, A. W. (1979). Species-specific predation on amphipod crustacea by the pinfish *Lagodon rhomboides*: mediation by macrophyte standing crop. *Marine Biology*, **55**, 201–7.

Stoner, A. W. (1980). Perception and choice of substratum by epifaunal amphipods associated with seagrasses. *Marine Ecology Progress Series*, **3**, 105–11.

Sushchenya, L. M. and Khmeleva, N. N. (1967). Consumption of food as a function of body weight in crustaceans. *Doklady (Proceedings) of the Academy of Sciences of the USSR, New York, Washington*, **172**, 559-62.

Tattersall, W. M. (1913). Clare Island Amphipoda. *Proceedings of the Irish Academy*, **31**, 1-24.

Tegner, M. J. and Dayton, P. K. (1987). El Niño effects on Southern California kelp forest communities. *Advances in Ecological Research*, **17**, 243-79.

Van Dolah, R. F. (1978). Factors regulating the distribution and population dynamics of the amphipod *Gammarus palustris* in intertidal salt-marsh communities. *Ecological Monographs*, **48**, 191-217.

Vince, S., Valiela, I., Backus, N., and Teal, J. M. (1976). Predation of the salt-marsh killifish *Fundulus heteroclitus* (L.) in relation to prey size and habitat structure: consequences for prey distribution and abundance. *Journal of Experimental Marine Biology and Ecology*, **23**, 255-66.

Virnstein, R. W. (1977). The importance of predation by crabs and fishes on benthic infauna in Chesapeake Bay. *Ecology*, **58**, 1199-1217.

Waldbauer, G. P., Cohen, R. W., and Friedman, S. (1984). Self-selection of an optimal nutrient mix from defined diets of larvae of the corn earworm, *Heliothis zea*. *Physiological Zoology*, **57**, 590-97.

Wallerstein, B. R. and Brusca, R. C. (1982). Fish predation: a preliminary study of its role in the zoogeography and evolution of shallow-water idoteid isopods (Crustacea: Isopoda: Idoteidae). *Journal of Biogeography*, **9**, 135-50.

Warner, G. (1977). *The biology of crabs*. Van Nostrand Reinhold, New York.

Warwick, R. M., Davey, J. T., Gee, J. M., and George, C. L. (1982). Faunistic control of *Enteromorpha* blooms: a field experiment. *Journal of Experimental Marine Biology and Ecology*, **56**, 23-31.

Welsh, B. L. (1975). The role of grass shrimp, *Palaemonetes pugio*, in a tidal marsh ecosystem. *Ecology*, **56**, 513-30.

Wenner, A. M. (1972). Sex ratio as a function of size in marine crustacea. *The American Naturalist*, **106**, 321-50.

Wildish, D. J. (1988). Ecology and natural history of aquatic Talitroidea. *Canadian Journal of Zoology*, **66**, 2340-59.

Williams, A. B. and Bynum, K. H. (1972). A ten year study of meroplankton in North Carolina estuaries: amphipods. *Chesapeake Science*, **13**, 175-92.

Willows, R. I. (1987). Population and individual energetics of *Ligia oceanica* (L.) (Crustacea: Isopoda) in the rocky supralittoral. *Journal of Experimental Marine Biology and Ecology*, **105**, 253-74.

Young, D. K., Buzas, M. A., and Young, M. W. (1976). Species densities of macrobenthos associated with seagrass: a field experimental study of predation. *Journal of Marine Research*, **34**, 577-92.

Zeller, D. C. (1988). Short-term effects of territoriality of a tropical damselfish and experimental exclusion of large fishes on invertebrates in algal turfs. *Marine Ecology Progress Series*, **44**, 85–93.

Zimmerman, R., Gibson, R., and Harrington, J. (1979). Herbivory and detritivory among gammaridean amphipods from a Florida seagrass community. *Marine Biology*, **54**, 41–7.

12. Ecological and physiological aspects of herbivory in benthic suspension-feeding molluscs

B. L. BAYNE and A. J. S. HAWKINS

The Plymouth Marine Laboratory, Prospect Place, West Hoe, Plymouth, England, UK

Abstract

Benthic suspension- (or filter-) feeding bivalve molluscs such as mussels, oysters, and clams act as important agents of sedimentation in estuaries and coastal systems. Some investigations suggest that suspension-feeding may be a dominant factor controlling phytoplankton biomass, although hydrodynamic and other factors may act to limit the availability of particulate seston to the feeding currents of the bivalves. These animals may also affect nutrient regeneration, directly via nitrogenous excretion or indirectly via microbial mineralization of their biodeposits. Many filter-feeding bivalves are opportunistic feeders, relying on phytoplankton, organic detritus and bacteria in proportions that vary with location and with season. We consider some key physiological aspects of these interactions and conclude that feeding behaviour is flexible in response to changes in food quantity and quality; rates of nitrogen excretion, although variable, may be modelled as a function of rates of nitrogen ingestion; and there are circumstances under which bivalves may be carbon- rather than nitrogen-limited by the available food. Better understanding of these physiological processes is necessary for further insights into the role of suspension-feeders in ecosystem processes

Introduction

Suspension-feeding molluscs are normally considered to be herbivores, relying on phytoplankton as food. Studies by Verwey in the Dutch

Plant–Animal Interactions in the Marine Benthos, (ed. D.M. John, S.J. Hawkins, and J.H. Price), Systematics Association Special Volume No. 46, pp. 265–88. Clarendon Press, Oxford, 1992. © The Systematics Association 1992.

Wadden Sea four decades ago (Verwey 1952) placed the feeding behaviour of cockles (*Cerastoderma edule*) and mussels (*Mytilus edulis*) within the context of ecosystem processes such as sedimentation (Haven and Morales-Alamo 1966) and the carbon and nutrient budgets of coastal environments (Dame et al. 1980; Jordan and Valiela 1982). More recently, the roles of such benthic suspension-feeders in the control of phytoplankton biomass and in nutrient cycling within coastal waters have received increased attention as key factors in coastal ecosystem dynamics (Officer et al. 1982; Dame et al. 1989). Such studies have led, in turn, to concepts of food limitation within natural populations of bivalve molluscs, including suggestions of self-limitation (Fréchette and Lefaivre 1990). There have also been detailed examinations of the true diet of bivalves in their natural habitats, including the roles of detritus and bacteria (Mann 1988; Langdon and Newell 1990).

To gain further quantitative understanding of the ecological consequences of such interrelationships requires improved knowledge of the factors which determine feeding behaviour, metabolism, and growth of the suspension-feeders. In this paper we discuss briefly some recent studies of the ecological role of suspension- (or filter-) feeding bivalve molluscs, and we then consider some physiological processes which are most relevant to this role.

The role of suspension-feeding bivalves in ecosystem processes

1. The control of phytoplankton biomass

Studies postulating a role for benthic suspension feeders in controlling phytoplankton biomass are normally based on models of the hydrodynamics of the system under consideration (e.g., San Francisco Bay, USA: Officer et al. 1982; Cloern 1982; Nichols 1985; Laholm Bay, Sweden: Loo and Rosenberg 1989; St Lawrence Estuary, Canada: Demers et al. 1989), and knowledge of the standing stock and production of the phytoplankton and of the benthic macrofauna. Extrapolations from laboratory determinations of the filtration rates of appropriate suspension-feeders (i.e., the volume of water filtered free of particles per unit time) allow gross budgeting of phytoplankton flux, which suggests that these animals may control phytoplankton biomass as a function of hydrodynamic residence time, the dynamics of phytoplankton growth, the mixing regime, and the biomass and filtration capacity of the filter-feeding fauna.

These 'budgeting' studies were acknowledged by their authors to be simplifications of reality; they relied on laboratory determinations of filtration rates using as food pure phytoplankton cultures, which were not representative of natural diets, and which overestimated natural rates of feeding (Loo and Rosenberg 1989). They disregarded effects of particle concentration on filtration rate (Nichols 1985) and they ignored the various factors that may limit the flux of particles actually available to the suspension-feeders within the benthic boundary layer (Wildish and Kristmanson 1984; Muschenheim 1987). Doering and Oviatt (1986) specifically challenged the application of various filtration-rate models to experimental data from mesocosm experiments and concluded (p. 274) that '. . . filtering rate models founded on other than natural suspensions of particulate matter are unlikely accurately to reflect processes in the field'. Fréchette and Bourget (1985a) concluded (p. 1158) that '. . . food is often depleted immediately above mussel populations, and water movement is critical in determining food availability to suspension feeders'.

Mesocosm and field experiments have thrown further light on the rather complex relationships between filter-feeding behaviour, the types and quantities of particles in suspension, and the net removal of suspended particulate organic matter (POM) and particulate inorganic matter (PIM). Cloern (1982) and Carlson et al. (1984) related the alterations in phytoplankton community structure to the activity of benthic filter-feeders, and Doering et al. (1986) concluded that benthic suspension-feeding may select against phytoplankton which sink rapidly (e.g., large diatoms) and promote the growth of smaller forms which are less likely to contribute to net benthic sedimentation. Riemann et al. (1988) considered that the impact of mussels as a controlling factor on phytoplankton biomass depended, *inter alia*, on the phytoplankton size structure; when the phytoplankton were dominated by picoplankton of 1–2 μm cell diameter, mussels were less effective at reducing phytoplankton biomass than when the phytoplankton cells were larger. Kemp et al. (1990) recorded the 'substantial capacity' of the mussel *Geukensia demissa* to select among particles in suspension and possibly to alter selective preferences over a short time-span; they suggested that factors other than size of available particles may be important in affecting the impact of suspension-feeding on the fate of the seston within the salt-marsh, since some of the changes they observed were independent of particle size.

Sessile benthic suspension-feeders depend upon water transport to bring food particles within range of their feeding currents; the flux of potential food items is a function of the interaction between particle concentrations, the potential particle settling velocities, current flow near the bed, and the mixing characteristics of the region. Hydrodynamic conditions within the near-bed benthic boundary layer are critical in defining the food environment (Wildish and Kristmanson

1984; Fréchette et al. 1989). The consequences of the feeding activity of the bivalves themselves in depleting the food supply to the population as a whole have been explored by Wildish and Kristmanson (1985), Fréchette and Bourget (1985b) and Petersen and Black (1987), among others. For example, Petersen and Black (1987) conducted a field experiment to test the effects of tidal elevation and bivalve density on growth and survivorship of filter-feeding bivalves on a tidal flat (Western Australia). Differences in density and in the time available for feeding (which declined with increased tidal elevation) could not explain growth differences. The authors concluded that depletion of food items as a result of feeding by bivalves low on the tidal-flat led to food limitation at the higher tidal sites; they also speculated that different particle preferences by different bivalve species imposed further complexity on interactions between the food environment and both distributions and growth of the bivalves.

In such studies it is necessary to discriminate between the effects of particle flux which act on population processes, such as 'food-driven self thinning' (Fréchette and Lefaivre 1990), and effects on the feeding behaviour of the individual organisms. Further, the two main components of particle flux (particle concentration and flow velocity) need to be uncoupled as factors affecting individual feeding behaviour (Grizzle and Lutz 1989). Cahalan et al. (1989) addressed this problem and concluded that the growth rates of juvenile bay scallops (*Argopecten irradians*) were predominantly a function of the algal cell concentrations, and much less dependent on flow velocity. Wildish et al. (1987) have postulated a 'reverse ramp function' to describe the relationship between flow velocity and growth; at low velocities, growth will be independent of flow, until a critical velocity is reached at which feeding is suppressed. Whatever processes may be acting to determine the concentration of particles within feeding reach of the bivalves (e.g., particle clearance by individuals elsewhere within the population; frictional bed stress; vertical mixing in the water column), it appears that seston concentration and nutritional quality are the key factors which directly affect feeding behaviour, and therefore the role of suspension-feeders in controlling phytoplankton biomass, with water-flow velocity *per se* becoming significant at high flow velocities. Selective feeding, in response both to particle size (Bayne and Newell 1983) and to chemical cues from the microplankton (Ward and Targett 1989) are important factors also.

2. Biodeposition and nutrient exchanges

Many suspension-feeding molluscs filter a greater mass of particulate material from suspension than is actually ingested; the filtered material is subjected to ciliary sorting mechanisms on the ctenidia and labial palps, much material being rejected from the mantle cavity as pseudofaeces (reviewed by Bayne and Newell 1983). The proportion of filtered material that is rejected varies between species and with environmental circumstances, but may often be as high as 80–90 per cent (Kuenzler 1961: for *Geukensia demissa*; Tenore *et al.* 1973: for *Mytilus edulis*, *Crassostrea virginica*, and *Mercenaria mercenaria*; Sornin *et al.* 1988: for *C. virginica*). Resuspended biodeposits, comprised both of pseudofaeces and the true faeces, may contribute significantly to total suspended loads in coastal environments (Kautsky 1981; Newell *et al.* 1982; Smaal *et al.* 1986). In the kelp ecosystem of the Cape Peninsula, South Africa, resuspended biodeposits comprise an important component of food for the suspension-feeders under both upwelling and downwelling conditions (Field *et al.* 1986). Kautsky and Evans (1987) suggested that biodeposition by mussels increases the total annual deposition of carbon, nitrogen and phosphorus by 10 per cent in the Asko region of the Baltic; remineralization of these biodeposits recycles nutrients back to the water column, accounting both for all of the computed nutrient demands of the benthic primary producers and for 12 per cent and 22 per cent, respectively, of the annual nitrogen and phosphorus demands of pelagic primary producers.

Finally, we refer to the role suspension-feeders may play in nutrient cycling as a result of their excretion of dissolved nitrogenous products, particularly ammonia. In their mesocosm experiments, Doering *et al.* (1986) concluded that the effects of filter-feeders on production and on the transport of carbon from the water column to the benthos were due in large part to their role in nutrient cycling. Measurements in various field situations have confirmed this role, for example: a salt-marsh (Jordan and Valiella 1982; Bertness 1984); oyster beds (Dame *et al.* 1985; Boucher and Boucher-Rodoni 1988); mussel beds (Dame and Dankers 1988; Asmus *et al.* 1990); a brackish lagoon (Nakamura *et al.* 1988). It is important to recognize, however, that although the measured rates of exchange for particles and for nutrients may be dominated by the activities of the suspension-feeders themselves, physical factors and other biotic components of the benthic communities will also be playing a part (Dame *et al.* 1989).

This brief survey of some of the main areas of influence of molluscan suspension-feeders on the dynamics of coastal ecosystems (see Fig. 12.1) highlights certain features of the feeding biology and metabolism of the bivalves that need to be addressed. We will discuss some of these below, but first we pose the question 'What material actually comprises the natural food which is digested and assimilated by filter-feeding bivalves?'

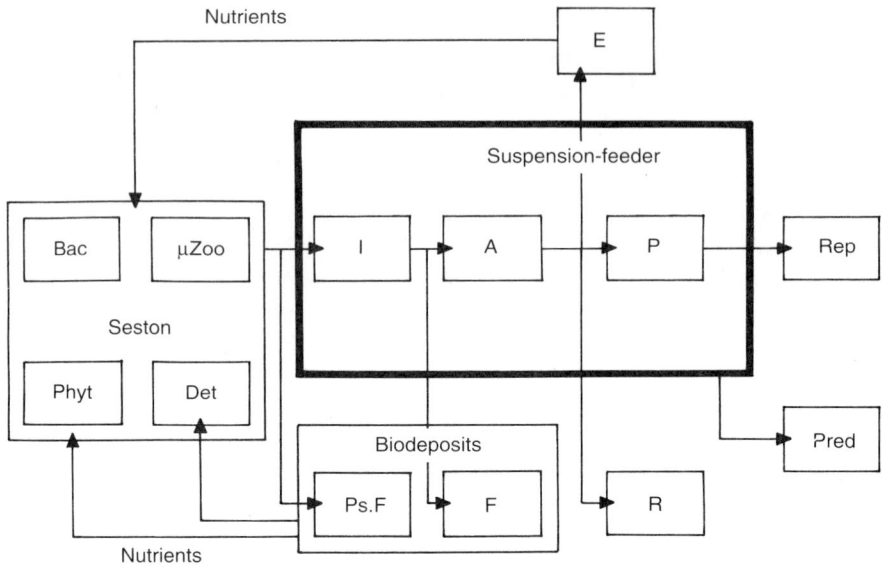

Fig. 12.1. A schematic diagram showing some of the ways in which suspension-feeding bivalves interact with ecological processes. I, ingestion; A, absorption; P, production; E, excretion; R, respiration; Rep, reproduction; Pred, predation; Bac, bacteria; µZoo, microzooplankton; Phyt, phytoplankton; Det, detritus; Ps. F, pseudofaeces; F, true faeces.

The diet of suspension-feeders

The 'food environment' for suspension-feeding bivalves comprises the particulate material (the seston) which is within reach of their feeding currents. The caveat, that it is the seston available to the feeding currents that is important, is significant (references and discussion above), although the dimensions of the normal feeding currents of bivalves, and how these may interact with hydrodynamic forces within the benthic boundary layer, are unknown. It is clear, nevertheless, that bivalves may have access to a great variety of particulate food items, including bacteria, phytoplankton of various species, detritus (which may vary from recently dead phytoplankton cells to long-degraded terrestrial vegetation), and microzooplankton. Various studies (e.g., Berg and Newell 1986) have demonstrated considerable spatial and seasonal variability in these seston components as potential food for suspension-feeders.

Stable-isotope tracers (such as ^{13}C, ^{15}N, and ^{34}S) have been used to trace the flows of such particulate organic matter through the biota of estuaries, salt-marshes, and offshore sites. By relating the isotopic 'signature' of source material (e.g., phytoplankton, *Spartina*-derived detritus, terrestrial plants) to measured isotopic depletions or enrichments within animal tissues, a quantitative measure may be gained of the animals' diet. Such studies have concluded that many bivalve filter-feeders are opportunistic as regards diet. For example, Peterson *et al.* (1985, 1986) examined a number of bivalve species within a salt-marsh and concluded (p. 873) that '. . . the bivalves ingest a mixture . . . of bacteria, algae and detritus, with the relative proportions varying according to location and season'. Fig. 12.2 shows some of their data, along with comparable measures made by Langdon and Newell (1990) on a different salt-marsh system. Similar conclusions regarding the dependence of diet on available suspended matter were reached by Stephenson and Lyon (1982) for a suspension-feeder (*Chione stutchburyi*) in an estuary in New Zealand. At an offshore site (Georges Bank in the North-west Atlantic), Fry (1988) recorded depth-related differences in isotopic composition (^{13}C) in *Placopecten magellanicus* which may be due to spatial differences in phytoplankton distribution. Some selection against available suspended particulates may nevertheless be apparent in some situations; Stephenson *et al.* (1986) concluded from isotopic composition determinations that mussels from

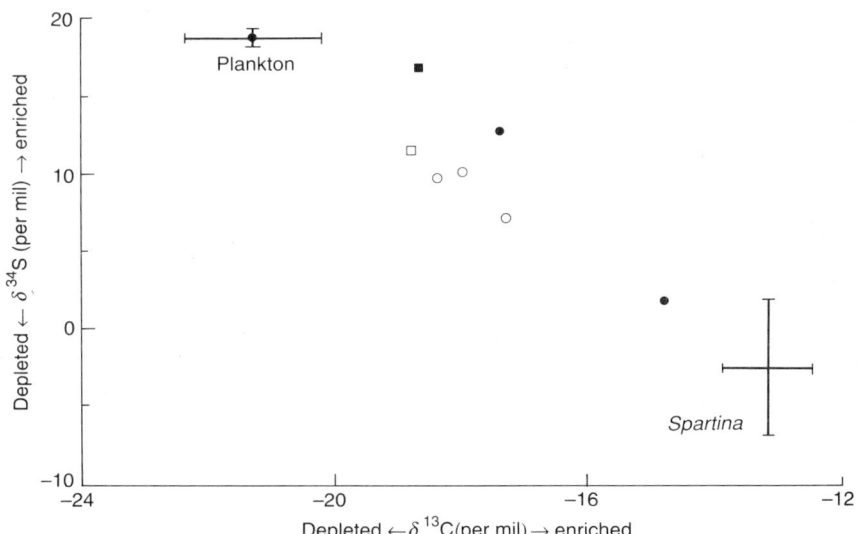

Fig. 12.2. Stable isotope signatures for plankton, *Spartina*, *Geukensia demissa* (○, ●), and *Mytilus edulis* (□, ■) from two salt-marshes in the USA. Filled symbols are from Peterson *et al.* (1985), open symbols from Langdon and Newell (1990). After Langdon and Newell (1990).

a seagrass bed were not assimilating *Zostera*-based detritus, relying instead on phytoplankton.

Other approaches have been used to deduce the natural diet of suspension-feeding bivalves. For example, Widdows *et al.* (1979) and Lucas *et al.* (1987) (for *Mytilus edulis* in a UK estuary) and Seiderer and Newell (1985) and Fielding and Davis (1989) (for *Choromytilus meridionalis* and *Aulacomya ater* in a South African kelp-bed ecosystem) related energy-balance studies on individual mussels to characteristics of the seston nominally available. Such studies have suggested that phytoplankton and organically rich detrital particles provide the dominant proportion of the diet, although bacteria, which are relatively enriched in nitrogen, may in some circumstances be a significant nitrogenous source. Laboratory experiments in which attempts have been made to reproduce the naturally available seston have also illuminated the dietary significance of both phytoplankton and various types of detrital particle (Kiørboe *et al.* 1980, 1981; Bayne *et al.* 1987; Cranford and Grant 1990). Using the results of laboratory feeding expriments, Newell and Field (1983) and Field *et al.* (1986) have modelled the flows of carbon and nitrogen in a kelp-bed ecosystem and have postulated that changes in the relative contributions of bacteria, phytoplankton, detritus, and resuspended faeces as food to the suspension-feeders are dependent on patterns of water transport through the kelp system.

The most powerful approach to elucidating the nutritional value of natural particulates to bivalve suspension-feeders has proved to be a detailed analysis of the animals' digestive and absorptive efficiencies when fed carefully prepared diets in laboratory experiments, combined with a budgetary approach to determine the percentage contributions of the dietary components to the animals' carbon and nitrogen requirements. Langdon and Newell (1990) have recently reviewed, with colleagues (Kreeger *et al.* 1990; Crosby *et al.* 1990), their experiments which address the roles of *Spartina*-detritus and bacteria in the diets of the oyster *Crassostrea virginica* and the mussel *Geukensia demissa*. They estimated that free-living bacteria and those attached to refractory detrital material may provide, in summer, 5·5 per cent and 31 per cent of the metabolic carbon requirements of the oysters and mussels, respectively. Similar calculations indicated that bacteria could contribute 27 per cent and 71 per cent of the metabolic nitrogen requirements of these two bivalves. Although these careful experiments highlight significant nutritional contributions from detritus and bacteria, they also demonstrate that other food sources, such as phytoplankton, nanozooplankton (which may feed on microbial organisms which themselves utilize detrital particles) and, possibly, dissolved organic matter (Manahan et al. 1983), are also potentially significant dietary elements.

Finally, we draw attention to some of the many studies that have demonstrated the influence of food availability in controlling the growth of bivalves in their natural habitats (e.g., in a variety of bivalve suspension-feeders: Peterson and Black 1984; *Corbicula fluminea*: Foe and Knight 1985; *Ostrea edulis* and *Pecten maximus* in suspended culture: Wilson 1987; *Crassostrea gigas*: Brown 1988; *Macoma balthica*: Thompson and Nichols 1988; *Mytilus edulis*: Bayne and Widdows 1978; Navarro et al. 1991). Such studies make plain that it is not only the quantity of available seston that is significant, but also its nutritional quality. A key physiological variable which determines the effects of dietary quality upon individual energy balance is absorption efficiency (Ae). Fig. 12.3 (from Navarro et al. 1991) illustrates the decline in Ae that is

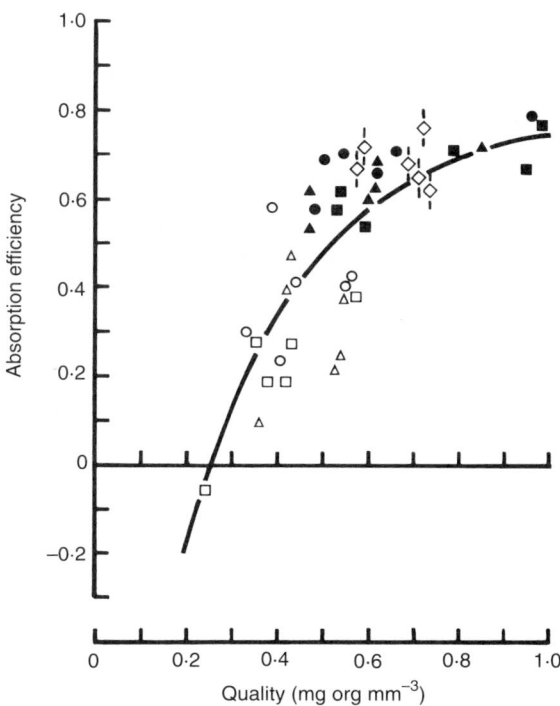

Fig. 12.3. Absorption efficiency in *Mytilus edulis*, related to food quality expressed as organic content per unit volume of particulate matter. The data are from studies in the Ria de Arosa, Galicia, Spain, and are based on natural diets; different symbols refer to different experiments. After Navarro et al. (1991).

consequent upon the ingestion of particles of reduced organic content. Indeed, below a certain threshold, Ae takes negative values as the animal is unable to recover from the diet sufficient energy-return to repay its 'investment' in the processes of digestion. These 'investments' include digestive enzymes as well as mucus and the discharged end-products of intracellular digestion within the cells of the digestive gland (Bayne and Hawkins 1990).

By utilizing different particles to different degrees, bivalves may impose qualitative as well as quantitative changes on the composition of the seston within estuarine and coastal systems. In addition, bivalves may themselves be subject to periods of food limitation on seasonal and possibly shorter time-scales, and on different spatial scales depending on the hydrodynamic mixing regime and the distribution of individuals within the populations. Therefore, interactions between all these sets of processes need to be considered when evaluating the ecological role of suspension-feeders.

Relevant physiological processes

Some of the likely complexities in the relationships between physiological traits as expressed by individual suspension-feeding bivalves and associated ecological processes emerge from this brief survey. Three aspects in particular may benefit from further research which draws on the results of recent studies: factors controlling the rates of feeding; processes governing rates of excretion of nitrogenous metabolic end-products; relative balances between carbon and nitrogen in the dietary requirements for growth.

1. Rates of feeding

In addition to the normal allometric relationship with body size, rates of filtration are known to be dependent on temperature, on seston quantity and quality, and on seasonal changes in the physiological condition of the individual animal (Winter 1976; Bayne and Newell 1983). Doering and Oviatt (1986), following earlier studies by Hibbert (1977) and others (see Bayne and Newell 1983), derived an empirical equation combining animal size and temperature as the best descriptor of filtration rate of *Mercenaria* in mesocosm experiments. Such descriptive equations could readily be combined with equally empirical descriptions of declining rates of filtration with increasing seston concentrations as measured using natural or closely simulated natural diets (Widdows et al. 1979; Bayne et al. 1989). The effects of changes in the nutritional quality of the seston are less easy to generalize; Bayne et al. (1987) observed a decline in filtration rates when mussels were fed a reduced-quality diet, but suggested that the more important physiological responses to such changes occur within the digestive tract (in terms of digestive and absorption efficiency) rather than in the rates of feeding *per se*.

As homeostatic systems, organisms show physiological adaptations to changes in dietary quality and quantity, and suspension-feeding organisms are no exception (see Jørgensen 1989 for an alternative view). In an attempt to explore the roles of both exogenous and endogenous factors in determining rates of filtration, Willows (1991) derived a model based on optimal foraging theory (e.g., Lehman 1976; Taghon 1981), but incorporating a role for 'digestive investment'. This model recognizes the interrelatedness of filtration rate, gut residence time, and the efficiency with which ingested material is digested and absorbed. Following Sibly (1981), the model also incorporates terms to describe the investment that the animal makes in digestion (e.g., in the form of digestive enzyme production and the products of intracellular digestion processes), and predicts the consequences for the optimal filtration rate and net energy-balance of changes in digestibility of the diet and in the rate of recovery, by intestinal absorption, of the by-products of this digestive investment. Fig. 12.4a illustrates the model formulation of the relationship between gut residence time and absorption efficiency. Negative Ae values exist at low gut residence times, when the digestive investment exceeds the energy return from the food.

Fig. 12.4b is a model output plot of net energy balance (EN) as a function of gut residence time (T) and the rate of digestive investment (S), and illustrates a region of optimal values for T and S where energy balance is maximized. In Fig. 12.4c, filtration rate (F) is shown as a function of T and S; the model predicts values for F, at maximal EN, which are themselves less than maximal. Indeed, at high filtration rates (and consequently low gut residence times), over a wide range of digestive investments, net energy balance is predicted to decline, as a result of reduced absorption efficiencies. In Fig. 12.4d a rather different aspect of this model is illustrated; here the optimal filtration rate is shown as a function of food (= seston) concentration for different values of a metabolic cost term which embraces the energy costs of feeding. The form of this curve is similar to some experimental data (e.g., Winter 1976;) and well illustrates the need, in ecological studies, to recognize the dynamic nature of the feeding behaviour of bivalves as they respond to changes in their food environment.

This model (Willows 1991) implicitly recognizes that the nutritional quality of a particle depends partly on its energy content per unit volume, and partly on features of the digestive system, which govern how nearly (and how quickly) the maximum energy gained approaches

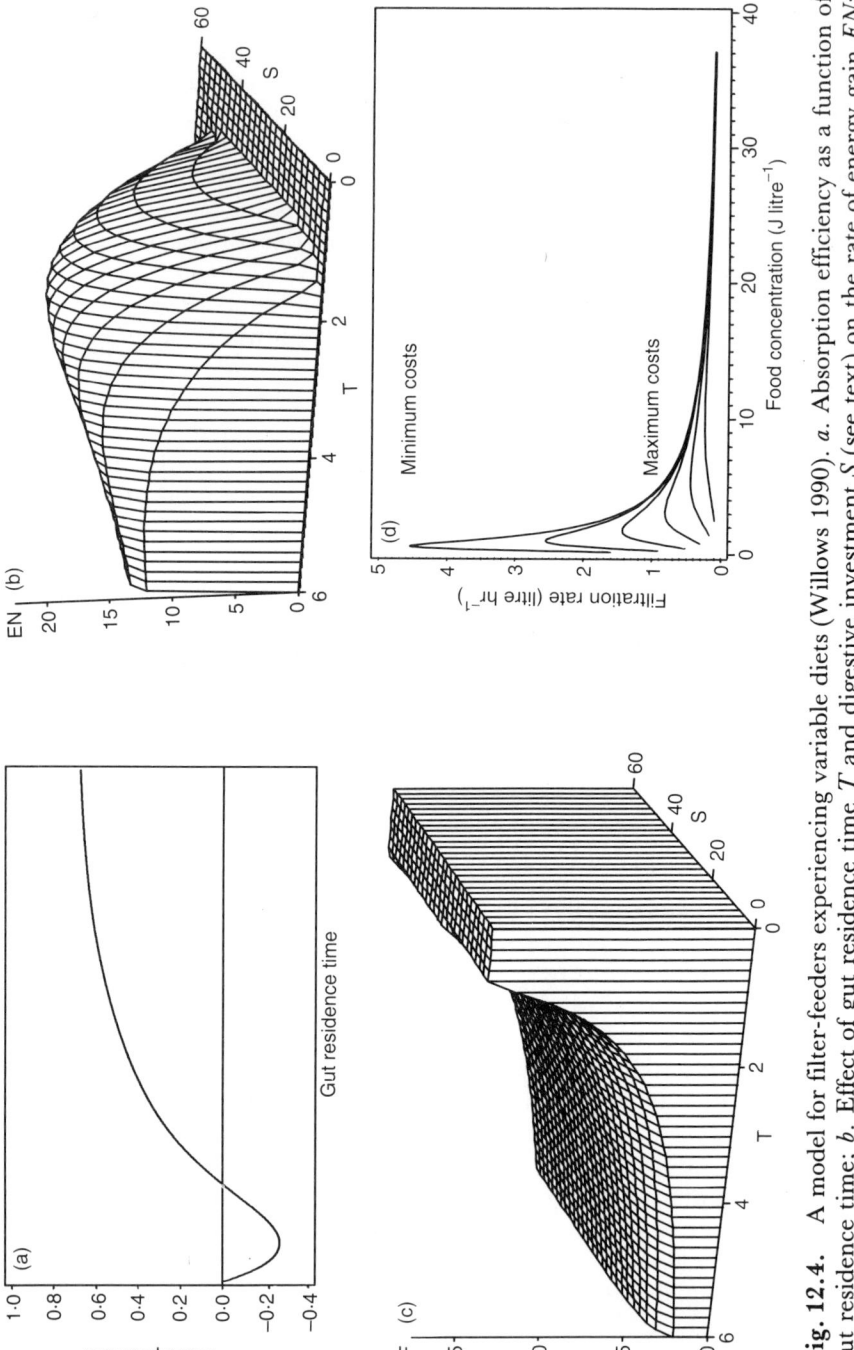

Fig. 12.4. A model for filter-feeders experiencing variable diets (Willows 1990). *a.* Absorption efficiency as a function of gut residence time; *b.* Effect of gut residence time T and digestive investment S (see text) on the rate of energy gain EN; *c.* The effect of gut residence time T and digestive investment S on filtration rate F; *d.* The effects of food concentration and a parameter describing the costs of filtration on optimal filtration rate. The different curves represent the relationship for different cost functions.

the total energy content of the particle. The important balances here are between the shortest gut residence times and digestive investments that are consistent with maximizing digestibility and absorption. When feeding on seston mixtures of different qualities, the animals are predicted to select against less favourable particles (Kiørboe et al. 1981; Newell and Jordan 1983; Bayne et al. 1989) above certain threshold concentrations. Under these circumstances, filtration rates and absorption efficiencies increase, and the quantity of the digestive investment declines. Although these model predictions have not yet been tested explicitly, they bear a qualitative resemblance to physiological compensations observed as *Mytilus edulis* individuals were subjected in laboratory experiments to changes in both the quantity and quality of the available food (Bayne et al. 1987).

We refer to this modelling approach in some detail to emphasize two points. Firstly, a relatively detailed understanding of the feeding behaviour of suspension-feeders, which is a necessary prerequisite to gaining further insight into the ecological role of these animals, is now available and has been subjected to modelling based on optimal foraging theory. But, secondly, further experimental studies are still neccessary, and these should challenge the predictions of the modelling approach, particularly with respect to the complex interplay of exogenous and endogenous factors in determining the rates of feeding in the natural environment.

2. Rates of excretion

Rates of nitrogen excretion (the primary end-product is ammonia) by filter-feeding bivalves have been subjected to less experimental analysis than have rates of feeding. Measurements have indicated that rates of excretion may be extremely variable on seasonal time-scales, and the form of this variability may change with body size (Bayne and Widdows 1978). Mussels (*Mytilus edulis*), for example, utilize stored protein as an energy-yielding substrate during certain periods in the seasonal reproductive cycle (Gabbott 1983; Hawkins and Bayne 1985). The contribution of protein relative to carbohydrate and lipid substrates, however, changes with body size, resulting in seasonally variable exponents relating rates of nitrogen excretion to tissue mass. No simple descriptive equation is available which embraces these effects of body size, temperature, and reproductive condition on total rates of ammonia release (compare feeding rates, discussed above). In large part, this variability is due to variability in the balance between rates of *exogenous* nitrogen excretion (i.e., excretion deriving directly from assimilated dietary nitrogen) and rates of *endogenous* excretion (i.e., excretion by the animal in fasting conditions), as we will discuss later.

Possibly the most useful way to formulate rates of ammonia release, as a contribution to understanding processes of nutrient cycling, is to consider nitrogen excretion as a proportion of nitrogen ingested or assimilated. In their measurements on a mussel bed, Asmus et al. (1990) estimated (p. 60) that '. . . about 37 per cent of the net uptake of particulate nitrogen is recycled by the mussel bed system'. This figure agrees rather well with the physiological data, although some of the recycled nitrogen will have been due to excretion by organisms other than mussels. Fig. 12.5 plots values from three studies on *Mytilus edulis*; as the amounts of nitrogen ingested increase, the percentage excreted per day declines, from 100 per cent at maintenance ration (by definition), to between 5 per cent and 15 per cent at ration

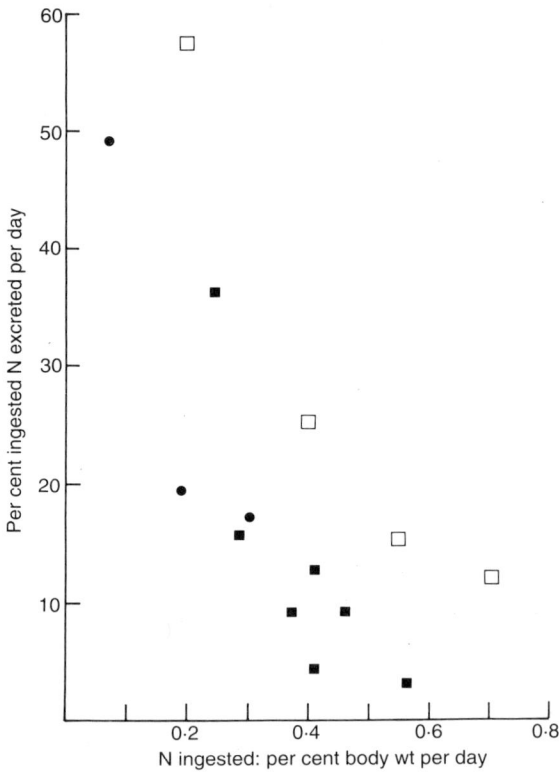

Fig. 12.5. The per cent of ingested nitrogen that is excreted as ammonia per day by *Mytilus edulis*, as a function of the rate of nitrogen ingestion expressed as the per cent of animal dry body weight per day. Values are from Bayne and Widdows (1978; ■), Hawkins and Bayne (1985; ●) and Hawkins et al. (1989; □).

levels from five to eight times the maintenance value. Some of the variability in this figure may be due to animal body size, with the smaller individuals (Hawkins *et al.* 1989) having higher proportional excretion rates. There is also recent evidence (see Hawkins and Bayne 1991) of an effect of dietary quality, with nitrogen losses increasing exponentially with increased dietary protein content.

These considerations, however, underestimate the role of bivalves in nutrient cycling, with regard to two further and related processes. The first concerns 'metabolic faecal loss', which represents material abraded or secreted into the gut lumen during the passage of food, and not subsequently reabsorbed (Bayne and Hawkins 1990). This is the material modelled by Willows (1991) and referred to above as the 'digestive investment'. Estimates of the metabolic faecal loss under normal feeding conditions vary (and may exceed the ingested energy or nitrogen equivalents; see Fig. 12.3 at low dietary quality); Bayne and Hawkins (1990) illustrate a value of 40 per cent of ingested ration in energy units, and Hawkins and Bayne (1985) measured metabolic faecal nitrogen at 23 ± 11 per cent of ingested nitrogen for mussels feeding on a unialgal diet. This faecal nitrogen, which is probably predominantly proteinaceous in nature, is then available for microbial remineralization within the sediments (Kautsky and Evans 1987).

3. Carbon: nitrogen relationships in the diet

We have occasionally referred above to 'dietary quality', a term which embraces many measurements, from the availability of specific fatty acids within phytoplankton cells to particulate energy content per unit volume of sea-water. Particular attention has been paid, however, to the carbon: nitrogen (C:N) ratio of the diet, in attempts to understand the relationships between the seston and the growth of filter-feeders within natural communities (e.g., Wickens and Field 1986). In such studies, the animal's metabolic requirements for carbon (or energy) and nitrogen are compared with the C:N ratio of specific components of the seston.

The nitrogen requirements for maintenance may be estimated as endogenous nitrogen excretion. Hawkins and Bayne (1991) recorded an allometric relationship for the maintenance protein requirement for *Mytilus* (related to body mass, W, in grams dry flesh) which is equivalent, in terms of nitrogen, to:

Maintenance nitrogen requirement $= 0 \cdot 16 (\pm 0 \cdot 02)$ mg N day^{-1}. $W^{0 \cdot 77 (\pm 0 \cdot 19)}$

From our studies we estimate a mean maintenance carbon requirement to be:

Maintenance carbon requirement = $2.58(\pm 0.27)$ mg C day^{-1} $W^{0.69(\pm 0.13)}$

These equations indicate a maintenance C:N ratio of 16:1, similar to Russell-Hunter's (1970) calculation of 17:1 for animals generally, and with no significant differences due to body size. The indication is that pure phytoplankton diets, with C:N ratios of 5:1 to 8:1 (Parsons *et al.* 1984) may satisfy the nitrogen requirements of the mussels more readily than their carbon requirements, when total phytoplankton biomass is low.

Recent evidence from studies of whole-body protein turnover in mussels (i.e., measures of rates of protein synthesis and breakdown using ^{15}N-labelled foods), reviewed by Hawkins and Bayne (1991), point to a marked 'protein sparing' effect in bivalve nutrition, whereby protein is conserved relative to energy, both with respect to seasonal changes in physiology and in response to changes in the diet. Fig. 12.6 plots net growth efficiencies for both protein and energy, related to rates of ingestion of an algal food; these mussels expressed higher efficiencies for protein. Hawkins and Bayne (1991) have suggested that component processes of protein turnover, such as the recycling of amino acids from protein breakdown to effect further protein synthesis, help to conserve nitrogen but are energetically expensive, to the extent that ATP-

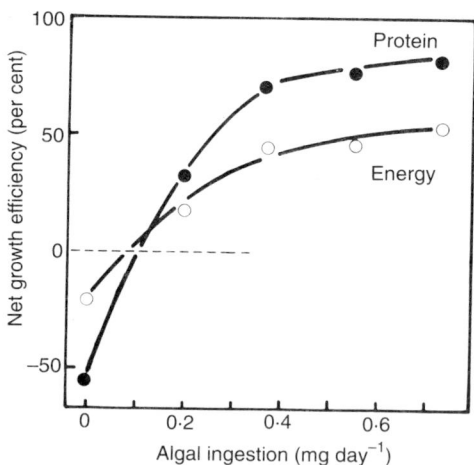

Fig. 12.6. The net growth efficiencies for protein and energy in *Mytilus edulis*, related to rates of ingestion of algal dry matter. The data are from Hawkins *et al.* (1989).

dependent protein turnover contributes at least 25 per cent to the energy requirements of maintenance (Hawkins et al. 1989). The primary reliance on carbohydrates as respiratory substrates, coupled with the high energy demands for protein (nitrogen) conservation, indicate a likelihood that carbon-rich rather than nitrogen-rich dietary sources may be limiting, particularly in winter months when metabolic demands are high relative to the supply of adequate food. This is not inconsistent with the evidence, discussed earlier, that identifies a significant detrital component to the diet of many bivalves in their natural habitats (Stuart et al. 1982; Langdon and Newell 1990).

We have not dealt in this paper with all known aspects of relationships between the functioning of bivalve suspension-feeders and the processes of benthic–pelagic coupling (e.g., the effects of reproductive products released to the pelagic environment; the input of organic materials from bivalve shells to the sediments). Nevertheless, it is clear that a full appreciation of ecosystem dynamics involving suspended seston and benthic bivalves must take account of, *inter alia*, physiological processes of feeding and metabolism in individual organisms, set in the context of the physical forces that control the delivery of seston to the benthos and the resuspension of bivalve biodeposits.

Acknowledgements

We are grateful to colleagues at the Plymouth Marine Laboratory for support and collaboration, in particular Bob Willows, Mandy Day, and John Widdows. Bob Willows kindly gave us permission to use the results of his unpublished feeding model. Our work has also benefited from collaborations with E. Navarro, J.I.P. Iglesias, I. Uriarte, and A. Farias.

References

Asmus, H., Asmus, R.M., and Reise, K. (1990). Exchange processes in an intertidal mussel bed: a Sylt-flume study in the Wadden Sea. *Berichte der Biologischen Anstalt Helgöland*, **6**, 1–79.

Bayne, B.L. and Hawkins, A.J.S. (1990). Filter-feeding in bivalve molluscs: controls on energy balance. In *Animal nutrition and transport processes. 1. Nutrition in wild and domestic animals*, (ed. J. Mellinger), *Comparative Physiology*, Basel, Karger, Vol. 5, pp. 70–83.

Bayne, B.L. and Newell, R.C. (1983). Physiological energetics of marine molluscs. In *The Mollusca*, Vol. 4, *Physiology*, Part 1, (ed. A.S.M. Saleuddin and K.M. Wilbur), pp. 407–515. Academic Press, New York.

Bayne, B. L. and Widdows, J. (1978). The physiological ecology of two populations of *Mytilus edulis*. L. *Oecologia*, Berlin **37**, 137–62.

Bayne, B. L., Hawkins, A. J. S., and Navarro, E. (1987). Feeding and digestion by the mussel *Mytilus edulis* L. (Bivalvia: Mollusca) in mixtures of silt and algal cells at low concentrations. *Journal of Experiment Marine Biology and Ecology*, **111**, 1–22.

Bayne, B. L., Hawkins, A. J. S., Navarro, E., and Iglesias, I. P. (1989). Effects of seston concentration on feeding, digestion and growth in the mussel *Mytilus edulis*. *Marine Ecology Progress Series*, **55**, 47–54.

Berg, J. A. and Newell, R. I. E. (1986). Temporal and spatial variations in the composition of seston available to the suspension feeder *Crassostrea virginica*. *Estuarine, Coastal and Shelf Science*, **23**, 375–86.

Bertness, M. D. (1984). Ribbed mussels and *Spartina alterniflora* production in a New England salt-marsh. *Ecology*, **65**, 1794–1807.

Boucher, B. and Boucher-Rodoni, R. (1988). In situ measurements of respiratory metabolism and nitrogen fluxes at the interface of oyster beds. *Marine Ecology Progress Series*, **44**, 229–38.

Brown, J. R. (1988). Multivariate analyses of the role of environmental factors in seasonal and site-related growth variation in the Pacific oyster *Crassostrea gigas*. *Marine Ecology Progress Series*, **45**, 225–36.

Cahalan, J. A., Siddall, S. E., and Luckenbach, M. W. (1989). Effects of flow velocity, food concentration and particle flux on growth rates of juvenile bay scallops *Argopecten irradians*. *Journal of Experimental Marine Biology and Ecology*, **131**, 43–60.

Carlson, D. J., Townsend, D. W., Hilyard, A. L., and Eaton, J. F. (1984). Effect of an intertidal mud-flat on plankton of the overlying water column. *Canadian Journal of Fisheries and Aquatic Science*, **41**, 1523–8.

Cloern, J. E. (1982). Does the benthos control phytoplankton biomass in South San Francisco Bay? *Marine Ecology Progress Series*, **9**, 191–202.

Cranford, P. J. and Grant, J. (1990). Particle clearance and absorption of phytoplankton and detritus by the sea scallop *Placopecten magellanicus* (Gmelin). *Journal of Experimental Marine Biology and Ecology*, **137**, 105–21.

Crosby, M. P., Newell, R. I. E., and Langdon, C. J. (1990). Bacterial mediation in the utilization of carbon and nitrogen from detrital complexes by the American oyster, *Crassostrea virginica*. *Limnology and Oceanography*, **35**, 625–35.

Dame, R. F., Zingmark, R. G., Stevenson, L. H., and Nelson, D. D. (1980). Filter feeding coupling between the estuarine water column and benthic subsystems. In *Estuarine perspectives*, (ed. V. Kennedy), pp. 521–6. Academic Press, New York.

Dame, R. F., Wolaver, T. G., and Libes, S. M. (1985). Summer uptake and release of nitrogen by an intertidal oyster reef. *Netherlands Journal of Sea Research*, **19**, 265–8.

Dame, R. F. and Dankers, N. (1988). Uptake and release of materials by a Wadden Sea mussel bed. *Journal of Experimental Marine Biology and Ecology*, **118**, 207-16.

Dame, R. F., Spurrier, J. D., and Wolaver, T. G. (1989). Carbon, nitrogen and phosphorus processing by an oyster reef. *Marine Ecology Progress Series*, **54**, 249-56.

Demers, S., Therriault, J.-C., Bourget, E., and Desilets, H. (1989). Small-scale gradients of phytoplankton productivity in the littoral fringe. *Marine Biology*, **100**, 393-9.

Doering, P. H. and Oviatt, C. A. (1986). Application of filtration rate models to field populations of bivalves: an assessment using experimental mesocosms. *Marine Ecology Progress Series*, **31**, 265-75.

Doering, P. H., Oviatt, C. A., and Kelly, J. R. (1986). The effects of the filter-feeding clam *Mercenaria mercenaria* on carbon cycling in experimental marine mesocosms. *Journal of Marine Research*, **44**, 839-61.

Field, J. G., Wickens, P. A., and Moloney, D. L. (1986). Modelling studies of material flows in a shallow ecosystem compared to the open ocean. In *Biogeochemical processes at the land-sea boundary*, (ed. P. Lasserre and J.-M. Martin), pp. 75-97. Elsevier, Amsterdam.

Fielding, P. J. and Davis, C. L. (1989). Carbon and nitrogen resources available to kelp bed filter feeders in an upwelling environment. *Marine Ecology Progress Series*, **55**, 181-9.

Foe, C. and Knight, A. (1985). The effect of phytoplankton and suspended sediment on the growth of *Corbicula fluminea* (Bivalvia). *Hydrobiologia*, **127**, 105-15.

Fréchette, M. and Bourget, E. (1985a). Energy flow between the pelagic and benthic zones: factors controlling particulate organic matter available to an intertidal mussel bed. *Canadian Journal of Fisheries and Aquatic Science*, **42**, 1158-65.

Fréchette, M. and Bourget, E. (1985b). Food-limited growth of *Mytilus edulis* L. in relation to the benthic boundary layer. *Canadian Journal of Fisheries and Aquatic Science*, **42**, 1166-70.

Fréchette, M. and Lefaivre, D. (1990). Discriminating between food and space limitation in benthic suspension feeders using self-thinning relationships. *Marine Ecology Progress Series*, **65**, 15-23.

Fréchette, M., Butman, C. A., and Geyer, W. R. (1989). The importance of boundary-layer flows in supplying phytoplankton to the benthic suspension feeder, *Mytilus edulis* L. *Limnology and Oceanography*, **34**, 19-36.

Fry, B. (1988). Food web structure on Georges Bank from stable C, N and S isotopic compositions. *Limnology and Oceanography*, **33**, 1182-90.

Gabbott, P. A. (1983). Development and seasonal metabolic activities in marine molluscs. In *The Mollusca*, Vol. **2**, (ed. P. W. Hochachka), pp. 165-217. Academic Press, New York.

Grizzle, R. E. and Lutz, R. A. (1989). A statistical model relating horizontal seston fluxes and bottom sediment characteristics to growth of *Mercenaria mercenaria*. *Marine Biology*, **102**, 95–105.

Haven, D. S., and Morales-Alamo, R. (1966). Aspects of biodeposition by oysters and other invertebrate filter feeders. *Limnology and Oceanography*, **11**, 487–98.

Hawkins, A. J. S. and Bayne, B. L. (1985). Seasonal variation in the relative utilization of carbon and nitrogen by the mussel *Mytilus edulis*: budgets, conversion efficiencies and maintenance requirements. *Marine Ecology Progress Series*, **25**, 181–8.

Hawkins, A. J. S. and Bayne, B. L. (1991). Nutrition of marine mussels: factors influencing the relative utilizations of protein and energy. *Aquaculture*, **94**, 177–96.

Hawkins, A. J. S., Widdows, J., and Bayne, B. L. (1989). The relevance of whole-body protein metabolism to measured costs of maintenance and growth in *Mytilus edulis*. *Physiological Zoology*, **62**, 745–63.

Hibbert, C. J. (1977). Energy relations of the bivalve *Mercenaria mercenaria* on an intertidal mud-flat. *Marine Biology*, **44**, 77–84.

Jordan, J. E. and Valiela, I. (1982). A nitrogen budget of the ribbed mussel, *Geukensia demisa*, and its significance in nitrogen flow in a New England salt-marsh. *Limnology and Oceanography*, **27**, 75–90.

Jørgensen, C. B. (1989). Water processing in filter-feeding bivalves. In *Behavioural mechanisms of food selection*, (ed. R. N. Hughes), pp. 615–35. Springer-Verlag, Berlin.

Kautsky, N. (1981). On the trophic role of the blue mussel (*Mytilus edulis* L.) in a Baltic coastal ecosystem and the fate of the organic matter produced by the mussels. *Kieler Meeresforschungen (Sonderh.)*, **5**, 454–61.

Kautsky, N. and Evans, S. (1987). Role of biodeposition by *Mytilus edulis* in the circulation of matter and nutrients in a Baltic coastal ecosystem. *Marine Ecology Progress Series*, **38**, 201–12.

Kemp, P. F., Newell, S. Y., and Krambeck, C. (1990). Effects of filter-feeding by the ribbed mussel *Geukensia demissa* on the water-column microbiota of a *Spartina alterniflora* salt-marsh. *Marine Ecology Progress Series*, **59**, 119–31.

Kiørboe, T., Møhlenberg, F., and Nøhr, O. (1980). Feeding, particle selection and carbon absorption in *Mytilus edulis* in different mixtures of algae and resuspended bottom material. *Ophelia*, **19**, 193–205.

Kirbøe, T., Møhlenberg, F., and Nøhr, O. (1981). Effect of suspended bottom material on growth and energetics in *Mytilus edulis*. *Marine Biology*, **61**, 283–8.

Kreeger, D. A., Newell. R. I. E., and Langdon, C. J. (1990). Effect of tidal exposure on utilization of dietary lignocellulose by the ribbed mussel *Geukensia demissa* (Dillwyn) (Mollusca: Bivalvia). *Journal of Experimental Marine Biology and Ecology*, **144**, 85–100.

Kuenzler, E.J. (1961). Structure and energy flow of a mussel population in a Georgia salt-marsh. *Limnology and Oceanography*, **6**, 191-204.

Langdon, C.J. and Newell, R.I.E. (1990). Utilization of detritus and bacteria as food sources by two bivalve suspension-feeders, the oyster *Crassostrea virginica* and the mussel *Geukensia demissa*. *Marine Ecology Progress Series*, **58**, 299-310.

Lehman, J.T. (1976). The filter-feeder as an optimal forager, and the predicted shapes of feeding curves. *Limnology and Oceanography*, **21**, 501-16.

Loo, L.-O. and Rosenberg, R. (1989). Bivalve suspension-feeding dynamics and benthic–pelagic coupling in an eutrophicated marine bay. *Journal of Experimental Marine Biology and Ecology*, **130**, 253-76.

Lucas, M.I., Newell, R.C., Shumway, S.E., Seiderer, L.J., and Bally, R. (1987). Particle clearance and yield in relation to bacterioplankton and suspended particulate availability in estuarine and open coast populations of the mussel *Mytilus edulis*. *Marine Ecology Progress Series*, **36**, 215-24.

Manahan, D.T., Wright, S.H., and Stephens, G.C. (1983). Simultaneous determination of net uptake of 16 amino acids by a marine bivalve. *American Journal of Physiology*, **244**, R832-8.

Mann, K.H. (1988). Production and use of detritus in various freshwater, estuarine and coastal marine systems. *Limnology and Oceanography*, **33**, 910-30.

Muschenheim, D.K. (1987). The dynamics of near-bed seston flux and suspension-feeding benthos. *Journal of Marine Research*, **45**, 473-96.

Navarro, E., Iglesias, J.I.P., Camacho, A.P., Labarta, U., and Beiras, R. (1991). The physiological energetics of mussels (*Mytilus galloprovincialis* Lmk.) from different cultivation rafts in the Ria de Arosa (Galicia, N.W. Spain). *Aquaculture*, **94**, 197-212.

Nakamura, M., Yamamuro, M., Ishitawa, M., and Nishimura, H. (1988). Role of the bivalve *Corbicula japonica* in the nitrogen cycle in a mesohaline lagoon. *Marine Biology*, **99**, 369-74.

Newell, R.C. and Field, J.G. (1983). The contribution of bacteria and detritus to carbon and nitrogen flow in a benthic community. *Marine Biology Letters*, **4**, 23-36.

Newell, R.C., Field, J.G., and Griffiths, C.L. (1982). Energy balance and significance of micro-organisms in a kelp bed community. *Marine Ecology Progress Series*, **8**, 103-13.

Newell, R.I.E. and Jordan, S.J. (1983). Preferential ingestion of organic material by the American oyster, *Crassostrea virginica*. *Marine Ecology Progress Series*, **13**, 47-53.

Nichols, F.H. (1985). Increased benthic grazing: an alternative explanation for low phytoplankton biomass in Northern San Francisco Bay during the 1976-1977 drought. *Estuarine, Coastal and Shelf Science*, **21**, 379-88.

Officer, C. B., Smayda, T. J., and Mann, R. (1982). Benthic filter feeding: a natural eutrophication control? *Marine Ecology Progress Series*, **9**, 203-10.

Parsons, T. R., Takahashi, M., and Hargrave, B. (1984). *Biological oceanographic processes*. Pergamon Press, Oxford.

Petersen, C. H. and Black, R. (1987). Resource depletion by active suspension feeders on tidal flats: influence of local density and tidal elevation. *Limnology and Oceanography*, **32**, 143-66.

Peterson, B. J., Howarth, R. W., and Garritt, R. H. (1985). Multiple stable isotopes used to trace the flow of organic matter in estuarine food webs. *Science*, **227**, 1361-3.

Peterson, B. J., Howarth, R. W., and Garritt, R. H. (1986). Sulfur and carbon isotopes as tracers of salt-marsh organic matter flow. *Ecology*, **67**, 865-74.

Riemann, B., Nielsen, T. G., Horsted, S. J., Bjørsen, P. K., and Pock-Steen, J. (1988). Regulation of phytoplankton biomass in estuarine enclosures. *Marine Ecology Progress Series*, **48**, 205-15.

Russell-Hunter, W. D. (1970). *Aquatic productivity*. Macmillan, London.

Seiderer, L. J. and Newell, R. C. (1985). Relative significance of phytoplankton, bacteria and plant detritus as carbon and nitrogen resources for the kelp bed filter-feeder *Choromytilus meridionalis*. *Marine Ecology Progress Series*, **22**, 127-39.

Sibly, R. M. (1981). Strategies of digestion and defecation. In *Physiological ecology: an evolutionary approach to resource use*, (ed. C. R. Townsend and P. Calow), pp. 109-39. Blackwell, Oxford.

Smaal, A. C., Verhagen, J. H. G., Coosen, J., and Haas, H. A. (1986). Interaction between seston quantity and quality and benthic suspension feeders in the Oosterschelde, The Netherlands. *Ophelia*, **26**, 385-99.

Sornin, J. M., Deslous-Paoli, J. M., and Hesse. O. (1988). Experimental study of the filtration of clays by the oyster *Crassostrea gigas* (Thunberg): adjustment of particle size for best retention. *Aquaculture*, **69**, 355-66.

Steen, J. (1988). Regulation of phytoplankton biomass in estuarine enclosures. *Marine Ecology Progress Series*, **48**, 205-15.

Stephenson, R. L. and Lyon, G. L. (1982). Carbon-13 depletion in an estuarine bivalve: detection of marine and terrestrial food sources. *Oecologia* (Berlin), **55**, 110-3.

Stephenson, R. L., Tan, F. C., and Mann, K. H. (1986). Use of stable carbon isotope ratios to compare plant material and potential consumers in a seagrass bed and a kelp bed in Nova Scotia, Canada. *Marine Ecology Progress Series*, **30**, 1-7.

Stuart, V. R., Field, J. G., and Newell, R. C. (1982). Evidence for absorption of kelp detritus by the ribbed mussel *Aulacomya ater* using a new ^{51}Cr-labelled microsphere technique. *Marine Ecology Progress Series*, **9**, 263-71.

Taghon, G. L. (1981). Beyond selection: optimal ingestion rate as a function of food value. *American Naturalist*, **118**, 202-14.

Tenore, K. R., Goldman, J. C., and Clarner, J. P. (1973). The food chain dynamics of the oyster, clam and mussel in an aquaculture food chain. *Journal of Experimental Marine Biology and Ecology*, **12**, 19-26.

Thompson, J. K. and Nichols, F. H. (1988). Food availability controls seasonal cycle of growth in *Macoma balthica* (L.) in San Francisco Bay, California. *Journal of Experimental Marine Biology and Ecology*, **116**, 43-61.

Verwey, J. (1952). On the ecology of distribution of cockle and mussel in the Dutch Wadden Sea, their role in sedimentation and the source of their food supply — with a short review of the feeding behaviour of bivalve molluscs. *Archives Neerlandaises*, **10**, 171-239.

Ward, J. E. and Targett, N. M. (1989). Influence of marine microalgal metabolites on the feeding behaviour of the blue mussel *Mytilus edulis*. *Marine Biology*, **101**, 313-21.

Wickens, P. A. and Field, J. G. (1986). The effects of water transport on nitrogen flow through a kelp-bed community. *South African Journal of Marine Science*, **4**, 79-92.

Widdows, J., Fieth, P., and Worrall, C. M. (1979). Relationship between seston, available food and feeding activity in the common mussel *Mytilus edulis*. *Marine Biology*, **50**, 195-207.

Wildish, D. J. and Kristmanson, D. D. (1984). Importance to mussels of the benthic boundary layer. *Canadian Journal of Fisheries and Aquatic Science*, **41**, 1618-25.

Wildish, D. J., Kristmanson, D. D., Hoar, R. L., DeCoste, A. M., McCormick, S. D., and White, A. W. (1987). Giant scallop feeding and growth responses to flow. *Journal of Experimental Marine Biology and Ecology*, **113**, 207-20.

Willows, R. I. (1991). Optimal digestive investment: a model for filter-feeders experiencing variable diets. *Limnology and Oceanography*. (In press).

Wilson, J. H. (1987). Environmental parameters controlling growth of *Ostrea edulis* L. and *Pecten maximus* L. in suspended culture. *Aquaculture*, **64**, 119-31.

Winter, J. E. (1976). A review of the knowledge of suspension-feeding in lamellibranchiate bivalves, with special reference to artificial aquaculture systems. *Aquaculture*, **13**, 1-33.

13. Foraging behaviour of marine benthic grazers

M.G. CHAPMAN and A.J. UNDERWOOD
Institute of Marine Ecology, Zoology Building, University of Sydney, NSW, Australia

Abstract

Benthic grazers exhibit a variety of different patterns of foraging. These vary among habitats and among taxonomic groups. Broadly, foraging behaviour has two components: firstly, the behaviour that brings animals into contact with their plant foods; secondly, the animals respond to the presence of food in different ways. Each can be affected by different physical or biological conditions. Here, foraging behaviour is examined in order to evaluate the evidence upon which our understanding of this behaviour is based. The roles of physical factors (such as shelter) and biological interactions (such as competition and predation) are considered. These are discussed in the light of some general models to account for foraging by different animals and with a view to outlining hypotheses of interest for further study.

Introduction

Foraging is the series of behaviours used by grazing animals in the search for and acquisition of their plant foods. Marine benthic grazers forage in a variety of habitats. These include hard substrata like intertidal and subtidal rocky reefs and coral reefs and soft-sediments like beaches and mud-flats. To date, much research has been on grazers feeding on hard substrata, often with an obvious cover of foliose or encrusting macroalgae. The impact of grazing on these algae is an important component of many ecological studies of these systems (e.g., Lubchenco and Gaines 1981; Underwood 1980, 1985; Underwood and Jernakoff 1984).

Grazers in marine habitats include polychaetes, molluscs (such as chitons and gastropods), echinoids and asteroids, crustaceans (including crabs, hermit crabs, isopods, and amphipods), some insect larvae, and chordates such as fish, reptiles, and mammals (Hawkins and Hartnoll 1983). Most research to date has examined foraging of gastropods and echinoids, the most abundant and obvious grazers in many systems. Thus, much of this review will concentrate on these two groups. Molluscs are by far the most abundant benthic grazers in most rocky intertidal systems. Their activity influences the distribution and abundance of algal cover (Underwood 1980; Lubchenco 1983) and can lead to predominance of crustose algae (Lubchenco and Cubit 1980) and exclusion of foliose macroalgae from mid-shore levels (Hay 1979; Underwood 1980) where grazing pressure is great. In many subtidal habitats with hard substrata, fish and echinoids are the dominant grazers. The relative importance of different groups may be correlated with latitude; echinoids and gastropods are more important in temperate systems, fish and echinoids in tropical systems (Choat 1982; Hawkins and Hartnoll 1983).

Marine animals feed on algae in one of four different ways. Microalgal grazers scrape or sweep the substratum, collecting up microalgae such as diatoms in addition to the sporelings of larger macroalgae (e.g., Nicotri 1977). Macroalgal grazers feed on pieces of macroalgae which they scrape, bite, or tear off the adult plants (e.g., Steneck and Watling 1982). Suspension feeders and deposit feeders will not be considered further in this review.

There have been many attempts to group the grazers into different categories. For example, Steneck and Watling (1982) used the morphology of the radula to distinguish between different feeding groups of molluscs. Raffaelli (1985) created functional feeding groups by associating together molluscs that had the same diet (as determined by their gut contents). He found no clear relationship between the structure of the radula and the molluscs' food, but gut and faecal contents may not reflect which food is actually absorbed and assimilated (Peterson and Bradley 1978; Fairweather and Underwood 1983; Santelices *et al.* 1983).

Chelazzi *et al.* (1988) and Little (1989) have identified common patterns of behaviour among different groups of intertidal molluscs in different intertidal habitats. Here, we examine different patterns of foraging activity across a wide spectrum of benthic grazers in subtidal and intertidal habitats. We identify spatial and temporal variability in this behaviour and consider what is known of the various causal factors. Throughout, discussion is restricted to grazers, avoiding merging the grazers with predators (as consumers, Lubchenco 1979). We believe the effects of grazing on algae, which often show rapid recovery by vegetative growth, to be potentially different from the activities of many marine predators (see Underwood and Fairweather 1991).

Types of foraging behaviour

Foraging can be divided into relatively distinct patterns of behaviour, overlapping to some extent. These patterns relate to:

(1) how to find food;
(2) when to feed;
(3) how often to feed;
(4) when to stop feeding.

For any marine grazer, only some patterns have been studied; no overall understanding of the various aspects is yet possible. In one of the more extensive studies, Cook and Cook (Cook 1969, 1971; Cook and Cook 1978, 1981) examined timing of foraging, spatial and temporal variation in timing and extent of foraging, and the mechanics of homing after feeding excursions in the limpets *Siphonaria normalis* and *S. alternata*. For most species, however, only very limited patterns have been studied.

Nearly all benthic grazers move in order to feed (possible exceptions will be discussed later) and their movement patterns during feeding depend on their distribution and the distribution and abundance of their food. Similar patterns of movement can be shown by quite different phyla living in different habitats. For example, subtidal echinoids (Fletcher 1987) and intertidal neritid snails (Levings and Garrity 1983) forage in limited areas around crevices in which they shelter. In contrast, very different patterns can be shown by very closely related species living in the same habitat (Levings and Garrity 1983).

1. Free-ranging foraging

The simplest form of foraging is shown by free-ranging grazers (e.g., many gastropods) which are surrounded by their food. Any movements they make are essentially foraging excursions; taped recordings of radular movements of a number of molluscs species have shown that while these animals are active they are feeding (Boyden and Zeldis 1979; Kitting 1979). Free-ranging animals do not necessarily have to search for food because they are surrounded by it. They include periwinkles (Petraitis 1982; Underwood and Chapman 1989), limpets (Branch 1971; Underwood 1977) and other intertidal gastropods (Underwood 1977). Microalgae are often considered to be (but are not always) ubiquitous and predictable food sources, because they are rapidly (or partially) renewed by the incoming tide each time (e.g., discussion in Underwood 1978; MacLulich 1986).

The foraging movements of many microalgal grazers are random in orientation and extent (Dexter 1943; Underwood 1977; Petraitis 1982; Underwood and Chapman 1985, 1989; Chapman 1986), although many limpets that eat microalgae do home to specific sites after feeding (Branch 1971; Cook 1971; Mackay and Underwood 1977). A species surrounded by food and not homing after feeding need not have developed the ability to orientate its foraging excursions (Hughes 1980) compared with a species that moves between a resting site and a feeding site (Chelazzi et al. 1987). Nevertheless, some species which sometimes move randomly when foraging can show strongly directional movements at other times (Chapman 1986; Underwood and Chapman 1989). Whether these movements are influenced by physical factors (such as water movement) or by the quality or quantity of microalgal food is not yet known. Castenholz (1963), Underwood (1984) and MacLulich (1987) have shown that density and/or variety of microalgae on intertidal shores are spatially and temporally very patchy. Microalgal grazers possibly respond to this patchiness by orientating and moving along gradients of food (quality and/or quantity). They may thus move randomly and disperse little when food is abundant, but may move further and show directional orientation when strong gradients exist (Mackay and Underwood 1977). Species like this still differ from those moving regularly between a feeding ground and a home site because they are responding to local feeding stimuli. Such responses would certainly explain some of the temporal and spatial variability in foraging behaviour that has been recorded for microalgal grazers (Mackay and Underwood 1977; Underwood and Chapman 1985; Chapman 1986).

There must be some limit to apparently random movements of free-ranging species because long-term studies of movements of intertidal gastropods have shown that although, over a short period, the grazers appear to move randomly, over longer periods they do tend to remain within a general area of the shore, usually within a particular vertical level (Frank 1964; Hamilton 1978). The level at which the population lives and forages is not necessarily fixed, but can vary (up and down shore) on a seasonal basis (reviewed by Underwood 1979) or under the influence of stimuli such as the height of tides (Taylor 1971) and the reach of waves (Ohgaki 1989). The factors which cause animals to remain in their preferred shore levels are not well understood (reviewed by Underwood 1979) but are often investigated by experiments involving displacement of animals out of their normal habitat in the field (Chelazzi and Vannini 1976, 1980; Gendron 1977; Byers and Mitton 1981; McQuaid, 1981; Doering and Phillips 1983), or by studying the roles of various stimuli on taxa in the laboratory (reviewed by Underwood 1979). It is not always clear how displacement experiments actually relate to natural foraging movements of the animals, particularly if the associated disturbance itself can induce directional orientation, as suggested for *Littorina littorea* (Petraitis 1982) and *Littorina unifasciata* (Chapman 1986). In field studies of manipulated animals, there are serious problems of lack of controls, inadequate analyses, and, sometimes, illogicality of relationships between hypotheses and tests (reviewed by Chapman 1986; Underwood and Chapman 1992; Underwood 1988). Underwood (1979) clearly identified some of the pitfalls in relating orientation mechanisms recognized in the laboratory to natural foraging activity in the field, especially where timing (with respect to daily or tidal cycles) of movement is not taken into account.

2. Foraging in response to a patchy distribution of food

Some grazers make directional foraging excursions because their food is uncommon and spatially patchy; others respond to patches of algae as shelter (see below). Random movements would be unlikely to bring these grazers into contact with their food and directional movement is therefore probably more efficient in exploiting patches of food (Hughes 1980). Many species of urchin move to feed on algae which are patchily distributed, such as drift algae or large species of macroalgae. Often urchins will cover large distances in search of this food and will form feeding aggregations around individual food items when these are located (Schiel 1982; Dean *et al*. 1984; Choat and Andrew 1986; Vadas *et al*. 1986). The influence of patchy distributions of macroalgae on the movements of grazing gastropods has not been extensively studied although the gastropods *Tegula funebralis* (Paine 1971) and *Littorina littorea* (Watson and Norton 1985) have been reported to move towards and feed on drift algae in upper intertidal habitats.

Chemical cues are thought to be important in the location and choice of preferred species of algae, because some grazer species can distinguish between the chemical extracts of different species of algae, showing distinct preference for some extracts over others (Imrie *et al*.

1989; Norton *et al.* 1990). Although chemical attraction may be effective in the still water of an aquarium, it is not clear how the responses may be used in the field, where there is considerable water turbulence and where preferred and other species of algae are mixed (Norton *et al.* 1990). Chemical attraction is more probably involved in the choice of food consumed after it has been found rather than in the location of, and movement towards, patches of food on the shore.

Many studies of distribution and abundance of grazing animals show a clear association between abundance of grazers and a source of food. This association does not distinguish between differential settlement and/or survival on patches of different food and orientation and movement towards these patches. Creese and Underwood (1982) reported that adults of the two siphonarian limpets *Siphonaria denticulata* and *S. virgulata* invade new patches of macroalgae as they appear on the shore and respond to macroalgae growing on the shells of a limpet, *Cellana tramoserica*, by grazing on these moving 'patches'. Although new recruits of these siphonarians preferentially settle on patches of macroalgae (Underwood 1980; Quinn 1988), the adults can change their foraging behaviour according to the patchiness of their source of food, in order to locate and exploit new patches. *Acmaea scutum* feeding on boulders with a mixed cover of algae are found foraging on different algal species independently of the relative abundance of these species. It has been suggested that different species of algae are selected as food to maintain a mixed diet (Kitting 1980).

Chelazzi *et al.* (1983), in their study of two sympatric species of chiton, showed that the species feeding higher on the shore tended to change its feeding ground more frequently than the closely related species that migrated further downshore to feed. They concluded that the high shore species feeding on a more uniform cover of food had no need to be site-specific while foraging, whereas the food lower on the shore was distributed in patches and it was advantageous for the chitons to remain feeding in patches which they had already located. The limpet *Cellana tramoserica* alternated between homing and free-ranging while foraging (Mackay and Underwood 1977). Where there was abundant food, most limpets homed to specific sites after each foraging excursion. Where food was scarce, the limpets changed to free-ranging, presumably to increase their chances of finding food. There have been few studies on the responses of grazers to differences in the dispersion of food.

Some grazer species live directly on their source of food and show a very limited foraging range. *Acmaea insessa* settled on to the kelp *Egregia* from the plankton, preferring stipes where adult limpets have already established feeding scars (Black 1976). The small animals possibly cannot make their own grazing scars and need the scars of adults in which to feed. The limpet *Patina pellucida* also apparently settles from the plankton on to *Laminaria* or *Saccorhiza*, on which the adult exclusively lives and feeds (Graham and Fretter 1947). This preference in settling behaviour is not necessarily the case; Vahl (1971) considered that juvenile *P. pellucida* did not settle directly on to kelps, juveniles being unable to use kelp as a source of food. Juvenile *Acmaea testudinalis* are found in greater numbers on a particular species of coralline crust in the field, even in habitats of patchy and mixed coralline cover (Steneck 1982). It has been suggested that this may indicate preference at settling, or selective mortality on different species of crusts. The possibility of post-settlement migration of limpets among the species of crust cannot, however, be disregarded. *Littorina obtusata* lives exclusively on fucoid algae (Barkman 1955). This species has no planktonic larval stage, the young hatchlings crawling away directly on to their food where they remain. It has been reported that these snails will not lay their eggs in the absence of these algae (Barkman 1955).

Some gastropods which live on macroalgae have patterns of behaviour which allow them to leave one plant and move on to another as conditions change. *Patella compressa* drop from its home scar on *Ecklonia* on to the substratum if the plant is torn from the substratum (Choat and Black 1979). Adult *Acmaea insessa* leave their home scars and move between rachises to colonize new *Egregia* plants (Black 1976). *Patina pellucida* is reported to move from the fronds of plants down on to the stipes during seasonal dieback of the host laminarian algae (Graham and Fretter 1947), although this does not appear true for all populations of these limpets (Kain and Svendsen 1969).

Many small, phytal, grazing amphipods also live almost exclusively on their preferred food source, e.g., *Hyale nilssoni* is found in greatest abundance on *Polysiphonia* in the field. This species of alga is also the one to which the amphipods are most attracted in the laboratory (McBane and Crocker 1983). Modification of food to supply a shelter is shown by the amphipod *Pseudoamphithoides incurvaria*, which encases itself in a valve it constructs from the fronds of *Dictyota* (in Steneck 1982).

3. Foraging in response to a patchy distribution of shelter

Many benthic grazers are limited in the range over which they can forage because of the need to return to shelter, or to a different microhabitat, after foraging. Some species, particularly limpets, 'home' to specific sites (Branch 1975; Cook 1971), whereas others simply return to particular habitats different from those in which they feed (Vannini and Chelazzi 1978; Levings and Garrity 1983; Chelazzi et al. 1985). The shelter may ameliorate harsh physical conditions, such as desiccation, or protect the animal from predation.

Littorina irrorata ascends stalks of the marsh grass *Spartina* during high tide, descending to feed on the ground during low tide (Hamilton 1977). This is thought to be a response to predation; crabs which move into the area on the incoming tide eat *Littorina* if the snails are prevented from ascending the grass while submersed (Warren 1985). The snails show very similar patterns of behaviour in different seasons of the year, even though major predators are absent during winter.

Closely related species living on the same shore can show remarkably different patterns of foraging, despite the fact that they eat very similar food and are subjected to the same physical and biological stresses. For example, in Panama, *Nerita scabricosta* feeds on nocturnal low tides by following the receding water down during the ebb and moving back up shore ahead of the flood, covering up to 5 m in each grazing excursion. On the same shore, *Nerita funiculata* feeds during ebb and flood tides within 25 cm of the crevices where it shelters, but returns to the shelter during low and high tides. Field experiments have shown that the inactivity of each species during diurnal low tides offers protection from desiccation; inactivity during high tide offers protection from predation by fish and *N. funiculata* is inactive during nocturnal low tides because of predation by crabs (Levings and Garrity 1983).

The range of foraging of many organisms is often clearly visible as a halo of denuded space around shelters from which the animals forage. For example, around crevices in which the urchin *Centrostephanus rodgersii* shelters, there was a cover of crustose algae (Fletcher 1987). Further away from the crevice, cover changed, first to filamentous algae, then (even further away) to foliose macroalgae. Experiments have shown that these patterns of algal distribution are maintained by feeding activity of the urchins sheltering within crevices (Fletcher 1987). Haloes of bare sand can also be found around patch reefs in many tropical seagrass beds. These are maintained by the activities of urchins (Ogden et al. 1973) or fish (Randall 1965), moving out from these reefs to feed on seagrasses before returning to the reefs for shelter. This activity has been demonstrated by transplanting seagrasses into haloes, by experimental exclusions and by constructing new shelters within the seagrass beds. Hay (1984) has shown that the relative importance of fish or urchins, in grazing these systems and in maintaining these haloes of bare sand, depends on the extent of overfishing on the reef. On overfished reefs, urchins were more abundant because their predators had been removed. Thus, at least some generalizations about the role of urchins in tropical habitats are based on an overassessment of their importance because of non-representative sampling areas. This underlines the importance of data being obtained from several representative localities before generalizations are made.

Algae themselves may provide shelter for many organisms, and it is important to separate their roles as food source and as shelter. Nicotri (1980) showed that the phytal isopod *Idotea* preferred to live among less nutritional algae when offered a choice, choosing the alga for value as shelter rather than value as food. Worthington and Fairweather (1989) showed that the large gastropod *Turbo undulatum* is found in greatest abundance around the turf-forming alga *Corallina*. It too appeared to select for value as shelter, remaining for weeks around artificial turfs which mimicked *Corallina*, although these did not supply any food. Amphipods in New Zealand have also been shown to be confined to living and feeding among *Corallina* because it offers shelter from predation by blennies (Coull and Wells 1983). Holmlund et al. (1990) showed that abundance of the amphipod *Amphithoe longimana* on different species of algae varied seasonally, gradually declining on the more highly branched genus *Hypnea* as the level of predation by fish increased. Host-plant morphology, in addition to host-plant palatability, influenced distribution of the amphipods.

Some grazers shelter in one species of alga but move to feed on another. Buschmann (1990) showed that, in the laboratory, phytal amphipods consumed the alga *Iridaea* more than other species and yet this alga supported some of the smallest densities of amphipods in the field. Further field studies showed a significant increase in amphipods on *Iridaea* during nocturnal low tides and no change in the densities on *Ulva* between nocturnal and diurnal low tides. Buschmann (1990) concluded that the amphipods appeared to shelter in the surrounding turfing algae during the day and moved on to the preferred food source to feed at low tide at night.

The limpet *Siphonaria thersites* is found living within crevices or clustered around the algal holdfasts within the band of *Iridaea cornucopiaea* on which it also feeds (Branch 1988). Limpets transplanted on to the open rock were extremely prone to desiccation and suffered considerable mortality; their limited tenacity made them vulnerable to dislodgement by waves. Although this limpet will readily eat other species like *Ulva*, it will not cross open rock to reach them. *S. thersites* may therefore be limited in field diet by the necessity for shelter. The data on mortality in response to desiccation are not convincing, however, because there were no experimental controls for disturbance and tranplantation. In addition, there are difficulties in interpreting data obtained from limpets removed from their scars or shelters, because any change in their activity and behaviour is confounded by their being within or outside their normal habitat (Chapman and Underwood 1992). Increased mortality may be due to a change in the behaviour of the animals, imposed by experimental manipulations rather than a change in their habitat.

Some species are restricted in their grazing because of competitive interactions for food. For example, Creese (1982) described how the limpet *Patelloida latistrigata* was restricted to areas of the shore occupied by barnacles. Experimental evidence indicated that *P. latistrigata* died when outcompeted for food by larger grazers on open surfaces. Thus, the smaller limpet was only found in areas where it was subject to considerable predation by whelks (Creese 1982; Fairweather and Underwood 1983), but had a food supply inaccessible to larger grazers (Jernakoff 1985).

An extreme category of shuttling between a resting and a feeding habitat is homing to a particular site after feeding excursions (reviewed by Branch 1971, 1975; Underwood 1979; Chelazzi *et al.* 1987; among others). Many homing limpets return to individual home sites (reviewed by Underwood 1979; Branch 1981) which are often shown by a home scar, or indentation or mark on the rock. Because homing has been extensively reviewed, we do not discuss it further.

Some homing limpets establish territories around their home scars (Stimson 1970; Branch 1971), usually on patches of their preferred algal food. They vigorously defend both territories and food against invading animals, pushing away and occasionally dislodging invaders. *Lottia gigantea* maintains its territory on a patch of *Ralfsia* around its home scar (Stimson 1970, 1973), defending this against encroachment by pushing out introduced *Lottia* and smaller limpets such as *Acmaea digitalis* and sessile animals like mussels. The territorial behaviour of *Lottia* is extremely ritualized (Wright 1982); undisturbed invaders are seldom dislodged and they tend much of the time to evade the territorial animal. The size of the territory of this species is directly related to the quantity of food available, increasing when food is in short supply, when the limpets presumably need to graze over a larger area to maintain adequate intake (Stimson 1973).

Patella longicosta also establishes territories around the home scars within beds of *Ralfsia*, on which they feed in a non-random manner (Branch 1971, 1981). It cuts straight, evenly-spaced grazing paths through *Ralfsia*, avoiding recently grazed paths on later foraging trips. This behaviour has two consequences; it prevents overgrazing and creates new edges on the crust of *Ralfsia*. The alga grows more rapidly along these edges, so that productivity is increased.

Some territorial, reef-dwelling fish are reported to maintain algal gardens in their territories. The damselfish *Pomacentrus lividus* occupies a distinct territory with a cover of green ephemeral algae absent elsewhere (Vine 1974). Caging experiments showed that newly settled algae are rapidly removed from plates outside the territory by the large number of grazing fish. The small amphipod *Erichthorius braziliensis* builds a tube within which it resides, grazing algae in close proximity (Connell 1963). The spacing between adjacent tubes is inversely related to size of the animals. The grazing area around the tube is vigorously defended against other amphipods which wander into the area.

Some snails may even carry their garden with them. There are reports that the sandy beach scavenger *Bullia digitalis* eats plants in addition to carrion. Examination of the contents of its digestive gland and the presence of appropriate enzymes in the gut provide evidence. The shells of these snails have a covering of boring algae with sporelings of *Ulva* and other ephemeral algae, but there are few other algae in this sandy beach habitat (Harris *et al.* 1986). *Bullia* has a long mobile proboscis, and has been reported to graze the algae growing on its own shell (Da Silva and Brown 1984).

Finally, it has been suggested that certain limpets following their own trails may 'garden' algae caught up, or growing more rapidly, in the mucus of these trails (Connor 1986). Although it was shown that mucus trails of the homing limpets *Lottia gigantea* and *Collisella scabra* could trap and stimulate growth of microalgae, there was no evidence that the limpets actually ingested their trails (and the trapped algae) when following them. If the limpets do indeed retrace their outward trail on each foraging excursion, it is difficult to see how growth of the algae could have been stimulated during such a short period of time because Connor's data indicate that stimulation of algal growth requires several days. Because old trails are often located and followed by limpets (Cook *et al.* 1969; Cook 1971), it is unlikely that trails are habitually ingested.

4. Sit-and-wait grazers

A few species may be considered as 'sit-and-wait' grazers because, rather than searching, they wait for food in the form of drift algae to come to them. Some species of sea-urchins and abalone are commonly considered to be such grazers. The data for some abalone are convincing because large individuals have been observed in the field remaining on their home scars and feeding on detached drift algae (Tutschulte and Connell 1988). Small individuals of the same species are more mobile and feed on microalgae (Tutschulte and Connell 1988). Momma and Sato (1969) showed interspecific variation in feeding mechanisms of cavern-inhabiting abalones. *Haliotis sieboldi* remained on scars overnight, while *H. discus hannai* undertook grazing excursions, some individuals homing and others not. It was suggested, without supporting data, that *H. sieboldi* probably fed on drifting algae. Studies were inadequately replicated, however, with the positions of less than 12 individuals recorded and data collected from only a few nights of observation. It is not yet possible, therefore, to determine whether each species does have a different foraging pattern (i.e., actively grazing vs sitting and waiting for drift), or whether each species shows individual and temporal variation in foraging. Longer studies of more animals are necessary before foraging patterns are clear. Leighton and Boolootian (1963) showed that the abalone *Haliotis cracherodii* ate mainly attached macroalgae, with distinct algal preferences when foraging. They assumed, without presenting data, that abalones living in caverns were relying on drift algae because there were no attached macroalgae present.

Burrowing sea urchins almost certainly rely on passing drift algae, and the capture and ingestion of algal particles of various sizes has been described for *Echinostrephus molaris* in the laboratory and the field (Campbell *et al.* 1973). The evidence demonstrating that mobile sea urchins rely on drift under some conditions is rather more circumstantial.

Mattison *et al.* (1977) turned over specimens of *Strongylocentrotus franciscanus* in the field and found that they were eating detached algae, although there was no evidence that they waited for the food to come to them rather than going in search of it. More often, there have been no direct observations on feeding. It has been assumed that urchins in crevices or aggregating in kelp forests rely on drift because there are few visible accessible attached macroalgae and movement of the urchins seems to be limited (Lowry and Pearse 1973; Cowen 1983; Harrold and Reed 1985). Most of the data are not, however, adequate to determine whether these animals are, in fact, eating drift, whether they are simply eating drift algae whenever they find them, or have the strategy of waiting for the food to come to them. In a study of *Strongylocentrotus franciscanus*, the urchins showed reduced movement when abundant drift algae were available (Mattison *et al.* 1977). In contrast, *Evechinus chloroticus* showed increased movement when drift algae were experimentally introduced into their habitat (Andrew and Stocker 1986). Thus, there is good evidence that some species of urchin do eat drift algae, and that the animals move around to find this food rather than sitting and waiting for it.

Timing of foraging

Temporal patterns of activity have recently been reviewed by Hawkins and Hartnoll (1983) and Little (1989). Intertidal animals living high on the shore usually show unpredictable patterns of movement, responding quickly to sporadic external stimuli like wave-wash (Boyden and Zeldis 1979; Wells 1980), the weather (Hamilton 1977), or periods of submergence (Barnes 1986). Some high-shore animals, such as the limpet *Patella granularis*, will feed more or less continuously as long as the substratum is wet (Branch 1971), whereas other high-shore limpets show circatidal or circadian patterns of movement (Warburton 1973; Boyden and Zeldis 1979; Barnes 1986).

Low-shore animals are often more predictable in their temporal patterns of movement, frequently showing circatidal rhythms which persist from day to day (Zann 1973; Chelazzi et al. 1988; Little 1989). Although most species are active when submerged (Hawkins and Hartnoll 1983; Little 1989), some species only feed on the rising and falling tide, remaining inactive at low and high tide (Cook 1971; Thomas 1973; Levings and Garrity 1983). *Siphonaria pectinata* has been reported to forage more on the ebbing than the rising tide (Thomas 1973). Other low-shore species, e.g. the limpet *Patella miniata*, feed independently of either the time of day or the tide (Branch 1971). *Siphonaria capensis*, which homes to a scar after foraging, is only active at low tide during the night (Branch and Cherry 1985). Its weak tenacity while active may explain why it returns to shelter before high tide. Most explanations for the timing of activity involve escape from physical factors, such as desiccation, or biological factors, such as predation (Little 1989).

Some subtidal grazers also show circadian patterns of foraging. The sea-urchins *Evechinus chloroticus* (Andrew 1988) and *Centrostephanus coronatus* (Vance 1979) are nocturnal grazers, remaining immobile during the day although not necessarily sheltering in crevices. The urchin *Centrostephanus rodgersii* aggregates in shelters during the day, moving out at night to feed (Fletcher 1987). This timing of activity may be a response to diurnally active fish predators, such as wrasses (Nelson and Vance 1979).

Often, the explanations proposed for patterns of foraging behaviour are offered in response to a series of preliminary observations. Rarely are hypotheses derived from these models examined in subsequent experimental studies. Levings and Garrity (1983) have demonstrated how useful it is to examine the possible consequences of alternative explanations for the behaviour of intertidal foraging snails. Previous reviews of temporal patterns of foraging have been extensive

(Hawkins and Hartnoll 1983; Little 1989) but relatively uncritical. Some of the proposed causes of different aspects of behaviour are, as yet, untested speculations and more experimental data are needed before syntheses are realistic (see also Underwood 1979).

Variability of foraging

1. Spatial and temporal variation

Foraging activity in many species appears to be extremely labile, varying from place to place and time to time (Hawkins and Hartnoll 1983; Little 1989). In some species, there can be considerable individual variation in the extent of foraging from day to day, possibly in response to variable local environmental conditions although these may not always be apparent (Hamilton 1977; Chapman 1986). Usually variation in the extent or timing of foraging from day to day is explained in terms of the timing of low or high tide or the height of the tide in the neap/spring cycle (Cook and Cook 1978; Little et al. 1988). Cook and Cook (1978) showed that patterns in the timing and extent of foraging in the limpets *Siphonaria normalis* and *S. alternata* were temporally and spatially variable, with reduced activity during spring tides at exposed sites. Little et al. (1988, 1990) also showed variation in the timing of activity in *Patella vulgata* according to the time of low tide, but this was confounded by spatial variability, with low-shore animals and high-shore animals foraging at different times. Chapman (1986) showed considerable variation in the extent and directionality of foraging between mid-shore and high-shore populations of *Littorina unifasciata*. These differences were themselves variable from time to time and this was not correlated with tidal height or the time of high tide.

Animals may show different patterns of foraging in different microhabitats on the same shore. For example, the intertidal limpet *Siphonaria capensis* living in pools is active during day and night, whereas animals living on open rock on the same shores are only active at night when the humidity is greater (Branch and Cherry 1985). Underwood and Chapman (1989) showed significant differences in directionality of foraging excursions and distances dispersed by the periwinkle *Littorina unifasciata*, apparently related to fine-scale topographical complexity of the substratum on which the animals were moving. There have been many studies on the movements of *Patella vulgata* in different localities, with considerable variation in timing and extent of foraging from place to place (Cook et al. 1969; Hartnoll and Wright 1977; Little et al. 1988, 1990). Even in the same site, populations living within a few metres of each other vary considerably in timing of their foraging behaviour according to the height at which they live (Little et al. 1988,

1990), although limited spatial replication makes it difficult to interpret these differences in behaviour. Studies replicated at different times and in different places, which are therefore capable of showing variation in foraging, indicate how dangerous it is to generalize about foraging behaviour from inadequate data (Underwood and Chapman 1985, 1989; Chapman 1986).

2. Changes in foraging behaviour

Many grazers have morphological limitations on the degree to which they could change feeding in response to changes in availability, types, or diversity of food. Gastropods, for example, have radular structures not very adaptable for different types of microalgae (Nicotri 1977). The division of grazers into functional feeding groups correlated with diet (Steneck and Watling 1982) suggests that morphology and function might be tightly linked, although this is open to dispute (Raffaelli 1985). Nevertheless, many species do show behavioural changes in response to alterations in environmental or intrinsic factors.

(a) Seasonality Some species change feeding patterns at different seasons, because of either changes in activity or changes in availability of food. For example, the urchins *Strongylocentrotus droebachiensis* and *S. franciscanus* feed on a range of attached algae but have a preference for *Nereocystis* (Vadas 1977). This alga is an annual and it dies back each winter. In summer, the urchins mainly eat attached *Nereocystis* and a variety of other plants. In winter, *Nereocystis* forms a smaller portion of their diet, and is often located and eaten as drift (Vadas 1977).

Some intertidal species have also been shown to change patterns of foraging at different times of the year. The periwinkle, *Littorina littorea*, moves downshore during winter, reducing feeding activity (Bertness *et al.* 1983). Other littorinids (*Littorina scutulata*) are forced to change methods of moving and foraging because of changes in dispersion of their microalgal foods. During cooler seasons, mats of diatoms are extensive at high intertidal sites and the snails must graze into the patches from the edges. During warmer seasons, the food-supply does not form thick mats but is scattered over the surface. As a result, the animals can apparently graze more freely (Castenholz 1961).

(b) Ontogenic variation Many grazing gastropods show changes in food and foraging behaviour as they get larger. Small *Collisella purpureus* feed on microalgae growing on the rock surface, but large specimens feed on the kelp *Ecklonia* (Choat and Black 1979). The limpet *Patella compressa* is an algal specialist which lives on and feeds on *Ecklonia* and *Laminaria* (Branch 1971). Small specimens feed on the epiphytes and fronds, with more than one feeding on the same frond. Large specimens move on to

the stipe where they form a home scar, are territorial and feed on the kelp (Branch 1971; Choat and Black 1979). Many limpets appear to free-range while small and become territorial when larger. Young *Patella longicosta* settle on to the shells of older limpets, where they feed on *Ralfsia*, often descending on to the substratum to feed. When they are large enough (about two years old) to survive on encrusting algae outside the adults' territories, they move down on to the substratum and set up territories on encrusting *Lithothamnion* outside the adults' gardens (Branch 1971, 1975). Until they can remove enough *Lithothamnion* to allow establishment of *Ralfsia* and their own territory, they continue to range and feed on *Lithothamnion* (Branch 1975).

Echinoids can also change foraging activity as they get larger. Small specimens of the subtidal *Evechinus chloroticus* are cryptic and remain in crevices, although they can graze around the edges. Larger specimens are free-ranging and move into the open (Choat and Andrew 1986; Andrew 1988).

Some animals show quite marked shifts in diet after recruitment into the adult habitat and subsequent growth. For example, the starfish *Stichaster australis* only settled on the encrusting coralline alga *Mesophyllum insigna* at low levels on the shore (Barker 1977). It stayed there for some time after settlement, feeding on the algae. Later it moved off and began to feed on mussels and other prey. It is interesting that the larval starfish were responding to specific chemical cues produced by similar algae (Morse *et al.* 1984) that cause settlement of such archaeogastropod grazers as abalone. Nudibranchs are very well-known for ontogenic shifts of diet from being larval phytoplanktivores to settlement, metamorphosis, and subsequent specialization on particular species of prey (Thompson 1962, 1964).

(c) Food supply Many marine grazers appear to be rather labile in food preference and take the opportunity to eat what is available rather than search for better food (Hawkins and Hartnoll 1983). The extent of movement can sometimes be directly related to the amount of food available in the environment. For example, the urchin *Evechinus chloroticus* increases its foraging range if the kelp *Ecklonia* is absent, or if drift algae are provided (Andrew and Stocker 1986). Many urchins, including *Strongylocentrus franciscanus* in California, alternate between aggregating in small, stationary clumps and forming large, motile feeding aggregations (Bernstein *et al.* 1981; Dean *et al.* 1984). The stationary clumps are common in kelp forests and these animals are thought to rely on drift algae. Increased starvation, through decline in drift algae, is thought to cause formation of the mobile feeding fronts, in which the density of the urchins is increased. These animals move

into barren grounds and the edges of kelp forests, where they feed on attached macroalgae. They then establish and maintain these barrens, in which foraging is far more extensive.

One of the intriguing studies of change in foraging behaviour is that by McKillup (1983). Mud-snails, *Nassarius pauperatus*, on different shores are confronted by different spatial supplies of food. On some shores, the animals are scavengers and feed on point-sources, aggregating around pieces of carrion. Elsewhere, they tend to eat microalgae scattered over the surface of the mud. The animals also occur in two varieties: 'twisters' and 'non-twisters', reflecting their differences in behaviour when feeding among a crowded group of animals. Twisting of the shell may confer some advantages in dislodging other feeding snails from a point-source of food and preventing the twisting snails from being dislodged. McKillup and Butler (1983) demonstrated that hunger of snails differed from shore to shore and that algae are relatively poor-quality food even though they may be abundant. McKillup (1983) was able to demonstrate that the twisting behaviour used during foraging was inherited and more widespread on shores with carrion.

Conclusions

It is difficult to draw any general conclusions about patterns of foraging in marine benthic grazers for various reasons. Firstly, methods used by investigators vary according to the precise questions being asked. So, where the main focus has been the impact of grazing on distribution, abundance, and diversity of algae, foraging patterns of grazers have not been recorded (Gaines 1985; Hannah and Santelices 1985). Many investigations of orientation of gastropods have been done in the laboratory (Underwood 1979), with no clear indication as to how this behaviour related to natural movement in the field. In some field studies, displacement experiments have been used to test behaviour in relation to preferred habitats or levels of distribution (Chelazzi and Vannini 1976, 1980; Gendron 1977; Byers and Mitton 1981; McQuaid 1981; Doering and Phillips 1983), again with no clear indication of relationship to natural movement and habitat choice. Even when quantitative data are collected on movements of relatively undisturbed animals under natural conditions (Underwood 1977; Cook and Cook 1978, 1981; Chapman 1986; Chelazzi *et al.* 1987; Underwood and Chapman, 1989; Little *et al.* 1990 and others), it is difficult to find common patterns among the different species studied because different methods are used by investigators and different types of data collected.

Secondly, there is often lack of temporal and spatial replication in studies of a species, so that generalizations cannot be drawn about its behaviour. The behaviour of *Patella vulgata* (Cook et al. 1969; Hartnoll and Wright 1977; Little et al. 1988, 1990) and *Littorina unifasciata* (Underwood and Chapman 1985, 1989; Chapman 1986) varies considerably among populations living on different shores and among populations living within a metre of each other on the same shore. This variation has been shown to be correlated with the height at which the animals live (Little et al. 1988, 1990), or with fine-scale topographical complexity of the substratum (Underwood and Chapman 1989). Nevertheless, such correlation cannot indicate the cause of these patterns. Cook and Cook (1978) showed variation in foraging activity in *Siphonaria normalis* and *S. alternata* from place to place, and concluded that this was correlated with the degree of wave exposure. The lack of replication within each area meant that it was impossible to separate spatial variation caused by wave exposure from spatial variation caused by other factors. Many studies are only done at one place (Thomas 1973; Mattison et al. 1977; Barnes 1986; Branch 1988) or over extremely short periods (Momma and Sato 1969; Buschmann 1990) and not replicated further.

Thirdly, most studies have only concentrated on one aspect of foraging, so there is no overall picture including spatial and temporal variability and the various factors which might influence it. For example, Worthington and Fairweather (1989) showed a close association between the coralline alga *Corallina* and the distribution of *Turbo undulatum*, apparently in response to the shelter that the alga offers. It is not clear how this relates to the foraging behaviour of the gastropod.

It is apparent that a complete study of foraging by any species of grazer should at least include accounts of the following (not all of which will be relevant for all species):

General patterns in populations:

(1) patterns of distribution and dispersion of the grazers;
(2) size-classes and stages of the life-history of the animals in different parts of their habitat;
(3) patterns of distribution and diversity of the food plants;
(4) tidal, daily, and seasonal changes in these distributions.

For each size-class or ontogenic stage (in each part of a heterogeneous habitat):

(1) associations of the grazers with crevices, other topographical features, algae, or other organisms that might provide shelter;
(2) consistency of home-sites, scars, 'roosts', or positions to which the animals return repeatedly after foraging;
(3) timing, frequency, and any patterns in the commencement of foraging;
(4) distance, direction, and directionality (i.e., degree of similarity of

different individuals within the population) of movements;
(5) orientation and cues used during foraging;
(6) choices, handling, and mixtures of plants consumed.

Three different components are needed in any study of foraging. Firstly, proper quantitative observations of the above components of foraging are an essential prerequisite to understanding the patterns of behaviour of a species. Ideally, observations should be gathered over a range of patterns of weather, tidal ranges, heights on the shore, as appropriate for the target organism. Most importantly, careful attention should be given to the requirements of replication, particularly at relevant spatial scales (see Chapman 1986, for examples). The description and analysis of patterns of movement in complex data sets are becoming easier with better statistical procedures (Mardia 1972; Underwood 1977; Batschelet 1981; Underwood and Chapman 1985; Chapman and Underwood 1992).

Secondly, alternative models to account for different patterns of behaviour during foraging need to be precisely framed (Underwood 1990). Usually, several quite different models might explain a particular behaviour. For example, animals may be confined to grazing near crevices because they provide refuge from predation or from physical stresses. There is no way of distinguishing between these two models, other alternatives, or some combination of explanations, without critical experimental tests (Medawar 1969; Underwood 1985, 1990). The first step is therefore careful formulation of distinguishable, testable, predictive hypotheses based on the various models and combinations of models. Logical construction of these is essential (see Connell 1972; Underwood and Denley 1984, for examples).

Finally, carefully designed experiments are needed to test the various hypotheses. These require attention to ensure that they will not contain the same problems as many previous experiments (Hurlbert 1984; Underwood 1986, 1988; Chapman and Underwood 1992).

Only when such a sequence of steps has been completed, as has been the case in only a subset of the studies considered above, is it possible to be sure of patterns and explanations of patterns of foraging in benthic marine grazers.

Acknowledgements

Preparation of this review was supported by funds from the Australian Research Council (to AJU) and from the Research Grant and the Institute of Marine Ecology of the University of Sydney. We are grateful for discussions and advice from colleagues at the Institute and at the Ross Street Marine Laboratories.

References

Andrew, N.L. (1988). Ecology of the sea-urchin *Evechinus chloroticus* in northern New Zealand: a review. *New Zealand Journal of Marine and Freshwater Research*, **22**, 415-26.

Andrew, N.L. and Stocker, L.J. (1986). Dispersion and phagokinesis in the echinoid *Evechinus chloroticus*. *Journal of Experimental Marine Biology and Ecology*, **100**, 11-23.

Barker, M.F. (1977). Observations on the settlement of the brachiolaria larvae of *Stichaster australis* (Verrill) and *Coscinasterias calamaria* (Gray) (Echinodermata: Asteroidea) in the laboratory and on the shore. *Journal of Experimental Marine Biology and Ecology*, **30**, 95-108.

Barkman, J.J. (1955). On the distribution and ecology of *Littorina obtusata* (L.) and its subspecific units. *Archives Néerlandiases de Zoologie*, **11**, 22-86.

Barnes, R.S.K. (1986). Daily activity rhythms in the intertidal gastropod *Hydrobia ulvae* (Pennant). *Estuarine, Coastal and Shelf Science*, **22**, 325-34.

Batschelet, E. (1981). *Circular statistics in biology*. Academic Press, London.

Bernstein, B.B., Williams, B.E, and Mann, K.H. (1981). The role of behavioral responses to predators in modifying urchins' (*Strongylocentrotus droebachiensis*) destructive grazing and seasonal foraging patterns. *Marine Biology*, **63**, 39-50.

Bertness, M., Yund, P.O., and Brown, A.F. (1983). Snail grazing and the abundance of algal crusts on a sheltered New England rocky beach. *Journal of Experimental Marine Biology and Ecology*, **71**, 147-64.

Black, R. (1976). The effects of grazing by the limpet *Acmaea insessa* on the kelp *Egregia laevigata* in the intertidal zone. *Ecology*, **57**, 265-77.

Boyden, C.R. and Zeldis, J.R. (1979). Preliminary observations using an attached microphonic sensor to study feeding behaviour of an intertidal limpet. *Estuarine and Coastal Marine Science*, **9**, 759-70.

Branch, G.M. (1971). The ecology of *Patella* Linnaeus from the Cape Peninsula, South Africa. 1. Zonation, movements and feeding. *Zoologica Africana*, **6**, 1-38.

Branch, G.M. (1975). Mechanisms reducing intraspecific competition in *Patella* spp.: migration, differentiation and territorial behaviour. *Journal of Animal Ecology*, **44**, 575-600.

Branch, G.M. (1981). The biology of limpets: physical factors, energy flow, and ecological interactions. *Oceanography and Marine Biology Annual Review*, **19**, 235-80.

Branch, G.M. (1988). Activity rhythms in *Siphonaria thersites*. In *Behavioural adaptation to intertidal life*, (ed. G. Chelazzi and M. Vannini), pp. 27-44. Plenum Press, New York.

Branch, G.M. and Cherry, M.I. (1985). Activity rhythms of the pulmonate limpet *Siphonaria capensis* Quoy and Gaimard as an adaptation to osmotic stress, predation and wave action. *Journal of Experimental Marine Biology and Ecology*, **87**, 153-68.

Buschmann, A. H. (1990). Intertidal macroalgae as refuge and food for Amphipoda in central Chile. *Aquatic Botany*, **36**, 237-45.

Byers, B. A. and Mitton, J. B. (1981). Habitat choice in the intertidal snail *Tegula funebralis*. *Marine Biology*, **65**, 149-54.

Campbell, A. C., Dart, J. K. G., Head, S. M., and Ormond, R. F. G. (1973). The feeding activity of *Echinostrephus molaris* (de Blainville) in the central Red Sea. *Marine Behaviour and Physiology*, **2**, 155-69.

Castenholz, R. W. (1961). The effect of grazing on marine littoral diatom populations. *Ecology*, **42**, 783-94.

Castenholz, R. W. (1963). An experimental study of the vertical distribution of littoral marine diatoms. *Limnology and Oceanography*, **8**, 450-62.

Chapman, M. G. (1986). Assessment of some controls in experimental transplants of intertidal gastropods. *Journal of Experimental Marine Biology and Ecology*, **103**, 181-201.

Chapman, M. G. and Underwood, A. J. (1992). Experimental designs for analyses of movements by gastropods. *Journal of Molluscan Studies*. (In press).

Chelazzi, G. and Vannini, M. (1976). Researches on the coast of Somalia. The shore and the dune at Sar Vanle. 9. Coastward orientation after displacement in *Nerita textilis* Dillwyn (Gastropoda: Prosobranchia). *Italian Journal of Zoology*, **4**, 161-78.

Chelazzi, G. and Vannini, M. (1980). Zonal orientation based on local visual cues in *Nerita plicata* L. (Mollusca: Gastropoda) at Aldabra Atoll. *Journal of Experimental Marine Biology and Ecology*, **46**, 147-56.

Chelazzi, G., Focardi, S., and Deneubourg, J. L. (1983). A comparative study on the movement patterns of two sympatric tropical chitons (Mollusca: Polyplacophora). *Marine Biology*, **74**, 115-26.

Chelazzi, G., Della Santina, P., and Vannini, M. (1985). Long-lasting substrate marking in the collective homing of the gastropod *Nerita textilis*. *Biological Bulletin of the Marine Biological Laboratory, Woods Hole*, **168**, 214-21.

Chelazzi, G., Della Santina, P., and Parpagnoli, D. (1987). Trail following in the chiton *Acanthopleura gemmata*: operational and ecological problems. *Marine Biology*, **95**, 539-46.

Chelazzi, G., Focardi S., and Deneubourg, J. L. (1988). Analysis of movement patterns and orientation mechanisms in intertidal chitons and gastropods. In *Behavioural adaptation to intertidal life*, (ed. G. Chelazzi and M. Vannini), pp. 173-84. Plenum Press, New York.

Choat, J. H. (1982). Fish feeding and the structure of benthic communities in temperate waters. *Annual Review of Ecology and Systematics*, **13**, 423-49.

Choat, J. H. and Andrew, N. L. (1986). Interactions amongst species in a guild

of subtidal benthic herbivores. *Oecologia* (Berlin), **68**, 387–94.
Choat, J. H. and Black, R. (1979). Life histories of limpets and the limpet-laminarian relationship. *Journal of Experimental Marine Biology and Ecology*, **41**, 25–50.
Connell, J. H. (1963). Territorial behavior and dispersion in some marine invertebrates. *Researches in Population Ecology*, **5**, 87–101.
Connell, J. H. (1972). Community interactions on marine rocky intertidal shores. *Annual Review of Ecology and Systematics*, **3**, 169–92.
Connor, V. (1986). The use of mucous trails by intertidal limpets to enhance food resources. *Biological Bulletin of the Marine Biological Laboratory, Woods Hole*, **171**, 548–64.
Cook, A., Bamford, O. S., Freeman, J. D. B., and Teideman, D. J. (1969). A study of the homing habit of the limpet. *Animal Behaviour*, **17**, 330–9.
Cook, S. B. (1969). Experiments on homing in the limpet *Siphonaria normalis*. *Animal Behaviour*, **17**, 679–82.
Cook, S. B. (1971). A study of homing behaviour in the limpet *Siphonaria alternata*. *Biological Bulletin of the Marine Biological Laboratory, Woods Hole*, **141**, 449–57.
Cook, S. B. and Cook, C. B. (1978). Tidal amplitude and activity in the pulmonate limpets *Siphonaria normalis* (Gould) and *S. alternata* (Say). *Journal of Experimental Marine Biology and Ecology*, **35**, 119–36.
Cook, S. B. and Cook, C. B. (1981). Activity patterns in *Siphonaria* populations: heading choice and the effects of size and grazing interval. *Journal of Experimental Marine Biology and Ecology*, **49**, 69–80.
Coull, B. C. and Wells, J. B. J. (1983). Refuges from fish predation: experiments with phytal meiofauna from the New Zealand rocky intertidal. *Ecology*, **64**, 1599–1609.
Cowen, R. K. (1983). The effect of sheephead (*Semicossyphus pulcher*) predation on red sea-urchin (*Strongylocentrotus franciscanus*) populations: an experimental analysis. *Oecologia* (Berlin, **58**, 249–55.
Creese, R. G. (1982). Distribution and abundance of the acmaeid limpet *Patelloida latistrigata*, and its interaction with barnacles. *Oecologia* (Berlin), **52**, 85–96.
Creese, R. G. and Underwood, A. J. (1982). Analysis of inter- and intraspecific competition amongst limpets with different methods of feeding. *Oecologia* (Berlin), **53**, 337–46.
Da Silva, F. M. and Brown, A. C. (1984). The gardens of the sandy-beach whelk *Bullia digitalis* (Dillwyn). *Journal of Molluscan Studies*, **50**, 64–5.
Dean, T. A., Schroeter, S. C., and Dixon, J. D. (1984). Effects of grazing by two species of sea-urchins (*Strongylocentrotus franciscanus* and *Lytechinus anamesus*) on recruitment and survival of two species of kelp (*Macrocystis pyrifera* and *Pterygophyta californicana*). *Marine Biology*, **78**, 301–14.

Dexter, R. W. (1943). Observations on local movements of *Littorina littorea* (L.) and *Thais lapillus* (L.). *Nautilus*, **57**, 6-8.
Doering, P. H. and Phillips, D. W. (1983). Maintenance of the shore-level size gradient in the marine snail, *Tegula funebralis* (A. Adams): importance of behavioural responses to light and sea star predators. *Journal of Experimental Marine Biology and Ecology*, **67**, 159-73.
Fairweather, P. G. and Underwood, A. J. (1983). The apparent diet of predators and biases due to different handling times of their prey. *Oecologia* (Berlin), **56**, 169-79.
Fletcher, W. J. (1987). Interactions among subtidal Australian sea-urchins, gastropods and algae: effects of experimental removals. *Ecological Monographs*, **57**, 89-109.
Frank, P. W. (1964). On home range of limpets. *American Naturalist*, **98**, 99-104.
Gaines, S. D. (1985). Herbivory and between-habitat diversity: the differential effectiveness of defenses in a marine plant. *Ecology*, **66**, 472-85.
Gendron, R. P. (1977). Habitat selection and migratory behaviour of the intertidal gastropod *Littorina littorea* (L.). *Journal of Animal Ecology*, **46**, 79-92.
Graham, A. and Fretter, V. (1947). The life history of *Patina pellucida* (L.). *Journal of the Marine Biological Association of the United Kingdom*, **26**, 590-601.
Hamilton, P. V. (1977). Daily movements and visual location of plant stems by *Littorina irrorata* (Mollusca: Gastropoda). *Marine Behaviour and Physiology*, **4**, 293-304.
Hamilton, P. V. (1978). Intertidal distribution and long-term movements of *Littorina irrorata* (Mollusca: Gastropoda). *Marine Biology*, **46**, 49-58.
Hannah, G. and Santelices, B. (1985). Ecological differences between the isomorphic reproductive phases of two species of *Iridaea* (Rhodophyta: Gigartinales). *Marine Ecology Progress Series*, **22**, 291-303.
Harris, S. A., Da Silva, F. M., Bolton, J. J., and Brown, A. C. (1986). Algal gardens and herbivory in a scavenging sandy-beach nassariid whelk. *Malacologia*, **27**, 299-305.
Harrold, C. and Reed, D. C. (1985). Food availability, sea-urchin grazing, and kelp forest community structure. *Ecology*, **66**, 1160-69.
Hartnoll, R. G. and Wright, J. R. (1977). Foraging movements and homing in the limpet *Patella vulgata* L. *Animal Behaviour*, **25**, 806-10.
Hawkins, S. J. and Hartnoll, R. G. (1983). Grazing of intertidal algae by marine invertebrates. *Oceanography and Marine Biology Annual Review*, **21**, 195-282.
Hay, C. (1979). Some factors affecting the upper limit of the southern bull kelp *Durvillaea antarctica* (Charisso) Hurst on two New Zealand shores. *Royal Society of New Zealand*, **9**, 279-89.

Hay, M. E. (1984). Patterns of fish and urchin grazing on Caribbean coral reefs: are previous results typical? *Ecology*, **65**, 446-54.

Holmlund, M. B., Peterson, C. H., and Hay, M. E. (1990). Does algal morphology affect amphipod susceptibility to fish predation? *Journal of Experimental Marine Biology and Ecology*, **139**, 65-83.

Hughes, R. N. (1980). Optimal foraging theory in the marine context. *Annual Review of Marine Biology and Oceanography*, **18**, 423-81.

Hurlbert, S. J. (1984). Pseudoreplication and the design of ecological field experiments. *Ecological Monographs*, **54**, 187-211.

Imrie, D. W., Hawkins, S. J., and McCrohan, C. R. (1989). The olfactory-gustatory basis of food preference in the herbivorous prosobranch, *Littorina littorea* (Linnaeus). *Journal of Molluscan Studies*, **55**, 217-25.

Jernakoff, P. (1985). Interactions between the limpet *Patelloida latistrigata* and algae on an intertidal rock-platform. *Marine Ecology Progress Series*, **23**, 71-8.

Kain, J. M. and Svendsen, P. (1969). A note on the behaviour of *Patina pellucida* in Britain and Norway. *Sarsia*, **38**, 25-30.

Kitting, C. L. (1979). The use of feeding noises to determine the algal foods being consumed by individual intertidal molluscs. *Oecologia* (Berlin), **40**, 1-18.

Kitting, C. L. (1980). Herbivore-plant interactions of individual limpets maintaining a mixed diet of intertidal marine algae. *Ecological Monographs*, **50**, 527-50.

Leighton, D. and Boolootian, R. A. (1963). Diet and growth in the black abalone, *Haliotis cracherodii*. *Ecology*, **44**, 227-38.

Levings, S. C. and Garrity, S. D. (1983). Diel and tidal movements in two co-occurring neritid snails: differences in grazing paterns on a tropical rocky shore. *Journal of Experimental Marine Biology and Ecology*, **67**, 261-78.

Little, C. (1989). Factors governing patterns of foraging activity in littoral marine herbivorous molluscs. *Journal of Molluscan Studies*, **55**, 273-84.

Little, C., Williams, G. A., Morritt, D., Perrins, J. M., and Stirling, P. (1988). Foraging behaviour of *Patella vulgata* L. in an Irish sea-lough. *Journal of Experimental Marine Biology and Ecology*, **120**, 1-21.

Little, C., Morritt, D., Paterson, D. M., Stirling, P., and Williams, G. A. (1990). Preliminary observations on factors affecting foraging activity in the limpet *Patella vulgata*. *Journal of the Marine Biological Association of the United Kingdom*, **70**, 181-95.

Lowry, L. F. and Pearse, J. S. (1973). Abalones and sea urchins in an area inhabited by sea urchins. *Marine Biology*, **23**, 213-19.

Lubchenco, J. (1979). Consumer terms and concepts. *American Naturalist*, **113**, 315-17.

Lubchenco, J. (1983). *Littorina* and *Fucus*: effects of herbivores, substratum heterogeneity and plant escapes during succession. *Ecology*, **64**, 1116-23.

Lubchenco, J. and Cubit, J. (1980). Heteromorphic life histories of certain marine algae as adaptations to variations in herbivory. *Ecology*, **61**, 676–87.
Lubchenco, J. and Gaines, S. D. (1981). A unified approach to marine plant–herbivore interactions. I. Population and communities. *Annual Review of Ecology and Systematics*, **12**, 405–37.
Mackay, D. A. and Underwood, A. J. (1977). Experimental studies on homing in the intertidal patellid limpet *Cellana tramoserica* (Sowerby). *Oecologia (Berlin)*, **30**, 215–38.
MacLulich, J. H. (1986). Colonization of bare rock surfaces by microflora in a rocky intertidal habitat. *Marine Ecology Progress Series*, **32**, 91–6.
MacLulich, J. H. (1987). Variations in the density and variety of intertidal epilithic microflora. *Marine Ecology Progress Series*, **40**, 285–93.
Mardia, K. V. (1972). *Statistics of directional data*. Academic Press, London.
Mattison, J. E., Trent, J., Shanks, A., Akin, T. B., and Pearse, J. S. (1977). Movement and feeding activity of red sea-urchins (*Strongylocentrotus franciscanus*) adjacent to a kelp forest. *Marine Biology*, **39**, 25–30.
McBane, C. D. and Crocker, R. A. (1983). Animal–algal relationships of the amphipod *Hyale nilssoni* (Rathke) in the rocky intertidal. *Journal of Crustacean Biology*, **3**, 592–601.
McKillup, S. C. (1983). A behavioural polymorphism in the marine snail *Nassarius pauperatus*: geographic variation correlated with food availability and differences in competitive ability between morphs. *Oecologia (Berlin)*, **56**, 58–66.
McKillup, S. C. and Butler, A. J. (1983). Measurement of hunger as a relative estimate of food available to populations of *Nassarius pauperatus*. *Oecologia (Berlin)*, **56**, 16–22.
McQuaid, C. D. (1981). The establishment and maintenance of vertical size gradients in populations of *Littorina africana knysnaensis* (Phillips) on an exposed rocky shore. *Journal of Experimental Marine Biology and Ecology*, **54**, 77–89.
Medawar, P. (1969). *Induction and intuition in scientific thought*. Methuen, London.
Momma, H. and Sato, R. (1969). The locomotion behaviour of the disc abalone, *Haliotis discus hannai* Ino, and the Siebold's abalone, *Haliotis sieboldi* Reeve, in the fishing grounds. *Tohuko Journal of Agricultural Research*, **20**, 1501–57.
Morse, A. N. C., Froyd, C. A., and Morse, D. E. (1984). Molecules from cyanobacteria and red algae that induce larval settlement and metamorphosis in the mollusc *Haliotis rufescens*. *Marine Biology*, **18**, 293–8.
Nelson, B. V. and Vance, R. R. (1979). Diel foraging patterns of the sea urchin *Centrostephanus coronatus* as a predator avoidance strategy. *Marine Biology*, **51**, 251–8.

Nicotri, M. E. (1977). Grazing effects of four marine intertidal herbivores on the microflora. *Ecology*, **58**, 1020–32.

Nicotri, M. E. (1980). Factors involved in herbivore food preference. *Journal of Experimental Marine Biology and Ecology*, **42**, 13–26.

Norton, T. A., Hawkins, S. J., Manley, N. L., Williams, G. A., and Watson, D. C. (1990). Scraping a living: a review of littorinid grazing. *Hydrobiologia*, **193**, 117–38.

Ogden, J. C., Brown, R. A., and Salesky, N. (1973). Grazing by the echinoid *Diadema antillarum* Philippi: formation of halos around West Indian patch reefs. *Science*, **182**, 715–17.

Ohgaki, S. (1989). Vertical movement of the littoral fringe periwinkle *Nodilittorina exigua* in relation to wave height. *Marine Biology*, **100**, 443–8.

Paine, R. T. (1971). Energy flow in a natural population of the herbivorous gastropod *Tegula funebralis*. *Limnology and Oceanography*, **16**, 86–98.

Peterson, C. H. and Bradley, B. P. (1978). Estimating the diet of a sluggish predator from field observations. *Journal of the Fisheries Research Board of Canada*, **35**, 136–41.

Petraitis, P. S. (1982). Occurrence of random and directional movements in the periwinkle, *Littorina littorea* (L.). *Journal of Experimental Marine Biology and Ecology*, **59**, 207–17.

Quinn, G. P. (1988). Effects of conspecific adults, macroalgae and height on the shore on recruitment of an intertidal limpet. *Marine Ecology Progress Series*, **48**, 305–6.

Raffaelli, D. (1985). Functional feeding groups of some intertidal molluscs defined by gut contents analysis. *Journal of Molluscan Studies*, **51**, 233–9.

Randall, J. E. (1965). Grazing effect on sea grasses by herbivorous reef fishes in the West Indies. *Ecology*, **48**, 255–60.

Santelices, B., Correa, J., and Avila, M. (1983). Benthic algal spores surviving digestion by sea-urchins. *Journal of Experimental Marine Biology and Ecology*, **70**, 263–70.

Schiel, D. R. (1982). Selective feeding by the echinoid *Evechinus chloroticus*, and the removal of plants from subtidal algal stands in northern New Zealand. *Oecologia* (Berlin), **54**, 379–88.

Steneck, R. S. (1982). A limpet–coralline alga association: adaptations and defenses between a selective herbivore and its prey. *Ecology*, **63**, 507–22.

Steneck, R. S. and Watling, L. (1982). Feeding capabilities and limitation of herbivorous molluscs: a functional group approach. *Marine Biology*, **68**, 299–319.

Stimson, J. (1970). Territorial behavior of the owl limpet *Lottia gigantea*. *Ecology*, **51**, 113–8.

Stimson, J. (1973). The role of the territory in the ecology of the intertidal limpet *Lottia gigantea* (Gray). *Ecology*, **54**, 1020–30.

Taylor, J. D. (1971). Intertidal zonation at Aldabra Atoll. *Philosophical Transactions of the Royal Society*, Series B, **260**, 173-213.

Thomas, R. F. (1973). Homing behaviour and movement rhythms in the pulmonate limpet *Siphonaria pectinata* Linnaeus. *Proceedings of the Malacological Society of London*, **40**, 303-11.

Thompson, T. E. (1962). Studies on the ontogeny of *Tritonia hombergi* Cuvier (Gastropoda, Opisthobranchia). *Philosophical Transactions of the Royal Society*, Series B, **245**, 171-218.

Thompson, T. E. (1964). Grazing and the life-cycles of British nudibranchs. In *Grazing in terrestrial and marine environments*, (ed. D. J. Crisp), pp. 275-97. Blackwell, Oxford.

Tutschulte, T. C. and Connell, J. H. (1988). Feeding behaviour and algal food of three species of abalones (*Haliotis*) in southern California. *Marine Ecology Progress Series*, **49**, 57-64.

Underwood, A. J. (1977). Movements of intertidal gastropods. *Journal of Experimental Marine Biology and Ecology*, **26**, 191-201.

Underwood, A. J. (1978). An experimental evaluation of competition between three species of intertidal prosobranch gastropods. *Oecologia* (Berlin), **33**, 185-208.

Underwood, A. J. (1979). The ecology of intertidal gastropods. *Advances in Marine Biology*, **16**, 111-210.

Underwood, A. J. (1980). The effects of grazing by gastropods and physical factors on the upper limits of distribution of intertidal macroalgae. *Oecologia* (Berlin), **46**, 201-13.

Underwood, A. J. (1984). The vertical distribution and seasonal abundance of intertidal microalgae on a rocky shore in New South Wales. *Journal of Experimental Marine Biology and Ecology*, **78**, 199-220.

Underwood, A. J. (1985). Physical factors and biological interactions: the necessity and nature of ecological experiments. In *The ecology of rocky coasts*, (ed. P. G. Moore and R. Seed), pp. 371-90. Hodder and Stoughton, London.

Underwood, A. J. (1986). The analysis of competition by field experiments. In *Community ecology: pattern and process*, (ed. J. Kikkawa and D. J. Anderson), pp. 240-68. Blackwells, Melbourne.

Underwood, A. J. (1988). Design and analysis of field experiments on competitive interactions affecting behaviour of intertidal animals. In *Behavioural adaptation to intertidal life*, (ed. G. Chelazzi and M. Vannini), pp. 333-58. Plenum Press, New York.

Underwood, A. J. (1990). Experiments in marine ecology and management: their logics, functions and interpretations. *Australian Journal of Ecology*, **15**, 365-389.

Underwood, A. J. and Chapman, M. G. (1985). Multifactorial analysis of directions of movement of animals. *Journal of Experimental Marine Biology and Ecology*, **91**, 17-43.

Underwood, A.J. and Chapman, M.G. (1989). Experimental analyses of the influences of topography of the substratum on movements and density of an intertidal snail, *Littorina unifasciata*. *Journal of Experimental Marine Biology and Ecology*, **134**, 175-96.

Underwood, A.J. and Chapman, M.G. (1991). Experiments on topographic influences on density and dispersion of *Littorina unifasciata* in New South Wales. *Journal of Molluscan Studies*. (In press).

Underwood, A.J. and Denley, E.J. (1984). Paradigms, explanations and generalizations in models for the structure of intertidal communities on rocky shores. In *Ecological communities: conceptual issues and the evidence*, (ed. D.R. Strong, D. Simberloff, L.G. Abele, and A. Thistle), pp. 151-80. Princeton University Press, New Jersey.

Underwood, A.J. and Fairweather, P.G. (1991). Marine invertebrates. In *Natural enemies: the population biology of predators, parasites and diseases*, (ed. M.J. Crawley). Blackwell, Oxford. (In press).

Underwood, A.J. and Jernakoff, P. (1984). The effects of tidal height, wave-exposure, seasonality and rock-pools on grazing and the distribution of intertidal macroalgae in New South Wales. *Journal of Experimental Marine Biology and Ecology*, **75**, 71-96.

Vadas, R.L. (1977). Preferential feeding — optimization strategy in sea-urchins. *Ecological Monographs*, **47**, 337-72.

Vadas, R.L., Elner, R.W., Garwood, P.E., and Babb, I.G. (1986). Experimental evaluation of aggregation behavior in the sea urchin *Strongylocentrotus droebachiensis*: a reinterpretation. *Marine Biology*, **90**, 433-48.

Vahl, O. (1971). Growth and density of *Patina pellucida* (L.) (Gastropoda: Prosobranchia) on *Laminaria hyperborea* (Gunnerus) from Western Norway. *Ophelia*, **9**, 31-50.

Vance, R.R. (1979). Effects of grazing by the sea urchin *Centrostephanus coronatus* on prey community composition. *Ecology*, **60**, 537-46.

Vannini, M. and Chelazzi, G. (1978). Field observations on rhythmic behaviour of *Nerita textilis* (Gastropoda: Prosobranchia). *Marine Biology*, **45**, 113-22.

Vine, P.J. (1974). Effects of algal grazing and aggressive behaviour of the fishes *Pomacentrus lividus* and *Acanthurus sohal* on coral-reef ecology. *Marine Biology*, **24**, 131-6.

Warburton, K. (1973). Solar orientation in the snail *Nerita plicata* (Prosobranchia: Neritacea) on a beach near Watamur, Kenya. *Marine Biology*, **23**, 93-100.

Warren, J.H. (1985). Climbing as an avoidance behaviour in the salt-marsh periwinkle, *Littorina irrorata* (Say). *Journal of Experimental Marine Biology and Ecology*, **89**, 11-28.

Watson, D.C. and Norton, T.A. (1985). Dietary preferences of the common periwinkle, *Littorina littorea*. *Journal of Experimental Marine Biology and Ecology*, **88**, 193-212.

Wells, R.A. (1980). Activity pattern as a mechanism of predator avoidance in two species of acmaeid limpet. *Journal of Experimental Marine Biology and Ecology*, **48**, 151-68.

Worthington, D.G. and Fairweather, P.G. (1989). Shelter and food: interactions between *Turbo undulatum* (Archaeogastropoda: Turbinidae) and coralline algae on rocky seashores in New South Wales. *Journal of Experimental Marine Biology and Ecology*, **129**, 61-79.

Wright, W.G. (1982). Ritualized behaviour in a territorial limpet. *Journal of Experimental Marine Biology and Ecology*, **60**, 245-52.

Zann, L.P. (1973). Relationships between intertidal zonation and circatidal rhythmicity in littoral gastropods. *Marine Biology*, **18**, 243-50.

14. Chemical mediation of seaweed–herbivore interactions

MARK E. HAY* and WILLIAM FENICAL[†]
*University of North Carolina at Chapel Hill, Institute of Marine Sciences, Morehead City, USA [†]Scripps Institution of Oceanography, University of California, San Diego, La Jolla, USA

Abstract

Seaweeds produce a variety of secondary metabolites that deter feeding by herbivores. The defensive value of a compound is, however, a specific function of the compound's structure and the species of herbivore attacking the plant; previous generalizations about the effectiveness and mode of action of different classes of compounds (e.g., qualitative vs quantitative chemical defences) are unfounded.

Herbivore size and mobility are often correlated with resistance to seaweed chemical defences. Small, sedentary herbivores (mesograzers) that live on the plants they consume often preferentially consume, or specialize on, seaweeds that are chemically defended from fishes. These mesograzers avoid or deter predators by associating with chemically noxious host plants and may use the compounds that deter fishes as specific feeding or host identification cues.

Numerous brown algae in the family Dictyotaceae and red algae in the genus *Laurencia* have chemical and morphological characteristics suggesting that these species may form groups of Mullerian or Batesian mimics. Mimicry could be:

(1) chemical and functional against nonvisually searching herbivores;
(2) morphological and functional against visually searching herbivores;
(3) both chemical and morphological.

Mimicry might also function against algal taxonomists and be responsible for the taxonomic uncertainties common in both the Dictyotaceae and the genus *Laurencia*.

Plant–Animal Interactions in the Marine Benthos, (ed. D. M. John, S. J. Hawkins, and J. H. Price), Systematics Association Special Volume No. 46, pp. 319–37. Clarendon Press, Oxford, 1992. © The Systematics Association 1992.

Introduction

Seaweeds are similar to terrestrial plants in that they produce polyphenolics, terpenes, acetogenins, aromatic compounds, and amino acid-derived substances as secondary metabolites. They differ from terrestrial plants by not producing alkaloids and by commonly incorporating halogens into their secondary compounds. Algal polyphenolics (called phlorotannins because of their biosynthetic differences from the tannins produced by terrestrial plants) have been reviewed by Ragan and Glombitza (1986) and by Steinberg (1992). Other types of algal secondary metabolites, as well as phlorotannins, have been reviewed by Faulkner (1984, 1986, 1987, 1988); Paul and Fenical (1987); Hay and Fenical (1988); Van Alstyne and Paul (1988); and Paul (1992). Although numerous older publications suggested that these compounds were chemical defences against herbivores, adequate tests of this hypothesis were not conducted until very recently (see reviews by Hay and Fenical 1988; Duffy and Hay 1990; Hay 1991; Paul 1992).

Can chemical defences be effective?

In the field, seaweeds are attacked by many species of herbivores; this is especially true on species-rich coral reefs where rates of herbivory are higher than for any other known habitat (Carpenter 1986; Hay 1991). Thus, to be advantageous in such a habitat, a defensive compound would have to function against a broad range of herbivores, and, on average, lower the probability that a given amount of plant will be consumed by the diverse group of herbivores attacking the seaweed. Several experimental field studies demonstrate that common seaweed metabolites are often able to do this (Hay and Fenical 1988; Hay 1991). As examples, when the following secondary metabolites were coated on to the palatable seagrass *Thalassia* and placed on shallow portions of Caribbean coral reefs, all significantly decreased the amount of plant material lost to herbivorous fishes: pachydictyol-A from brown algae in the Dictyotaceae, cymopol from the green alga *Cymopolia*, stypotriol from the brown alga *Stypopodium*, and elatol and isolaurinterol from red algae in the genus *Laurencia* (Hay et al. 1987b).

To date, over 40 pure compounds from a variety of seaweeds have been tested in the field on Caribbean and Pacific coral reefs or in the laboratory against fishes, urchins, crabs, amphipods, or polychaetes (Hay 1991, 1992; Paul 1992; Steinberg 1992). Although many seaweed secondary-metabolites are broad-spectrum deterrents, several have no known effects against herbivores, and few, if any, are deterrent to all herbivores. As one of several possible examples, pachydictyol-A produced by several genera of brown algae in the Dictyotaceae:

(1) deters feeding by Caribbean reef fishes in field experiments and by four species of Pacific and Atlantic fishes against which it has been tested (Fig. 14.1);
(2) deters feeding by the Caribbean sea urchin *Diadema* but not by the temperate Atlantic urchin *Arbacia* (Hay et al. 1987a, b); and
(3) either does not affect, or significantly stimulates, feeding by the amphipods *Ampithoe longimana* and *Pseudamphithoides incurvaria* and the polychaete *Platynereis dumerilii* (Hay et al. 1987a, 1988d, 1990a).

Just as this single compound can vary markedly in its effects on the feeding of different herbivores, very similar compounds can differ markedly in their effects on the same or similar herbivores (Fig. 14.1). As an example, pachydictyol-A and dictyol-E are both diterpene alcohols produced by several species of *Dictyota*. Although their structures differ only by the substitution of one hydroxyl, the compounds differ considerably in their effects on feeding by the fishes, urchins, and mesograzers against which they have been tested (Fig. 14.1). It appears that neither the chemical structure (Hay and Fenical 1988) nor the pharmacological activity (Hay et al. 1987b) of a compound can be used to confidently predict its effect on feeding by any given herbivore. Effects of seaweed secondary-metabolites on a plant's susceptibility to herbivores appear to be determined by the specific effects of individual compounds on individual species of herbivore; few if any broad generalizations about the activities of different classes of compounds seem to be justified (Hay and Fenical 1988).

Although terrestrial ecologists have often conducted investigations of plant chemical defences by lumping compounds to simplify quantification (i.e., measuring total phenolics, total terpenes, etc.) and assuming that these classes of compounds had similar biological activities, this is clearly not a useful approach for marine plants and herbivores. The approach is probably invalid for terrestrial systems as well (Zucker 1983; Bernays et al. 1989).

What selects for resistance to seaweed chemical defences?

Herbivore resistance to plant chemical defences has been more thoroughly studied in terrestrial than in marine communities, with the majority of terrestrial studies focused on insect herbivores (Rosenthal and Janzen 1979; Futuyma 1983; Futuyma and Morano 1988; Bernays et al. 1989). Most species of terrestrial herbivores are insects and as many as 90 per cent of these herbivores may be relatively specialized

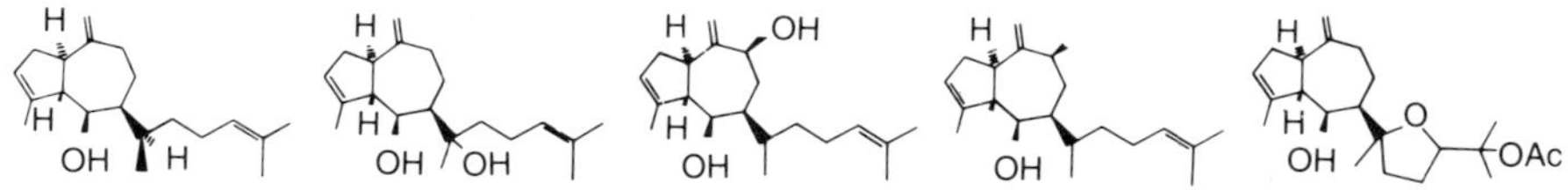

		Pachydictyol A	Dictyol E	Dictyol B	Dictyol B acetate	Dictyol H
Fish	Caribbean reef fish	−(1)	+(5)			
	Siganus doliatus	−(2)	NS(2)	−(5)	NS/−(5,6)	NS(5)
	Lagodon rhomboides	−(3)	−(3)	−(2)		−(2)
	Diplodus holbrooki	−(4)	−(4)			
Amphipods	Ampithoe longimana	NS/+(4)	NS(4)	NS(5)	NS(5)	NS(5)
	Ampithoe valida	NS(5)	−(5)	NS(5)	−(5)	NS(5)
	Gammarus mucronatus	−(5)	−(5)	−(5)	−(5)	−(5)
	Hyale macrodactyla	−(5)	NS(5)	NS(5)	−(5)	NS(5)
Worm	Platynereis dumerilii	NS/+(3)	NS(3)			
Urchins	Arbacia punctulata	NS(4)	−(4)			
	Diadema antillarum	−(1)				

Fig. 14.1. Structures of several dictyols produced by brown seaweeds in the Dictyotaceae and the activities of these compounds when tested as feeding deterrents againsts several herbivores. Key to symbols: − indicates a significant deterrent effect, + indicates that the compound significantly stimulated feeding, NS indicates there was no significant effect. If two symbols are given, then multiple assays have produced variable results. Numbers following these symbols give the reference for each assay.

[1] Hay et al. 1987b
[2] Hay et al. 1988a
[3] Hay et al. 1988c
[4] Hay et al. 1987a
[5] M.E. Hay, J.E. Duffy, and W. Fenical-unpublished data
[6] Hay et al. 1990a

feeders (Bernays 1989). The evolutionary reasons for this high degree of specialization are debated among terrestrial ecologists, with little agreement being evident (see Bernays and Graham 1988 and the immediately following papers in the same issue of *Ecology*). However, once adults place their eggs on a host plant, the factors selecting for the much less mobile larvae to be able to tolerate that plant's chemical defences are obvious (Futuyma and Morano 1988).

In contrast to patterns seen in terrestrial systems, marine communities are dominated by generalist herbivores such as fishes (Hay 1991), urchins (Lawrence 1975), and gastropods (Steneck and Watling 1982; Hawkins and Hartnoll 1983) that feed broadly from many plant types and families. The potential for reciprocal coevolution between these herbivores and any species of seaweed is very small, because few single species of seaweeds are likely to make up an essential portion of their diet. Thus, if there is selection for resistance to seaweed chemical defences among marine herbivores, it is most likely to be:

(1) for broad resistance to the many different metabolites encountered by the herbivore due to its diverse diet, or
(2) caused by ecological constraints other than the direct need to acquire food from specific plants (Hay 1992).

Partial support for the first point (1) comes from recent studies by Steinberg (1992) and Steinberg and Van Altena (in press), showing that feeding by several invertebrate herbivores in Australia and New Zealand is unaffected by the high levels of diverse phlorotannins found in numerous brown seaweeds from temperate Australasia. In contrast, brown algal phlorotannins at much lower concentrations are effective defences against invertebrate herbivores in North America. Abundant brown seaweeds in Australasia may have three times the phlorotannin content of similar seaweeds from North America; this apparently has selected for broad resistance to phlorotannins among Australasian herbivores. These herbivores are not, however, resistant to smaller organic molecules (e.g., certain terpenes) that are produced by some temperate Australasian seaweeds.

In support of the second point above (2), numerous studies indicate that predation by omnivorous and predatory fishes may select for herbivores that can counter seaweed chemical defences that deter fishes (reviewed by Hay 1992). These herbivores reduce their risk of predation by living on plants that are chemically defended from, and thus rarely visited by, their predators. The limited number of studies presently available suggest that this resistance most commonly occurs among small herbivores that use plants as both food and habitat — examples include amphipods (Hay *et al.* 1987a, 1988b, 1990a; Duffy and

Hay 1991a, b), polychaetes (Hay et al. 1988d), crabs (Hay et al. 1989, 1990b), and ascoglossans (Paul and Van Alstyne 1988b; Hay et al. 1989, 1990b). Although resistance to certain seaweed chemical defences occurs among both generalist (Hay et al. 1987a, 1988b, d, Duffy and Hay 1991a) and specialist (Hay et al. 1989, 1990a, b) mesograzers, the ability of this resistance to diminish predation on mesograzers is best studied for specialists. Examples include:

(1) ascoglossan gastropods that specialize on toxic green algae, sequester the algal toxins, and effectively use these for their own defence (Paul and Van Alstyne 1988b; Hay et al. 1990b);
(2) an amphipod that eats and builds a mobile domicile from a chemically defended species of *Dictyota*; when in this domicile, which it uses like a hermit crab does its shell, the amphipod is rejected by predatory fish — if removed from the domicile, the amphipod is rapidly eaten (Hay et al. 1990a); and
(3) tropical crabs in the Pacific and Caribbean that are camouflaged and relatively immune from fish predation when on their algal hosts, which are chemically defended from fishes; if removed from their hosts, they are immediately eaten (Hay et al. 1989, 1990b).

In each of these examples, deterrence or avoidance of predation appears to be the major advantage resulting from feeding specialization. More extensive reviews on how predation may affect the evolution of feeding specialization are available for both marine (Hay 1992) and terrestrial communities (Bernays 1989).

Do models of plant–herbivore interactions and chemical defences developed for terrestrial systems fit the marine data?

The plant apparency model (Feeney 1976; Rhoades and Cates 1976) has provided a conceptual framework for much of the plant–herbivore research conducted during the last 15 years. This model noted that those plant species dominating a community were 'bound to be found' by herbivores (i.e., were apparent) and would thus have to invest heavily in defences that were generalized and effective against a broad range of herbivores. Polyphenolics (tannins) were thought to fill this role by acting as digestibility reducers that allowed little, if any, possibility of counteradaptation by herbivores. Tannins were termed *quantitative* defences because they were thought to function in a dose-dependent manner.

In contrast, fast-growing plants that were unpredictable in space or time were termed unapparent plants because these plants would be more likely to escape detection by many herbivores. Because they allocated more of their resources to rapid growth, reproduction, and dispersal, unapparent plants were thought to be defended by inexpensive toxins that were effective in low concentrations against generalist herbivores but were more easily circumvented by specialist herbivores that had evolved detoxification mechanisms. These compounds were called *qualitative* defences because of their alleged toxicity and effectiveness at low concentrations. Thus, apparent plants were thought to differ from unapparent plants in three basic ways (Fox 1981):

(1) the amount of resources invested in defence;
(2) the types and modes of action of chemicals used for defence (i.e., digestibility reducers vs toxins); and
(3) the potential for coevolution of plants and their herbivores.

A more recent model (Coley *et al.* 1985) focuses on plant resource availability rather than apparency to explain most of the above patterns but retains as its foundation many of the same distinctions regarding the differences in costs and effects of qualitative vs quantitative defences. The ecological and evolutionary generalities upon which these models are based suggest they should apply equally well to plant–herbivore interactions in both marine and terrestrial communities.

Evidence available to date from marine communities suggests that these models may be flawed in their assumptions of differences between apparent and unapparent plants in the types of compounds produced (polyphenolics vs smaller organic molecules), differences in costs (polyphenolics expensive — toxins cheap), differences in modes of action (polyphenolics functioning as digestibility reducers vs smaller organic molecules functioning as toxins), and possibilities for herbivores to evolve a tolerance to these different chemical defences (Hay and Fenical 1988; Steinberg 1992; Steinberg and Van Altena, in press). Similar questions also have been raised regarding the robustness of these models for terrestrial communities (see review by Bernays *et al.* 1989). A few examples of these problems from marine communities are noted below (for a more detailed discussion, see Hay and Fenical 1988; Steinberg 1992).

1. Does production of polyphenolics correlate with apparency?

The apparency theory would predict that extremely apparent seaweeds such as kelps should have high levels of phlorotannins relative to the less apparent Fucales; available data show the opposite pattern (Steinberg 1985, 1992; Estes and Steinberg 1988; Steinberg and Van Altena, in press). The resource availability model would predict that seaweeds in habitats with high light but low nutrient availability (e.g., shallow coral reefs) would have higher phlorotannin levels than seaweeds in habitats experiencing less nutrient limitation (e.g., shallow temperate seaweed beds); this is also not the case (Steinberg 1989). Both within and among genera, phlorotannins are common and in relatively high abundance in temperate brown algae but almost completely absent from similar tropical species (Steinberg 1989, 1992; Van Alstyne and Paul 1990). Even comparisons between temperate regions can be striking (Steinberg 1989; Steinberg and Van Altena, in press).

Additionally, terrestrial comparisons between temperate and tropical forests show greater herbivory and greater tannin concentrations in tropical as opposed to temperate forests (Coley and Aide 1990); in contrast, although herbivory is extreme on tropical coral reefs (often removing 60–100 per cent of primary production, Carpenter 1986; Hay 1991), seaweeds in these habitats produce little, if any, phlorotannins (Steinberg 1989, 1992) even though phlorotannins appear to be effective defences against tropical herbivores (Van Alstyne and Paul 1990). On tropical reefs, apparent green (e.g., *Halimeda*), red (e.g., *Laurencia*), and brown (e.g., *Dictyota*) seaweeds produce small organic molecules (e.g., terpenes) as defences and are not known to produce significant concentrations of phlorotannins. Thus, apparent seaweeds on shallow tropical reefs are attacked by a large variety of generalist grazers (Hay 1991) and grow in a habitat where nutrients, not carbon or light, often limit their rate of growth. Both the plant-apparency model and the resource availability model would predict that these plants should be defended by phlorotannins rather than qualitative defences; the documented patterns are opposite to those predicted by the models. Thus, phlorotannin production appears to be more predictable on the basis of taxonomic relationships or geographical distribution than on the basis of apparency.

2. Costs of chemical defences

Both of the above models assume that high concentrations of polyphenolics are more costly than low concentrations of more toxic compounds. Clearly, more energy is stored in large concentrations (often 3–13 per cent of plant dry mass) of phlorotannins than in lesser concentrations of qualitative defences (rarely much above 1 per cent). It is not at all clear, however, that there is any significant difference in the total energy invested in each type of defence, once the costs of synthesis, storage, and functions other than herbivore deterrence are considered (Hay and Fenical 1988). Additionally, it is doubtful that energy is the currency in which costs should be measured because seaweeds are often nutrient rather than carbon limited, and they may leach large amounts of carbon when nutrients are low and the excess carbon cannot be used for growth (Mann 1982). It is possible that phlorotannins serve at times as simply a way to store excess carbon. If this is the case, then energy invested directly in the compounds may constitute no cost at all. For seaweeds, there is at present no evidence that there is any important difference in the costs of what are often termed qualitative and quantitative defences.

3. Do qualitative and quantitative defences differ in mode of action or the degree to which herbivores can resist the compounds?

Apparency theory proposes that polyphenolics are non-toxic digestibility reducers and that qualitative defences are toxins; additionally, it is assumed that it is more difficult for herbivores to evolve tolerance for polyphenolics than for terpenes or other qualitative defences (Feeney 1976; Fox 1981). The very limited data available on the physiological effects of seaweed metabolites on marine herbivores do not clearly support this distinction.

As would be predicted by the apparency theory, algal phlorotannins are broadly deterrent against generalist invertebrate herbivores in North America (Steinberg 1985). In Australasia, however, generalist invertebrate herbivores are largely unaffected by phlorotannin levels 2 to 3 times higher than those that deter marine herbivores in North America. When grown on seaweeds or artificial diets differing in phlorotannin concentration, the digestive efficiency of Australasian herbivores was unaffected by phlorotannins (Steinberg and Van Altena, in press). Thus, the common herbivores in temperate Australasia are obviously resistant to algal tannins and these tannins do not act as digestibility reducers when fed to these herbivores. Recent terrestrial studies also show that numerous herbivores have evolved resistance to tannins and that tannins may often function as cell toxins rather than as digestibility reducers (Bernays et al. 1989).

Less is known about the long-term effects of smaller organic molecules that function as algal defences. When the diterpene alcohol pachydictyol-A, from the brown seaweed *Dictyota*, was fed to a generalist fish, it significantly reduced the growth rate compared with that of control fish that ate the same amount of food (Hay et al. 1987a). In contrast to the prediction that qualitative defences function as toxins, no effect on survivorship could be demonstrated.

Numerous herbivores have been shown to be resistant to qualitative defences produced by several seaweeds (Hay 1992); however, several herbivores are also resistant to phlorotannins (Steinberg and Van Altena, in press). Thus, to date, there is no obvious difference in the probability of herbivores evolving resistance to these different types of defences. Similar patterns also are known for terrestrial communities; many herbivores eat tannin-rich plants and a few even grow better when tannins are added to their diets (Bernays *et al.* 1989).

At present, the only clear difference between phlorotannins and other chemical defences appears to be that phlorotannins are often produced in high concentrations (3–14 per cent of plant dry mass: Steinberg 1989; Steinberg and Van Altena, in press) while other defensive compounds usually occur as no more than 1–2 per cent of plant dry mass (Hay and Fenical 1988).

New directions

Above we have summarized what we presently know about seaweed chemical defences and how the data from marine systems compare with the models developed for terrestrial plants. But the more interesting focus is not what we know; it is what we should know but do not. In this section we will discuss briefly what we see as fertile areas for future investigations.

1. Does Batesian or Mullerian mimicry occur among seaweeds?

Batesian mimicry occurs when palatable species mimic distasteful ones. Given that palatable seaweeds of one morphology growing near unpalatable ones of a different morphology experience significantly less herbivory than they do when growing alone (Hay 1985, 1986; Littler *et al.* 1986; Pfister and Hay 1988), it seems reasonable to suspect that this associational refuge would become even more pronounced if the palatable species was morphologically similar to the unpalatable one. Although mimicry is unstudied so far among seaweeds, several brown seaweeds in the Dictyotales might be good candidates for investigation. As one of several possible examples, three of the four morphologically similar seaweeds shown in Fig. 14.2 produce compounds demonstrated to deter feeding by reef fishes (*Dictyota bartayresii* and *Dilophus* produce dictyols, including pachydictyol-A — see Fig. 14.1, and *Dictyopteris* produces the C_{11} hydrocarbons dictyopterenes A and B, see (Hay *et al.* 1988*b*). The fourth species, *Dictyota divaricata*, produces dolastanes instead of dictyols and the few dolastane-class compounds that have been tested are not deterrent (Hay *et al.* 1988*a*). The *Dictyota* species producing dolastane-class compounds appear to be more susceptible to herbivores than the dictyol-producing species (see references in Hay *et al.* 1990*a*). *Dictyota* species like *D. cervicornis*, *D. divaricata*, *D. furcellata*, *D. indica*, and *D. linearis*, which produce dolastanes and are not known to produce dictyols (Faulkner 1984, 1986, 1987, 1988), often occur intermixed with other morphologically similar *Dictyota* species that are

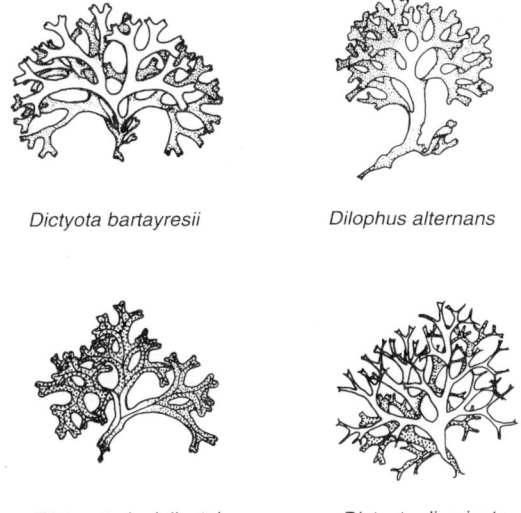

Fig. 14.2. An illustration of four morphologically similar, and in some cases chemically similar (*Dictyota* and *Dilophus*), seaweeds that commonly co-occur on Caribbean reefs. Illustrations are modified from Taylor (1960).

chemically defended from reef herbivores. Thus, the more palatable species could be functioning as Batesian mimics.

A similar situation could exist among several morphologically similar species of *Laurencia*; some are chemically rich and resistant to grazers, others are chemically bland and more susceptible. For investigators interested in pursuing this idea, the reviews of Faulkner (1984, 1986, 1987, 1988) provide a thorough background to the chemistry produced by each species. However, it should be borne in mind that if these algae are confounding herbivores by using mimicry, then they will almost certainly confound chemists and taxonomists (none of whom have to eat according to their taxonomic expertise). Thus, there are very likely to be incorrect reports of some compounds for some species. In fact, we know of one case where a compound characteristic of *Laurencia* species was incorrectly identified as coming from a *Chondria* species even though voucher specimens of the plant were identified by an excellent taxonomist (see Fenical *et al.* 1973). The potential for palatable species of *Chondria* to be mimicking unpalatable species of *Laurencia* should be obvious.

Seaweeds might also lessen losses to herbivores by forming groups of Mullerian mimics (multiple unpalatable species having a similar form, or similar chemistry, and thus minimizing the burden of the herbivore learning experience). Again, brown algae in the Dictyotaceae may be the best potential example of this. As shown in Fig. 14.2, chemically defended species of *Dictyota, Dilophus,* and *Dictyopteris* may have very similar morphologies. Because these genera often co-occur, grow in intermixed clumps, and are abundant, visually orienting herbivores might quickly learn to avoid brown-bladed species, thus lessening the losses that each seaweed might incur due to feeding by inexperienced fishes.

Although the above arguments are focused on morphological similarities, similar chemical defences could have a similar effect on herbivores such as sea urchins that search using chemical cues. A possible Mullerian complex employing both chemical and morphological mimicry is suggested by the fact that *Dictyota, Dilophus, Pachydictyon,* and *Glossophora* are all morphologically similar brown seaweeds that produce the feeding deterrent pachydictyol-A, and often other dictyols (see Fig. 14.1). Within the Dictyotales, the advantages of mimicry may have slowed, or prevented, morphological and chemical divergence among several of the species. With this in mind, it is interesting to note that *Dictyopteris delicatula* is defended by small organic molecules (Hay *et al.* 1988*b*) and resembles species of *Dictyota* with similar defences (Fig. 14.2). In contrast, most other species of *Dictyopteris* in the Caribbean are morphologically different from the *Dictyota* species and appear to be defended by highly acidic thalli (K. Van Alstyne and J. Norris, pers. comm.) rather than small organic molecules.

Although warning coloration has not been explored for seaweeds, chemically defended species might be advantaged by advertising their distastefulness. This would be especially critical when they were small and in greater danger of being killed by one errant bite from a large herbivore such as a parrotfish. We suspect that the striking blue iridescence of numerous tropical species in the Dictyotales, especially when they are small, may serve this function. If so, then this characteristic might also be mimicked by more palatable species.

The above hypotheses are all untested, but could easily play an important role in seaweed–herbivore interactions across a wide variety of habitats. Those wishing to pursue some of these ideas can find chemical information on the various algal species in the reviews by Faulkner (1984, 1986, 1987, 1988) and some information on the susceptibility of various algae to herbivores and how this may be affected by seaweed chemistry in Littler *et al.* (1983), Hay (1984, 1991), Steinberg (1985, 1992), Paul and Hay (1986), Paul (1987), Hay *et al.* (1988*a*), Paul *et al.* (1988), Wylie and Paul (1988), and Steinberg and Van Altena (in press). However, the most powerful studies will probably need to be conducted intensely and in a few localized habitats by

2. Population bottle-necks for chemically defended plants

Although adequately defended plants may be relatively safe from most consumers once they are large enough to be seen and appropriately identified, chemical defences may provide little protection when plants are very small and likely to be completely consumed in one bite by larger herbivores (e.g., fishes and urchins). Thus, chemically defended species may need to escape spatially or temporally while small. Studies assessing the importance of refuges from predation and how these change with plant size for defended vs undefended seaweeds have not been conducted. In the field, we have commonly seen well-defended plants like *Halimeda* or *Dictyota* establish first in the interstices of topographically complex substrates that may have provided refuge from large scraping fishes. As the plants became larger, they spread vegetatively over less complex substrates. This suggests that both clonal growth and temporary spatial refuges may be critically important for chemically defended seaweeds.

The iridescent coloration of several chemically rich seaweeds in the Dictyotales, and the bright red tips on some species of chemically defended *Laurencia* spp. seem to be especially prominent when individuals are small; this may allow herbivores to see and avoid them at a smaller size. Additionally, young plants or newly produced portions of *Halimeda* have been shown to have extremely high concentrations of the most deterrent metabolites (Hay *et al.* 1988*c*; Paul and Van Alstyne 1988*a*). This excessive defence of young tissues could occur among other seaweeds as well and may encourage herbivores to avoid chemically defended species at a smaller size than they would do otherwise.

3. Alternative and additional functions of secondary metabolites

Although many seaweed secondary-metabolites have been demonstrated to function as herbivore deterrents, this does not diminish their potential role in other processes. Their importance as allelopathic agents, as pathogen inhibitors, and in internal physiological processes of the plant has not been systematically studied. In preliminary laboratory assays run against larvae of several benthic invertebrates, Lindquist and Hay (personal observation) found that at concentrations of $5 \mu g\, m^{-1}$ several seaweed secondary metabolites either killed larvae or prohibited their metamorphosis. To realistically evaluate the potential ecological significance of such results we need to know more about where and how compounds are stored in the plant and if metabolites are released to the plant surface where they could act as antifouling agents. There are no rigorously quantified studies demonstrating the release of secondary metabolites under natural conditions (however, see Norton et al. 1990 for an indication that this probably occurs) and, thus, no studies assessing concentrations near the plants, how often these are turned over, which plant morphologies could most effectively maintain compounds in the immediate vicinity, etc.

4. Compound allocation within and among individuals

The allocation of compounds to different plant portions, to tissues of differing ages, among plants within a population, and between plants in different populations has rarely been addressed (see Hay and Fenical 1988 and Steinberg 1992 for references). This is an area in need of much additional attention. The few studies presently under way (J. Lubchenco, pers. comm., V.J. Paul, pers. comm., M.E. Hay, personal observations) show substantial variation at all of these levels. Although important, these studies will almost certainly progress slowly because they require chemical skills beyond the abilities of most ecologists and a sampling intensity (i.e., large numbers of replicate analyses) far beyond the patience levels of most accomplished chemists.

5. Complex interactions of chemical defences with the physical and biological history of plants

Neither seaweeds nor the animals that eat them are static organisms with immutable characteristics. Herbivore tolerance to chemical defences may change with increased exposure to a compound or plant (Brattsten 1979; Futuyma and Morano 1988), and with amounts of defences and types of compounds produced. Seaweed susceptibility to herbivores also may be altered by recent physical (e.g., desiccation) or biological (e.g., grazing) stresses (see Renaud et al. 1990 as an example and for references to the terrestrial literature on this topic).

Although there is a rich literature on how consumer attack may induce increased levels of chemical or morphological defences in terrestrial plants or in marine and freshwater invertebrates (Harvell 1990), this process is relatively unstudied for marine plants. Van Alstyne (1988) documented a modest, but ecologically significant, induction of increased phlorotannin levels in *Fucus distichus* following attack by gastropods, and Renaud et al. (1990) demonstrated that *Padina* plants attacked by urchins (or physically damaged to mimic urchin attack) were less susceptible to urchin grazing in later assays; however, this pattern was temporally variable and its underlying chemical basis not rigorously investigated. Other, and sometimes very extensive, attempts to demonstrate induction of chemical defences in seaweeds following real or artificial attack have not substantiated its occurrence in several tropical green (V.J. Paul, pers. comm.) or temperate brown seaweeds (P.D. Steinberg, pers. comm.).

Several green seaweeds in the tropics appear to store defensive compounds in a less biologically active form and immediately convert this to a more bioactive form when they are attacked by herbivores. This is best documented for species of *Halimeda* (Paul and Van Alstyne submitted) but also appears to occur in other genera of chemically defended tropical greens (V.J. Paul, pers. comm.).

We hope that the above summary of what we know, and would like to know, about chemical mediation of seaweed–herbivore interactions will facilitate the initiation of students into this area of research where so much remains to be done.

Acknowledgements

Support for this paper was provided by grants OCE 89-11872 and 89-00131 (to MEH) and CHE 86-20217 and 90-08621 (to WF) from the US National Science Foundation. Comments by Robert Steneck and an anonymous reviewer improved the manuscript.

References

Bernays, E.A. (1989). Host range in phytophagous insects: the potential role of generalist predators. *Evolutionary Ecology*, **3**, 299–311.

Bernays, E.A. and Graham, M. (1988). On the evolution of host specificity in phytophagous arthropods. *Ecology*, **69**, 886–92.

Bernays, E.A., Driver, G.C., and Bilgener, M. (1989). Herbivores and plant tannins. *Advances in Ecological Research*, **19**, 263–302.

Brattsten, L.B. (1979). Biochemical defense mechanisms in herbivores against plant allelochemicals. In *Herbivores: their interaction with secondary plant metabolites*, (ed. G.A. Rosenthal and D.H. Janzen), pp. 200–70. Academic Press, New York.

Carpenter, R.C. (1986). Partitioning herbivory and its effects on coral reef algal communities. *Ecological Monographs*, **56**, 345–65.

Coley, P.D. and Aide, T.M. (1990). A comparison of herbivory and plant defenses in temperate and tropical broad-leaved forests. In *Plant–animal interactions: evolutionary ecology in tropical and temperate regions*, (ed. P.W. Price, T.M. Lewinsohn, G.E. Fernandes, and W.W. Benson), pp. 25–49. John Wiley, New York.

Coley, P.D., Bryant, J.P., and Chapin, F.S. III (1985). Resource availability and plant antiherbivore defense. *Science*, **230**, 895–9.

Duffy, J. E. and Hay, M. E. (1990). Seaweed adaptations to herbivory. *BioScience*, **40**, 368-75.

Duffy, J. E. and Hay, M. E. (1991*a*) Host plants as food and shelter: determinants of food choice in an herbivorous marine amphipod. *Ecology*, **72**, 1286-98.

Duffy, J. E. and Hay, M. E. (1991*b*). All amphipods are not created equal: a reply to Bell. *Ecology*, **72**, 354-8.

Estes, J. A. and Steinberg, P. D. (1988). Predation, herbivory, and kelp evolution. *Paleobiology*, **14**, 19-36.

Faulkner, D. J. (1984). Marine natural products: metabolites of marine algae and herbivorous marine molluscs. *Natural Product Reports*, **1**, 251-80.

Faulkner, D. J. (1986). Marine natural products. *Natural Product Reports*, **3**, 2-33.

Faulkner, D. J. (1987). Marine natural products. *Natural Product Reports*, **4**, 539-76.

Faulkner, D. J. (1988). Marine natural products. *Natural Product Reports*, **5**, 613-63.

Feeney, P. (1976). Plant apparency and chemical defense. *Recent Advances in Phytochemistry*, **10**, 1-40.

Fenical, W., Sims, J. J., and Radlick, T. (1973). Chondriol: a novel acetylenic metabolite from the red alga *Chondria oppositiclada*. *Tetrahedron Letters*, **1973**, 313-16.

Fox, L. R. (1981). Defense and dynamics in plant-herbivore systems. *American Zoologist*, **21**, 853-64.

Futuyma, D. J. (1983). Evolutionary interactions among herbivorous insects and plants. In *Coevolution*, (ed. D. J. Futuyma and M. Slatkin), pp. 207-31. Sinauer Associates Inc., Sunderland, Massachusetts.

Futuyma, D. J. and Morano, G. (1988). The evolution of ecological specialization. *Annual Review of Ecology and Systematics*, **19**, 207-33.

Harvell, C. D. (1990). The ecology and evolution of inducible defenses. *The Quarterly Review of Biology*, **65**, 323-40.

Hawkins, S. J. and Hartnoll, R. G. (1983). Grazing of intertidal algae by marine invertebrates. *Oceanography and Marine Biology Annual Review*, **21**, 195-282.

Hay, M. E. (1984). Predictable spatial escapes from herbivory: how do these affect the evolution of herbivore resistance in tropical marine communities? *Oecologia* (Berlin), **64**, 396-407.

Hay, M. E. (1985). Spatial patterns of herbivore impact and their importance in maintaining algal species richness. In *Proceedings of the 5th International Coral Reef Congress*, 4, pp. 29-34. EPHE, Tahiti.

Hay, M. E. (1986). Associational plant defenses and the maintenance of species diversity: turning competitors into accomplices. *American Naturalist*, **128**, 617-41.

Hay, M. E. (1991). Fish–seaweed interactions on coral reefs: effects of herbivorous fishes and adaptations of their prey. In *The ecology of coral reef fishes*, (ed. P. F. Sale), pp. 96–119. Academic Press, New York.

Hay, M. E. (1992) Seaweed chemical defenses: their role in the evolution of feeding specialization and in mediating complex interactions. In *Ecological roles of marine natural products*, (ed. V. J. Paul). Comstock Publishing Associates, Ithaca. (In press).

Hay, M. E. and Fenical, W. (1988). Marine plant–herbivore interactions: the ecology of chemical defense. *Annual Review of Ecology and Systematics*, **19**, 111–45.

Hay, M. E., Duffy, J. E., Pfister, C. A., and Fenical, W. (1987a) Chemical defenses against different marine herbivores: are amphipods insect equivalents? *Ecology*, **68**, 1567–80.

Hay, M. E., Fenical, W., and Gustafson, K. (1987b). Chemical defense against diverse coral-reef herbivores. *Ecology*, **68**, 1581–91.

Hay, M. E., Duffy, J. E., and Fenical, W. (1988a). Seaweed chemical defenses: among-compound and among-herbivore variance. In *Proceedings of the 6th International Coral Reef Symposium*, **3**, pp. 43–8.

Hay, M. E., Duffy, J. E., Fenical, W., and Gustafson, K. (1988b). Chemical defense in the seaweed *Dictyopteris delicatula*: differential effects against reef fishes and amphipods. *Marine Ecology Progress Series*, **48**, 185–92.

Hay, M. E., Paul, V. J., Lewis, S. M., Gustafson, K., Tucker, J., and Trindell, R. N. (1988c). Can tropical seaweeds reduce herbivory by growing at night? Diel patterns of growth, nitrogen content, herbivory, and chemical versus morphological defenses. *Oecologia* (Berlin), **75**, 233–45.

Hay, M. E., Renaud, P. E., and Fenical, W. (1988d). Large mobile versus small sedentary herbivores and their resistance to seaweed chemical defenses. *Oecologia* (Berlin), **75**, 246–52.

Hay, M. E., Pawlik, J. R., Duffy, J. E., and Fenical, W. (1989). Seaweed-herbivore-predator interactions: host-plant specialization reduces predation on small herbivores *Oecologia* (Berlin), **81**, 418–27.

Hay, M. E., Duffy, J. E., and Fenical, W. (1990a). Host-plant specialization decreases predation on a marine amphipod: an herbivore in plant's clothing. *Ecology*, **71**, 733–43.

Hay, M. E., Duffy, J. E., Paul, V. J., Renaud, P. E., and Fenical, W. (1990b). Specialist herbivores reduce their susceptibility to predation by feeding on the chemically defended seaweed *Avrainvillea longicaulis*. *Limnology and Oceanography*, **35**, 1734–43.

Lawrence, J. M. (1975). On the relationship between marine plants and sea urchins. *Oceanography and Marine Biology Annual Review*, **13**, 213–86.

Littler, M. M., Taylor, P. R., and Littler, D. S. (1983). Algal resistance to herbivory on a Caribbean barrier reef. *Coral Reefs*, **2**, 111–18.

Littler, M. M., Taylor, P. R., and Littler, D. S. (1986). Plant defense associations in the marine environment. *Coral Reefs*, **5**, 63-71.

Mann, K. H. (1982). *Ecology of coastal waters: a systems approach*. Studies in Ecology, Vol. 8. University of California Press, Berkeley.

Norton, T. A., Hawkins, S. J., Manley, N. L., Williams, G. A., and Watson, D. C. (1990). Scraping a living: a review of littorinid grazing. *Hydrobiologica*, **193**, 117-38.

Paul, V. J. (1987). Feeding deterrent effects of algal natural products. *Bulletin of Marine Sciences*, **41**, 514-22.

Paul, V. J. (ed.) (1992). *Ecological roles of marine natural products*. Comstock Publishing Associates, Ithaca. (In press).

Paul, V. J. and Fenical, W. (1987). Natural products chemistry and chemical defense in tropical marine algae of the phylum Chlorophyta. In *Bioorganic marine chemistry*, (ed. P. J. Scheuer), Vol. 1, pp. 1-29. Springer-Verlag, Berlin.

Paul, V. J. and Hay, M. E. (1986). Seaweed susceptibility to herbivory: chemical and morphological correlates. *Marine Ecology Progress Series*, **33**, 255-64.

Paul, V. J. and Van Alstyne, K. L. (1988*a*). Chemical defense and chemical variation in some tropical Pacific species of *Halimeda* (Halimedaceae; Chlorophyta). *Coral Reefs*, **6**, 263-70.

Paul, V. J. and Van Alstyne, K. L. (1988*b*). Use of ingested algal diterpenoids by *Elysia halimedae* Macnae (Opisthobranchia: Ascoglossa) as antipredator defenses. *Journal of Experimental Marine Biology and Ecology*, **119**, 15-29.

Paul, V. J., Wylie, C. R., and Sanger, H. R. (1988). Effects of algal chemical defenses toward different coral-reef herbivorous fishes. *Proceedings of the 6th International Coral Reef Congress*, **3**, 73-8.

Pfister, C. A. and Hay, M. E. (1988). Associational plant refuges: convergent patterns in marine and terrestrial communities result from differing mechanisms. *Oecologia* (Berlin), **77**, 118-29.

Ragan, M. A. and Glombitza, K. W. (1986). Phlorotannins, brown algal polyphenolics. In *Progress in Phycological Research*, (ed. F. E. Round and D. J. Chapman), Vol. 4, pp. 129-241. BioPress, Bristol.

Renaud, P. E., Hay, M. E., and Schmitt, T. M. (1990). Interactions of plant stress and herbivory: intraspecific variation in the susceptibility of a palatable versus an unpalatable seaweed to sea urchin grazing. *Oecologia* (Berlin), **82**, 217-26.

Rhoades, D. and Cates, R. (1976). Toward a general theory of plant antiherbivore chemistry. *Recent Advances in Phytochemistry*, **10**, 168-213.

Rosenthal, G. A. and Janzen, D. H. (1979). *Herbivores: their interaction with secondary plant metabolites*. Academic Press, New York.

Steinberg, P. D. (1985). Feeding preferences of *Tegula funebralis* and chemical

defenses of marine brown algae. *Ecological Monographs*, **55**, 333-49.
Steinberg, P.D. (1989). Biogeographical variation in brown algal polyphenolics and other secondary metabolites: comparison between temperate Australasia and North America. *Oecologia* (Berlin), **78**, 373-82.
Steinberg, P.D. (1992). Geographical variation in the interaction between marine herbivores and brown algal secondary metabolites. In *Ecological roles of marine natural products*, (ed. V.J. Paul). Comstock Publishing Associates, Ithaca. (In press).
Steinberg, P.D. and Altena, I. Van (1991). Tolerance of marine invertebrate herbivores to brown algal phlorotannins in temperate Australia. *Ecological Monographs*.
Steneck, R.S. and Watling, L. (1982). Feeding capabilities and limitations of herbivorous molluscs: a functional group approach. *Marine Biology*, **68**, 299-319.
Taylor, W.R. (1960). *Marine algae of the eastern tropical and subtropical coasts of the Americas.* University of Michigan Press, Ann Arbor.
Van Alstyne, K.L. (1988). Herbivore grazing increases polyphenolic defenses in the intertidal brown alga *Fucus distichus*. *Ecology*, **69**, 655-63.
Van Alstyne, K.L. and Paul, V.J. (1988). The role of secondary metabolites in marine ecological interactions. In *Proceedings of the 6th International Coral Reef Symposium*, **1**, 175-86.
Van Alstyne, K.L. and Paul, V.J. (1990). The biogeography of polyphenolic compounds in marine macroalgae: temperate brown algal defenses deter feeding by tropical herbivorous fishes. *Oecologia* (Berlin), **84**, 158-63.
Wylie, C.R. and Paul, V.J. (1988). Feeding preferences of the surgeonfish *Zebrasoma flavescens* in relation to chemical defenses of tropical algae. *Marine Ecology Progress Series*, **45**, 23-32.
Zucker, W.V. (1983). Tannins: does structure determine function? An ecological perspective. *American Naturalist*, **121**, 335-65.

15. Herbivorous fishes: feeding and digestive mechanisms

MICHAEL H. HORN
Department of Biological Science, California State University, Fullerton, California, USA

Abstract

Feeding and digestion are sequential processes that seem to be especially tightly linked in herbivores, which are faced with a food resource not only low in protein and locked inside indigestible cell walls but also often chemically or morphologically defended. Herbivorous fishes browse or graze (if sediment is ingested) on their algal or seagrass foods with varying degrees of selectivity using a closely-spaced set of cropping or scraping teeth. These teeth may be conical or spatulate, smooth-edged, or cuspidate, flexible or rigidly attached, hinged or entire, separate and chisel-like or fused to form a beak-like structure. Before the ingested food is acted upon by an apparently conventional set of enzymes, the fishes gain access to the consumed material by using one or more of four main digestive mechanisms:

(1) lysing the cells in a highly acidic stomach;
(2) triturating the material in pharyngeal jaws;
(3) triturating the material in a muscular stomach;
(4) harbouring microbes that ferment the food in a hindgut region or compartment.

Convergences in these digestive mechanisms are apparent among phylogenetically unrelated groups of herbivorous fishes.

Introduction

Feeding and digestion are two interdependent stages of a single process that determines an animal's net rate of energy and nutrient gain (Penry and Jumars 1986). This premise implies that the entire process is optimal only if digestion follows an optimal path constrained by the food items actually consumed. Animals will show a digestive flexibility that tends to match the variety of foods eaten, in order to maximize the net rate of energy and nutrient gain. Thus, accurate assessment of the nutritional physiology of the animal and its impact on its food source requires that both the feeding and digestive stages be considered. These stages seem to be especially closely linked in herbivores because their food supply is not only low in protein and locked inside indigestible cell walls but also often chemically or morphologically defended. Presumably because of this low-quality, refractory diet, a variety of morphological and physiological specializations characterize the digestive tracts of herbivorous fishes. The poor-quality foods and generally low digestive efficiencies of herbivorous fishes may select for maximum feeding effort and assimilation efficiency in these fishes (Montgomery et al. 1989).

The purpose of this paper is to review recent studies of the consumption and digestion of benthic algae and seagrasses by marine herbivorous fishes. I will first describe the types of feeding mechanisms found among herbivorous fishes, and then examine the principal digestive mechanisms they use and attempt to show that feeding and digestion are integrated components of a single energy- and nutrient-acquiring process in these species.

Feeding mechanisms

Herbivorous fishes may be classified in an ecological sense according to their feeding behaviour or according to their feeding mechanisms. Fishes that consume benthic algae feed as solitary individuals, as members of roving groups or as defenders of territories containing their food supply (Ogden and Lobel 1978; Horn 1989). They usually have short, blunt snouts with closely set teeth that form a cropping or scraping edge (Ogden and Lobel 1978). These jaw teeth often undergo morphological changes with age (Christensen 1978; Stoner 1980; Orton 1989), changes that sometimes are associated with an ontogenetic shift from carnivory to herbivory (Christensen 1978; Stoner 1980), which commonly occurs in animals that are herbivorous as adults (White 1985). In terms of obtaining food, herbivorous fishes can be classified as either browsers or grazers (Hiatt and Strasburg 1960; Jones 1968; Bakus 1969; Ogden and Lobel 1978; Russ 1984; Clements and Bellwood 1988; Clements et al. 1989). Browsers bite or tear pieces from upright macroalgae and rarely ingest any inorganic material, whereas grazers pick up sediment in the process of feeding, by scraping or sucking (Jones 1968). Because browsing herbivores take whole or parts of individual seaweeds, they tend to be selective feeders; grazers, on the other hand, feed less selectively because their algal food is small and closely attached to the substratum (Lobel 1981; Choat 1982). Herbivorous fishes are concentrated in only about 19 (5 per cent) of the 409 recognized (Nelson 1984) families of teleostean fishes, 15 of which belong to the order Perciformes (Horn 1989). The total number of species of marine herbivorous fishes is unknown and unestimated, partly because the feeding habits of many species in the larger families (e.g., Blenniidae, Gobiidae) have been insufficiently studied. Herbivorous fishes are more diverse in tropical than in temperate waters, and few strictly herbivorous species are found beyond 40° N and S latitudes (Horn 1989).

Browsers and grazers do not always fall into mutually exclusive categories either at the family, generic, or even species level. For example, the tropical family Acanthuridae (surgeonfishes) contains both browsers and grazers (Table 15.1), as does the genus *Acanthurus* (see Horn 1989), and some surgeonfish species such as *Acanthurus dussumieri* and *A. nigroris* have been observed to feed either as browsers or grazers (Jones 1968). Girellids, in general, appear to be capable of feeding either as browsers or grazers (see Horn 1989), and they are discussed below as combination browsers/grazers. Nevertheless, these categories are useful designations that can lead to predictions about the ecological impacts and digestive physiology of the fishes.

1. Representative browsers

Browsing species belong to the tropical families Acanthuridae, Pomacentridae, and Siganidae, the tropical/warm temperate families Kyphosidae and Sparidae and the temperate-zone families Girellidae, Odacidae, and Stichaeidae, among others (Table 15.1). Although these fishes usually feed selectively, as a group they crop bites from a variety of red, green, and brown algae of both filamentous and foliose growth forms. The following brief descriptions of feeding in four different species of browsers help to illustrate the diversity of foraging modes and feeding mechanisms in this category of herbivorous fishes.

(a) The brown surgeonfish Acanthurus nigrofuscus (Fig. 15.1) This is a diurnal browser that forages in groups in shallow inshore areas of tropical reefs (Montgomery *et al.* 1989). It spends 55–70 per cent of the time in active foraging and, except for selective feeding on green algal blooms in winter, feeds non-selectively on a variety of red algae during most of the year (Montgomery *et al.* 1989). The jaw teeth are small (2–3 mm in length) and like those in most species of *Acanthurus* have

Table 15.1. Nineteen of the principal teleostean fish families containing herbivorous species with feeding type and distribution based on information compiled in Horn (1989).

Family	Feeding type(s)	Family distribution
Acanthuridae	Browsers and grazers	Tropical
Aplodactylidae	Grazers	Temperate
Balistidae	Browsers and grazers	Tropical/temperate
Blenniidae	Grazers	Tropical/temperate
Chanidae	Grazer	Tropical
Girellidae	Browsers and grazers	Temperate
Gobiidae	Grazers	Tropical/temperate
Hemiramphidae	Browsers	Tropical
Kyphosidae	Browsers	Tropical/temperate
Mugilidae	Grazers	Tropical/temperate
Odacidae	Browsers	Temperate
Pomacanthidae	Browsers and grazers	Tropical
Pomacentridae	Browsers and grazers	Tropical
Scaridae	Mostly grazers, few browsers	Tropical
Scorpididae	Browsers and grazers	Tropical/temperate
Siganidae	Browsers	Tropical
Sparidae	Browsers and grazers	Tropical/temperate
Stichaeidae	Browsers	Temperate
Tetraodontidae	Grazers	Tropical

denticulations on the margins (Jones 1968; Fig. 15.1A). Compared with *Acanthurus* grazers, browsers in this genus have fewer, broader teeth, which are better suited for taking bites of filamentous algae; filaments caught between the cusps of the teeth are partly cut and partly snapped off by a quick jerk of the head (Jones 1968).

(b) The monkeyface prickleback Cebidichthys violaceus (Fig. 15.1B) This is a sluggish, bottom-dwelling fish of mainly Californian rocky intertidal and subtidal habitats (Eschmeyer *et al*. 1983) that browses selectively on a variety of red and green algae and avoids encrusting, calcareous, and brown seaweeds (Horn *et al*. 1982). This stichaeid fish feeds solitarily, and its intertidal foraging apparently occurs on incoming tides, either during the day or night (Ralston and Horn 1986). The small, pointed, partially embedded teeth (Fig. 15.1B) are used to grasp algal thalli, which are then torn off by a head jerk or a rapid body spin (personal observations).

(c) The butterfish Odax pullus (Fig. 15.3B) This is a fish endemic to coastal waters of New Zealand (Gomon and Paxton 1985). It is a selective diurnal browser, which, like the closely related parrot-fishes (Liem and Greenwood 1981), feeds using its fused jaw teeth (Fig. 15.3B). Unlike parrotfishes (see below), it feeds mainly on laminarian (*Ecklonia radiata*) and fucalean (*Carpophyllum* spp.) brown seaweeds, and its feeding action differs according to the alga being consumed (Clements 1985; Clements and Bellwood 1988). When the fish feeds on *Ecklonia*, the oral surface is applied to the lamina and held there by opercular suction, and a disc of algal tissue is excised. The bite is usually taken from the tips and centres of the secondary laminae where the sori are located during the reproductive season. When feeding on *Carpophyllum*, the butterfish removes the reproductive bunches and thalli with a bite and a sideways jerk of the head. Feeding bouts, comprising two or fewer bites per minute, are separated by intervals of searching behaviour, and feeding rates are highest during the first two hours after dawn. After each feeding bout, the fish hangs in a tail-down position and masticates its food with the pharyngeal teeth, as evidenced by the raising and lowering of the hyoid apparatus.

(d) The silver drummer Kyphosus sydneyanus (Fig. 15.4) This is a schooling kyphosid fish of Australian (Coleman 1980) and New Zealand waters (Ayling and Cox 1982), is a diurnal or perhaps crepuscular (Russell 1983) browser that feeds primarily on brown algae (Rimmer and Wiebe 1987) including *Carpophyllum* and *Ecklonia* (Russell 1983), also eaten by *Odax pullus* (see above). It has small, cuspate teeth (Fig. 15.4), which serve to take cleanly bitten pieces of algae (Russell 1983).

2. Representative grazers

Grazing species belong to the tropical families Acanthuridae, Pomacentridae, and Scaridae, the tropical/warm temperate families Blenniidae and Mugilidae and the temperate-zone families Aplodactylidae and Girellidae, among others (Table 15.1). The following brief descriptions of feeding in three different species of grazers and the Scaridae, in general, exemplify the range of foraging modes and feeding mechanisms in this category of herbivorous fishes.

(a) The striped bristlemouth Ctenochaetus striatus (Fig. 15.2B) This is an acanthurid of Indo-Pacific reefs (Randall 1986a), a grazer that feeds on fine detritus and diatoms or other algal material by a combination of whisking with its flexible, comb-like teeth (Fig. 15.2B) and sucking (Jones 1968; Randall 1986a). This fish has also been classified more

specifically as a sucker of fine sediments (Russ 1984) and as a detritivore (Sutton and Clements 1988; Clements et al. 1989). Montgomery et al. (1989) found that C. striatus spends about 30 per cent of its time in active feeding on reefs and sandy substrata within its weakly defended home range and that it ingests both algae and large amounts of inorganic grit.

(b) The striped mullet Mugil cephalus (Fig. 15.2A) This is an active, schooling fish found world-wide in tropical and warm temperate coastal waters (Smith and Smith 1986), which feeds mainly by sucking up the surface layer of mud or by grazing on submerged rock or plant surfaces (Odum 1970). Its diet consists primarily of benthic or epiphytic microalgae and plant detritus (Thomson 1954), but sand or other sediment particles may make up >50 per cent of the stomach contents by weight (Collins 1981). The jaw teeth of *M. cephalus* are minute, and the gill rakers form a sieve-like filtering apparatus (Fig. 15.2A).

(c) The redlip parrotfish Scarus rubroviolaceus (Fig. 15.3A) This is a wide-ranging inhabitant of Indo-Pacific coral reefs (Rosenblatt and Hobson 1969), which feeds by scraping the surface of turf-covered substrata (Clements and Bellwood 1988) using the fused jaw teeth (Fig. 15.3A) that characterize this and other scarid species. Each bite produces a pair of narrow parallel scrapes marked by dislodged algae, but substratum scarring occurs only occasionally; bites are taken at the rate of 15 to 20 $\min^{-1}$ and are grouped into short feeding bouts (Clements and Bellwood 1988).

Although a few parrotfishes browse on seagrasses (Randall 1967; Lobel and Ogden 1981), most graze algae from reef surfaces (e.g., Randall 1986b). Bellwood and Choat (1990) have recently challenged the assumption that the grazers form a relatively uniform group of reef herbivores based on their functional analysis of grazing in 24 species of parrot-fishes from the Great Barrier Reef. Their findings led them to reject this assumption and to recognize two distinct functional groups of grazing parrotfishes: excavators and scrapers. Excavating species (e.g., *Scarus sordidus*) use their powerful jaw apparatus to take relatively few but large bites that leave distinct scars on the substratum. In contrast, scraping species (e.g., *S. rubroviolaceus*) have relatively weak jaws, take numerous small bites per minute and usually leave no scars on the substratum. As a result of these differences, Bellwood and Choat (1990) point out that excavators and scrapers will make dissimilar contributions to reef bioerosion and habitat modification and exert different impacts on benthic communities.

3. Representative browsers/grazers

The luderick, *Girella tricuspidata*, a girellid fish of Australian (Coleman 1980) and New Zealand (Ayling and Cox 1982) coastal waters, is a roving diurnal forager that browses mainly on large brown algae and grazes on smaller red algal turfs (Russell 1983). The teeth are spatulate, straight-edged, or cuspate and each hinged with a ball-and-socket joint; this dental morphology appears to allow a ripping action for effective browsing on foliose algae and a cropping action for efficient grazing on rock surfaces (see Orton 1989).

The Girellidae (nibblers), as a group of 15 species of warm temperate herbivorous fishes, exhibits a complex ontogeny of dental polymorphisms (Orton 1989), and most species probably combine browsing and grazing in their foraging activities (see Horn 1989). Two little-known, highly derived South Pacific species, *Girella fimbriata* from the Kermadec Islands and *G. nebulosa* from Easter Island, provide exceptions to the patterns of dentition, and probably of foraging, common to most girellids (Orton 1989). The jaw teeth of these species (Fig. 15.2c) lack hinges and form a single row of massive, chisel-shaped, almost perfectly occluded structures that take on an appearance approaching that of the fused teeth of parrotfishes. The tooth features and limited observations on one of the species (*G. fimbriata*) suggest that these fishes graze rock surfaces.

Digestive mechanisms

The foregoing section provided examples of some of the kinds of feeding behaviour and jaw dentition found among herbivorous fishes. Clearly, these fishes use several mechanisms to ingest their algal food. Once the food is ingested, the question becomes how do the fishes acquire sustenance from this refractory diet. In other words, what are the morphological and physiological specializations of the gut by which herbivorous fishes gain access to the nutrients locked inside algal cells? With some important exceptions, herbivorous fishes have longer guts and shorter gut transit times than carnivorous or omnivorous species, and they apparently employ a standard set of vertebrate enzymes once they have broken down the algal cells (Horn 1989).

Lobel (1981) observed that cell breakage can occur by either:

(1) lysis, resulting from gastric acidity;
(2) mechanical action, resulting primarily from trituration in the pharyngeal jaws or a gizzard-like stomach.

Lobel's conclusions were based on the assumption that fishes do not produce cellulolytic enzymes or harbour a fermentative microflora. The former assumption still holds, but at least two herbivorous fishes are now known (Rimmer and Wiebe 1987) to contain gut microbes that can digest algal material. Based on Lobel's (1981) results and those of more recent studies, Horn (1989) described four main types of alimentary canal in herbivorous fishes. These gut types are presented here in modified form as digestive mechanisms each identified by the dominant process involved. The same species discussed in the previous section are considered in this section so as to depict the linkage between feeding and digestion as clearly as possible.

1. Type I: acid lysis in a thin-walled stomach

Acid lysis appears to be one of the primary mechanisms whereby some herbivorous fishes gain access to the nutrients inside algal cell walls. It has been demonstrated in laboratory tests simulating the acidic gastric pH environment of several marine herbivorous fishes (Lobel 1981; Urquhart 1984). The stomach pH values of certain acanthurids, pomacanthids, and pomacentrids were found to range from 2·4–4·3, acidities that seemed to be as effective as trituration in releasing the cell contents of some algal species (Lobel 1981). The mechanism by which low pH causes cell lysis is poorly understood but may involve a weakening of the hydrogen bonds of the structural units in the algal cell walls. As has been shown in terrestrial plants (Wilkins 1984), this process would in effect loosen or expand the cell walls. Once released by acid lysis, nutrients are then presumably broken down by the conventional battery of enzymes found in the digestive tracts of fishes.

Fishes that depend on acid lysis as the predominant mechanism for breaking down algal cells tend to have thin-walled stomachs (Lobel 1981) and long (Horn 1989) or moderately long (Horn and K.S. Messer, unpublished data) intestines but otherwise show few morphological specializations for herbivory. These fishes are mainly browsers and according to Lobel (1981) should feed primarily on green and red algae with large cells. Lobel argued that a browsing rather than a grazing habit minimizes sand ingestion during feeding and thereby prevents the rapid buffering of stomach contents by calcium carbonate material.

Two species mentioned in the previous section on feeding mechanisms fit relatively well into the Type I category of digestive mechanisms. They are discussed in turn below.

(a) The brown surgeonfish Acanthurus nigrofuscus (Fig. 15.1A) This fish browses on red and green algae (Montgomery *et al.* 1989), and it has a thin-walled (Jones 1968), acidic (Montgomery and Pollak 1988*a*)

Feeding and digestion of herbivorous fishes 347

Type I: Acid lysis in a thin-walled stomach

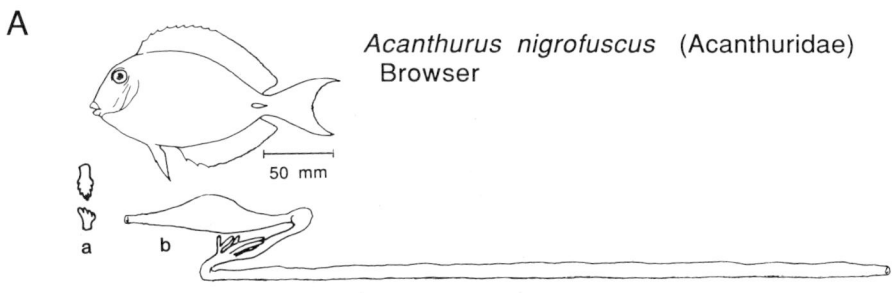

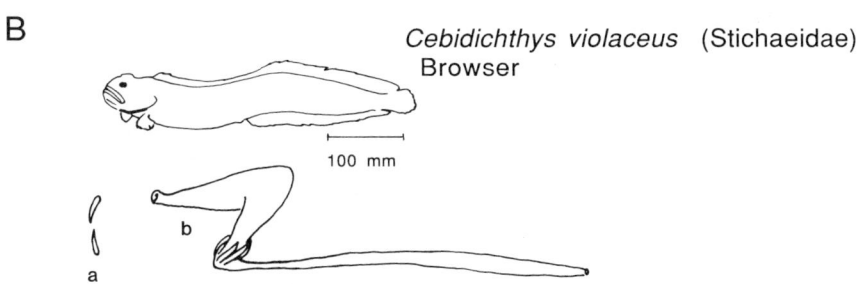

Fig. 15.1. Digestive tracts of two herbivorous fishes representing the Type I digestive mechanism: acid lysis in a thin-walled stomach. Gut lengths but not other dimensions are drawn to the same scale as the fish lengths; salient features are lettered consecutively for each fish. A. *Acanthurus nigrofuscus*: a = upper and lower jaw teeth; b = stomach. B. *Cebidichthys violaceus*: a = upper and lower jaw teeth: b = stomach (*A. nigrofuscus* illustration modified from fig. 2 in Horn (1989); information for drawing of *C. violaceus* from Barton (1982) and personal observations).

stomach. The digestive tract measures about 4 × the standard length (Jones 1968) but decreases in length during short periods of starvation (Montgomery and Pollak 1988a). Gastric pH can be as low as 2·9 but also can be only mildly acidic depending on the time of day and starvation state (Montgomery and Pollak 1988a).

(b) The monkeyface prickleback Cebidichthys violaceus (Fig. 15.1B) This species browses on red and green algae, and it has a thin-walled (personal observation), acidic stomach (Edwards and Horn 1982; Urquhart 1984). The digestive tract is only moderately long, ranging from $1 \cdot 1 \times$ (Barton 1982) to at least $1 \cdot 7 \times$ (Horn, unpublished data) the standard body length. Stomach pH values of $2 \cdot 2 - 2 \cdot 5$ were obtained for this fish by Urquhart (1984) who showed that acid lysis of the cell walls of the green alga *Ulva lobata* significantly increased the concentration of carbohydrate and protein in the medium compared with control solutions. After the alga had been treated with acidic solutions simulating those in the stomach, carbohydrates leached into these solutions, whereas proteins leached into solutions simulating the slightly alkaline environment of the intestine of the fish.

It seems clear that both *A. nigrofuscus* and *C. violaceus* gain sufficient nutrition for growth and reproduction by acid lysis of algal cells. The acanthurid may obtain digestive assistance from the microbial symbionts in its gut, but the degree and type of assistance has yet to be determined (see Montgomery and Pollak 1988*a*; Sutton and Clements 1988; Clements *et al.* 1989). On the other hand, *C. violaceus* appears to harbour a sparse microbial flora (J. A. Kandel and Horn, unpublished data) and yet assimilates nutrients efficiently from algae (Edwards and Horn 1982; Horn and Neighbors 1984; Horn *et al.* 1986) while leaving the algal cell wall largely intact (Urquhart 1984). Moreover, recent experiments (M.B. Fris and Horn, unpublished data) have shown that this temperate-zone herbivore can grow on a strictly seaweed diet.

Other fishes that belong in the Type I category include probably all of the acanthurids that are considered to be browsers and also to have thin-walled stomachs (Jones 1968; Clements *et al.* 1989; see Horn 1989). The stomach pH has been determined for several of these species and was found to be acidic ($3 \cdot 2 - 4 \cdot 3$) in each case (Lobel 1981; see Horn 1989). In addition, the following species all have thin-walled, acidic (pH $1 \cdot 9 - 3 \cdot 1$) stomachs and belong in this category (Horn 1989): *Chanos chanos* (Chanidae), *Apolemichthys xanthopunctatus* and *Centropyge flavissimus* (Pomacanthidae), *Eupomacentrus nigricans* and *E. planifrons* (Pomacentridae), and *Archosargus probatocephalus* (Sparidae). Rabbitfishes (Siganidae) have thin-walled stomachs and long intestines (Bryan 1975), but gastric pH has not been measured in these tropical herbivores. Herbivorous blenniids have moderately long to long intestines but lack a true stomach (Al-Hussaini 1947; Goldschmid *et al.* 1984); these fishes may also have a Type I digestive mechanism but, again, their gut pH has apparently not been determined.

The luderick *Girella tricuspidata* seems to be the only fish studied to date with the ability to assimilate constituents of algal cell walls (Anderson 1987). The highly acidic (pH as low as 2) and moderately muscular stomach of this fish (Anderson 1986) presumably contributes to this ability. In having both of these gastric traits, *G. tricuspidata* appears to combine the features of fishes with thin-walled, highly acidic stomachs and those with thick-walled, less acidic stomachs discussed below. Whether the bristle-like pharyngeal teeth borne on the lower elements of the gill arches (Orton 1989) aid in rupturing cell walls remains unknown, although Russell (1983) observed abrasion of algal stomach contents that he attributed to the possible action of pharyngeal teeth.

2. Type II: trituration in a gizzard-like stomach

A thick-walled, muscular stomach represents one of the two physical mechanisms by which certain herbivorous fishes break down algal cell walls (Lobel 1981). This gizzard-like organ serves mainly to triturate bacteria, blue-green algae, diatoms, and macroalgae, especially filamentous red and green forms, that have been ingested with quantities of sand or other sedimentary material. The fishes in this category are therefore mostly grazers. That the thick-walled stomach is a grinding organ is based upon observations of the condition of food before and after passage through the stomach (Al-Hussaini 1947; Jones 1968; Odum 1968, 1970; Randall 1974; Payne 1978; Nelson and Wilkins 1988). Fishes with gizzard-like stomachs tend to have slightly acidic to slightly alkaline gastric fluids (see Horn 1989).

Two species discussed in the previous section on feeding mechanisms fit relatively well into the Type II category. They are discussed in turn below.

(a) The striped mullet Mugil cephalus (Fig. 15.2A) It has an extremely thick-walled pyloric stomach and a long intestine (digestive tract 5–6 × the standard length). The fish feeds selectively on small, nutrient-rich particles in the sediments (Odum 1970; Marais 1980) using its finely divided pharyngeal filtering device (Odum 1968; Fig. 15.2A). It apparently relies solely (Payne 1978) on the muscular, grinding action of the pyloric portion of the stomach (Fig. 15.2A) and the abrasion of ingested sand grains to lyse the cells of ingested bacteria and blue-green algae. Gastric pH ranges from 3·5 to 8·5, but most values are in the higher parts of this range (see Horn 1989); an intestinal pH of 8·5 has been recorded (Payne 1978).

(b) The striped bristlemouth Ctenochaetus striatus (Fig. 15.2B) It has a thick-walled stomach and a gastro-intestinal length of about 3·5 × the

Type II: Trituration in a gizzard-like stomach

A

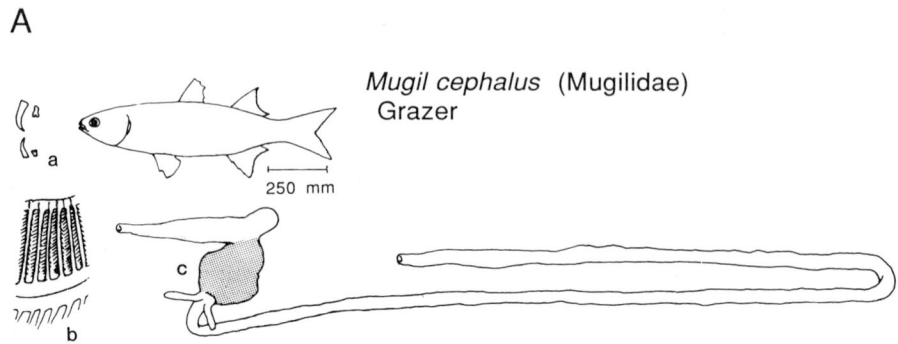

Mugil cephalus (Mugilidae)
Grazer

B

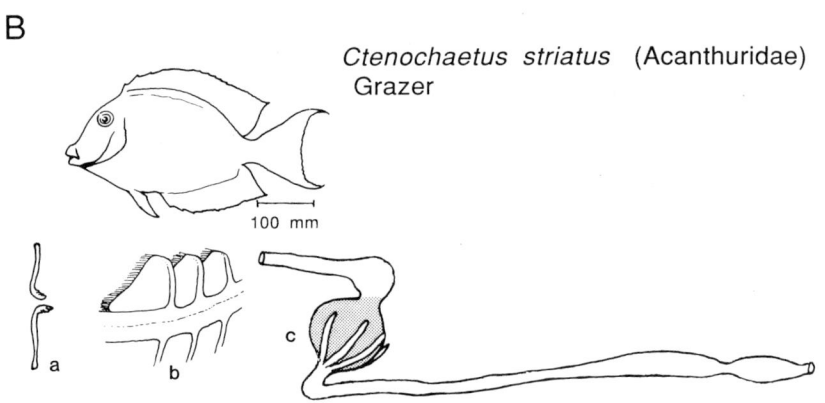

Ctenochaetus striatus (Acanthuridae)
Grazer

C

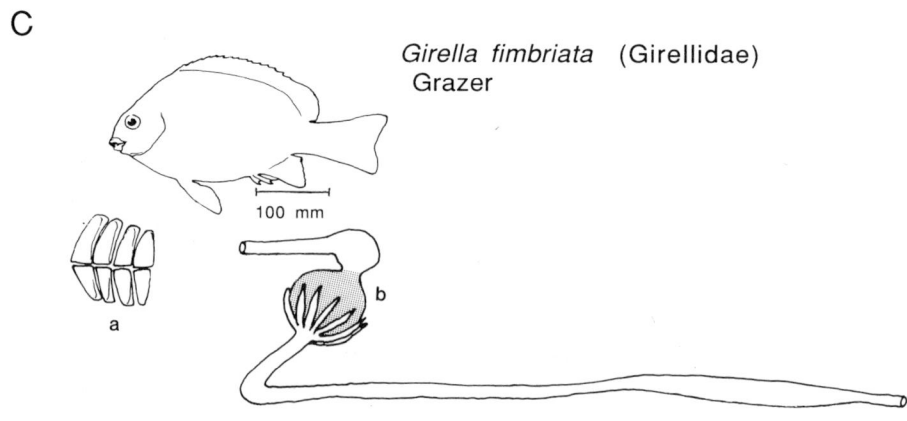

Girella fimbriata (Girellidae)
Grazer

standard body length based on measurements taken on two congeneric species with similar food habits (Jones 1968). This acanthurid sucks and sweeps detritus, diatoms, and inorganic particles from the substratum, passes them across gill rakers bearing many fine teeth and bristles and into the gizzard-like stomach where the particles are triturated. It has relatively low assimilation efficiencies for total organic matter (20 per cent) and nitrogen (37 per cent) but may compensate for them with high feeding rates (Nelson and Wilkins 1988). The grinding action of the stomach apparently reduces the size of the carbonate particles because the average particle size in the rectum is smaller than that in the stomach (Nelson and Wilkins 1988). Consequently, *C. striatus* probably has a local impact on the particle-size distributions of sediments within reef and lagoon habitats. Lobel (1981) found gastric pH in this species to be nearly neutral (7·1), thus supporting his hypothesis that browsers and not grazers should have acidic stomachs.

Other fishes that belong in the Type II category include other species of mullets, perhaps two species of girellids (see below) and those species of surgeon-fishes (in the genera *Acanthurus* and *Ctenochaetus*) that ingest sedimentary particles and have thick-walled stomachs with approximately neutral gastric fluids (see Horn 1989). The nibbler *Girella fimbriata* (Fig. 15.2c) has an extremely thick-walled, gizzard-like stomach and, among girellids, an exceptionally long gut (4·5 × the standard body length), which in the one specimen dissected was largely packed with calcareous red algae (Orton 1989; Horn, personal observations). Thus, in its digestive morphology and its suspected habit of ingesting large amounts of inorganic material, *G. fimbriata* converges on *Mugil cephalus* and *Ctenochaetus striatus*. Further study of *G. fimbriata* and its close relative, *G. nebulosa*, is required, however, before any firm conclusions can be drawn about feeding and digestion in these species.

3. Type III: trituration in pharyngeal jaws

A specialized pharyngeal apparatus represents the second of the two physical mechanisms by which certain herbivorous fishes break down

Fig. 15.2. Digestive tracts of three herbivorous fishes representing the Type II digestive mechanism: trituration in a gizzard-like stomach. A. *Mugil cephalus*: a = minute upper and lower jaw teeth; b = close-set, sieve-like gill rakers; c = thick-walled, muscular pyloric region (shaded) of stomach. B. *Ctenochaetus striatus*: a = mobile, loosely attached upper and lower jaw teeth; b = finely divided gill rakers; c = thick-walled, muscular pyloric region (shaded) of stomach. C. *Girella fimbriata*: a = close-set, unhinged, chisel-like upper and lower jaw teeth; b = thick-walled, muscular pyloric region (shaded) of stomach. Relative dimensions as in Fig. 15.1. (*M. cephalus* illustration modified from fig. 2 in Horn (1989); information for drawing of *C. striatus* from Jones (1968) and of *G. fimbriata* from Orton (1989)).

Type III: Trituration in pharyngeal jaws

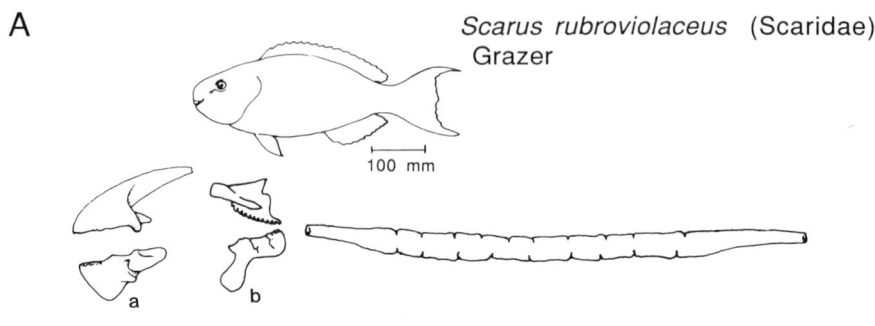

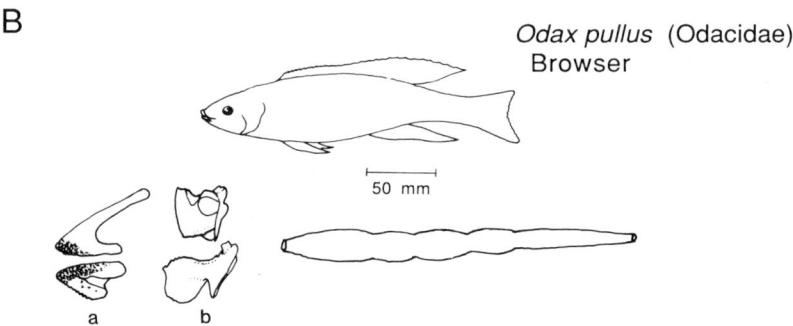

Fig. 15.3. Digestive tracts of two herbivorous fishes representing the Type III digestive mechanism: trituration in pharyngeal jaws. A. *Scarus rubroviolaceus*: a = fused upper and lower oral jaw teeth; b = lateral view of the upper and lower pharyngeal bones. B. *Odax pullus*: a = fused upper and lower oral jaw teeth; b = lateral view of the upper and lower pharyngeal bones. Relative dimensions as in Fig. 15.1. *S. rubroviolaceus* illustration modified from fig. 2 in Horn (1989); information for drawing of *O. pullus* from Clements and Bellwood (1988)).

algal cell walls (Lobel 1981). Inorganic material may or may not be ingested with the food material; therefore, fishes in this category may be either grazers or browsers. The pharyngeal jaws may function either to grind or to shred the algal material before it reaches the intestine. Fishes with a Type III digestive mechanism also have slightly acidic to alkaline digestive tracts and no stomach (see Horn 1989). The relative gut lengths of fishes with specialized pharyngeal structures are generally shorter than either those with highly acidic, thin-walled stomachs or those with nearly neutral or slightly alkaline, gizzard-like stomachs (Lobel 1981; Horn 1989).

Two species discussed in the previous section on feeding mechanisms represent the Type III category. Their digestive mechanisms are described below.

(a) The redlip parrotfish Scarus rubroviolaceus (Fig. 15.3A) This species scrapes algae from reef surfaces and then grinds the ingested material between the broad, flat, dentigerous surfaces of the highly mobile and powerful pharyngeal apparatus (Clements and Bellwood 1988). The intestinal contents consist of finely ground algal material and calcium carbonate particles, attesting to the pulverizing efficiency of the pharyngeal jaws. The intestine is moderately long (2·1 × the standard body length) and sacculated, a modification that may increase its holding capacity and also its retention times by reducing laminar flow rates (Clements and Bellwood 1988). The intestinal pH of *S. rubroviolaceus* has not been measured, but that of other parrotfishes ranges from slightly acidic to slightly alkaline (Horn 1989).

(b) The butterfish Odax pullus (Fig. 15.3B) This fish takes bites from upright, foliose brown algae and then shreds the ingested material between the narrow cutting edges of the dentigerous surfaces of the highly mobile pharyngeal apparatus (Clements and Bellwood 1988). Thus, the algal food is sliced into small pieces but left uncrushed before it is passed into the intestine. The intestine is moderately long (*c.* 1·5 × the standard body length) and has a dorsal swelling; intestinal pH is >8·1 (Clements and Bellwood 1988).

Other fishes with a Type III digestive system include other parrot-fishes, other herbivorous odacids, and the garfish *Hyporhamphus melanochir*. *H. melanochir* triturates ingested seagrass material in a pharyngeal mill before passing the material into the extremely short (0·5 × total body length) intestine, which has a pH ranging from slightly acidic to neutral (Klumpp and Nichols 1983).

4. *Type IV: microbial fermentation in the hindgut*

A fourth type of digestive mechanism found in herbivorous fishes involves gut micro-organisms with the apparent ability to digest algal cell walls. Lobel (1981) expected the possibility of such a mechanism, but strong evidence for microbial gut fermentation was not provided until Rimmer and Wiebe (1987) showed that the digestive tracts of two Australian kyphosids, *Kyphosus sydneyanus* and *K. cornelii*, contained a complex microflora and high concentrations of volatile fatty acids (VFAs). Food taken from the hindgut caecal chambers of these two fishes contained an abundant and diverse microflora of bacteria and ciliated and flagellated protozoans, which were undetectable in the foregut. VFAs, the assimilable, anaerobic degradation products resulting from hydrolysis of polysaccharides by microbial symbionts (see Smith and Douglas 1987), were recorded in the material sampled from the caecal pouch of both fishes and the rectum of *K. sydneyanus*. Rimmer and Wiebe (1987) considered this finding to be conclusive evidence of fermentative digestion in these fishes.

The two kyphosids are, at this writing, the only fishes that have convincingly been shown to possess the Type IV digestive mechanism. *K. sydneyanus*, the species discussed in the previous section, is illustrated here (Fig. 15.4) and, because of their apparent similarities, it and *K. cornelii* are discussed together below.

These two fishes are large (60–80 cm in length), strictly herbivorous species that swallow cleanly bitten pieces of algae without trituration. Although the gastric fluids are strongly acidic in both *K. sydneyanus* (pH 2·8–3·0) and *K. cornelii* (pH 2·9–3·9), lysis of algal cells was not apparent to Rimmer and Wiebe (1987). Algae were softened but retained their natural colour and intact appearance before passing from the stomach to the intestine. Both species have long

Type IV: Microbial fermentation in the hindgut

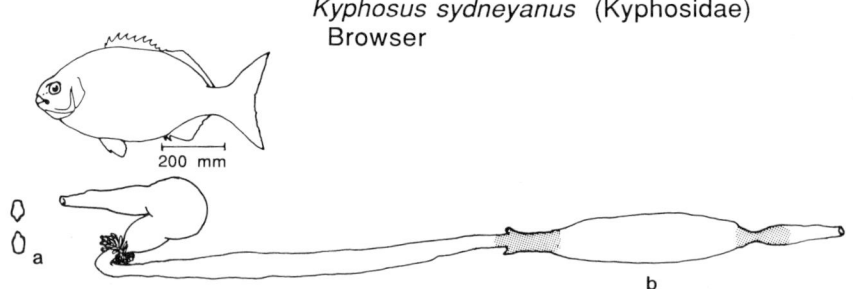

Fig. 15.4. Digestive tract of a herbivorous fish representing the Type IV digestive mechanism: microbial fermentation in the hindgut. *Kyphosus sydneyanus*: a = upper and lower jaw teeth; b = hindgut compartment separated by valves (shaded) from the intestine and rectum. Relative dimensions as in Fig. 15.1. (modified from fig. 2 of Horn (1989)).

intestines (4·0–4·6 × fork length of body) and thin-walled, caecum-like pouches near the posterior end of the gut. *K. sydneyanus* has a single-lobed pouch or enlargement (Fig. 15.4), whereas *K. cornelii* has two blind, lateral sacs connected to a median lobe; in both species the pouches are separated by valves from the adjacent parts of the gut.

K. sydneyanus and *K. cornelii* seem to meet most of the basic requirements (Bjorndal 1987) for a vertebrate to maintain an efficient gut microflora: based on Rimmer and Wiebe's (1987) work, the two species have:

(1) a reasonably constant food supply of red and brown seaweeds;
(2) a relatively long gut transit time (21 hours);
(3) expected anaerobic conditions in the caecal pouches;
(4) a fairly narrow pH range in these caecal pouches (6·3–6·7 for *K. sydneyanus*, 6·1–6·2 for *K. cornelii*);
(5) possible evidence of waste product removal in the form of much larger numbers of lysed bacteria in the rectum than in anterior regions of the gut.

The final requirement, a constant, preferably elevated, body temperature, is perhaps the most difficult criterion to meet because temperatures would be expected to vary seasonally in the warm temperate to subtropical habitats of these fishes. Water temperatures, however, fluctuated by only 4°C (22–26°C) over the year at Rimmer and Wiebe's (1987) collection site in coastal waters of Western Australia (Rimmer 1986). This temperature range is apparently narrow enough to allow effective microbial fermentation to occur in the digestive tracts of the two species.

Information is slowly emerging to indicate that complex, symbiotic microfloras are characteristic of other herbivorous fishes and that these symbionts probably play important roles in the digestive processes of these fishes. Acanthurids appear to have particularly rich and distinctive microfloras. In a survey of 26 species of surgeon-fishes on the Great Barrier Reef, Clements *et al.* (1989) found protistan symbionts, in a range of forms, to be present only in the herbivorous and detritivorous species. These symbionts were absent from all members of the families Kyphosidae, Pomacentridae, Scaridae, Zanclidae, Siganidae, and Blenniidae examined. *Acanthurus nigrofuscus* is perhaps the best-studied surgeon-fish in terms of symbiotic microfloras. The gut of this species was found by Fishelson *et al.* (1985) to contain bacteria, flagellates, and high densities of a large (30–500 μm), cigar-shaped protistan, which has been subsequently described (Montgomery and Pollak 1988*b*) as a new genus and species of micro-organism, *Epulopiscium fishelsoni*. Little is known about the functional role of these gut symbionts, but two recent studies provide some intriguing clues. Montgomery and Pollak (1988a) showed that if *A. nigrofuscus* was starved for two or three days, the protistan symbiont was eliminated, and that if the starved fish was released back into the wild to forage with other surgeon-fishes, it did not re-establish the protist in its gut. Sutton and Clements (1988) studied the aerobic and facultatively anaerobic bacteria in the digestive tracts of four Great Barrier Reef fishes with different food habits and found that the intestine of only the herbivorous *A. nigrofuscus* was dominanted by agar-digesting non-*Vibrio* bacteria. The authors suggested that these bacteria may have a role in the breakdown of algal structural carbohydrates.

A study on the microbial floras in the halfmoon *Medialuna californiensis* (Scorpididae), a warm-temperate fish common in southern California waters (Eschmeyer *et al.* 1983), has revealed numerous rod-shaped bacteria in the intestine and a range of volatile fatty acids (VFAs), especially in the hindgut, similar to that observed in the rumen of cattle (J. S. Kandel, W. Van Antwerp, and Horn, unpublished data). In the same study, on the other hand, the monkeyface prickle-back *Cebidichthys violaceus*, a cool-temperate fish common in central California coastal waters (see above), had somewhat fewer such bacteria and only acetic acid in its gut. These results strongly suggest that microbial fermentation plays a role in the digestive processes of *M. californiensis* but probably not of *C. violaceus*. Research in progress by the same investigators indicates that features and processes similar to those found in *M. californiensis* (Scorpididae) occur in the guts of the opaleye *Girella nigricans* (Girellidae) and the zebra-perch *Hermosilla azurea* (Kyphosidae), also warm-temperate fishes found in southern california coastal waters (Eschmeyer *et al.* 1983). The three families mentioned here are related to an uncertain degree (Johnson and Fritzsche 1989; Orton 1989; Stevens *et al.* 1989); further functional analysis of their alimentary tracts, combined with analysis of their phylogenetic relationships, may provide insights into the evolution of digestive mechanisms in these fishes.

Conclusions on feeding and digestive mechanisms

The four digestive mechanisms discussed above provide a framework for the dominant processes found among marine herbivorous fishes for gaining access to the nutrients in algae and seagrasses. These mechanisms involve fundamental mechanical and physiological specializations, and serve to emphasize the refractory nature of a herbivorous diet and to underscore the close relationship between feeding and digestion in herbivores. Many browsers appear to require only a stomach pH just slightly more acidic than that of a carnivore in order to leach enough nutrients from algal cells without really breaking down the cell wall. Yet, others such as the kyphosids and their relatives (some girellids and scorpidids) have acidic stomachs as well as apparently fermentative microbial digestion. Interestingly, these fishes commonly eat species of brown seaweeds that are considered to be (e.g., Littler *et al.* 1983; Estes and Steinberg 1988) among the most chemically defended of marine macrophytes. *Odax pullus* also eats some of the same heavily defended brown algae and, even though it has a short, seemingly unspecialized gut, Clements (1990) has found dense concentrations of prokaryotic and eukaryotic microbes in its posterior intestine. Whether gut microfloras play a role in the tolerance to, or detoxification of, such algae remains to be determined. Whatever their exact roles, micro-organisms seem likely to become increasingly recognized as being involved in the digestive processes of herbivorous fishes.

Grazers frequently have either a pharyngeal or gastric grinding mechanism as a triturating process. These species usually have non-acidic stomachs or lack stomachs altogether. Gizzard-like stomachs are markedly convergent features in taxonomically unrelated grazers such as the Mugilidae and certain members of the Acanthuridae and Girellidae. Several groups of herbivorous fishes, e.g., the Acanthuridae, Girellidae, and Kyphosidae, exhibit more than one of the four digestive mechanisms discussed in this review. Even a single species such as *Girella tricuspidata* may use three (types I, II, and IV) of the afore-mentioned digestive mechanisms. At least some of these fishes consume large, tough and apparently chemically defended brown algae as a major part of their diets. Multiple digestive mechanisms may be required to extract sufficient nutrients from such resistant, unpalatable, and toxic seaweeds.

Of the numerous ecological and physiological questions that remain to be answered concerning marine fish herbivory (see Horn 1989), most, of course, involve feeding and digestion. These questions centre on factors that determine food choice; the physiological effects of seaweed secondary compounds on herbivore feeding and digestion; the influences of gut pH on digestive processes; the constraints on growth in herbivorous fishes; the ecological impacts of herbivorous fishes on algal communities; the greater diversity of herbivorous fishes in tropical that in tempere habitats; and the evolution of herbivory (and carnivory) in monophyletic groups. Arguably, research progress in all these areas would be enhanced by recognizing that feeding and digestion are integrated parts of a single process of energy gain.

Acknowledgements

I thank Randal Orton and Diane Waugh for generously making a specimen of the remarkable *Girella fimbriata* available to me; Richard Feeney for a quick loan of surgeon-fish from the Natural History Museum of Los Angeles County; Kathleen Burton for rapid delivery of these specimens; Kristin Mailhiot for ably keeping track of important references; and once again, Candace Irelan for expertly preparing the figures. Michael Fris and Candace Irelan made valuable comments on the manuscript. My research on herbivorous fishes has been supported by the National Science Foundation, most recently by Grant OCE-8716368.

References

Al-Hussaini, A. H. (1947). The feeding habits and the morphology of the alimentary tract of some teleosts living in the neighborhood of the Marine Biological Station, Ghardaqa, Red Sea. *Publications of the Marine Biological Station, Ghardaqa, Red Sea*, **5**, 1–61.

Anderson, T. A. (1986). Histological and cytological structure of the gastrointestinal tract of the luderick, *Girella tricuspidata* (Pisces, Kyphosidae), in relation to diet. *Journal of Morphology*, **190**, 109–19.

Anderson, T. A. (1987). Utilization of algal cell fractions by the marine herbivore the luderick, *Girella tricuspidata* (Quoy and Gaimard). *Journal of Fish Biology*, **31**, 221–8.

Ayling, T. and Cox, G. J. (1982). *Collins guide to the sea fishes of New Zealand*. Collins, Auckland, New Zealand.

Bakus, G. J. (1969). Energetics and feeding in shallow marine waters. *International Review of General and Experimental Zoology*, **4**, 275–369.

Barton, M. G. (1982). Intertidal vertical distribution and diets of five species of central California stichaeoid fishes. *California Fish and Game*, **68**, 174–82.

Bellwood, D. R. and Choat, J. H. (1990). A functional analysis of grazing in parrot-fishes (family Scaridae): the ecological implications. *Environmental Biology of Fishes*, **28**, 189–214.

Bjorndal, K. A. (1987). Digestive efficiency in a temperate herbivorous reptile, *Gopherus polyphemus*. *Copeia*, **1987**, 714–20.

Bryan, P. G. (1975). Food habits, functional digestive morphology, and assimilation efficiency of the rabbitfish *Siganus spinus* (Pisces, Siganidae) on Guam. *Pacific Science*, **29**, 269–77.

Choat, J. H. (1982). Fish feeding and the structure of benthic communities in temperate waters. *Annual Review of Ecology and Systematics*, **13**, 423–49.

Christensen, M. S. (1978). Trophic relationships in juveniles of three species of sparid fishes in the South African marine littoral. *Fishery Bulletin*, **76**, 389-401.

Clements, K.D. (1985). Feeding in two New Zealand herbivorous fish, the butterfish *Odax pullus* and the marblefish *Aplodactylus arctidens*. Unpublished M.Sc. Thesis. University of Auckland, New Zealand.

Clements, K.D. (1990). The endosymbiotic communities of two herbivorous labroid fishes, *Odax cyanomelas* (Richardson) and *O. pullus* Schneider. *Marine Biology*, **109**, 223-9.

Clements, K.D. and Bellwood D.R. (1988). A comparison of the feeding mechanisms of two herbivorous labroid fishes, the temperate *Odax pullus* and the tropical *Scarus rubroviolaceus*. *Australian Journal of Marine and Freshwater Research*, **39**, 87-107.

Clements, K.D., Sutton, D.C., and Choat, J.H. (1989). Occurrence and characteristics of unusual protistan symbionts from surgeonfishes (Acanthuridae) of the Great Barrier Reef, Australia. *Marine Biology*, **102**, 403-12.

Coleman, N. (1980). *Australian sea fishes south of 30° south*. Doubleday, Sydney, Australia.

Collins, M.R. (1981). The feeding periodicity of striped mullet, *Mugil cephalus* L., in two Florida habitats. *Journal of Fish Biology*, **19**, 307-15.

Edwards, T.W. and Horn, M.H. (1982). Assimilation efficiency of a temperate zone intertidal fish (*Cebidichthys violaceus*) fed diets of macroalgae. *Marine Biology*, **67**, 247-53.

Eschmeyer, W.N., Herald, E.S., and Hammann, H. (1983). *A field guide to Pacific Coast fishes of North America from the Gulf of Alaska to Baja California*. Houghton Mifflin, Boston.

Estes, J.A. and Steinberg, P.D. (1988). Predation, herbivory and kelp evolution. *Paleobiology*, **14**, 19-36.

Fishelson, L., Montgomery, W.L., and Myrberg, A.A. (1985). A unique symbiosis in the gut of tropical herbivorous surgeon-fish (Acanthuridae: Teleostei) from the Red Sea. *Science*, **229**, 49-51.

Goldschmid, A., Kotrschal, K., and Wirtz, P. (1984). Food and gut length of 14 Adriatic blenniid fish (Blenniidae; Percomorpha; Teleostei). *Zoologischer Anzeiger*, **213**, 145-50.

Gomon, M.F. and Paxton, J.R. (1985). A revision of the Odacidae, a temperate Australian-New Zealand labroid fish family. *Indo-Pacific Fishes*, **8**, 57 pp.

Hiatt, R.W. and Strasburg, D.W. (1960). Ecological relationships of the fish fauna on coral reefs of the Marshall Islands. *Ecological Monographs*, **30**, 65-127.

Horn, M.H. (1989). Biology of marine herbivorous fishes. *Oceanography and Marine Biology Annual Review*, **27**, 167-272.

Horn, M.H. and Neighbors, M.A. (1984). Protein and nitrogen assimilation as as factor in predicting the seasonal macroalgal diet of the monkeyface prickleback. *Transactions of the American Fisheries Society*, **113**, 388-96.

Horn, M. H., Murray, S. N., and Edwards, T. W. (1982). Dietary selectivity in the field and food preferences in the laboratory for two herbivorous fishes (*Cebidichthys violaceus* and *Xiphister mucosus*) from a temperate intertidal zone. *Marine Biology*, **67**, 237-46.

Horn, M. H., Neighbors, M. A., and Murray, S. N. (1986). Herbivore responses to a seasonally fluctuating food supply: growth potential of two temperate intertidal fishes based on the protein and energy assimilated from their macroalgal diets. *Journal of Experimental Marine Biology and Ecology*, **103**, 217-34.

Johnson, G. D. and Fritzsche, R. A. (1989). *Graus nigra*, an omnivorous girellid, with a comparative osteology and comments on relationships of the Girellidae (Pisces: Perciformes). *Proceedings of the Academy of Natural Sciences of Philadelphia*, **141**, 1-27.

Jones, R. S. (1968). Ecological relationships in Hawaiian and Johnston Island Acanthuridae (surgeonfishes). *Micronesica*, **4**, 309-61.

Klumpp, D. W. and Nichols, P. D. (1983). Nutrition of the southern sea garfish *Hyporhamphus melanochir*: gut passage rate and daily consumption of two food types and assimilation of seagrass components. *Marine Ecology Progress Series*, **12**, 207-16.

Liem, K. F. and Greenwood, P. H. (1981). A functional approach to the phylogeny of the pharyngognath teleosts. *American Zoologist*, **21**, 83-101.

Littler, M. M., Taylor, P. R., and Littler, D. S. (1983). Algal resistance to herbivory on a Caribbean barrier reef. *Coral Reefs*, **2**, 111-18.

Lobel, P. S. (1981). Trophic biology of herbivorous reef fishes: alimentary pH and digestive capabilities. *Journal of Fish Biology*, **19**, 365-97.

Lobel, P. S. and Ogden, J. C. (1981). Foraging by the herbivorous parrotfish *Sparisoma radians*. *Marine Biology*, **64**, 173-83.

Marais, J. F. K. (1980). Aspects of food intake, food selection, and alimentary canal morphology of *Mugil cephalus* (Linnaeus, 1758), *Liza tricuspidens* (Smith, 1935), *L. richardsoni* (Smith, 1846) and *L. dumerili* (Steindachner, 1869). *Journal of Experimental Marine Biology and Ecology*, **44**, 193-209.

Montgomery, W. L. and Pollak, P. E. (1988a). Gut anatomy and pH in a Red Sea surgeonfish, *Acanthurus nigrofuscus*. *Marine Ecology Progress Series*, **44**, 7-13.

Montgomery, W. L. and Pollak, P. E. (1988b). *Epulopiscium fishelsoni* n. gen., n. sp., a protist of uncertain taxonomic affinities from the gut of an herbivorous reef fish. *Journal of Protozoology*, **35**, 565-9.

Montgomery, W. L., Myrberg, A. A. Jr., and Fishelson, L. (1989). Feeding ecology of surgeonfishes (Acanthuridae) in the northern Red Sea, with particular reference to *Acanthurus nigrofuscus* (Forsskål). *Journal of Experimental Marine Biology and Ecology*, **132**, 179-207.

Nelson, J. S. (1984). *Fishes of the world*. 2nd edn. Wiley, New York.

Nelson, S. G. and Wilkins, S. De C. (1988). Sediment processing by the surgeon-fish *Ctenochaetus striatus* at Moorea, French Polynesia. *Journal of Fish Biology*, **32**, 817–24.

Odum, W. E. (1968). The ecological significance of fine particle selection by the striped mullet *Mugil cephalus*. *Limnology and Oceanography*, **13**, 92–8.

Odum, W. E. (1970). Utilization of the direct grazing and plant detritus food chains by the striped mullet *Mugil cephalus*. In *Marine food chains*, (ed. J. H. Steele), pp. 222–40. University of California Press, Berkeley.

Ogden, J. C. and Lobel, P. S. (1978). The role of herbivorous fishes and urchins in coral reef communities. *Environmental Biology of Fishes*, **3**, 49–63.

Orton, R. D. (1989). The evolution of dental morphology in the Girellidae (Acanthopterygil: Perciformes), with a systematic revision of the Girellidae. Unpublished Ph.D. Thesis. University of California, Los Angeles.

Payne, A. I. (1978). Gut pH and digestive strategies in estuarine grey mullet (Mugilidae) and tilapia (Cichlidae). *Journal of Fish Biology*, **13**, 627–9.

Penry, D. L. and Jumars, P. A. (1986). Chemical reactor analysis and optimal digestion. *BioScience*, **36**, 310–15.

Ralston, S. L. and Horn, M. H. (1986). High tide movements of the temperate-zone herbivorous fish *Cebidichthys violaceus* (Girard) as determined by ultrasonic telemetry. *Journal of Experimental Marine Biology and Ecology*, **98**, 35–50.

Randall, J. E. (1967). Food habits of reef fishes of the West Indies. *Studies in Tropical Oceanography*, **5**, 665–847.

Randall, J. E. (1974). The effects of fishes on coral reefs. In *Proceedings of the Second International Coral Reef Symposium*, Vol. 1 (ed. Great Barrier Reef Committee), pp. 159–66. Brisbane, Australia.

Randall, J. E. (1986a). Acanthuridae. In *Smiths' sea fishes*, (ed. M. M. Smith and P. C. Heemstra), pp. 811–23. Springer-Verlag, Berlin.

Randall, J. E. (1986b). Scaridae. In *Smiths' sea fishes*, (ed. M. M. Smith and P. C. Heemstra), pp. 706–14. Springer-Verlag, Berlin.

Rimmer, D. W. (1986). Changes in diet and the development of microbial digestion in juvenile buffalo bream, *Kyphosus cornelii*. *Marine Biology*, **92**, 443–8.

Rimmer, D. W. and Wiebe, W. J. (1987). Fermentative microbial digestion in herbivorous fishes. *Journal of Fish Biology*, **31**, 229–36.

Rosenblatt, R. H. and Hobson, E. S. (1969). Parrotfishes (Scaridae) of the eastern Pacific, with a generic rearrangement of the Scarinae. *Copeia*, **1969**, 434–53.

Russ, G. (1984). Distribution and abundance of herbivorous grazing fishes in the central Great Barrier Reef. I. Levels of variability across the entire continental shelf. *Marine Ecology Progress Series*, **20**, 23–34.

Russell, B.C. (1983). The food and feeding habits of rocky reef fish of northeastern New Zealand. *New Zealand Journal of Marine and Freshwater Research*, **17**, 121–45.

Smith, D.C. and Douglas, A.E. (1987). *The biology of symbiosis*. Edward Arnold, London.

Smith, M.M. and Smith, J.L.B. (1986). Mugilidae. In *Smiths' sea fishes*, (ed. M.M. Smith and P.C. Heemstra), pp. 714–20. Springer-Verlag, Berlin.

Stevens, E.G., Watson, W., and Moser, H.G. (1989). Development and distribution of larvae and pelagic juveniles of three kyphosid fishes (*Girella nigricans, Medialuna californiensis,* and *Hermosilla azurea*) off California and Baja California. *Fishery Bulletin*, **87**, 745–68.

Stoner, A.W. (1980). The feeding ecology of *Lagodon rhomboides* (Pisces: Sparidae): variation and functional responses. *Fishery Bulletin*, **78**, 337–52.

Sutton, D.C. and Clements, K.D. (1988). Aerobic, heterotrophic gastrointestinal microflora of tropical marine fishes. In *Proceedings of the Sixth International Coral Reef Symposium*, 3, pp. 185–90. Townsville, Queensland, Australia.

Thomson, J.M. (1954). The organs of feeding and the food of some Australian mullet. *Australian Journal of Marine and Freshwater Research*, **5**, 469–85.

Urquhart, K.A.F. (1984). Macroalgal digestion by *Cebidichthys violaceus*, a temperate marine fish with highly acidic stomach fluids. Unpublished MA Thesis. California State University, Fullerton.

White, T.C.R. (1985). When is a herbivore not a herbivore? *Oecologia* (Berlin), **67**, 596–7.

16. Digestion survival in seaweeds: an overview

B. SANTELICES
Departamento de Ecología, Facultad de Ciencias Biológicas, Pontificia Universidad Católica de Chile, Santiago, Chile

Abstract

Several types of seaweeds survive passage through the digestive tract of various types of grazers. Originally described as a type of escape mechanism, survival depends upon three types of algal responses:

(1) regeneration and growth of partially digested vegetative tissues;
(2) germination and growth of non-digested algal propagules; and
(3) grazing-induced cell and protoplast release after cell wall digestion.

Buccal morphology and digestion efficiency of the grazer, algal life history, and seaweed morphology determine the type of algal response. Opportunistic algal forms survive digestion more frequently and through a more diverse array of grazers than late successionists. The interaction seems ecologically important. Observational and experimental field evidence suggests that grazer-mediated propagule dispersal is widespread in benthic systems. Algal propagules surviving in faecal pellets resist more a extended desiccation periods, attach equally well, and both sink and grow faster than non-ingested propagules. The number of propagules released from pellets is within known field densities of settling propagules, but their relative importance is likely to increase when seaweed biomass available for reproduction is reduced by grazing. Overall, the phenomenon here reviewed shares several characteristics with animal-mediated seed-dispersal processes of land plants. However, the grazer-induced cell and propagule production response has no parallel in land producers.

Introduction

The possibility that benthic sexual or asexual algal propagules may survive digestion by herbivores was not considered prior to 1983. Although incompletely digested algal remains in faecal pellets of marine invertebrates had been noticed since the early fifties (Lasker and Giese 1954; Farmanfarmaian and Phillips 1962; Cloakie and Norton 1974; Prim and Lawrence 1975; Lobel 1980), most reports assumed complete tissue destruction after ingestion or limited attachment capabilities of the few surviving cells. However, results gathered over the last seven years have drastically changed this view. Experimental incubation of faecal pellets showed, firstly for sea urchins (Santelices et al., 1983) and later for a variety of other invertebrates and fishes, that incompletely digested algal-remains often are able to originate new thalli. Field experiments have shown that some of these surviving fragments and propagules can reattach. Such experiments have compared digestion survival of seaweeds with some of the animal-mediated dispersal processes of land plants (e.g., McKey 1975; Janzen 1983; Herrera, 1985; Cobo and Andreu 1988) and with community effects derived from differential survival of some species in planktonic communities (Gibor 1956; Porter 1973, 1977; Nicotri 1977). In addition, studies on protoplast production have contributed useful data to understand grazer-related aspects of the process. Today, survival of seaweeds to digestion is considered to be an ecologically important phenomenon in benthic communities (Duffy and Hay 1990). The main purpose of this study is to review the accumulated evidence on this process comparing, whenever pertinent, the known characteristics of seaweed survival to digestion with general trends described for animal-mediated seed dispersal processes of land plants.

The nature of seaweed resistance to digestion

1. Types of algal responses

Although seaweed resistance to digestion was originally described as a type of escape mechanism (Santelices et al. 1983), the accumulated evidence suggests that the phenomenon involves at least three different types of algal response.

(a) Regeneration and growth of partially digested vegetative tissues This is the most common response (Santelices and Ugarte 1987). The faecal pellets of many invertebrates often contain algal fragments, with regeneration capabilities, of a variety of morphologically different seaweeds (Figs 16.1–16.4). This type of survival is wholly dependent

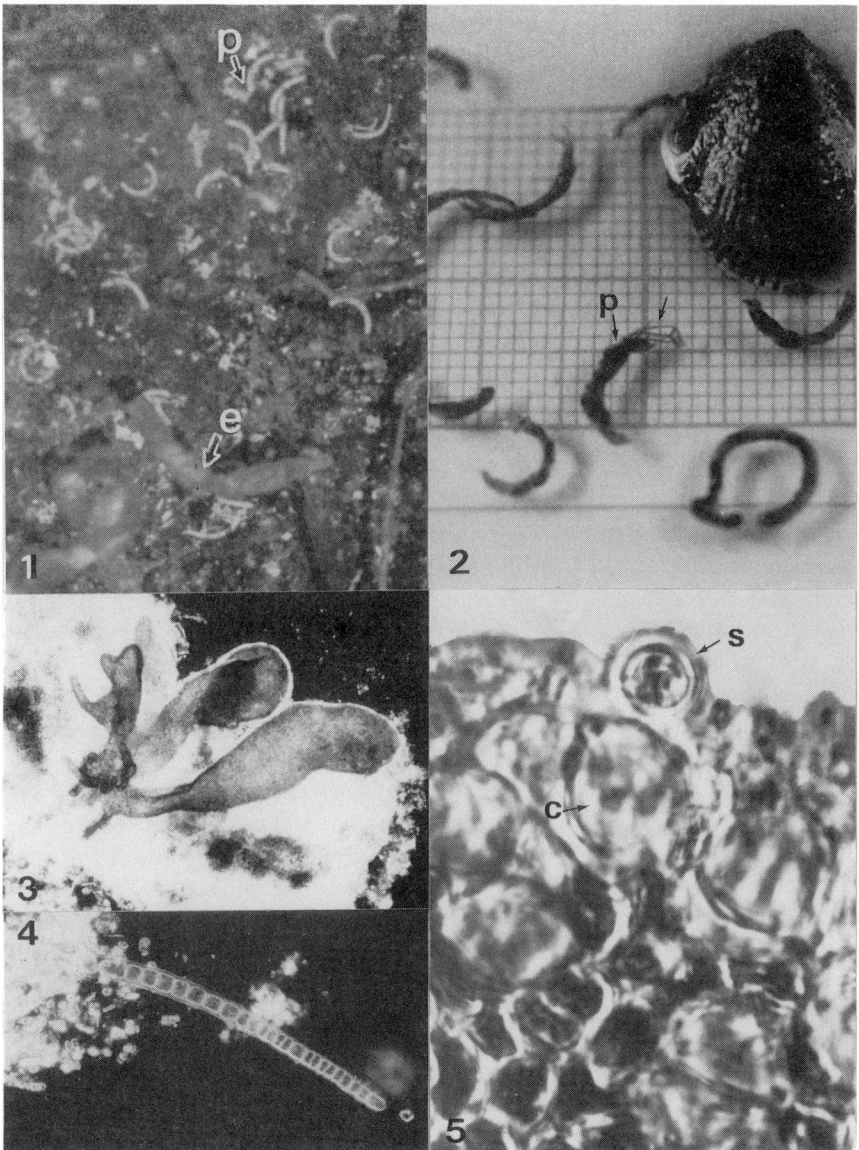

Fig. 16.1. Plants of *Enteromorpha* sp. (*e*) growing in intertidal pools in close spatial proximity with faecal pellets (*p*) of *Collisella* sp. **Fig. 16.2.** Faecal pellets (*p*) recently ejected by a grazer (*Collisella* sp.) filled with undigested fragments (arrow) of species of *Enteromorpha* and *Chaetomorpha*. **Fig. 16.3.** Small thalli of *Enteromorpha compressa* regenerating from a fragment in the faecal pellet of *Fissurella crassa* incubated in the laboratory. **Fig. 16.4.** Small thalli of *Chaetomorpha* sp. regenerating from fragments in the faecal pellet of *Fissurella crassa*. **Fig. 16.5.** Swarmer (*s*) being released from a fragment of *Enteromorpha* recovered from faecal pellets of *Siphonaria lessonii*. Note the empty cells (*c*).

upon the capability of meristematic cells to pass alive through the digestive tract of grazers. Therefore morphological differentiation and location of meristematic tissues over the seaweed thalli are of overall importance. Species with regeneration capacity in most cells throughout the thallus (e.g., opportunistic foliose forms such as *Ulva* or *Enteromorpha*) should exhibit higher regeneration capabilities than species with more localized meristems. Filamentous forms with many apical cells (e.g., densely branched species) would have a higher survival potential than sparsely branched forms, with fewer apical cells. Intercalary meristems are likely to survive the action of digestive enzymes more frequently than apical meristems.

To be ecologically important from a population point of view, surviving fragments should be able to reattach to the substratum. Breeman and Hoeksema (1987) found this to be the case for *Audouinella purpurea*, as in the field, small tufts of this species found on bare substratum originated from fragments contained in sticky, detritus-rich envelopes identified as faecal pellets. Under laboratory conditions, uningested fragments of *A. purpurea* exhibited fast and high regeneration capacities, first forming one or more adhesive rhizoids and subsequently new filamentous shoots (Pearlmutter and Vadas 1978). In this species, rhizoids and erect filaments may differentiate within three days after fragmentation (Breeman and Hoeksema 1987).

Similar findings were reported by Jernakoff (1985), working with *Patelloida latistrigata* in south-eastern Australia. He transferred faeces of this invertebrate, containing blue-green algae and germlings of ulvoids, to clean sandstone plates covered with filtered seawater. Within two weeks after transfer, two of the four original germlings in the faeces had reattached.

Audouinella purpurea and ulvoids are by no means unique among seaweeds with respect to their reattachment capabilities. Many frondose and filamentous seaweeds are known to be able to fragment, float, drift for a period, reattach to the substratum, and generate new individuals. This is common in the Ectocarpaceae (Russell 1967), several filamentous, thin-frondose and cylindrical green, brown, and red algae (Norton and Mathieson 1983), and even in some thick-frondose, normally attached species such as *Solieria chordalis* (Floc'h et al. 1987). It is not yet clear whether fragment size could limit regeneration and reattachment capabilities of any of these species. However, under calm-water conditions, many fragments should be able to reattach to the substratum after surviving digestion.

(b) Germination and growth of non-digested algal propagules Seaweed spores lack protective outer coverings and are expected to be readily digested

when consumed. However, these spores are often consumed together with vegetative cells that can protect them during digestion. For example, gametangia and sporangia in the Chlorophyta are unicellular and are often produced intermixed with vegetative cells. Even in species where the spores are grouped into more or less extensive fertile areas (= sori), they are often in close spatial proximity to vegetative cells that protect them during digestion. Experimental evidence (Santelices and Ugarte 1987) shows that these surviving spores are released from the sporogenous tissues, germinate, and are able to attach themselves to the substratum. Nevertheless, such spores often showed reduced germination when compared with uningested controls, suggesting some mortality due to digestion.

Free spores also have been found to survive ingestion, at least by two different types of grazers. The amphipod *Hyale media* grazes on cystocarpic tissues as well as on carpospores of *Iridaea laminarioides*. Cultivation of 100 faecal pellets from individuals fed with mature cystocarpic tissues showed germinating carpospores in 67 such pellets (Buschmann and Santelices, 1987). Although the germination of ingested spores was significantly lower than uningested spores, spores viably surviving ingestion became attached to the culture dishes and showed significantly higher growth rates than the control, uningested spores.

Free spores also survive ingestion by mussels (Santelices and Martínez 1988). Gut content analysis of *Perumytilus purpuratus*, a mid-intertidal, belt-forming species of mussel abundant in wave-exposed habitats of central Chile, indicated that macroalgal spores were one of the most frequent food items. Experimental incubation demonstrated survival of eight to ten taxa growing in faecal cultures.

Given that in all these experiments the surviving spores were able to attach themselves to the culture containers, originating new individuals, in nature they may be expected to settle and recruit much as any uningested spore.

(c) Grazer-induced cell and protoplast release A third type of algal response to digestion has been observed in the Chlorophyta and in members of the Bangiales, Rhodophyta (Santelices and Ugarte, 1987). In some fragments of these species recovered from faecal pellets, the cytoplasm of the surviving cells is concentrated towards one side of the cell. When the cell wall disintegrates, the tissue losses integrity, with cells and protoplasts being set free in the culture medium (Fig. 16.5). The green algal cells or protoplasts develop flagella, swim for a while and settle down at the bottom of the culture vessels to form new thalli. Depending on the reproductive condition of the surviving cells, protoplasts from the

frondose phase of Bangiales tested developed into a callus, a filamentous phase, or reconstituted the frondose phase.

The induction of cells or protoplast release by incomplete digestion, and their ecological roles in generating new individuals and therefore acting as accessory means of reproduction form a unique process, with no described parallel in plant–herbivore interactions within terrestrial or planktonic communities. In a broad sense, the process approaches some of the dispersal mechanisms described for some land plants, where seed germination is stimulated by passage through the digestive tract of seed predators (McKey 1975; Janzen 1983; Cobo and Andreu 1988). However, in this seaweed–herbivore interaction there is *de novo* grazer-induced propagule production, a process significantly different from that of enhanced germination by passage of seeds through an animal.

Experiments with faecal pellets have shown that grazer-induced propagule production is common only among the Chlorophyta and the Bangiales in the Rhodophyta. However, biotechnological studies on protoplast release (Liu *et al.* 1984; Polne-Fuller and Gibor 1984; Cheney *et al.* 1986; Tokuda and Kawashima 1988; Buttler *et al.* 1989; Kloareg *et al.* 1989; Boyen *et al.* 1990) have revealed that digestive enzymes of sea urchins and molluscs are also effective in releasing cells or protoplasts of other brown and red algae. Thus it seems that, depending on the herbivore, grazer-induced propagule production could be a widespread process affecting all types of seaweeds. It is not expected, however, that these single cells or protoplasts directly regenerate new individuals in the field since in laboratory cultures they form calluses prior to further development (Polne-Fuller and Gibor 1987). This probably explains the lack of growth of these types of seaweed in cultures started from faecal pellets.

2. The interacting species

Data documenting survival of seaweeds after ingestion by several types of grazers are summarized in Table 16.1. Although to be regarded as tentative due to the limited number of interacting pairs so far studied, a few general statements could be advanced from the comparative data in this table.

Benthic algal survival through digestive tracts has been documented in several ecologically important groups of herbivores, including sea urchins (Santelices *et al.* 1983), molluscs (Jernakoff 1985; Santelices and Correa 1985), amphipods (Breeman and Hoeksema 1987; Buschmann and Santelices 1987), fishes (Paya and Santelices 1989), and filter-feeders consuming algal spores (Santelices and

Table 16.1. Comparison between the presence (+) or absence (−) of algal species in gut contents of different grazers (G) and their presence or absence in culture (C) started from the respective faecal pellets. Algal diversity contained in the gut contents of filter-feeders is unknown; species therefore have not been included in calculations

	Fish	Grazing molluscs							Filter feeder	Sea urchin	Amphipod	
	Sicyases sanguineus	*Littorina peruviana*	*Chiton granosus*	*Siphonaria lessonii*	*Collisella zebrina*	*Collisella ceciliana*	*Fissurella crassa*	*Fissurella limbata*	*Perumytilus purpuratus*	*Tetrapygus niger*	*Hyale* spp.	% survival ($C/G \times 100$) through various herbivores
	GC	GC	GC	GC	GC	GC	GC	GC	GC	GC	GC	
Chlorophyta												
Opportunistic species												
Enteromorpha sp.	++	++	++	++	++	++	++	++	++	++	−−	100
Ulva rigida	+−	++	++	++	++	++	++	++	+−	++	−−	88
Ulvella sp.	−−	++	−−	++	++	−−	−−	−−	−−	−−	+−	100
Urospora sp.	−−	−−	−−	−−	−−	−−	−−	−−	−−	++	−−	100
Cladophora sp.	+−	−−	+−	−−	−−	−−	−−	+−	−−	++	−−	25
Chaetomorpha sp.	−−	−−	−−	+−	++	++	++	++	−−	++	−−	80
Entocladia sp.	+−	−−	−−	−−	++	++	−−	++	+−	++	−−	80
Late successional forms												
Codium dimorphum	+−	+−	−−	−−	−−	−−	+−	++	−−	+−	−−	20
Phaeophyta												
Opportunistic species												
Ectocarpus sp.	+−	+−	++	++	++	++	−−	−−	+−	−−	−−	71
Pilayella littoralis	−−	−−	−−	−−	++	−−	−−	−−	−−	+−	−−	50
Petalonia fascia	−−	−−	+−	−−	−−	++	++	++	+−	−−	−−	66.6
Late successional forms												
Ralfsia sp.	−−	+−	−−	−−	+−	−−	−−	+−	−−	+−	−−	0
Lessonia nigrescens	−−	−−	−−	−−	−−	−−	+−	+−	−−	++	−−	50
Glossophora kunthii	−−	−−	−−	−−	−−	−−	−−	−−	−−	+−	−−	0

Table 16.1. cont'd

	Fish	Grazing molluscs							Filter feeder	Sea urchin	Amphipod	
	Sicyases sanguineus	Littorina peruviana	Chiton granosus	Siphonaria lessoni	Collisella zebrina	Collisella ceciliana	Fissurella crassa	Fissurella limbata	Perumytilus purpuratus	Tetrapygus niger	Hyale spp.	% survival ($C/G \times 100$) through various herbivores
Rhodophyta												
Opportunistic species												
Erythrotrichia sp.	−−	−−	+−	++	++	++	++	+−	−	++	−−	71.4
Erythrocladia sp.	+−	−−	−−	++	−−	−−	−−	−−	−	++	−−	100
Acrochaetium sp.	+−	+−	−−	−−	−−	−−	−−	−−	−	+−	−−	0
Bangia fuscopurpurea	−−	−−	−−	−−	−−	−−	++	++	−	++	−−	100
Antithamnion sp.	+−	−−	−−	−−	−−	−−	−−	++	−	+−	−−	50
Pterothamnion sp.	+−	−−	−−	++	−−	−−	−−	−−	−	++	−−	100
Centroceras clavulatum	+−	−−	+−	−−	−−	−−	+−	−−	−	−−	−−	0
Porphyra columbina	+−	−−	−−	++	−−	−−	++	−−	+	+−	+−	40
Polysiphonia sp.	+−	−−	−−	−−	−−	−−	−−	+−	+	+−	−−	0
Pterosiphonia dendroidea	−−	−−	−−	−−	−−	−−	−−	−−	−	+−	−−	0
Herposiphonia sp.	−−	−−	−−	−−	−−	−−	−−	−−	+	−−	−−	0
Ceramium rubrum	−−	−−	−−	−−	−−	−−	−−	−−	−	−−	−−	0
Late successional forms												
Placamium pacificum	−−	−−	−−	−−	−−	−−	−−	−−	−	+−	−−	0
Gelidium chilense	+−	−−	−−	−−	−−	−−	−−	−−	−	+−	−−	0
Iridaea laminarioides	++	+−	+−	−−	+−	−−	++	+−	−	+−	+−	40
Schottera nicaeensis	−−	−−	−−	−−	−−	−−	+−	−−	−	++	++	40
Crustose corallines	+−	+−	+−	+−	−−	+−	+−	+−	−	−−	−−	0
Nothogenia fastigiata	−−	−−	−−	−−	−−	−−	−−	+−	−	+−	−−	0
Corallina officinalis	+−	−−	−−	−−	−−	−−	−−	−−	−	+−	−−	0
Gymnogongrus furcellatus	−−	−−	−−	−−	−−	−−	−−	−−	−	+−	+−	0
Total number of species	16 2	9 3	9 3	9 7	8 6	9 8	12 8	14 7	? 8	27 12	5 1	
% species surviving	12.5	33.3	33.5	77.7	75	88.8	63.6	50.0	?	44.4	20.0	
Oppor. vs late succ. forms												
—in gut contents	11/5	5/4	1/2	8/1	6/2	8/1	9/4	10/4	?	17/10	2/3	
—among surviving forms	1/1	3/0	3/0	6/0	6/0	8/0	7/1	7/0	7/0	10/2	0/1	

Martínez 1988). Thus, the phenomenon seems to be widespread among most types of marine herbivores.

However, although it is a general phenomenon, similar proportions of algal species do not necessarily survive the passage through the digestive tracts of every grazer species. Algal survival through the only fish studied is much reduced in comparison with sea urchins and molluscs. Within-group differences are also evident. For example, as many as 83 per cent of the algal species found in the digestive tract of the gastropod *Collisella ceciliana* may pass through alive, whereas only 25–30 per cent of the essentially similar algal species consumed by *Littorina peruviana* and *Chiton granosus* do so.

The accumulated evidence suggests that the above differences result from several factors related to the biological attributes of the seaweeds, the grazing capabilities of the herbivores, and the grazing process.

(a) Biological attributes of the seaweeds Not all seaweed species show similar regeneration capabilities, have equivalent numbers of vegetative cells intermixed with their reproductive propagules, or have similar capabilities to regenerate new individuals from recently released protoplasts. Therefore, their digestion survival probabilities are different.

The way that a given seaweed species apportions resources to reproduction and other functions, and the degree of cellular differentiation and division of labour reached among the cells in a given thallus, are also important. Opportunist species usually have little division of labour and the cells with either reproductive, regenerative, or both capacities develop over much of the thallus. Generally, when tissues from these algal taxa survive digestion, they should have a much higher chance of regenerating new individuals than tissues from late successional forms. Experimental results (Table 16.1) show that some opportunistic forms (e.g., *Enteromorpha, Ulva*) are able to regenerate from faecal pellets of almost all generalist grazers. Survival in late successionists is more reduced and depends on the possibility that meristematic or reproductive cells, (in these species more localized in space and time than in opportunistic forms,) are ingested by the grazers and survive digestion. For example, *Gelidium lingulatum* and *Iridaea laminarioides* could grow only when the faecal cultures contained either surviving spores or meristematic tissues. In the case of *G. lingulatum*, only those fragments with apical cells, or with cortical cells able to differentiate into apical cells, regenerate new growth. In the case of *Iridaea laminarioides*, fragments from the holdfast and basal parts of the stipes were the only tissues with regenerative capabilities (Santelices and Ugarte 1987).

Certain late successional forms have shown no survival capacity to digestion, despite being very frequent food items of several of the grazers studied. This is the case with encrusting coralline species. Vegetative tissues of these forms have always failed to grow from faecal cultures and the spores have evidenced high sensitivity and reduced survival to digestion. Perhaps most grazers are consuming only the external, non-meristematic layers of these calcareous forms.

As suggested in the previous section, morphological organization of the seaweed consumed is also important. Some of the most obviously important morphological characteristics include the number and position of the meristems and the number of layers and size of vegetative cells intermixed with reproductive propagules, which may protect these latter during digestion. Several other characteristics are perhaps important, but experimental data are lacking. As in land plants and fruit traits (Janzen, 1983), there is an immense amount of work still to be done on the integration of morphological characteristics of seaweeds with their functions in relation to marine animals. For example, it seems evident that shape, size, position on the plant, palatability, toughness, surface contour, and other characteristics of the seaweed reproductive structures (sori, cystocarps, sporophylls) will all influence consumption, and probably transit and survival of these tissues through animal guts. The great diversity in these algal traits has not been sufficiently considered in relation to herbivory.

(b) Grazing capabilities of the herbivore Experimental results so far indicate that the differential survival of algae to ingestion and digestion depends on at least two traits of the herbivore, the structure of the feeding apparatus, and the digestive efficiency. Structurally different feeding apparatus may produce different degrees of damage to the algal tissue. For example, the polyplacophoran radula has as many as 17 teeth per row, at least four of these apparently used for grazing (Steneck and Watling 1982). During grazing these 'dominant' teeth scrape the substratum and are followed by a light inward-sweeping action of the marginal teeth that, according to Steneck and Watling (1982), probably picks up debris and algal filaments. The combined action of dominant and marginal teeth probably affects a large number of algal cells in each thallus grazed by chitons, allowing only a reduced number to remain with intact cell walls and thus explaining the low numbers of algal species surviving passage through chiton guts. On the other hand, docoglossan radulae (such as those present in the species of *Collisella*) have few, short, and small mobile teeth placed in such a way that they often leave unreached spaces between them (Newell 1979; Branch 1981). These teeth have good excavating capabilities on tough substrata but are poorly suited to grazing on filamentous and foliose algae. Small

plants or small pieces of algal tissue may slip between the teeth during grazing, thus entering the digestive tract without suffering much damage by these radular teeth.

Although the structure and functioning of feeding apparatus are important, not all results can be explained on that basis alone. Perhaps the best example here is the low number of algae surviving digestion by the fish *Sicyases sanguineus*. Only two of the 16 algal species consumed by this fish were shown to survive digestion (Paya and Santelices 1989). However, the teeth of this animal are not sufficiently numerous or close together to macerate filamentous or thin, foliose algal forms (Cancino and Castilla 1988). Therefore, low algal survival in this case seems related to digestive efficiency and enzymatic action.

Enzymatic action does not, however, always kill the algae. As commented previously, digestion by some grazers (e.g., *Siphonaria*) stimulates protoplast release more than by others (e.g., *Chiton*). Perhaps, as was suggested by Santelices and Ugarte (1987), seaweed survival involves the action of functionally different types of digestive enzymes, some killing cells, others digesting the cell wall. The different abundance or levels of activity of each of these enzymes in the gastric juices of a given invertebrate could determine greater of lesser algal mortality or varying abundances of propagule production.

Evidence gathered in protoplast production studies supports the above hypotheses. Crude digestive juices from grazing molluscs and sea urchins contain mixtures of cellulases, fuconases, alginate lyases, and proteases (Liu *et al*. 1984; Boyen *et al*. 1990). Some of these compounds, such as cellulases and fuconases, have been very useful in studies of protoplast isolation, whereas others, such as proteases, lipases, and ribonucleases, are harmful to plant protoplasts (Boyen *et al*. 1990). Some of these harmful fractions may affect some algal species but not others. For example, partially purified alginate lyases from *Haliotis* are harmful to species of *Laminaria* but induce protoplast release in species of *Fucus, Sphacelaria*, and *Macrocystis* (Boyen *et al*. 1990). The abundance of harmful or non-harmful fractions may vary from one to another species of grazer or during different physiological states of a given grazer. All these variations may explain inter-grazer differences in capability to induce protoplast release and lead to the possibility of specific interactions between a given seaweed and a given grazer.

(c) The grazing process Data on digestion survival of algae and protoplast production suggest that several factors related to the digestive process may be particular important in determining abundance and diversity of the different types of algae surviving. The state of hunger of the herbivore at the time of feeding is one such factor. Generally,

fewer algal cells or fragments are likely to survive digestion in a hungry than in a recently-fed animal. Experimental data (Santelices and Paya 1989) indicate that regeneration frequency and propagule production are linear functions of the total abundance of algal fragments in the faecal pellets. Regeneration of the chlorophytan *Enteromorpha* sp., surviving through the digestive tract of the gastropod *Siphonaria lessonii*, could be as low as zero in faecal pellets containing less than 20 fragments. Above this threshold abundance, all pellets showed at least a few fragments able to regenerate, the quantity increasing linearly with increasing numbers of fragments in the pellets. The total density of new germlings arising from recently released protoplasts showed a similar trend. All cultures containing pellets with more than 40 algal fragments per pellet showed germling growth, density increasing with increasing number of fragments in the pellets.

Previous feeding also seems particularly important. Experimental studies of protoplast-production (Polne-Fuller and Gibor 1987) recommend feeding an invertebrate on a specific diet of a particular seaweed, before extracting its gut enzymes, as a means to increase the specificity of enzymes for that seaweed. This process also results in an increase in protoplast yield. These findings are consistent with what seem to be specialized examples of digestion survival of algae. In central Chile the amphipod *Hyale media* has a marked trophic preference for mature cystocarpic tissues of the middle–late successionist red alga *Iridaea laminarioides*. Although under laboratory conditions this invertebrate may consume fertile tetrasporic tissues of *I. laminarioides*, and or vegetative or reproductive tissues of a few other seaweeds (*Gymnogongrus, Ulva*), when given simultaneous offers, the amphipod's consumption is normally highly specific and restricted to cystocarpic walls and carpospores of *I. laminarioides* (Buschmann and Santelices 1987; Luxoro and Santelices 1989). Carpospores are the only type of reproductive unit of *I. laminarioides* to survive passage through the digestive tract of these amphipods, thus revealing a specialized relationship. More recently, Buschmann and Bravo (1990) documented germination of carpospores of *I. laminarioides* in faecal pellets of *Hyale hirtipalma*. The pellets were collected in the laboratory after experimental feeding of the amphipod with *I. laminarioides* and, in the field, in patches of *Iridaea* showing abundance of these amphipods.

Most grazers consume a mixed diet of algae. However, little is known about the effects of these mixed diets on the survival capacities of seaweeds. Laxative chemicals produced by the seaweed may speed up transit through the animal; mildly astringent or otherwise annoying chemicals in the frond may cause the invertebrate to consume small amounts of a given type of alga at a particular feeding and large amounts at other feedings; allelochemicals produced by some algal species in a mixed diet may affect the germination rates of the released protoplasts of other species. Even though as many as 500 to 600 secondary compounds are estimated (Hay and Fenical 1988) to have been isolated from marine algae, their ecological functions under natural conditions have been addressed only recently and none has yet been studied from the perspective of seaweed survival to digestion.

Ecological consequences of seaweed survival to digestion

1. Dispersal

The most obvious ecological consequence of seaweed resistance to digestion is dispersal of seaweed propagules. This is an especially important function because many seaweeds have both short-lived propagules and restricted (a few metres) dispersal shadows (see Santelices, 1990 for review). Given that several seaweed species can be grown in cultures started from faecal pellets excreted even 3 to 4 days after the last feeding by an invertebrate (Santelices et al. 1983), the short-distance dispersal effect of the grazer may be very significant. The importance of this phenomenon in the long-distance dispersal of seaweeds is more difficult to assess due to methodological problems involved in studying dispersal of microscopic propagules in marine conditions. However, since the largest dispersal shadow so far known for a seaweed propagule is no greater than 35 km (Amsler and Searles 1980; Zechman and Mathieson 1985), it is likely that highly mobile dispersers, such as fish, could also favour larger dispersal potential for seaweeds.

The studies referred to above and describing *Audouinella purpurea* growing in the field from faecal pellets of the amphipod *Gammarus salinus* in the northern Netherlands (Breeman and Hoeksema 1987) and also ulvoids and blue-green algae growing from pellets of the gastropod *Patelloida latistrigata* (Jernakoff 1985), both document the ecological roles of these herbivores as seaweed dispersers. Further, recent experimental evidence suggests that animal-mediated dispersal could help propagules to cross natural barriers that otherwise limit their dispersal. Knowing that faecal pellets of the amphipod *Hyale hirtipalma* contain live spores of the red seaweed *Iridaea laminarioides* and that this amphipod uses patches of *Ulva rigida* as homing and sheltering areas, Buschmann and Vergara (1992) placed coverslips in the field beneath the canopy of *U. rigida*. They found that the amphipod may disperse live carpospores of *I. laminarioides* via faecal pellets, transporting such propagules through the *Ulva* canopy which otherwise limits their settlement and attachment.

These experimental results suggest, in addition, that animal-generated propagule shadows are very likely not homogeneous, monotonically declining functions of distance from the parental plant. Existence of favoured defaecation sites, either as in the case of the above amphipods or as have been documented for fishes, probably introduces much heterogeneity in the grazer-mediated dispersal shadow of seaweeds. Likewise, specific differences in seaweed transit times through the herbivores, differences in animal mobility and presence or absence of homing or territorial behaviour, should all affect the resultant dispersal pattern.

2. Algal settlement and growth

The surviving seaweed propagules may be ejected to either intertidal or subtidal habitats. Experimental evidence suggests that algal propagules excreted alive in faecal pellets have some ecological advantages in both habitats when compared with free spores (Santelices and Paya 1980). When falling on intertidal bottoms, the propagules inside pellets are protected from desiccation by the sticky nature of the pellet. Significant differences in desiccation resistance exist between propagules of the same species (*Enteromorpha compressa*) depending on whether or not they are included in a faecal pellet. This protection is important because in intertidal habitats desiccation-tolerance can be of great significance in determination of the vertical distribution of several algal species (Edwards 1977; Schoenbeck and Norton 1978), germlings generally having more limited exposure-tolerance to desiccation than do adult plants.

The sticky surfaces of the pellets also allow them to remain attached to intertidal rocky surfaces, as described by Jernakoff (1985) for some ulvoids and by Breeman and Hoeksema (1987) for *Audouinella purpurea* (= *Rhodochorton purpureum*). Experimental comparisons of adhesive capacities of *Enteromorpha compressa* swarmers suggest conspicuous differences depending on whether or not they are included in a faecal pellet (Santelices and Paya 1989). The adhesive capacity of free spores decreases exponentially with increased desiccation, due to raised mortality among air-exposed swarmers. By contrast, adhesive capacity of faecal pellets increases significantly and exponentially to a maximum with air exposure. As desiccation increases, the more superficial parts of the pellet lose water and its mucilaginous envelope becomes firmly attached to the substratum. Some irreversible change in the mucilaginous layer seems to occur with increasing dehydration, as the pellets are not released from the substratum upon rehydration.

The very different patterns of adhesion of spores and pellets containing *Enteromorpha* tissues under air emersion conditions suggest two different strategies of attachment to the substratum for a given algal species, and two contrasting but perhaps complementary roles for free spores and for propagules inside pellets. Under short emersion times, spores can readily attach. As their survival decreases with increased desiccation, propagule attachment via faecal pellets becomes effective and comparatively more important.

When falling to subtidal habitats, the pellets sink much faster than free algal propagules. Faecal pellets of the small-sized grazing mollusc *Siphonaria lessonii* sink 8 to 22 times faster than the fastest sinking seaweed propagule and 40 to 100 times faster than common seaweed propagules (Santelices and Ugarte 1989). Thus, algal propagules inside pellets would remain for less time in the water column, reaching the bottom faster than propagules not so contained. It should be remembered that the problems faced by these propagules are, in general, the opposite of those faced by truly planktonic algae. While fast sinking of phytoplanktonic cells is disadvantageous because of removal from the euphotic zone, fast sinking of seaweed propagules is advantageous due to reduced viability with time and temporally limited fixation abilities of these reproductive bodies.

Sinking within a faecal pellet may also involve arriving together with other conspecific propagules at a given site. In some seaweed species this may be beneficial. For example, in species with heteromorphic alternation of generations and microscopic gametophytes, having spores sinking together increases the probability of gamete encounters and fertilization as gametophytes develop in close proximity. Density of gamete-producing individuals per unit area and the resulting distance between them are known to limit fertilization and sporophyte production in some kelps (e.g., *Macrocystis*, North 1971). Species such as *Pelagophycus* and *Nereocystis* release fertile segments of their blades rather than individual spores (Coon *et al.* 1972); these blades have the capacity to sink quickly, thus releasing large number of spores which generate gametophytes in close proximity. Such a role could also be played by fragments of sporophylls contained in pellets. Interestingly, Tugwell and Branch (1989) recently reported very low polyphenol levels in sporogenous tissues of South African populations of *Laminaria pallida*, as compared with equivalent tissues in the local representatives of *Macrocystis angustifolia* and *Ecklonia maxima*. Given that spores remain *in situ* on *L. pallida* for a significantly longer period than on the other two species, relative availability to herbivores could not be invoked as an explanation for the low polyphenol levels. Tugwell and Branch (1989) suggested that the spores of *L. pallida* may be resistant to or may escape digestion, increasing their dispersal by grazer activity. Perhaps the grazers also facilitate germination of gametophytes in close proximity.

All experiments so far performed with seaweed spores indicate that germination levels of ingested spores are significantly lower than for uningested spores, thus pointing to some mortality or reduced viability due to digestion. However, most ingestion-surviving tissues and some types of spores have shown, in laboratory cultures, growth rates and regeneration capabilities significantly faster than non-ingested control thalli. The nature of any stimulatory factors contained in the faecal pellets and enhancing seaweed growth is still unknown. Perhaps these correspond with substances that can be used by seaweeds as nutrients or growth factors.

3. Quantitative contribution of the propagules surviving digestion

Given the abundance of uningested algal propagules in the water column, it is important to estimate the quantitative contribution made by algal propagules that survive digestion by grazers. Such quantification is difficult and somewhat uncertain as it involves combining laboratory data (number of propagules released in faecal pellets) with field data (number of pellets produced by an average grazer density). Further, many of these comparisons involve extrapolations of results gathered from mm^2 or cm^2 areas in culture dishes to m^2 of rocky surface in the field, and the results have to be aligned with highly variable phenomena such as spore production and settlement. Such comparisons remain necessary to evaluate ecological importance and relative contribution of algal digestion-survival to propagule supply at a given site.

These comparisons have involved quantification of grazer densities at a given site, pellet production per grazer per unit time, average number of algal fragments, fragment area per pellet, and average number of cells and propagules produced per fragment area. Results (Santelices and Paya 1989) indicate that, on average, the quantities of propagules produced by some species compare well with the recorded abundance of propagules settling in experimental plots in the field and could constitute significant additional source of propagules, especially for opportunist taxa. However, in many habitats, density, and biological activity of herbivores fluctuate in space and time. Thus, the quantitative importance of seaweed digestion survival in acting as an additional source of propagules is variable. It should be especially important at sites and seasons in which herbivore consumption is intense. It is precisely at those sites and seasons that the resources allocated to seaweed reproduction are reduced through direct consumption by the herbivore. Thus, digestion survival of seaweed tissues and propagules appears to form a process directly complementing the free reproductive capabilities of seaweeds.

The evolutionary nature of the relationship

All the observational and experimental data gathered so far tend to characterize survival of algae to grazer-digestion as a coarse, diffuse adjustment between seaweed and herbivores in benthic habitats. None of the seaweed responses, including regeneration after fragmentation and cell and protoplast release, is unique to seaweed–herbivore interactions. None of the seaweed species so far studied relies exclusively on grazer-mediated dispersal mechanisms. None of the grazers characterized as disperser consumes exclusively one species or one type of seaweed. Further, survival to digestion as a process depends on a number of factors (discussed previously) some of which seem to have certain random components. The most specialized relationship known to date seems to be the interaction between some amphipods and some Rhodophyta, where the invertebrate shows a marked feeding preference for the algal spores which, in turn, depend to some extent on the feeding activities of the amphipod for release and dispersal.

Several alternative explanations could be advanced for this apparent lack of close evolutionary relationship. Perhaps the general and specific constraints known in terrestrial systems to reduce the likelihood of close coevolution between particular plants and dispersers (Wheelwright and Orians 1982; Howe 1984; Janzen 1984; Herrera 1985; Jordano 1987) are also operating in these benthic systems. Alternatively, perhaps truly specialist marine herbivores have not yet, or only inadequately, been studied. It should be remembered, however, that specialist herbivores are scarce in marine habitats (Lubchenco and Gaines 1981; Hawkins and Hartnoll 1982; Steneck 1982). The few specialists known often tend also to have limited mobility and to live on vegetative tissues of the seaweeds that they consume (e.g., limpets on kelps). Therefore, they could be expected to be rather poor dispersers. Perhaps a closer relationship may be found in the ecological situations known as gardens (Branch 1975, 1976; Branch and Griffiths 1988), but these interactions remain unexplored from the perspective of algal propagules surviving digestion.

It is concluded, therefore, that as far as we know today, the evolutionary importance of digestion survival of seaweeds lies in the possibilities of reducing the damage done to the seaweed, transforming grazing into dispersal. Because herbivory has been one of the important selective factors in the evolutionary history of seaweeds, algal adaptations that may reduce or counterbalance damage resulting from herbivory should be particularly important.

Acknowledgements

This study was suppported by Dirección de Investigación, Pontificia Universidad Católica de Chile (Grants DIUC E-20 and 8903-E), Fundación Andes (Convenio C-51284) and Fondo Nacional de Investigación Científica y Tecnológica-Chile (FONDECYT 803/90). My gratitude goes to all these institutions for their support. I thank Drs G. Collantes, F. Jaksic, A. Hoffmann, and M. Kalin-Arroyo for critically reading the manuscript and A. Buschmann for letting me use his studies, currently in press.

References

Amsler, C. D. and Searles, R. B. (1980). Vertical distribution of seaweed spores in a water column offshore of North Carolina. *Journal of Phycology*, **16**, 617–19.

Boyen, C., Kloareg, B., Polne-Fuller, M., and Gibor, A. (1990). Preparation of alginate lyases from marine molluscs for protoplast isolation in brown algae. *Phycologia*, **29**, 173–81.

Branch, G. M. (1975). Mechanisms reducing intraspecific competiton in *Patella* spp.: migration, differentiation and territorial behaviour. *Journal of Animal Ecology*, **44**, 575–600.

Branch, G. M. (1976). Intraspecific competition experienced by South African *Patella* species. *Journal of Animal Ecology*, **45**, 507–29.

Branch, G. M. (1981). The biology of limpets: physical factors, energy flow and ecological interactions. *Oceanography and Marine Biology Annual Review*, **19**, 235–380.

Branch, G. M. and Griffiths, C. L.(1988). The Benguela ecosystem. Part V. The coastal zone. *Oceanography and Marine Biology Annual Review*, **26**, 395–486.

Breeman, A. M. and Hoeksema, B. W. (1987). Vegetative propagation of the red alga *Rhodochorton purpureum* by means of fragments that escape digestion by herbivores. *Marine Ecology Progress Series*, **35**, 197–201.

Buschmann, A. and Bravo, A. (1990). Intertidal amphipods as potential dispersal agents of carpospores of *Iridaea laminarioides* (Gigartinales: Rhodophyta). *Journal of Phycology*, **26**, 417–20.

Buschmann, A. and Santelices, B. (1987). Micrograzers and spore release in *Iridaea laminarioides* Bory (Rhodophyta, Gigartinales). *Journal of Experimental Marine Biology and Ecology*, **108**, 171–9.

Buschmann, A. and Vergara, P. A. (1992). A field study on the role of rocky

intertidal amphipods as algal spore dispersal agents in southern Chile. *Journal of Phycology*, **27**. (In press).
Buttler, D. M., Ostgaard, K., Boyen, C., Evans, L. V., Jensen, A., and Kloareg, B. (1989). Isolation conditions for high yields of protoplasts from *Laminaria saccharina* and *L. digitata* (Phaeophyceae). *Journal of Experimental Botany*, **40**, 1237–46.
Cancino, J. M. and Castilla, J. C. (1988). Emersion behaviour and foraging ecology of the common Chilean clingfish *Sycases sanguineus* (Pisces: Gobiosocidae). *Journal of Natural History*, **22**, 249–261.
Cheney, D. P., Mar, E., Saga, N., and van der Meer, J. (1986). Protoplast isolation and cell division in the agar producing seaweed, *Gracilaria* (Rhodophyta). *Journal of Phycology*, **22**, 238–43.
Cloakie, J. J. P. and Norton, T. A. (1974). The effects of grazing on algal vegetation of pebbles from the Firth of Clyde. *British Phycological Journal*, **9**, 216.
Cobo, M. and Andreu, A. C. (1988). Seed consumption and dispersal by the spurthighed tortoise *Testudo graeca*. *Oikos*, **51**, 267–73.
Coon, D., Neushul, M., and Charters, A. C. (1972). The settling behavior of marine algal spores. *Proceedings of the International Seaweed Symposium*, **7**, 237–42.
Duffy, J. E. and Hay, M. E. (1990). Seaweed adaptations to herbivory. *BioScience*, **40**, 368–75.
Edwards, P. (1977). An investigation of the vertical distribution of selected marine algae with a tide-simulating machine. *Journal of Phycology*, **13**, 62–8.
Farmanfarmaian, A. and Philips, J. H. (1962). Digestion, storage and translocation of nutrients in the purple sea urchin (*Strongylocentrotus purpuratus*). *Biological Bulletin (Woods Hole, Massachusetts)*, **123**, 105–20.
Floc'h, J. Y., Deslandes, E., and Le Gall, Y. (1987). Evidence for vegetative propagation of the carrageenophyte *Solieria chordalis* (Solieriaceae, Rhodophyceae) on the coast of Brittany (France) and in culture. *Botanica Marina*, **30**, 315–21.
Gibor, A. (1956). Some ecological relationships between phyto and zooplankton. *Biological Bulletin (Woods Hole, Massachusetts)*, **111**, 230–4.
Hawkins, S. J. and Hartnoll, R. G. (1983). Grazing of intertidal algae by marine invertebrates. *Oceanography and Marine Biology Annual Review*, **21**, 195–282.
Hay, M. E. and Fenical, W. (1988). Marine plant–herbivore interactions: the ecology of chemical defense. *Annual Review of Ecology and Systematics*, **19**, 111–45.
Herrera, C. M. (1985). Determinants of plant–animal coevolution: the case of mutualistic dispersal of seeds by vertebrates. *Oikos*, **44**, 132–41.

Howe, H. F. (1984). Constraints on the evolution of mutualisms. *American Naturalist*, **123**, 764-77.

Janzen, D. H. (1983). Dispersal of seeds by vertebrate guts. In *Coevolution*, (ed. D. Futuyma and M. Slatkin), pp. 232-62. Sinauer, Sunderland.

Jernakoff, P. (1985). Interactions between the limpet *Patelloida latistrigata* and algae on an intertidal rock platform. *Marine Ecology Progress Series*, **23**, 71-8.

Jordano, P. (1987). Patterns of mutualistic interactions in pollination and seed dispersal: connectance, dependence, asymmetries and coevolution. *American Naturalist*, **129**, 657-77.

Kloareg, B., Polne-Fuller, M., and Gibor, A. (1989). Mass production of viable protoplasts from *Macrocystis pyrifera* (L.) C. Ag. (Phaeophyta). *Plant Science*, **69**, 105-12.

Lasker, R. and Giese, A. (1954). Nutrition of the sea urchin, *Strongylocentrotus purpuratus*. *Biological Bulletin (Woods Hole, Massachusetts)*, **106**, 328-40.

Liu, W. S., Tang, Y. L., Liu, X. W., and FANG, T. C. (1984). Studies on the preparation and on the properties of sea snail enzymes. *Hydrobiologia*, **116/117**, 319-20.

Lobel, P. S. (1980). Herbivory by damselfishes and their role in coral reef community ecology. *Bulletin of Marine Sciences*, **30**, 273-89.

Lubchenco, J. and Gaines, S. D. (1981). A unified approach to marine plant-herbivore interactions. I. Population and communities. *Annual Review of Ecology and Systematics*, **12**, 405-37.

Luxoro, C. and Santelices, B. (1989). Additional evidence for ecological differences among isomorphic reproductive phases of *Iridaea laminarioides* (Rhodophyta: Gigartinales). *Journal of Phycology*, **25**, 206-12.

McKey, D. (1975). The ecology of coevolved seed dispersal systems. In *Coevolution of animal and plants*, (ed. L. E. Gilbert and P. H. Raven), pp. 159-191. University of Texas Press, Austin, Texas.

Newell, R. C. (1979). *The biology of intertidal animals* (3rd edn). Marine Ecology Surveys Ltd., Faversham, Kent, UK.

Nicotri, M. E. (1977). Grazing effects of four marine intertidal herbivores on the micro-flora. *Ecology*, **58**, 1020-32.

Norton, T. A. and Mathieson, A. C. (1983). The biology of unattached seaweeds. In *Progress in Phycological Research*, Vol. 2, (ed. F. E. Round and D. J. Chapman), pp. 333-86.

North, W. J. (1971). Introduction and background. In *The biology of giant kelp beds (Macrocystis) in California* (ed. W J. North). *Nova Hedwigia*, No. 32, pp. 1-97.

Paya, I. and Santelices, B. (1989). Macroalgae survive digestion by fishes. *Journal of Phycology*, **25**, 186-8.

Pearlmutter, N. L. and Vadas, R. L (1978). Regeneration of thallus fragments of *Rhodochorton purpureum* (Rhodophyceae, Nemalionales). *Phycologia*, **17**, 186-190.

Polne-Fuller, M. and Gibor, A. (1984). Developmental studies in *Porphyra*. Blade differentiation in *Porphyra perforata* as expressed by morphology, enzymatic digestion and protoplast regeneration. *Journal of Phycology*, **20**, 609-16.

Polne-Fuller, M. and Gibor, A. (1987). Tissue culture in seaweeds. In *Developments in aquaculture and fisheries science. 16. Seaweed cultivation for renewable resources*, (ed. K. T. Bird and P. M. Benson), pp. 219-39. Elsevier, Amsterdam.

Porter, K. G. (1973). Selective grazing and differential digestion of algae by zooplankton. *Nature*, **244**, 179-80.

Porter, K. G. (1977). The plant-animal interface in freshwater ecosystems. *American Scientist*, **65**, 159-70.

Prim, P. and Lawrence, J. M. (1975). Utilization of marine plants and their constituents by bacteria isolated from the gut of echinoids (Echinodermata). *Marine Biology*, **33**, 167-73.

Russell, G. (1967). The ecology of some free-living Ectocarpaceae. *Helgoländer Wissenschaftliche Meeresuntersuchungen*, **15**, 155-62.

Santelices, B. (1990). Patterns of reproduction, dispersal and recruitment in the seaweed. *Oceanography and Marine Biology Annual Review*, **28**, 177-276.

Santelices, B. and Correa, J. (1985). Differential survival of macroalgae to digestion by intertidal herbivore molluscs. *Journal of Experimental Marine Biology and Ecology*, **38**, 183-91.

Santelices, B. and Martínez, E. (1988). Effects of filter-feeders and grazers on algal settlement and growth in mussel beds. *Journal of Experimental Marine Biology and Ecology*, **118**, 281-306.

Santelices, B. and Paya, I. (1989). Digestion survival of algae: some ecological comparisons between free spores and propagules in fecal pellets. *Journal of Phycology*, **25**, 693-9.

Santelices, B. and Ugarte, R. (1987). Algal life-history strategies and resistance to digestion. *Marine Ecology Progress Series*, **35**, 267-75.

Santelices, B., Correa, J. and Avila, M. (1983). Benthic algal spores surviving digestion by sea urchins. *Journal of Experimental Marine Biology and Ecology*, **70**, 263-9.

Schoenbeck, M. W. and Norton, T. A. (1987). Factors controlling the upper limits of fucoid algae on the shore. *Journal of Experimental Marine Biology and Ecology*, **31**, 303-13.

Steneck, R. S. (1982). A limpet-coralline algal association: adaptations and defences between a selective herbivore and its prey. *Ecology*, **63**, 507-22.

Steneck, R. S. and Watling, L. (1982). Feeding capabilities and limitations of herbivorous molluscs: a functional group approach. *Marine Biology*, **68**, 299-319.

Tokuda, H. and Kawashima, Y. (1988). Protoplast isolation and culture of a brown alga *Undaria pinnatifida*. In *Algal biotechnology*, (ed. T. Stadler), pp. 151-9. Elsevier Applied Science, London and New York.

Tugwell, S. and Branch, G. M. (1989). Differential polyphenolic distribution among tissues in the kelps *Ecklonia maxima, Laminaria pallida* and *Macrocystis angustifolia* in relation to plant-defense theory. *Journal of Experimental Marine Biology and Ecology*, **129**, 219-30.

Wheelwright, N. T. and Orians, G. H. (1982). Seed dispersal by animals: contrast with pollen dispersal, problems of terminology, and constraints on coevolution. *American Naturalist*, **119**, 402-413.

Zechman, F. W. and Mathieson, A. C. (1985). The distribution of seaweed propagules in estuarine coastal and offshore waters of New Hampshire, USA. *Botanica Marina*, **28**, 283-94.

17. Role of algae in the recruitment of marine invertebrate larvae

AILEEN N.C. MORSE
Marine Science Institute, University of California, Santa Barbara, California, USA

Abstract

Recruitment of marine invertebrate larvae is a dynamic feature of the structuring of shallow-water marine communities. It has recently been recognized that a clear understanding of early settlement events is essential for a more complete assessment and prediction of the effect of recruitment on the structure of these communities. The availability and nature of suitable settlement and recruitment substrata must be considered as important factors in this process. Both direct and indirect evidence suggest that algae provide critical substrata for the recruitment of a number of invertebrate larvae in the form of essential settlement cues, space, shelter from predators, and a reliable food supply for further development and growth. The wide distributions (and in many instances extensive stands) of a variety of algal species apparently involved in these recruitment processes serve to emphasize the important contribution of algae to overall structuring of shallow-water benthic communities. Where possible, the roles of algae as suitable settlement substrata are discussed in relation to other factors perceived as important to the recruitment processes of marine invertebrate larvae.

Introduction

The need to re-examine recruitment of sessile marine invertebrates in the light of factors which contribute to variation in initial settlement events has been recently more widely recognized (Dayton 1979; Caffey 1985; Underwood and Denley 1984; Watanabe 1984). It has been

Plant–Animal Interactions in the Marine Benthos, (ed. D.M. John, S.J. Hawkins, and J.H. Price), Systematics Association Special Volume No. 46, pp. 385–403. Clarendon Press, Oxford, 1992. © The Systematics Association 1992.

suggested that processes which effect larval recruitment on marine hard substrata may be determined in part by variation in larval delivery to a particular site (Gaines *et al.* 1985) and, on very small scales, by the availability of unoccupied settlement space (Gaines and Roughgarden 1985). Connell (1985) recently reviewed the possible effects of initial variation in settlement of marine invertebrate larvae (specifically barnacles) on their later distribution and abundance. He concluded that densities of recruits may be direct indicators of initial settlement if mortality between the two stages is density-independent. Moreover, the available data collected at different sites suggest that mechanisms affecting recruitment at a particular site are fairly consistent and that it should be expected that variation in recruitment patterns between sites will be largely environment-specific.

It is well recognized that for many marine invertebrates selection of an initial settlement (*sensu* Keough and Downes 1982) site is not random and passive but rather an active process. Planktonic larvae of most specialist sessile invertebrates have evolved complex mechanisms which ensure correct selection of specific benthic substrata on which to settle, partially or completely attach, and metamorphose to the early juvenile stages (Crisp 1974; Chia and Rice 1978).

Settlement and metamorphosis in a variety of marine invertebrates have been demonstrated in laboratory experiments to be controlled by larval sensory recognition of exogenous mechanical, chemical, and other environmental stimuli, associated with different kinds of settlement surfaces (for reviews see Crisp 1974, 1984; Scheltema 1974; Hadfield 1978, 1986; Chia 1978; Burke 1983, 1986; A. Morse, 1985; D. Morse, 1985, 1990; Rittschof and Bonaventura 1986). In many of these instances, settlement cues, such as chemical cues associated with algae, are substratum-specific. Laboratory studies have provided information on the role of several groups of algae in the settlement and metamorphosis of a number of important invertebrate species. Field studies on recruitment of several of these species suggest that a number of criteria identified in laboratory studies of settlement may have important consequences for field recruitment.

Several groups of algae, including representatives of the Rhodophyta, Phaeophyta, and Chlorophyta, have been identified as potential sources of invertebrate larval settlement cues. Non-geniculate coralline red algae provide one of the most important sources of settlement cues among the algae investigated (Barnes and Gonor 1973; D. Morse *et al.* 1979, 1988; Heslinga 1981; Sebens 1983*b*; D. Morse 1984; Morse and Morse 1984*a*; A. Morse 1988; Rowley 1989; Pearce and Scheibling 1990). As suggested earlier (Morse and Morse 1984*a*), this is not too surprising for, unlike other fleshy red, brown or green macroalgae (generally ephemeral), the non-geniculate coralline algae persist (unless subjected to extreme conditions) as permanent encrusting forms cemented to benthic hard substrata. They are represented by numerous genera, widely distributed (and often dominant) in all the oceans (cf. Johansen 1981). Several fleshy red macrophytes have been identified as settlement substrata important for large herbivorous gastropods (Kriegstein *et al.* 1974; Switzer-Dunlap and Hadfield 1977; Davis *et al.* 1990). Numbers of *Sargassum* spp. have been reported to support extensive colonies of hydroids; these associations, involving large numbers of flora and fauna, are fairly species–species specific (Nishihira 1965). Their role in promoting settlement has been extensively documented (e.g., Nishihira 1968; Kato *et al.* 1975). Fucoids have been reported to be important in settlement of spirorbids (cf. Williams 1964). There are several reports on the possible role of other phaeophytes as both specialized habitats and recruitment substrata for colonization by a number of mobile invertebrates such as amphipods, copepods and gastropods (e.g., Gunnill 1984; Watanabe 1984; see Williams and Seed, chapter 9 this volume). So far, data available for green algae are limited to a single genera–species interaction (Switzer-Dunlap and Hadfield 1977).

Inducers of larval settlement and metamorphosis associated with algae

1. Non-geniculate coralline algae

Species exhibiting selective settlement of larvae on coralline algae in the laboratory include the tubeworm *Spirorbis rupestris* (Gee 1965); the chitons *Tonicella lineata* (Barnes and Gonor 1973), *Mopalia muscosa* (D. Morse *et al.* 1979), and *Katharina tunicata* (Rumrill and Cameron 1983); the asteroids *Acanthaster planci* (Henderson and Lucas 1971) and *Stichaster australis* (Barker 1977); the echinoderms *Strongylocentrotus purpuratus* (Rowley 1989) and *S. droebachiensis* (Pearce and Scheibling 1990); the temperate soft coral *Alcyonium siderium* (Sebens 1983b); the scleractinian corals *Agaricia humilis*, *A. tenuifolia*, and *A. agaricites* (D. Morse *et al.* 1988); the gastropod mollusc *Trochus niloticus* (Heslinga 1981); the gastropod mollusc *Haliotis rufescens* (D. Morse *et al.* 1979; Morse and Morse 1984a) and eleven of its congeners (cf. D. Morse 1984). The data obtained from these controlled laboratory experiments demonstrate that specificity of larval recognition of a settlement cue is high, except in the two echinoderm species. The fact that these echinoderms also settle in response to other types of substrata indicates that their specificity is relatively relaxed. Stringency of requirement for an exogenous cue is, however, high for all of these species.

These studies have examined several factors involved in the process:

(1) differential settlement rates and responses of larvae to different substrata;
(2) factors associated with particular algae apparently responsible for induction of the larval settlement response;
(3) the potential for differential post-settlement survival on different habitat-specific substrata, in the absence of disturbance (although very few studies present data for longer than a week after settlement);
(4) differential mortality in the presence of habitat-specific substrata likely to be encountered by planktonic larvae during their benthic explorations; and
(5) the consequences of delayed metamorphosis in the absence of a suitable substratum.

For a number of the species examined, larval chemosensory mechanisms facilitate recognition of a required algal substratum. There have been very few indications that texture of an algal surface is important for such species. Although larval contact with the substratum is generally required for the initiation of settlement processes *in situ*, molecules extracted into solution from these algae alone are sufficient to induce larval settlement, attachment, and normal developmental metamorphosis. Several of these inducers have been biochemically characterized.

The chemical nature of the settlement cue of larvae of the Pacific red abalone, *Haliotis rufescens*, and the molecular processing of this cue by the larva, have been the most extensively investigated and characterized. In the laboratory, *H. rufescens* larvae demonstrate specific settlement on any of a number of non-geniculate corallines commonly found within juvenile and adult habitats in the field (D. Morse *et al.* 1979, 1980c; Morse and Morse 1984a) as well as on other genera in this group collected world-wide (cf. Morse and Morse 1984a; A. Morse 1988). Larvae continued their normal larval swimming behaviour in the presence of seawater alone, clean inert substrata, marine bacteria, diatoms and microalgae, animal cohabitants found in adult and juvenile habitats (D. Morse *et al.* 1979, 1980c), representative red, brown, and green macroalgae from the same habitats (D. Morse *et al.*, 1979, 1980c; Morse and Morse 1984a, 1984b), and several strains of Cyanobacteria (blue-green algae) (A. Morse *et al.* 1984). Larvae are also induced to settle in response to water-soluble extracts of these algae. Inducers have been biochemically characterized and purified. They are small peptides with unusual composition, complexed to proteins at the surfaces of coralline algae (Morse and Morse, 1984a, 1984b; A. Morse et al. 1984; A. Morse 1985, 1988).

This chemosensory induction of *Haliotis* larvae at algal surfaces is contact dependent (D. Morse et al. 1980c; Morse and Morse 1984a). Observation (with the microscope) of the settling process reveals that a single random contact event with a coralline surface initiates the beginning of settlement behaviour; competent larvae do not resume their planktonic swimming behaviour (personal observations). Experimental demonstration of the unique availability of inducers complexed to coralline surfaces for larval detection has been hypothesized to determine, at least in part, the substratum-specific recruitment of *H. rufescens* larvae from the plankton (Morse and Morse 1984a).

The findings that 12 species (distributed world-wide) of the genus *Haliotis* are induced by a number of different coralline species, as well as by algal extracts and by the mammalian neurotransmitter gamma-aminobutyric acid GABA (cf. D. Morse 1984), suggest that members of this genus all recognize the same class of chemical cue. As has been shown for one congener, *H. rufescens*, this class of molecule seems to be ubiquitously distributed in non-geniculate coralline algae (cf. Morse and Morse 1984a; A. Morse 1988). When this is considered alongside field observations by Tegner and Dayton (1977) (see D. Morse et al. 1980c) that young *H. rufescens* recruits are found exclusively on coralline algae, and recent controlled field experiments which have demonstrated early recruitment only on coralline algae (*H. discus hannai*: Saito 1981; *H. laevigata* and *H. scalaris*: Shepherd and Turner 1985, Shepherd 1990; *H. cracherodii*: Douros 1985; *H. rubra*: Prince et al. 1987), it seems reasonable to hypothesize that members of this genus show the same specificity of settlement cue recognition. Laboratory experiments (D. Morse et al. 1980c) and field studies (Shepherd and Turner 1985; Douros 1985; Prince et al. 1987) both exclude any significant contribution by either juvenile or adult conspecifics to recruitment.

The stringency and specificity of *Haliotis rufescens* persists for several weeks (cf. D. Morse et al. 1979, 1980a), thus enhancing its capability for dispersal. Estimates have been made for the settlement consequences for this species as it is carried by different water masses with different velocities in the vicinity of kelp beds (Jackson 1986).

Gee (1965) reported the inducer of *Spirorbis* settlement to be a low molecular-weight compound present in an aqueous extract of *Lithothamnion* sp. Larvae were induced to settle on experimental panels which had been soaked in an extract of this alga. Similar results were found with inducers of *Tonicella lineata* (Barnes and Gonor 1973) present in extracts of *Lithothamnion* sp. and *Lithophyllum* sp., except that the inducer was much larger (c. 60–100 000 daltons). This difference, however, may be largely an effect of differential extraction procedures, as was demonstrated with the well-characterized inducer of *Haliotis rufescens* (Morse and Morse 1984a; A. Morse *et al.* 1984). *Tonicella* larvae responded to this macromolecular inducer after it had been given time to adsorb on to slate panels. The idea, current at the time, that an inducer must be adsorbed to a surface to facilitate larval recognition of the substance (Crisp and Meadows 1963) almost certainly dictated the design of these experiments. It was only after the discovery by D. Morse *et al.* (1979) that *H. rufescens* larvae could be induced to settle in response to an algal extract in solution, and that this response could be mimicked by a defined chemical (GABA), that the larval response of *Tonicella* was re-examined. D. B. Bonar (unpublished observations) found that larvae were induced to settle in the presence of GABA, suggesting that these larvae can respond to a soluble inducer. Similarly, both extracts of *Lithothamnion* spp. and GABA induce larvae of *Trochus niloticus* (Heslinga 1981; Heslinga and Hillman 1981) and of all the *Haliotis* species thus far tested (cf. D. Morse 1984, 1990). *Mopalia muscosa* was induced by intact coralline algae and by GABA (D. Morse *et al.* 1979). The specificity of substratum selection exhibited by *Katharina tunicata* is very similar to that seen with *H. rufescens*, except that ciliary activity in this species seems to be sensitive to GABA (Rumrill and Cameron 1983). There are no reports of studies of recruitment for either of these species. The grazing activities of *K. tunicata* have been linked to the persistence of established coralline algal species (Paine 1980). Rumrill and Cameron (1983) reported that they collected adults of *K. tunicata* from coralline algae-encrusted rocks. The evidence obtained from laboratory experiments on preferential settlement suggests that *M. muscosa* and *K. tunicata* may also recruit specifically to corallines in the field.

The results of these experiments demonstrate that a group of invertebrates, found in association with coralline algae in the natural environment, responds similarly to molecules extracted from these algae and to pure GABA. This suggests that larvae of all these species may be responding to the same or similar settlement cues present as a unique class of molecules in coralline algae. The available data on the characterization of the algal inducers identified in the above studies, alone, would not be sufficient to support this hypothesis. In *Haliotis rufescens*, the larval receptors controlling settlement and metamorphosis exhibit a fairly stringent stereochemical requirement for activation (see D. Morse *et al.* 1979, 1980b). A number of criteria (for review, see D. Morse 1990) support the hypothesis that these receptors are a unique class of GABA receptors similar to, but pharmacologically distinct from, any of the known GABA receptors found in the mammalian brain and CNS. These receptors have recently been radioactively labelled. The fact that both GABA and the natural inducer purified from *Lithothamnion* sp. compete specifically for binding of this label to these receptors demonstrates that both GABA and the natural inducer are acting at the same larval receptors. The receptors have been identified, by other criteria, to be the larval receptors controlling metamorphosis (Trapido-Rosenthal and Morse 1986). These data thus invoke the hypothesis that other invertebrate larvae, which are induced to settle specifically in response both to GABA and to extracts of coralline algae, may be responding to the same or similar chemical cues found in this group of algae; and that the larval processes and mechanism for detection of settlement cues in these species may be very similar, if not the same (D. Morse, 1985; A. Morse 1988).

This clearly does not apply to settlement of echinoderms. The results of settlement experiments in the laboratory with larvae of *Strongylocentrotus purpuratus* demonstrated that settlement on bare rock was significantly lower and much more variable than on either coralline algae or algal turf substrata. Larvae, however, did not show a preference for either of these algal surfaces; initiation of metamorphosis was delayed on turf (1–3 hours) compared with that observed on coralline algae (just minutes) (Rowley 1989). Although larvae settled in response to the algal peptide inducer of *Haliotis rufescens* larvae (cf. A. Morse *et al.* 1984), the response was not concentration dependent, indicating that a receptor–ligand interaction is not involved. Field sampling of newly settled recruits in the field, in this same study, revealed that the densities of recruits were similarly high on coralline algae plus rock from the 'barrens' and kelp bed, and on red algal turf from the kelp bed, but marginally lower on rock. *Strongylocentrotus franciscanus* showed similar size–frequency patterns in the field, but at much lower densities.

The green sea urchin, *Strongylocentrotus droebachiensis*, was recently reported to settle and metamorphose in response to three species of coralline algae, and to algal extracts prepared from these algae, as well as partially to GABA, but only at very high (non-physiological) concentrations (Pearce and Scheibling 1990). Subsequent experiments, testing a wide range of other possible settlement substrata, indicated that this species does not significantly discriminate between algae and many of the other substrata tested (Pearce, pers. comm.). Results of field and laboratory experiments with *S. droebachiensis* suggest that recruitment to coralline algal substrata may result in reduced post-settlement growth rates compared with the availability of kelp as a food source (Raymond and Scheibling 1987). These studies support previous findings with the same and other echinoderm species (Cameron and

Hinegardner 1974; Cameron and Schroeter 1980) which indicate that specificity of substratum-selection is relaxed.

Recent evidence suggests that some non-geniculate coralline algae may harbour more than one class of inductive chemical cue. Three shallow-water Caribbean agariciid corals settle specifically in response to a water-insoluble fraction associated with algal cell walls. They all have a strict requirement for an environmental cue. Evidence suggests that this inducer is a cross-linked polysaccharide (D. Morse *et al.* 1988). These larvae settle neither in response to the peptide inducer of *Haliotis rufescens* prepared from either Pacific or Caribbean algae, nor in response to GABA. Also, further experiments revealed that metamorphosis in these corals is controlled by a different pathway of signal transduction. Moreover, unlike *Haliotis* larvae, which respond equally well to all species of corallines, larvae of *Agaricia humilis* discriminate between different algae commonly found within the adult range. Preliminary evidence suggests that this is similarly so for *A. tenuifolia* and *A. agaricites*. This specificity of discrimination was found to be independent of the (randomly selected) larval batch. The percentages of individuals metamorphosed on coralline algae in the laboratory were not significantly different from the percentages of young recruits distributed on these algae in the field for either *A. humilis* or *A. tenuifolia*. Larval metamorphosis and recruitment of *A. humilis* favoured substrata in direct light, whereas these of *A. tenuifolia* were highly specific for shade.

These agariciids have been reported to be among the most important opportunistic colonizers of shallow reefs throughout the Caribbean (Bak and Engel 1979; van Moorsel 1989). Controlled field experiments on recruitment of *A. humilis* and *A. agaricites* (van Moorsel 1989) revealed that no recruits appeared on settlement plates until various coralline algae became established on those plates. The spatial distribution of corallines affected recruitment; density of algal cover was directly correlated with density of recruitment.

Sebens (1983*a*, *b*) reported similar findings for the temperate octocoral *Alcyonium siderium*, although in this case there was not such strict correlation between dependence on coralline substrata in field and laboratory. However, metamorphosis in the laboratory was unaffected by GABA and strongly influenced by light. In all these species, larvae can delay metamorphosis for up to 10 days without loss of specificity or requirement for a particular cue, although the data suggest that *Agaricia* may be less affected by delay than *Alcyonium*.

2. Other Rhodophyta

Veligers of the opisthobranch mollusc *Aplysia californica* preferentially crawled on *Laurencia pacifica* when offered a variety of other red algae (species of *Plocamium, Laurencia, Polysiphonia, Dasya, Chondrus*) and a green alga (*Ulva* sp.); they metamorphosed only on *Laurencia* (Kriegstein *et al.* 1974). The requirement for an exogenous cue was demonstrated to be stringent. Metamorphosis can be delayed for up to four weeks without loss of stringency or specificity of requirement (although only *Laurencia* was offered). Young individuals in the field were found almost exclusively on *Laurencia* and *Plocamium* (Audesirk 1979). Laboratory experiments have demonstrated the dietary advantages of settlement on *Plocamium*. Acquisition of anti-predatory defences and enhancement of growth were found to be correlated with specialization on this alga (Pennings 1990). Several species of red algae (*Callithamnion halliae, Laurencia* sp., *Chondrococcus hornemanni*) and a green alga (*Ulva* sp.) induce species-specific settlement of several other herbivorous opisthobranchs (Aplysiidae) (Switzer-Dunlap and Hadfield 1977).

Veligers of the queen conch *Strombus gigas* are induced to settle and metamorphose in response to intact *Laurencia poitei* and to a macromolecular extract of this alga (Siddall 1983). The fact that this process can be mimicked by KCl (Davis *et al.* 1990) indicates that an externally accessible chemosensory membrane is involved in the processing of the environmental stimulus in this species (cf. Yool *et al.* 1986).

3. Phaeophyta

Spirorbis borealis occurs most abundantly on *Fucus serratus* and shows a significant preference for this alga in laboratory studies. Williams (1964) demonstrated that an extract of this alga adsorbed to a clean surface was all that was required for induction of larval metamorphosis. The larval response was both time- and concentration-dependent. There was no attraction of larvae towards extracts in solution, neither were larvae able to respond to a solution of this extract in seawater; induction of settlement was strictly dependent upon larval detection of this inducer on or very close to a surface. Contact with this chemical cue affected the free-swimming and exploratory phases, and the phototrophic response of larvae. Settlement rates in the presence of the chemical cue varied with larval age although, significantly, discrimination of the required settlement cue did not diminish with age. Although the cue was only very crudely characterized biochemically, chemoreception as a mechanism for induction was clearly satisfied.

In both field surveys and transplant experiments, Nishihira (1967) established that the hydroid *Sertularella miurensis* recruits preferentially to *Sargassum tortile*; in contrast, *Coryne uchidai* recruited to a number of different algae. Extracts of *S. tortile*, several of its congeners, and related species were all found to promote settlement of the hydroid *C. uchidai*, although this species is not usually found on *S. tortile* in the field (Nishihira 1968). An active component of an extract of this alga also induced settlement and metamorphosis of *C. uchidai*; this steroid-like compound has been identified as an epoxide of delta-tocotrienol (Kato *et al*. 1975). These results suggest that inducers of the many hydroids examined in these studies may very well form a class of inducer compounds with the same or similar structures as that inducer partially characterized by Kato *et al*. (1975). This class may be ubiquitously distributed in *Sargassum* spp. and in related genera (cf. Nishihira 1968). As Nishihira suggests, other factors apparently account for the absence of *C. uchidai* from *S. tortile*. An analogous situation has been reported for *Haliotis rufescens*. Although larvae did not settle on intact foliose red macroalgae, extracts of such species were found to contain potent inducers (Morse and Morse 1984a, b). Biochemical characterization of these molecules reveal them to be the same class of molecules as the natural inducer from coralline algae (Morse and Morse, 1984a, b; A. Morse *et al*. 1984). It was experimentally demonstrated that inducer molecules were not present in surface materials of foliose red algae; they were found only in surface materials of coralline algae (Morse and Morse 1984a). This has been hypothesized to account for the substrata-specific recruitment of abalone in the field to coralline algae only (Morse and Morse 1984a).

Conclusions

For most sessile benthic marine invertebrates, settlement from the plankton represents an irreversible transition from larval to early juvenile life stages, not merely a change of habitat. For most specialist invertebrate species examined, sensory recognition of species-specific environmental cues has been shown (in laboratory experiments) to be required for activation of programmed behavioural and developmental processes required for facilitation of the transition from developmentally arrested larval stage to actively developing juvenile (cf. Chia and Rice 1978; A. Morse 1990). For such larvae, available settlement space is not enough; it has to be the right kind of space. Similarly, laboratory experiments indicate that, irrespective of the degree of larval abundance in the plankton, if the required settlement substrata are not readily available to larvae then they will not settle but remain either temporarily or permanently (Williams 1964) arrested in the larval stage. After an excessive time in the plankton, a few species have been reported to settle, apparently randomly; usually this has resulted in abnormal metamorphosis (cf. Crisp 1974). On the other hand, the demonstrated ability of a number of species successfully to delay settlement and metamorphosis for varying lengths of time after onset of developmental competence has been postulated to increase genetic fitness of particular species by increasing their dispersal capabilities (Highsmith and Emlet 1986).

For specialist species, efficiency of detection of the required cue could be expected to be an important contributing (but by no means exclusive) factor in determining initial settlement rates. The availability of particular cue-associated settlement substrata could be expected directly to affect larval detection efficiency. Thus the distribution and abundance of particular settlement substrata within habitats will have an important influence on estimates of potential settlement rates of those invertebrates having highly tuned, fail-safe mechanisms for the detection of specific cues.

In species with relatively relaxed stringency and specificity, it might be expected that other factors (e.g., availability of competent planktonic larvae and physical parameters affecting larval delivery) would be more important in determining settlement rates. These factors, either alone or in combination with other sensory cues associated with substrata, could also be expected to be important for species in which settlement on surfaces seems to be determined by larval response to physical and mechanical properties of the water column directly over hard substrata (see Rittschof et al. 1984; Wethey 1986; Butman et al. 1988).

There is only one study in which it has been directly demonstrated that a defined substance, identified as a settlement cue in the laboratory, does enhance settlement and metamorphosis in the ocean. This one instance has recently been demonstrated for the gregarious polychaete *Phragmatopoma californica* (Jensen and Morse 1990). As these authors have pointed out, there is great need to improve our understanding of the actual contribution of settlement cues identified in laboratory studies to settlement events in the ocean, where undoubtedly an array of interactive effects serves to control individual settlement events.

Algal assemblages are recognized as being structurally dominant in subtidal and intertidal communities throughout the world. Although the relative abundance of species within a particular community varies considerably, particular dominant patterns within communities give rise to variation in patterns between communities (Connor and Adey 1977; Chapman and Johnson 1990; Foster 1990; Santelices 1990; Schiel 1990; Underwood and Kennelly 1990). The results of numerous field studies indicate that invertebrate recruitment to algae within these diverse communities is common and sometimes prevalent

(*Spirorbis* spp.: Williams 1964, Crisp 1974; *Acanthaster planci*: Lucas 1975; *Stichaster australis*: Barker 1977; *Aplysia californica*: Audesirk 1979, Pennings 1991; *Strongylocentrotus* spp.: Cameron and Schroeter 1980, Raymond and Scheibling 1987, Rowley 1989; *Haliotis* spp.: Saito 1981, Tegner and Dayton 1977, Shepherd and Turner 1985, Douros 1985, Prince et al. 1987, Shepherd 1990; *Acmaea testudinalis*: Steneck 1982; *Alcyonium siderium*: Sebens 1983b; *Hyale* spp. and *Ampithoe* spp.: Gunnill 1984; *Tegula* spp.: Watanabe 1984; *Agaricia* spp.: D. Morse et al. 1988). None of these field studies has evaluated initial settlement events. The chemical and physical requirements for settlement on algal substrata have, however, been determined in laboratory studies (in the absence of disturbance) for a number of these larvae. Both rates of settlement responses of larvae to living algae and their associated chemical cues, and rates of subsequent development of metamorphic processes, have been documented (e.g., Kriegstein et al. 1974; D. Morse et al. 1980a, 1988; Morse and Morse 1984a; Rowley 1989; Davis et al. 1990; Pearce and Scheibling 1990). Recent studies at the molecular level point to a second level of fine-scale regulation of the larval primary response to a required chemical cue. In response to certain conditions in the water column, settlement rates can be modified by secondary environmental stimuli (D. Morse 1990).

Differential post-settlement mortality (in the absence of disturbance) on a variety of plant, animal, microbial, and 'bare rock' substrata commonly found within the habitat range of young recruits and adults, has been investigated. For taxa with high specificity of cue recognition associated with a particular species or group of algae (e.g., *Haliotis*), laboratory data suggest that recruitment will be independent of this parameter. For taxa (e.g., *Strongylocentrotus*) with relaxed specificity, recruitment will vary as a function of this parameter. Pre-settlement differential mortality in the presence of substrata likely to be encountered by larvae during normal exploratory behaviour has also been examined. The significance of this variable in recruitment seems, except in a few instances, to be rather low. The density of a prey species on a settlement substratum, however, appears to be indirectly correlated with early post-settlement survival (e.g., serpulid worms/ *Haliotis*, personal observations).

I would like to suggest that carefully controlled laboratory experiments, carried out in the absence of disturbance factors (including grazers), will provide us with guidelines for potential parameters which might describe the dynamics of initial settlement events, on a species-specific basis, in the field. Such dynamics are hypothesized to be a direct reflection of the limits of physiological response of a particular invertebrate species to a substratum within its normal distribution range.

References

Audesirk, T. E. (1979). A field study of growth and reproduction in *Aplysia californica*. *Biological Bulletin (Woods Hole, Massachusetts)*, **157**, 407-21.

Bak, R. P. M. and Engel, M. S. (1979). Distribution, abundance and survival of juvenile hermatypic corals (Scleractinia) and the importance of life history strategies in the parent coral community. *Marine Biology*, **54**, 341-52.

Barker, M. F. (1977). Observations on the settlement of the Brachiolaria larvae of *Stichaster australis* (Verrill) and *Coscinasterias calamaria* (Gray) (Echinodermata: Asteroidea) in the laboratory and on the shore. *Journal of Experimental Marine Biology and Ecology*, **30**, 95-108.

Barnes, J. R. and Gonor, J. J. (1973). The larval settling response of the lined chiton *Tonicella lineata*. *Marine Biology*, **201**, 259-64.

Burke, R. D. (1983). The induction of metamorphosis of marine invertebrate larvae: stimulus and response. *Canadian Journal of Zoology*, **61**, 1701-19.

Burke, R. D. (1986). Pheromones and the gregarious settlement of marine invertebrate larvae. Stimulus and response. *Bulletin of Marine Science*, **39**, 323-31.

Butman, C. A., Grassle, J. P., and Webb, C. M. (1988). Substrate choices made by marine larvae settling in still water and in a flume flow. *Nature*, **333**, 771-3.

Caffey, H. M. (1985). Spatial and temporal variation in settlement and recruitment of intertidal barnacles. *Ecological Monographs*, **55**, 313-32.

Cameron, R. A. and Hinegardner, R. T. (1974). Initiation of metamorphosis in laboratory cultured sea urchins. *Biological Bulletin (Woods Hole, Massachusetts)*, **146**, 335-42.

Cameron, R. A. and Schroeter, S. C. (1980). Sea urchin recruitment: effect of substrate selection on juvenile distribution. *Marine Ecology Progress Series*, **2**, 243-7.

Chapman, A. R. O. and Johnson, C. R. (1990). Disturbance and organization of macroalgal assemblages in the Northwest Atlantic. *Hydrobiologia*, **192**, 77-121.

Chia, F.-S. (1978). Perspectives: settlement and metamorphosis of marine invertebrate larvae. In *Settlement and metamorphosis of marine invertebrate larvae*, (ed. F.-S. Chia and M. E. Rice), pp. 283-5. Elsevier, New York.

Chia, F.-S. and Rice, M. E. (ed.) (1978). *Settlement and metamorphosis of marine invertebrate larvae*. Elsevier, New York.

Connell, J. H. (1985). The consequences of variation in settlement vs. post-settlement mortality in rocky intertidal communities. *Journal of Experimental Marine Biology and Ecology*, **93**, 11-45.

Connor, J. L. and Adey, W. H. (1977). The benthic algal composition, standing crop, and productivity of a Caribbean algal ridge. *Atoll Research Bulletin*, **211**, 1-15.

Crisp, D.J. (1974). Factors influencing the settlement of marine invertebrate larvae. In *Chemoreception in marine organisms*, (ed. P.T. Grant and A.M. Mackie), pp. 177-265. Academic Press, New York.

Crisp, D.J. (1984). Overview of research on marine invertebrate larvae, 1940-1980. In *Marine biodeterioration*, (ed. J.D. Costlow and R.C. Tipper), pp. 103-26. Naval Institute Press, Annapolis.

Crisp, D.J. and Meadows, P.S. (1963). Adsorbed layers: the stimulus to settlement in barnacles. *Proceedings of The Royal Society*, Series B, **158**, 364-87.

Davis, M., Heyman, W.D., Harvey, W., and Withstandley, C.A. (1990). A comparison of two inducers, KCl and *Laurencia* extracts, and techniques for the commercial scale induction of metamorphosis in queen conch *Strombus gigas* Linnaeus, 1758 larvae. *Journal of Shellfish Research*, **9**, 67-73.

Dayton, P.K. (1979). Ecology: science or a religion? In *Ecological processes in coastal and estuarine systems*, (ed. R. Livingston), pp. 3-17. Plenum Press, New York.

Douros, W.J. (1985). Density, growth, reproduction and recruitment in an intertidal abalone: effects of intraspecific competition and prehistoric predation. Unpublished MA Dissertation. University of California, Santa Barbara. 112 pp.

Foster, M.S. (1990). Organization of macroalgal assemblages in the Northeast Pacific: the assumption of homogeneity and the illusion of generality. *Hydrobiologia*, **192**, 21-33.

Gaines, S. and Roughgarden, J. (1985). Larval settlement rate: a leading determinant of structure in an ecological community of the marine intertidal zone. *Proceedings, US National Academy of Sciences*, **82**, 3707-11.

Gaines, S., Brown, S., and Roughgarden, J. (1985). Spatial variation in larval concentrations as a cause of spatial variation in settlement for the barnacle, *Balanus glandula*. *Oecologia* (Berlin), **67**, 267-72.

Gee, J.M. (1965). Chemical stimulation of settlement of larvae of *Spirorbis rupestris* (Serpulidae). *Animal Behavior*, **13**, 181-6.

Gunnill, F.C. (1984). Differing distributions of potentially competing amphipods, copepods and gastropods among specimens of the intertidal alga *Pelvetia fastigiata*. *Marine Biology*, **82**, 277-91.

Hadfield, M.G. (1978). Metamorphosis in marine molluscan larvae: an analysis of stimulus and response. In *Settlement and metamorphosis of marine invertebrate larvae*, (ed. F.-S. Chia and M.E. Rice), pp. 165-75.

Hadfield, H.G. (1986). Settlement and recruitment of marine invertebrates: a perspective and some proposals. *Bulletin of Marine Science*, **39**, 418-25.

Henderson, J. A. and Lucas, J. S. (1971). Larval development and metamorphosis of *Acanthaster planci* (Asteroidea). *Nature* (London), **232**, 655-7.

Heslinga, G. A. (1981). Larval development, settlement and metamorphosis of the tropical gastropod *Trochus niloticus*. *Malacologia*, **20**, 349-57.

Heslinga, G. A. and Hillman, A. (1981). Hatchery culture of the commercial top shell *Trochus niloticus* in Palau, Caroline Islands. *Aquaculture*, **22**, 35-43.

Highsmith, R. C. and Emlet, R. B. (1986). Delayed metamorphosis: effect on growth and survival of juvenile sand dollars (Echinoidea: Clypeasteroida). *Bulletin of Marine Science*, **39**, 347-61.

Jackson, G. A. (1986). Interaction of physical and biological processes in the settlement of planktonic larvae. *Bulletin of Marine Science*, **39**, 202-12.

Jensen, R. A. and Morse, D. E. (1990). Chemically induced metamorphosis of polychaete larvae in both the laboratory and ocean environment. *Journal of Chemical Ecology*, **16**, 911-30.

Johansen, H. W. (1981). *Coralline algae, a first synthesis*. CRC Press, Boca Raton, Florida.

Kato, T. *et al.* (1975). Active components of *Sargassum tortile* effecting the settlement of swimming larvae of *Coryne uchidai*. *Experientia*, **31**, 433-4.

Keough, M. J. and Downes, B. J. (1982). Recruitment of marine invertebrates: the role of active larval choices and early mortality. *Oecologia* (Berlin), **54**, 348-52.

Kriegstein, A. R., Castellucci, V., and Kandel, E. R. (1974). Metamorphosis of *Aplysia californica* in laboratory culture. *Proceedings, US National Academy of Sciences*, **71**, 3654-8.

Lucas, J. S. (1975). Environmental influences on the early development of *Acanthaster planci* (L.). In *Crown-of-thorns starfish seminar proceedings, Brisbane*, pp. 109-21. Australian Government Publication Service, Canberra.

Morse, A. N. C. (1985). Characterization of inducers of molluscan metamorphosis. Unpublished Ph.D. Thesis. University of California, Santa Barbara. 249 pp.

Morse, A. N. C. (1988). The role of algal metabolites in the recruitment process. In *Marine biodeterioration, advanced techniques applicable to the Indian ocean*, (ed. M.-F. Thompson, R. Sarojini, and R. Nagabhushanam), pp. 463-73. Oxford and IBH Publishing, New Delhi.

Morse, A. N. C. and Morse, D. E. (1984a). Recruitment and metamorphosis of *Haliotis* larvae are induced by molecules uniquely available at the surfaces of crustose red algae. *Journal of Experimental Marine Biology and Ecology*, **75**, 191-15.

Morse, A. N. C. and Morse, D. E. (1984b). GABA-mimetic molecules from *Porphyra* (Rhodophyta) induce metamorphosis of *Haliotis* (Gastropoda) larvae. *Hydrobiologia*, **116**, 155-8.

Morse, A. N. C., Froyd, C., and Morse, D. E. (1984). Molecules from Cyanobacteria and red algae that induce larval settlement and metamorphosis in the mollusc *Haliotis rufescens*. *Marine Biology*, **81**, 293-8.

Morse, D. E. (1984). Biochemical and genetic engineering for improved production of abalones and other valuable molluscs. *Aquaculture*, **39**, 263-83.

Morse, D. E. (1985). Neurotransmitter-mimetic inducers of larval settlement and metamorphosis. *Bulletin of Marine Science*, **37**, 697-706.

Morse, D. E. (1990). Recent progress in larval settlement and metamorphosis: closing the gaps between molecular biology and ecology. *Bulletin of Marine Science*, **46**, 465-83.

Morse, D. E., Hooker, N., Duncan, H., and Jensen, L. (1979). γ-Aminobutyric acid, a neurotransmitter, induces planktonic abalone larvae to settle and begin metamorphosis. *Science*, **204**, 407-10.

Morse, D. E., Duncan, H., Hooker, N., Baloun, A., and Young, G. (1980*a*). GABA induces behavioral and developmental metamorphosis in planktonic molluscan larvae. *Federation Proceedings*, **39**, 3237-41.

Morse, D. E., Hooker, N., and Duncan, H. (1980*b*). GABA induces metamorphosis in *Haliotis*, V: Stereochemical specificity. *Brain Research Bulletin*, **5**, Suppl. 2, 381-7.

Morse, D. E., Tegner, M., Duncan, H., Hooker, N., Trevelyan, G., and Cameron, A. (1980*c*). Induction of settling and metamorphosis of planktonic molluscan (*Haliotis*) larvae. III: Signaling by metabolites of intact algae is dependent on contact. In *Chemical signals in vertebrate and aquatic animals*, (ed. D. Muller-Schwarze and R. M. Silverstein), pp. 67-86. Plenum Press, New York.

Morse, D. E., Hooker, N., Morse, A., and Jensen, R. A. (1988). Control of larval metamorphosis and recruitment in sympatric agariciid corals. *Journal of Experimental Marine Biology and Ecology*, **116**, 193-217.

Nishihira, M. (1965). The association between Hydrozoa and their attachment substrata with special reference to algal substrata. *Bulletin of the Marine Biological Station of Asamushi*, **12**, 75-92.

Nishihira, M. (1967). Observations on the selection of algal substrata by hydrozoan larvae, *Sertularella miurensis* in nature. *Bulletin of the Marine Biological Station of Asamushi*, **13**, 35-48.

Nishihira, M. (1968). Brief experiments on the effect of algal extracts in promoting the settlement of the larvae of *Coryne uchidai* Stechow (Hydrozoa). *Bulletin of the Marine Biological Station of Asamushi*, **13**, 91-101.

Nishihira, M. (1971). Colonization patterns of Hydrozoa on several species of *Sargassum*. *Bulletin of the Marine Biological Station of Asamushi*, **14**, 99-108.

Paine, R. T. (1980). Food webs: linkage, interaction strength and community infrastructure. *Journal of Animal Ecology*, **49**, 667–85.
Pearce, C. M. and Scheibling, R. E. (1990). Induction of metamorphosis of larvae of the green sea-urchin, *Strongylocentrotus droebachiensis*, by coralline red algae. *Biological Bulletin (Woods Hole, Massachusetts)*, **179**, 304–11.
Pennings, S. C. (1990). Multiple factors promoting narrow host range in the sea hare, *Aplysia californica*. *Oecologia* (Berlin), **82**, 192–200.
Pennings, S. C. (1991). Temporal and spatial variation in the recruitment of the sea hare, *Aplysia californica* at Santa Catalina island. In *Recent advances in California island research*, Proceedings of the third California island symposium, in press. Santa Barbara Museum of Natural History.
Prince, J. D., Sellers, T. L., Ford, W. B., and Talbot, S. R. (1987). Experimental evidence for limited dispersal of haliotid larvae (genus *Haliotis*; Mollusca: Gastropoda). *Journal of Experimental Marine Biology and Ecology*, **106**, 243–63.
Raymond, B. G. and Scheibling, R. E. (1987). Recruitment and growth of the sea urchin *Strongylocentrotus droebachiensis* (Müller) following mass mortalities off Nova Scotia, Canada. *Journal of Experimental Marine Biology and Ecology*, **108**, 31–54.
Rittschof, D. and Bonaventura, J. (1986). Macromolecular cues in marine systems. *Journal of Chemical Ecology*, **12**, 1013–23.
Rittschof, D., Branscomb, E. S., and Costlow, J. (1984). Settlement and behaviour in relation to flow and surface in larval barnacles, *Balanus amphitrite* Darwin. *Journal of Experimental Marine Biology and Ecology*, **82**, 131–46.
Rowley, R. J. (1989). Settlement and recruitment of sea urchins (*Strongylocentrotus* spp.) in a sea urchin barren ground and a kelp bed: are populations regulated by settlement or post-settlement processes? *Marine Biology*, **100**, 485–94.
Rumrill, S. S. and Cameron, R. A. (1983). Effects of gamma-aminobutyric acid on the settlement of larvae of the black chiton *Katharina tunicata*. *Marine Biology*, **72**, 243–7.
Saito, K. (1981). The appearance and growth of 0-year-old *Ezo* abalone. *Bulletin of Japanese Society for Science and Fisheries*, **47**, 1393–400.
Santelices, B. (1990). Patterns of organization of intertidal and shallow subtidal vegetation in wave exposed habitats of central Chile. *Hydrobiologia*, **192**, 35–57.
Scheltema, R. S. (1974). Biological interactions determining larval settlement of marine invertebrates. *Thalassia Jugoslavica*, **10**, 263–96.
Schiel, D. R. (1990). Macroalgal assemblages in New Zealand: structure, interactions and demography. *Hydrobiologia*, **192**, 59–76.
Sebens, K. P. (1983a). The larval and juvenile ecology of the temperate octocoral *Alcyonium siderium* Verrill. I. Substratum selection by benthic

larvae. *Journal of Experimental Marine Biology and Ecology*, **71**, 73–89.

Sebens, K. P. (1983*b*). Settlement and metamorphosis of a temperate soft-coral (*Alcyonium siderium* Verrill): induction by crustose algae. *Biological Bulletin (Woods Hole, Massachusetts)*, **165**, 286–304.

Shepherd, S. A. (1990). Studies on Southern Australian abalone (genus *Haliotis*). XII. Long-term recruitment and mortality dynamics of an unfished population. *Australian Journal of Marine and Freshwater Research*, **41**, 475–92.

Shepherd, S. A. and Turner, J. A. (1985). Studies on Southern Australian abalone (genus *Haliotis*). VI. Habitat preference, abundance and predators of juveniles. *Journal of Experimental Marine Biology and Ecology*, **93**, 285–98.

Siddall, S. E. (1982). Biological and economic outlook for hatchery production of juvenile queen conch. *Gulf and Caribbean Fisheries Institute Proceedings*, **35**, 46–52.

Steneck, R. S. (1982). A limpet-coralline alga association: adaptations and defenses between a selective herbivore and its prey. *Ecology*, **63**, 507–22.

Switzer-Dunlop, M. and Hadfield, M. G. (1977). Observations on development, larval growth and metamorphosis of four species of Aplysiidae (Gastropoda: Opistobranchia) in laboratory culture. *Journal of Experimental Marine Biology and Ecology*, **29**, 245–61.

Tegner, M. J. and Dayton, P. K. (1977). Sea urchin recruitment patterns and implications of commercial fishing. *Science*, **196**, p. 324.

Trapido-Rodenthal, H. G. and Morse, D. E. (1986). Availability of chemosensory receptors is down-regulated by habituation to a morphogenetic signal. *Proceedings, US National Academy of Sciences*, **83**, 7658–62.

Underwood, A. J. and Denley, E. J. (1984). Paradigms, explanations and generalizations in models for the structure of intertidal communities on rocky shores. In *Ecological communities: conceptual issues and the evidence*, (ed. D. Strong, D. Simberloff, L. G. Abele, and A. B. Thistle), pp. 151–80. Princeton University Press, New Jersey, USA.

Underwood, A. J. and Kennelly, S. J. (1990). Ecology of marine algae on rocky shores and subtidal reefs in temperate Australia. *Hydrobiologia*, **192**, 3–20.

Van Moorsel, G. W. N. M. (1989). Juvenile ecology and reproductive strategies of reef corals. Unpublished Ph.D. Thesis. Rijksuniversiteit Gronigen, Holland. 103 pp.

Watanabe, J. M. (1984). The influence of recruitment, competition, and benthic predation on spatial distributions of three species of kelp forest gastropods (Trochidae: *Tegula*). *Ecology*, **65**, 920–36.

Wethey, D. S. (1986). Ranking of settlement cues by barnacle larvae: influence of surface contour. *Bulletin of Marine Science*, **39**, 393–400.

Williams, G. B. (1964). The effect of extracts of *Fucus serratus* in promoting the settlement of larvae of *Spirorbis borealis* [Polychaeta]. *Journal of the Marine Biological Association of the United Kingdom*, **44**, 397–414.

Yool, A. J., Grau, S. M., Hadfield, M. G., Jensen, R. A., Markell, D. A., and Morse, D. E. (1986). Excess potassium induces larval metamorphosis in four marine invertebrate species. *Biological Bulletin (Woods Hole, Massachusetts)*, **170**, 255–66.

18. Algal 'gardening' by grazers: a comparison of the ecological effects of territorial fish and limpets

G.M. BRANCH, J.M. HARRIS, C. PARKINS,
R.H. BUSTAMANTE, and S. EEKHOUT
Department of Zoology and Marine Biology Research Institute, University of Cape Town, Rondebosch, South Africa

Abstract

Algal 'gardening' by grazers modifies macroalgal assemblages by selectively enhancing particular species and increasing the food value of the algae. Three main groups of grazers are involved: fish, limpets, and a few polychaetes. Gardening shifts algal assemblages towards more productive and/or easily consumed species such as filamentous or finely branched algae and non-coralline crusts. Species richness and biomass of gardens depend on the intensity of grazing in gardens relative to background levels. Grazers potentially influence their gardens by altering grazing intensity (usually by territorial defence), nutrient enhancement, 'weeding' or 'sowing' selectively, and modifying availability of substratum. Gardening grazers seem 'prudent', but this may relate to changes in the productivity, food value, or anti-herbivore defences of the algae in their gardens. Territorial defence of gardens is most intense against competitors, and gardening is likely to be most frequent where algal productivity is low and/or grazing intense, as can be shown by geographical gradients. Thus gardening has both evolutionary and ecological implications.

Introduction

Grazers which influence the composition or growth of plants are often referred to as 'gardeners', but the phenomenon means different things to different people. We define it as: 'modification of plant assemblages, caused by the activities of an individual grazer within a fixed site, which selectively enhances particular plant species and increases the food value of the plants for the grazer'.

The fact that it needs precise definition is instructive, for gardening forms part of a grazing continuum and is often difficult to separate from other effects that grazers have on food plants. For example, grazing herds enhance the productivity and food quality of grasses on African plains (McNaughton 1984), and grazing by zooplankton increases the net growth of phytoplankton (Sterner 1986). In neither case, however, does the individual grazer benefit from its own actions. Gardening at a fixed point does permit this and, potentially, leads to tightly coevolved interactions between particular grazers and algae.

In this review we focus only on algal gardening by grazers, particularly fish and limpets, while appreciating that comparable interactions occur between bacteria and detritivores (Hylleberg 1975; Branch and Pringle 1985) and between diatoms and micrograzers (Bianchi et al. 1989). We concentrate on three facets: the effects that gardening has on the algae themselves, its influence on community structure and dynamics, and some of the evolutionary implications of the phenomenon.

Examples

Three major groups of animals have been recorded as gardeners. The case of tropical reef fish (mainly pomacentrids) is especially well documented (e.g., Kaufman 1977; Vine 1974; Brawley and Adey 1977; Lobel 1980; Lassuy 1980; Montgomery 1980; Williams 1980; Sammarco and Williams 1982; Hixon and Brostoff 1983; Sammarco 1983; Klumpp et al. 1987; Russ 1987; Klumpp and Polunin 1989). Individual fish defend patches of algae, aggressively repelling other fish and urchins. The result is the development of small algal assemblages quite different from those in the surrounds.

Some limpets also garden. In California *Lottia gigantea* maintains a fine algal film and excludes other grazers. It also hinders invasion of its gardens by sessile organisms (Stimson 1970, 1973). More specialized associations are exhibited by South African limpets. *Patella longicosta* and *Patella tabularis* have specific associations with *Ralfsia verrucosa* and also defend their gardens against grazers (Branch 1975b, 1976, 1981). *Patella cochlear*, which lives in dense aggregations, has a narrow fringing garden around each limpet of fine red algae, usually *Herposiphonia heringii* or *Gelidium micropterum*; the latter apparently occurring only in these gardens (Branch 1975a). Although previously unrecorded as gardening, *Patella barbara* and *Patella miniata* present particularly interesting cases because they garden only over a portion of their geographical range.

The final example comprizes nereid polychaetes. *Platynereis bicanaliculata* and *Nereis vexillosa* catch drifting fragments of green algae and attach them to their tubes which are embedded in soft sediments. The attached algae provide a predictable food supply. Overgrazing is avoided in most cases because the worms crop the seaweeds distally (Woodin 1977). As these nereids interact aggressively, their tubes are spaced out (Woodin 1974), helping to restrict use of the algae to individuals that garden them. Apart from supplying a supplementary source of food, the algae reduce temperature and salinity stress, but diminish oxygen levels in the sediments below, thus tending to exclude some species such as maldanid polychaetes (Woodin 1977).

Some urchins' activities approach a condition of gardening. *Diadema antillarum* concentrates its feeding in a foraging range and has the capacity to modify algal assemblages (Carpenter 1981; Williams and Carpenter 1988). As yet, however, no urchin has been shown to garden in the sense defined above.

Effects on algae

One of the striking features of gardens is the similarity between the functional algal types present, even though the grazers responsible differ radically in terms of taxonomy, morphology, and mobility. Most gardens consist of opportunistic, fast-growing species — often filamentous, delicate red or green algae, but sometimes encrusting forms. Common finely-branched or filamentous genera include *Polysiphonia* and related genera, *Gelidiopsis, Centroceras*, and *Ceramium* (Vine 1974; Branch 1975a; Montgomery 1980; Lassuy 1980; Sammarco 1983; Klumpp and Polunin 1989), while crusts include *Ralfsia* (Branch 1975b).

Gardens increase local productivity relative to that in adjacent areas (Fig. 18.1) (Montgomery 1980; Russ 1987; Klumpp *et al.* 1987). This shift is brought about, firstly, by the nature of the algae involved and, secondly, because they are maintained in an early, rapid phase of growth by continual grazing. For example, grazing by the giant blue damselfish *Microspathodon dorsalis* keeps its algal garden (a near monoculture of *Polysiphonia*) in an early stage of growth. Removal of a territorial fish allows the alga to almost quadruple the size of its filaments within 10 days (Montgomery 1980). Similarly, the limpet *Patella longicosta* maintains its garden of *Ralfia verrucosa* in a young, fast-growing stage (Branch 1981). Such effects are, of course, not

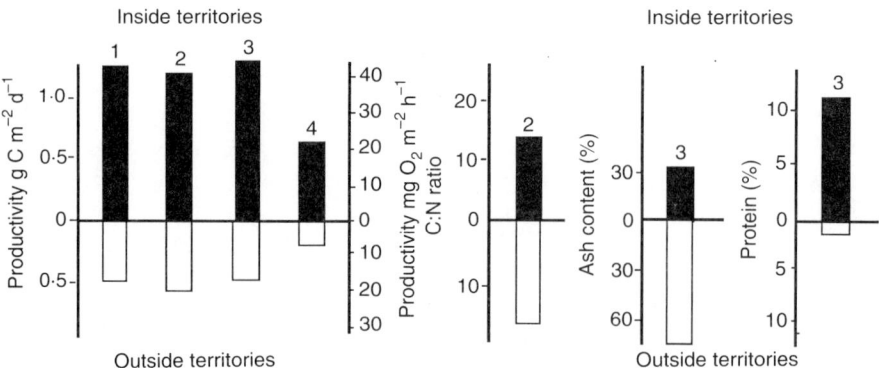

Fig. 18.1. Comparisons of algal productivity, C:N ratio, ash, and protein content for algae inside and outside territorial gardens of fish. Data from: 1. Klumpp *et al.* (1987); 2. Russ (1987); 3. and 4. Montgomery (1980). Production in g C for 1 and 2; in mg O_2 for 3 and 4.

specific to gardens. Carpenter (1981) has shown that intermediate levels of grazing by the urchin *Diadema antillarum* yield maximal production, reducing biomass relative to ungrazed situations but increasing production/biomass. We believe that territorial gardeners achieve an intermediate level of grazing, thus optimizing production of their food. They may accomplish this by either decreasing or increasing grazing intensity, depending on whether background levels of grazing are high or low.

Apart from their higher productivity, algae in the gardens of fish have a higher proportion of protein and lower ash content and C:N ratios than algae outside territories (Montgomery 1980; Russ 1987; Klumpp *et al.* 1987). Coupled with their fragile structure, this increases the value of algae in gardens. Many fish lack the necessary morphology and physiology to digest complex algae. Pomacentrids do have a low gut pH, which allows them to lyse algal cells, but they are only capable of digesting delicate red algae and blue-greens, not robust algae (Lassuy 1980). Klumpp and Polunin (1989) have shown that territorial fish selectively feed on algae with a high organic content and low C:N ratios.

Processes establishing and maintaining gardens

The key effect that grazers appear to have on their gardens is to modify the intensity of grazing within their territories. Pomacentrid fish reduce the amount of grazing inside their territories by restricting access by other grazers. Outside damselfish territories, algal biomass is low

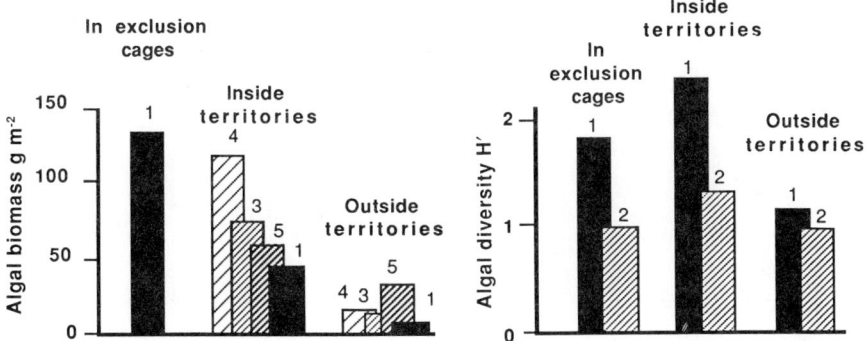

Fig. 18.2. Algal biomass and diversity inside and outside fish territories and in fish-exclusion cages: 1. *Hemiglyphododon plagiometopon* (Sammarco 1983); 2. *Stegastes fasciolatus* (Hixon and Brostoff 1983); 3 and 4. *Eupomacentrus lividus* and *H. plagiometopon* (Lassuy 1980); 5. *S. fasciolatus* (Russ 1987; biomass estimated by doubling his C values).

(Fig. 18.2) and dominated by encrusting corallines, although large grazer-resistant foliose algae (e.g., *Halimeda, Lobophora, Dictyota,* and *Padina*) are also present. Inside territories, biomass rises (Brawley and Adey 1977; Lassuy 1980; Sammarco 1983; Hixon and Brostoff 1983; Klumpp et al. 1987; Russ 1987) and becomes dominated by filamentous forms. Total exclusion of fish results in a biomass even higher than in territories (Sammarco 1983). Only in one species (*Microspathodon dorsalis*) has algal biomass been shown to be low inside territories relative to background levels (Montgomery 1980).

Direct tests of whether grazing really is reduced inside fish territories are few. Hixon and Brostoff (1983) produce convincing data that the number of bite-marks is lower inside territories of *Stegastes fasciolatus* than outside. This is not in itself conclusive evidence. As Russ (1987) points out, the depth of the bite and the amount of material removed need to be considered. Russ's own direct measurements on the same species revealed no significant difference between the rates of removal in and out of territories. The issue thus remains in need of further rigorous testing. This does not, however, alter the general pattern of high biomass in gardens relative to undefended areas.

Limpets present a different picture. They do react aggressively against intruding conspecifics and other invertebrate grazers which present a competitive threat (Stimson 1970, 1973; Branch 1975*b*, 1976, 1981). Despite this, they appear to intensify grazing pressure within their gardens by concentrating their grazing on a small area. Algal biomass is invariably lower in their gardens than in adjacent areas

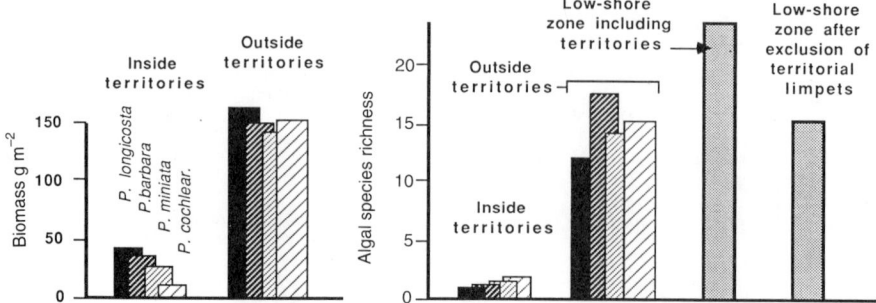

Fig. 18.3. Algal biomass (g ash-free dry mass m^{-2}) and species richness (per 0·25 m^2) in and out of limpet gardens, and species richness in low-shore zones from which territorial limpets had or had not been removed 1 year previously (Branch, unpublished data).

(Fig. 18.3). They do, however, enhance productivity. For example, gardens of *Ralfsia verrucosa* which are grazed by *Patella longicosta* are about 30 per cent more productive than ungrazed controls (Branch 1981).

Increases or decreases in grazing may have direct effects on algal composition. Montgomery (1980) has developed a general model which predicts that as grazing intensifies, an increasing number of species will be excluded, leaving only those productive enough to compensate for grazing. Montgomery's own evidence shows grazed gardens to be a staggering 47 times more productive per unit biomass than surrounding turfs.

Part of the effect of grazing is not the intensity *per se* but the selectivity. Although Montgomery's (1980) model depends on damselfish being generalized grazers they may selectively 'weed out' particular species. Lassuy (1980) demonstrated this by introducing 'strings' of algae into gardens or into undefended sites. Strings inside gardens were rapidly stripped of large, tough, inedible species. Conversely, if the strings lay outside territories, fine filamentous algae were soon removed by non-territorial fish.

Another factor potentially influencing algal biomass and productivity is the input of nutrients from territorial individuals. Meyer *et al.* (1983) have shown that haemulid fish excrete sufficient dissolved and particulate nutrients to increase the growth rate of the coral heads in which they rest by day. Algae grazed by the urchin *Diadema antillarum* are 2–10 times more productive than ungrazed algae (Carpenter 1981). Accumulation of ammonia beneath each urchin may contribute to this, potentially meeting up to 19 per cent of the nitrogen needs of the algae

(Williams and Carpenter 1988). Neither of these situations relates directly to gardening. Polunin and Koike (1987) have described localized nutrient enrichment in the territories of *Plectroglyphidodon lacrymatus*, although they point out that defaecation (the main agent of enrichment) takes place at the edge of the territory, separate from the feeding areas, so that there is probably no direct link with the productivity of the garden. Thus it remains an open question whether territorial fish contribute sufficiently to nutrients to boost the production of their gardens. In terms of this function, limpets are more likely candidates, for their excreta will accumulate beneath each animal and remain concentrated in the vicinity of the garden, particularly during low tide. Whether this happens remains untested.

Finally, grazers may actively create space for algal recruitment. Both Kaufman (1977) and Wellington (1982) record how damselfish damage and kill portions of corals, creating space in which algae can settle. In a different way, nereids also increase the substratum available for algae: on the surface of soft sediments their tubes provide some of the few firm bases on which algae can attach (Woodin 1977).

Community effects

By enhancing algal productivity, grazers potentially increase overall rates of energy flow. The extent to which gardening *per se* contributes to this effect depends largely on the density of gardens. *Patella cochlear* has contiguous gardens and dominates the low shore on moderately wave-exposed coasts in southern Africa (Branch 1975a), and *P. longicosta* gardens cover up to 70 per cent of the substratum in the zone immediately above *P. cochlear*. In several localities gardens of territorial fish are almost contiguous. *Stegastes apicalis* covers 40 to 50 per cent of shallow reefs with its territories, and a high proportion of the productivity generated in the gardens is used locally and rapidly (Klumpp and Polunin 1989). The territorial fish themselves consume 25 to 38 per cent and resident cryptofauna in the algal gardens eat 31 per cent, while invading fish account for about 14 per cent. Gardens of *Eupomacentrus planifrons* contribute 70 to 80 per cent of the productivity of fore and back-reefs at St Croix in the US Virgin Islands (Brawley and Adey 1977).

Coupled with their elevated productivity, it has been suggested that gardens are sites of high nitrogen fixation (Brawley and Adey 1977). Lobel (1980) noted that the gardens of *Eupomacentrus planifrons* and *Eupomacentrus nigricans* have much higher biomasses of blue-green algae than areas outside gardens. However, Sammarco (1983) and Wilkinson and Sammarco (1983) recorded that N-fixation increased as the intensity of fish grazing increased, being highest outside gardens, intermediate in gardens, and lowest in fish-exclusion cages.

The development of gardens influences algal species richness or diversity in three ways: at the levels of the gardens themselves (equivalent to α diversity), within zones in which the gardens occur ($\sim \beta$ diversity) and, finally, when one compares different zones on the shore or at different depths ($\sim \gamma$ diversity).

Within gardens of territorial reef fish, species richness tends to be high (Fig. 18.2) because grazing is generally considered to be moderate (Hixon and Brostoff 1983; Sammarco 1983; but see Montgomery 1980). Outside gardens, intense grazing yields low-biomass, low-diversity assemblages dominated by a thin turf and encrusting corallines, although in refuges where grazers are scarce or cannot graze effectively, high algal biomass and diversity can be found (Hay 1981; Lewis 1986; Morrison 1988). If all fish are experimentally excluded from gardens, (or even areas outside gardens) dense algal mats with a high diversity may develop initially (Montgomery 1980), but ultimately result in a low diversity because a small number of species come to dominate and less-competitive species are displaced (Hixon and Brostoff 1983). Experimental manipulation of urchin densities has also shown that intermediate levels of grazing are associated with maximal levels of species richness (Carpenter 1981).

Compared with fish, territorial limpets consistently depress algal species richness in their gardens (Fig. 18.3). This supports our earlier contention that limpets increase the amount of grazing in their gardens in comparison with background levels.

Within zones where limpet gardens occur, overall diversity is often boosted: species which occur in gardens are often not found outside the gardens and vice versa. A small-scale mosaic of different algal assemblages is created. Experimental removal of gardening limpets from the low-shore in southern Africa diminished species richness as the assemblages converged to resemble those that previously lay outside gardens (see Fig. 18.3). In part this was due to the invasion of gardens by competitively superior algae after the removal of the limpets. Parallel mosaic patchiness is created on tropical reefs within zones where gardens are established by fish.

Where limpets have dense or contiguous gardens, they may radically alter algal composition. For example, *Patella cochlear* dominates the low shore on wave-exposed shores in South Africa, severely reducing biomass and richness (McQuaid *et al.* 1985). The composition of algae there bears little relationship to that in zones above or below. Rate of change of species composition between zones is consequently increased, influencing γ diversity. Similarly, differences in algal composition at different depths on tropical reefs are attributable to differences in grazing pressure and to the presence or absence of territorial gardens (Hay 1981; Lewis 1986; Hatcher 1988).

Apart from direct interactions between grazers and algae, gardening has impacts on other organisms. Small infaunal species increase inside fish gardens and may contribute to the diet of the fish (Lobel 1980). On the other hand, limpet gardens are typified by a low abundance of cryptofauna because of their simple structure and low biomass (unpublished data). Macrofaunal grazers are largely excluded from the gardens of both limpets and fish (Vine 1974; Myberg and Thresher 1974; Stimson 1970, 1973; Branch 1975a, b, 1981; Brawley and Adey 1977; Ebersole 1977; Williams 1980; Hixon and Brostoff 1983). Experimental exclusion of the limpet *Patella cochlear* leads to substantial increases in the densities and diversity of other invertebrate grazers (Fig. 18.4).

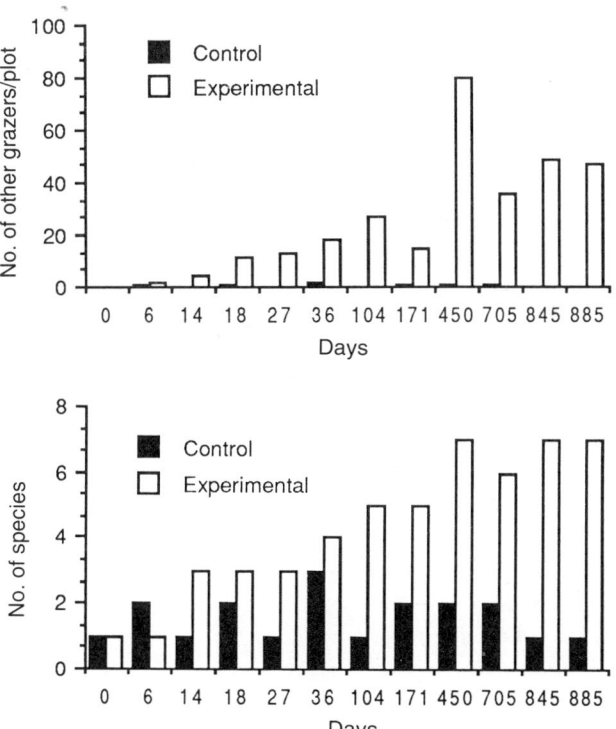

Fig. 18.4. Changes in the numbers and species richness of grazers (per 0·5 m^2 plot) in the cochlear zone, in control plots and after experimental exclusion of *Patella cochlear* (J. Harris and G. M. Branch, unpublished data).

A more complicated case is the interplay between *Eupomacentrus planifrons* and urchins. This damselfish attacks *Diadema antillarum* aggressively and excludes it from its territories. Removal of damselfish results in a rapid invasion by *Diadema* and an almost immediate elimination of the gardens. *E. planifrons* is more tolerant of *Echinometra viridis*, which does occur within the damselfish's territories. The urchins themselves compete, *Diadema* being a superior competitor in terms of its ability to remove food. *E. planifrons* thus contributes to spatial partitioning and coexistence of the urchins, leading Williams (1980) to describe it as a 'non-predatory keystone species'.

Gardening also affects other space-occupiers. Growth of algae in the gardens of *Eupomacentrus lividus* inhibits settlement of corals and other sessile invertebrates such as serpulids (Vine 1974). *Eupomacentrus planifrons* territories are sites of high mortality for coral spat. Despite this, some corals successfully compete with algae and survive better inside fish territories because urchins decimate coral recruits outside the territories. Fish territoriality thus increases coral diversity (Sammarco and Williams 1982). Territorial fish may have profound effects on the zonation and abundance of corals. In the Gulf of Panama, *Eupomacentrus acapulcoensis* kills areas on corals, particularly massive corals, to create algal gardens. As the fish are concentrated in shallow waters, massive corals are rare there. *E. acapulcoensis* also chases away other coral-eating fish (e.g., acanthurids and scarids) and thus helps to maintain delicate branching corals as the dominant forms in shallows. In deeper waters, there are few damselfish and other fish readily attack the branching corals, leaving only a sparse cover of corals, all of which are massive (Wellington 1982).

Territorial fish may have more indirect effects on corals. Risk and Sammarco (1982) noted that corals are weaker inside fish territories, and postulated that a reduction of grazing in the territories allows more boring organisms to settle and subsequently erode the coral. Sammarco *et al.* (1987) failed, however, to detect any differences between internal bioerosion in and out of territories. On the other hand, grazing by fish (which also erodes coral) is higher outside territories than inside (Sammarco *et al.* 1986). One final complication is that N-excretion by fish may strengthen corals by increasing their growth inside territories (Polunin and Koike 1987).

A generalized model

Despite variations between species, general patterns do emerge regarding the effects of gardening (Fig. 18.5). At high levels of grazing, diversity and biomass of algae are low. Both increase as grazing declines,

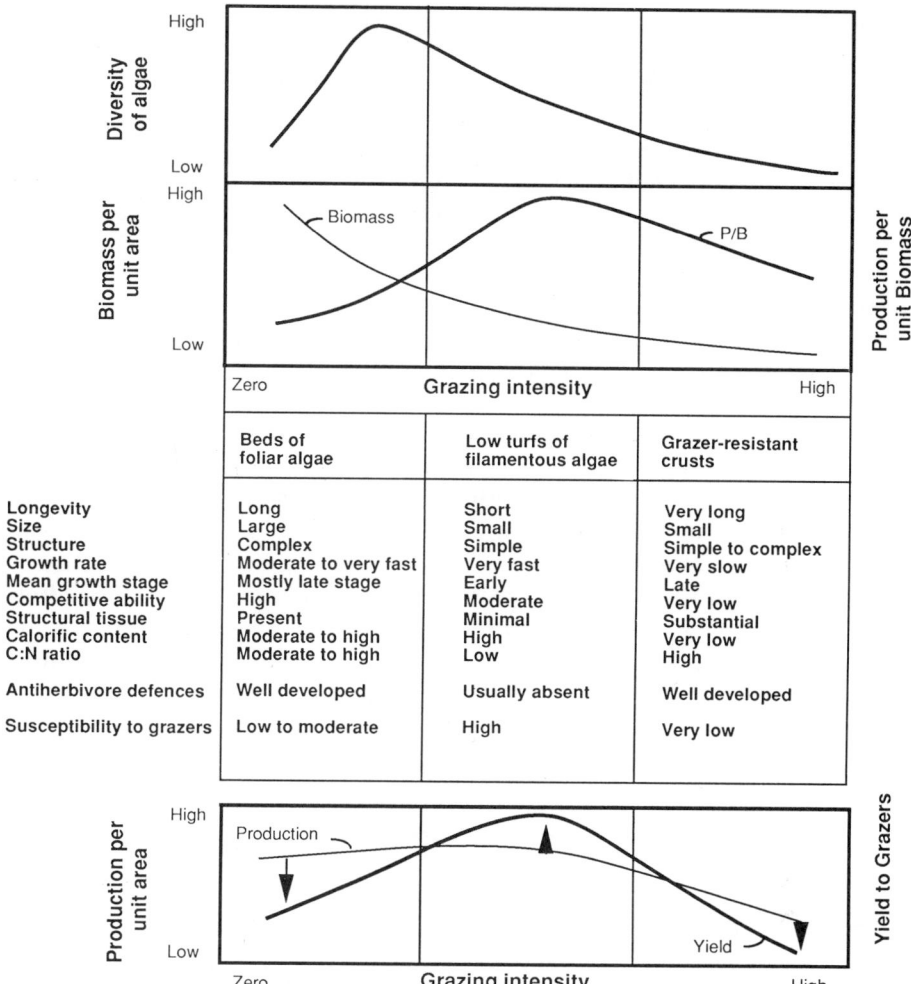

Fig. 18.5. A generalized model for algal diversity, biomass, and production per unit biomass in relation to grazing intensity. Characteristics of the major functional groups of algae typically associated with each grazing regime are summarized in relation to their productivity per unit area. Yield reflects how edibility of different algal types increases or decreases the gains that grazers may obtain from production.

although diversity peaks at a relatively low level of grazing and then drops as competition eliminates some species at very low levels of grazing. Production per unit biomass peaks at an intermediate level of grazing. From the perspective of the grazer, two issues impact on the value of gardens as sources of food. Productivity per unit area determines the amount of food available. The quality of the food determines how available and usable it is.

The relative proportions of three major functional types of algae can be related to grazing intensity. At very low levels of grazing, dense beds of large foliose algae become established. At high intensities, grazer-resistant coralline crusts dominate, a condition often called 'the barrens' or 'overgrazed'. At intermediate levels of grazing (as, we believe, is experienced in gardens), low turfs of fine or filamentous algae are frequent and brown or red algal crusts have also been recorded. In all cases, gardens of such algae are more productive when consumed by their territorial grazers than when they are not — whether this grazing increases or decreases biomass relative to background levels.

These three functional types of algae hold differing attractions for grazers. Turfs of filamentous or fine algae tend to comprise species which are small, short-lived, opportunistic, fast-growing, and highly productive. They also have minimal structural tissue, high energy, and nitrogen content and few anti-herbivore defences. Their value to grazers is high. Conversely, coralline crusts and large foliose algae often have properties which diminish their yield to herbivores (see Fig. 18.5).

Fig. 18.6 summarizes our views on where territorial gardening fits into this gradient of grazing intensity, and the fact that it results in algal types which are both productive and of high value, increasing yield to the grazer. It also emphasizes our belief that most gardening

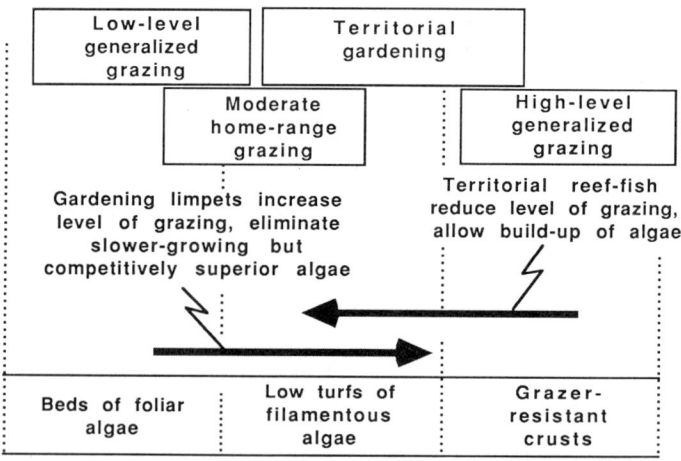

Fig. 18.6. Generalized predictions of how different types of grazing relate to different algal types, and how limpet and fish territoriality influence algal composition.

fish create this condition by decreasing the intensity of grazing (relative to that experienced outside territories), whereas most (if not all) limpets increase the level of grazing.

An evolutionary perspective

Several questions still challenge our understanding of gardening. One is whether algae benefit from being grazed. Much has been written about the potential for coevolution between terrestrial plants and their grazers and the putative benefits that plants gain from being grazed. Westoby (1989) concludes that while individual plants do not benefit from being grazed, they may benefit 'from living in environments where most plants are being grazed'. So, for example, some algae benefit from living where grazing prevents competitively superior species from becoming dominant (e.g., Lubchenco 1978). Encrusting corallines seem to be a clear case (Vine 1974; Brawley and Adey 1977; Morrison 1988). In relation to algae in gardens, a clear dichotomy emerges between those associated with limpets and those with fish. Limpets increase the level of grazing in their gardens, and maintain a relatively small suite of competitively inferior species which would otherwise be overgrown. Most fish decrease the level of grazing in their territories, benefiting algae which would be eliminated by more intense grazing (Fig. 18.6). If algae in gardens appear to 'compensate' for grazing by increased productivity, this does not imply increased fitness, merely that they are being held in an early and productive stage of growth.

What prevents territorial animals from overgrazing their gardens? If non-territorial species gain unfettered access to gardens they may rapidly eliminate them (Lassuy 1980; Branch 1981; Hixon and Brostoff 1983). Yet both territorial fish and limpets maintain gardens for long periods of time without overgrazing. In part the answer may depend on the fact that territorial animals can reap the benefits of an apparently 'prudent' pattern of feeding. The question is particularly interesting in the case of *Patella longicosta*, for if it is placed in ungrazed patches of *Ralfsia* it grazes rapidly at first, but once territories are established and 'paths' cut through the alga, its feeding rate drops substantially, closely balancing the replacement rate of the alga (Branch 1981 and unpublished data). This seemingly 'prudent' reduction of grazing rate may be due to a number of factors, including an improvement of the quality of the food-plant once it is grazed. Grazing may also be inhibited if it induces an increase in polyphenols (cf. Van Alstyne 1988) or if high polyphenol levels are associated with the appearance of fresh, young growths of *Ralfsia*. Young tissues of some algae do possess high levels of anti-herbivore chemicals (Hay *et al.* 1988; Paul and Van Alstyne 1988; but see Ragan and Glombitza 1986; Coley *et al.* 1985).

Our measurements of polyphenols (Fig. 18.7) reveal exactly the opposite: levels are twice as high in ungrazed plants as in gardens. This apparently correlates with the age of the tissues, for young (ungrazed) plants have low and old tissues have high concentrations. Furthermore, polyphenol levels in old plants can be reduced by artificial grazing, and elevated levels can be induced in gardens by removing the resident limpets and allowing the plant to develop to a more mature stage. Possibly reductions of polyphenols in gardens improve the digestibility of *Ralfsia* and allow resident limpets to reduce their grazing rate without reducing yield.

What circumstances favour the development of territorial gardens? Gardening enhances the local production of a food source. But for gardening to evolve, the yield from a garden must be realizable by the individual responsible. This implies the need for a fixed feeding site and the means to defend that site against intruding competitors, both well-documented phenomena. Defence should be most intense against species which present a competitive threat, as Myberg and Thresher (1974) and Ebersole (1977) have shown for two pomacentrid fish.

But this does not in itself provide a reason for territoriality to arise. A second factor may be the relative need for it to develop if background productivities are low, or the demands of grazers are high, or both. This may be the clue as to why particular limpets and fish have embarked on gardening. Around the coasts of southern Africa, for example, there is a strong gradient of nutrients, with upwelling well

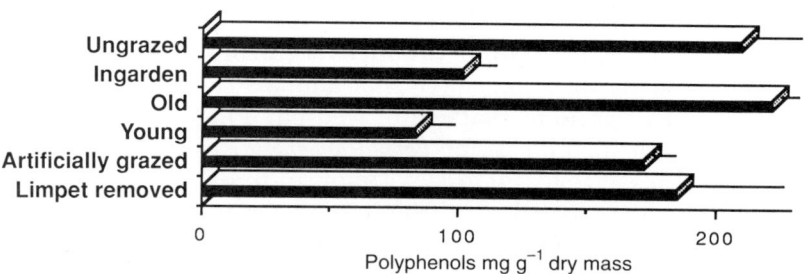

Fig. 18.7. Polyphenol levels in the tissues of *Ralfsia verrucosa*. Data are for plants in gardens of *Patella longicosta*, ungrazed plants outside gardens, old and young ungrazed plants, old plants which were artificially grazed, and gardens from which territorial limpets were removed 12 weeks previously (C. Parkins and G. M. Branch, unpublished data).

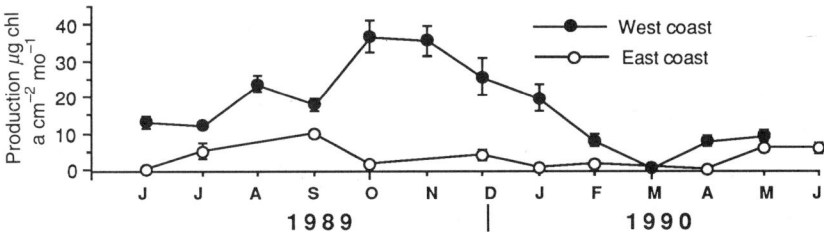

Fig. 18.8. Microalgal productivity (units of chlorophyll a cm^{-2} month^{-1}) on the west and east coasts of South Africa. Data were obtained from, respectively, Groenrivier (R. Bustamante, S. Eekhout, and G. M. Branch, unpublished data) and Dwesa, Transkei (A. Dye and D. White, unpublished data).

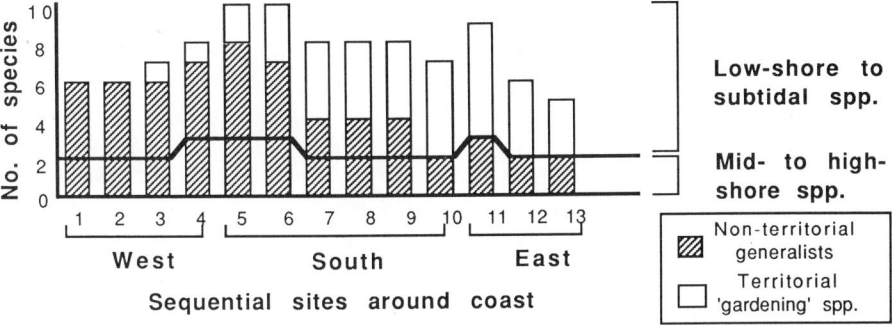

Fig. 18.9. Number of *Patella* spp. and the proportion of species which have territorial gardens at consecutive sites around the coast of southern Africa, from Lüderitz to Cape Vidal (G. M. Branch, unpublished data).

developed on the west coast and virtually absent on the east coast. We do not yet know if this results in an equivalent gradient in intertidal productivity, but measurements at two sites (Fig. 18.8) suggest that this is so. Correlated with this, there is a gradient in the proportion of limpet species that are gardeners (Fig. 18.9). There are also two species (*Patella barbara* and *Patella miniata*) which do not garden on the west coast but do so on the east coast. These facts support the idea that gardening becomes increasingly important in nutrient-poor waters. Gardening by fish occurs predominantly in tropical waters, long recognized as not only nutrient-poor but also regions of intense herbivory (Vine 1974; Hay 1981; Lewis 1986; Morrison 1988; Hatcher 1988).

Finally, for gardening to be feasible, the physical environment must allow continuous localized productivity sufficient to meet the needs of the grazer. In southern Africa, no limpets that live in the high- to mid-shore, where physical stresses seasonally limit algal growth, are territorial gardeners (Fig. 18.9).

These patterns are not merely of local interest, for they illustrate how large-scale driving forces (such as productivity) may influence local-scale biological interactions. As Menge and Olson (1990) have stressed, a major challenge confronting ecologists is the need to understand the relative importance of local and geographical processes before we can develop predictive models of how communities are regulated.

References

Bianchi, T.S., Jones, C.G., and Shachak, M. (1989). Positive feedback of consumer population density on resource supply. *Trends in Ecology and Evolution*, **4**, 234–8.

Branch, G.M. (1975*a*). Intraspecific competition in *Patella cochlear* Born. *Journal of Animal Ecology*, **44**, 263–82.

Branch, G.M. (1975*b*). Mechanisms reducing intraspecific competition in *Patella* spp.: migration, differentiation and territorial behaviour. *Journal of Animal Ecology*, **44**, 575–600.

Branch, G.M. (1976). Interspecific competition experienced by South African *Patella* species. *Journal of Animal Ecology*, **45**, 507–29.

Branch, G.M. (1981). The biology of limpets: physical factors, energy flow, and ecological interactions. *Oceanography and Marine Biology Annual Review*, **19**, 235–380.

Branch, G.M. and Pringle, A. (1985). The impact of the sand prawn *Callianassa kraussi* Stebbing on sediment turnover and on bacteria, meiofauna, and benthic microflora. *Journal of Marine Biology and Ecology*, **107**, 219–35.

Brawley, S.H. and Adey, W.H. (1977). Territorial behaviour of threespot damselfish (*Eupomacentrus planifrons*) increases algal biomass and productivity. *Environmental Biology of Fishes*, **2**, 45–51.

Carpenter, R.C. (1981). Grazing of *Diadema antillarum* (Philippi) and its effects on the benthic algal community. *Ecological Monographs*, **39**, 749–65.

Coley, P.D., Bryant, J.P., and Chapin, F.S. (1985). Resource availability and plant antiherbivore defence. *Science*, **230**, 895–99.

Ebersole, J.P. (1977). The adaptive significance of interspecific territoriality in the reef fish *Eupomacentrus leucosticus*. *Ecology*, **58**, 914–20.

Hatcher, B.G. (1988). Coral reef primary productivity: a begger's banquet. *Trends in Ecology and Evolution*, **3**, 106–11.

Hay, M.E. (1981). Herbivory, algal distribution, and the maintenance of between-habitat diversity on a tropical reef. *American Naturalist*, **118**, 520–40.

Hay, M.E., Paul, V.J., Lewis, S.M., Gustafson, K., Tucker, J., and

Trindell, R. N. (1988). Can tropical seaweeds reduce herbivory by growing at night? Diel patterns of growth, nitrogen content, herbivory, and chemical versus morphological defences. *Oecologia* (Berlin), **75**, 233-45.

Hixon, M. A. and Brostoff, W. N. (1983). Damselfish as keystone species in reverse: intermediate disturbance and diversity of reef algae. *Science*, **220**, 511-13.

Hylleberg, J. (1975). Selective feeding by *Abarenicola pacifica* with notes on *Abarenicola vagabonda* and a concept of gardening in lugworms. *Ophelia* (Berlin), **14**, 113-37.

Kaufman, L. (1977). The three-spot damselfish: effects on benthic biota of Carribean coral reefs. In *Proceedings of the Third International Coral Reef Symposium*, 1, pp. 559-64.

Klumpp, D. W. and Polunin, N. V. C. (1989). Partitioning among grazers of food resources within damselfish territories on a coral reef. *Journal of Experimental Marine Biology and Ecology*, **125**, 145-69.

Klumpp, D. W., McKinnon, D., and Daniel, P. (1987). Damselfish territories: zones of high productivity on coral reefs. *Marine Ecology Progress Series*, **40**, 41-51.

Lassuy, D. R. (1980). Effects of 'farming' behaviour by *Eupomacentrus lividus* and *Hemiglyphidodon plagiometopon* on algal community structure. *Bulletin of Marine Science*, **30**, 304-12.

Lewis, S. M. (1986). The role of herbivorous fishes in the organization of a Caribbean reef community. *Ecological Monographs*, **56**, 183-200.

Lobel, P. S. (1980). Herbivory by damselfishes and their role in coral reef community ecology. *Bulletin of Marine Science*, **30**, 273-89.

Lubchenco, J. (1978). Plant species diversity in a maritime intertidal community: importance of herbivore food preferences and algal competitive abilities. *American Naturalist*, **112**, 23-39.

McNaughton, S. J. (1984). Grazing lawns: animals in herds, plant forms and coevolution. *American Naturalist*, **124**, 863-86.

McQuaid, C. D., Branch, G. M., and Crowe, A. A. (1985). Biotic and abiotic influences on rocky intertidal biomass and richness in the southern Benguela region. *South African Journal of Zoology*, **20**, 115-22.

Menge, B. A. and Olson, A. M. (1990). Role of scale and environmental factors in regulation of community structure. *Trends in Ecology and Evolution*, **5**, 52-7.

Meyer, J. C., Schultz, E. P., and Helfman, G. S. (1983). Fish shoals: an asset for corals. *Science*, **220**, 1047-9.

Montgomery, W. L. (1980). The impact of non-selective grazing by the giant blue damselfish, *Microspathodon dorsalis*, on algal communities in the Gulf of California, Mexico. *Bulletin of Marine Science*, **30**, 290-303.

Morrison, D. (1988). Comparing fish and urchin grazing in shallow and deeper coral reef algal communities. *Ecology*, **69**, 1367-82.

Myberg, A. A. and Thresher, R. E. (1974). Interspecific aggression and its relevance to the concept of territoriality in reef fish. *American Zoologist*, **14**, 81-96.

Paul, V. T. and Van Alstyne, K. L. (1988). Chemical defences and chemical variation in some tropical Pacific species of *Halimeda* (Halimedaceae: Chlorophyta). *Coral Reefs*, **6**, 263-9.

Polunin, N. V. C. and Koike, I. (1987). Temporal focussing of nitrogen release by a periodically-feeding herbivorous reef-fish. *Journal of Experimental Marine Biology and Ecology*, **111**, 286-96.

Ragan, M. A. and Glombitza, K. W. (1986). Phlorotannins, brown algal phenols. *Progress in Phycological Research*, **4**, 129-241.

Risk, M. J. and Sammarco, P. W. (1982). Bioerosion of corals and the influence of damselfish territoriality: a preliminary study. *Oecologia* (Berlin), **52**, 376-80.

Russ, G. R. (1987). Is the rate of removal of algae by grazers reduced inside territories of the tropical damselfish? *Journal of Experimental Marine Biology and Ecology*, **110**, 1-17.

Sammarco, P. W. (1983). Effects of fish and damsel-fish territoriality on coral reef algae I. Algal community structure. *Marine Ecology Progress Series*, **13**, 1-14.

Sammarco, P. W. and Williams, A. H. (1982). Damselfish territoriality: influence on *Diadema* distribution and implications for coral community structure. *Marine Ecology Progress Series*, **8**, 53-9.

Sammarco, P. W., Carlton, J. H., and Risk, M. J. (1986). Effects of grazing and damselfish territoriality on bioerosion of dead corals: direct effects. *Journal of Experimental Marine Biology and Ecology*, **98**, 1-19

Sammarco, P. W., Risk, M. J., and Rose, C. (1987). Effects of grazing and damselfish territoriality on internal bioerosion of dead corals: indirect effects. *Journal of Experimental Marine Biology and Ecology*, **112**, 185-99.

Sterner, R. W. (1986). Herbivores' direct and indirect effects on algal populations. *Science*, **231**, 605-6.

Stimson, J. (1970). Territorial behaviour of the owl limpet, *Lottia gigantea*. *Ecology*, **51**, 113-18.

Stimson, J. (1973). The role of the territory in the ecology of the intertidal limpet, *Lottia gigantea* (Gray). *Ecology*, **54**, 1020-30.

Vine, P. J. (1974). Effects of algal grazing and aggressive behaviour of the fishes *Pomacentrus lividus* and *Acanthurus sohal* on coral-reef ecology. *Marine Biology*, **24**, 131-6.

Van Alstyne, K. L. (1988). Herbivore grazing increases polyphenolic defences in the intertidal brown alga *Fucus distichus*. *Ecology*, **69**, 655-63.

Wellington, G. M. (1982). Depth zonation of corals in the Gulf of Panama:

control and facilitation by resident reef fishes. *Ecological Monographs*, **52**, 223-41.

Westoby, M. (1989). Selective forces exerted by vertebrate herbivores on plants. *Trends in Ecological Evolution*, **4**, 115-17.

Wilkinson, C. R. and Sammarco, P. W. (1983). Effects of fish grazing and damselfish territoriality II. Nitrogen fixation. *Marine Ecology Progress Series*, **13**, 15-19.

Williams, A. H. (1980). The threespot damselfish: a noncarnivorous keystone species. *American Naturalist*, **116**, 138-42.

Williams, S. C. and Carpenter, R. C. (1988). Nitrogen limited primary production of coral reef algal turfs: potential contribution of ammonium excreted by *Diadema antillarum*. *Marine Ecology Progress Series*, **47**, 145-52.

Woodin, S. A. (1974). Polychaete abundance patterns in a marine soft-sediment environment: the importance of biological interactions. *Ecological Monographs*, **44**, 171-87.

Woodin, S. A. (1977). Algal 'gardening' behavior by nereid polychaetes: effects on soft-bottom community structure. *Marine Biology*, **44**, 39-42.

19. Grazing and succession in marine algae

WAYNE P. SOUSA* and JOSEPH H. CONNELL[†]
*Department of Integrative Biology, University of California, Berkeley, California, USA [†]Department of Biological Sciences, University of California, Santa Barbara, California, USA

Abstract

Assemblages of sessile organisms on rocky seashores and nearshore subtidal hard bottoms experience a variety of space-clearing disturbances. The order and rate of recolonization of cleared space by algae and invertebrates depend on characteristics of the initial disturbance, the pool of available colonists, effects exerted by prior occupants on subsequent colonists, and the actions of grazers and predators. Grazers may affect succession in diverse ways depending on the model(s) of succession that is operating (Connell and Slatyer 1977), the effect of the particular grazer on species of differing successional status, and the intensity of grazing. Experimental studies have shown that browsing herbivores, such as crabs, coiled snails, and dipteran larvae, that prefer early-colonizing algae can accelerate succession by breaking the inhibitory effect of these early species on later colonists. In contrast, herbivores such as limpets and sea urchins that damage plants in a less discriminating fashion, or preferentially feed on later successional species, tend to stall succession or to shift the climax state to one composed of species that are well-defended against intense rasping or bulldozing. Lastly, some studies have found that, late in a successional sequence, grazers have little effect on the replacement of one species by another.

Introduction

Ecological succession refers to the sequence of species replacements that occurs in patches of habitat from which a disturbance has removed some or all resident individuals and opened space for recolonization.

Assemblages of sessile organisms on rocky seashores, subtidal hard substrata, and coral reefs experience a variety of space-clearing disturbances (Connell and Keough 1985; Dayton 1985; Huston 1985; Sousa 1985). Open space is a key resource for many sessile marine organisms. It provides sites for attachment as well as access to other essential resources such as light and suspended food. The areal extent and intensity of disturbance determines how large the open patches are and whether there are survivors that can grow to partially or completely fill the space. In this paper, we examine the influence of grazing on the rate and direction of succession after disturbance on marine hard substrates. The models of Connell and Slatyer (1977) provide the framework for our discussion, as do the predictions of Farrell (1991) concerning the effect of different patterns of grazing on the rate of species replacement under each of the Connell–Slatyer models. Against this background, we review the results of experimental studies of the role of grazers in successional dynamics.

Models of succession

Connell and Slatyer (1977) proposed three alternative models of successional change in assemblages of sessile organisms: facilitation, tolerance, and inhibition. All three models agree that a subset of species in the pool of available colonists will characteristically be the first to reoccupy the newly opened space. These species do so by virtue of their broad powers of dispersal and rapid growth to maturity. In some systems, these early successional species exhibit less seasonal patterns of recruitment than later arriving species (Sousa 1979), but in others a different set of opportunistic species appears in each season (e.g., Paine 1977; Turner 1983a). The traditional facilitation model of succession assumes additionally that these first colonists are the only species that are able to recruit to the site and persist there under the supposedly harsh physical conditions that occur immediately after the space has been opened. The tolerance and inhibition models assume that any species, including those that usually appear later in a sequence, can successfully recruit at the beginning if its propagules are dispersed to the site.

The three models differ in the mechanisms that are proposed to drive subsequent species invasions and replacements. In the facilitation model, the early successional species modify the environment so that it is more suitable for later successional species to invade and grow to maturity, while making the site less suitable for the local recruitment of members of their own species. Assuming that facilitation characterizes each replacement in a sequence, succession continues until the resident species that dominates the site no longer modifies it in ways that facilitate the invasion and growth of a different species.

Under the tolerance model, the presence of early successional species neither increases nor decreases the rate of recruitment and growth to maturity of species that attain peak abundance later in succession. These later species colonize either immediately after the disturbance or more commonly some time later. They tend to grow to maturity more slowly than early species, but are more efficient competitors for limited resources. Later successional species are able to survive and grow in the presence of established populations of early species which reduce the availability of resources such as light and space. Each successive species in a sequence is more efficient at exploiting resources than its predecessor. The end point of a sequence of tolerance-mediated replacements is reached when the site is occupied by the species most efficient at resource use. This species reduces the available resources to such low levels that no other species can invade. The mechanism of replacement is competitive exclusion; as populations of later species colonize and grow, they deprive earlier species of resources. For example, a shade-tolerant, later successional algal species would grow up in the shade cast by the canopy of early species, overtop them, and shade them out.

The inhibition model predicts that well-established populations of early successional species pre-empt patch resources preventing the invasion of subsequent colonists or suppressing the growth of those species that colonized soon after the disturbance. Unlike the tolerance model, early species are not killed by competition with later species. The former can resist invasion as long as they remain healthy and undamaged. The observed replacement of early successional species by later species is a simple product of differential longevity of the species that comprise each stage. Species characteristic of later successional stages replace early species by outliving them (often in a suppressed condition beneath the canopy), gradually accumulating as early species succumb to physical stress and to natural enemies including grazers, predators, or parasites.

A number of authors (e.g., Dean and Hurd 1980; Day and Osman 1981; Turner 1983*a*; Quinn and Dunham 1983; Breitburg 1985; Pickett *et al.* 1987; Walker and Chapin 1987) have commented that the Connell–Slatyer models ignore several sources of variability in succession including seasonal and spatial variation in recruitment, growth, and mortality; density-dependence in the quality (sign) of interspecific interactions; heterogeneous effects of different species within the same successional stage; and indirect interactions among species. The models were also faulted for assuming that the same mechanism accounts for all species replacements in a sequence. Such sources of variation are indeed important, but it is unrealistic to expect a general model to incorporate all of them. Although not explicitly stated in Connell and Slatyer (1977), no mechanism was intended to apply to an entire successional sequence and the models focused only on the net effects of earlier species on later colonists (Connell et al. 1987). While there are some successional sequences in which the same general mechanism accounts for each species replacement (Sousa 1979), in other successions several mechanisms may operate (e.g., Turner 1983a, 1983b; Farrell 1991). In this paper, we consider each species replacement in time as a separate event to which the proposed mechanisms of Connell and Slatyer (1977) may or may not apply. We recognize that the three models represent extremes along a continuum of effects of earlier on later species (Dean and Hurd 1980; Connell et al. 1987).

Grazing and the rate of succession

It is not the aim of this paper to review the evidence for or against these successional models. Rather, we wish to examine the consequences of overlaying different patterns of grazing on an underlying model of successional change. In other words, given that a particular mechanism of species replacement operates in the absence of grazing, what effects do different patterns of grazing have on the direction and rate of change? To dissociate grazing from the successional models is clearly artificial, particularly with respect to the inhibition model, since differential mortality due to grazing is one of a suite of mechanisms that could break the inhibitory effect of early species and cause their replacement by less-vulnerable, longer-lived, later successional species. For the sake of clarity, however, we will treat grazing and the successional models as independent processes. Farrell (1991) has recently made a set of predictions concerning the influence of consumers on the rate of species replacement under the three Connell–Slatyer models of succession. This influence depends not only on the model of succession but also on the relative damage or mortality caused by the consumer to members of the different successional stages. Farrell's predictions are general in that they may apply to the effects of consumers on succession in any assemblage of sessile organisms. Our discussion of his predictions focuses more narrowly on the effects of grazing herbivores on plant or plant/animal successions in marine benthic environments.

Farrell (1991) considers the influence of three qualitatively different patterns of grazing (Fig. 19.1). Grazers differentially damage or remove:

		Successional status of algal species that grazers damage most		
		Earlier	Neither	Later
Model of succession	Facilitation	−	−	−
	Tolerance	0	−	−
	Inhibition	+	0	−

Fig. 19.1. Effect of different patterns of damage or mortality from grazing on the rate of species replacement under the Connell–Slatyer models of sucession. + = rate increase, − = rate decrease, 0 = little or no effect. Column headings indicate whether net damage by grazers falls more heavily on earlier successional species, later successional species, or neither (i.e., equal damage to both). Modified from Farrell (1991).

(1) earlier successional species;
(2) later successional species;
(3) neither category of species, i.e. the net loss of cover by early and late species to grazing is equivalent.

Regardless of the successional model, differential damage or mortality of later colonizing species slows the rate at which they replace species that dominate earlier in succession. Predicted effects of grazing vary with the underlying model of succession when grazing is more harmful to early species or equally damaging to populations of all species. Under a facilitation model, greater loss of early successional species to grazers will slow the rate at which they are replaced, since the cover of these species makes the environment more favourable for the establishment of later species. Non-selective damage will also slow species replacement under this model, since the facilitating influence is reduced as well as the survival of later species that have colonized.

The tolerance model represents an idealized relationship that lies midway between the facilitation and inhibition models. It represents one of the two extreme cases of asymmetrical competition between early and later-species. Early species are assumed to have no negative effect on later species. The inhibition model represents the opposite extreme in which earlier species are completely dominant over later species and can exclude them indefinitely as long as early species hold the space. While recognizing that in reality there is a continuum of relative competitive abilities, we will focus on the extremes for the purpose of discussion.

Under the tolerance model, differential removal of the cover of early species will not affect the rate of their replacement since the establishment and growth of later species is unaffected by earlier colonists. When grazing damage is non-selective, species replacement should slow; the removal of early species has no effect on the rate, but the loss of later colonists will slow the rate at which they accumulate on the site.

If early species inhibit the recruitment and growth of later colonists (inhibition model), removal of the former should speed their replacement. Non-selective damage will have little net effect on the rate of replacement; the loss of early species will enhance the recruitment of later species, but those that do colonize will be removed at roughly the same rate.

The above predictions concern qualitative changes in the rate of succession when damage due to grazing varies among species of different successional status. In situations predicted to produce a rate change, the more intense the grazing damage, the more markedly the rate will be altered.

Experimental evidence

Very few published studies have been explicitly designed to evaluate which, if any, of the Connell–Slatyer models apply to succession on marine hard substrata. Ideally, we would want to know the underlying successional mechanism before embarking on a test of Farrell's (1991) predictions. The three successional models are distinct in the predicted effect of earlier species on the recruitment and growth of later species. However, only a few studies (Sousa 1979; Sousa *et al.* 1981; Lubchenco 1983; Turner 1983*a*, *b*; Johnson and Mann 1988; Farrell, 1991) have directly manipulated the abundance of earlier successional species, an experimental protocol recommended by Connell and Slayter (1977). Other studies have adopted the indirect approach of manipulating the density of grazers (treatments are usually complete removal of grazers vs ambient density) and inferring the model of succession from observed patterns of species replacement with and without grazing. There is obviously an inherent circularity in using this procedure as a test of Farrell's predictions, since one must assume that the predictions are correct before an underlying successional mechanism can be inferred. In those manipulative studies of successional dynamics that are of this indirect sort, we cannot rigorously test the predictions. Instead, we will examine cases in which grazing accelerates, slows, or has little or no effect on the rate of successional species replacement, and will evaluate whether the patterns revealed are consistent with Farrell's predictions. We only consider the results of studies that have conducted replicated experiments in the field or, in one case, a realistic laboratory microcosm.

1. Cases in which grazing accelerates succession

A relatively large number of studies have found that grazing by particular herbivores causes more rapid replacement of certain early species by later ones (Table 19.1). As noted earlier, the first and third studies in Table 19.1 (Sousa 1979; Lubchenco 1983) independently manipulated the cover of the earlier successional species and thereby demonstrated that, in the absence of grazing, earlier species inhibit the establishment of later species. These same studies, as well as Irvine's (1983) investigation of algal succession in damselfish territories on a Caribbean coral reef, have shown, by laboratory and field experiments, that the grazer preferentially feeds on the early successional species.

It is doubtful that the earlier species in these successions could prevent the gradual accumulation of later species, even in the absence of grazers, since the earlier species listed in Table 19.1 are generally more vulnerable to physical stresses such as desiccation and water motion. Clearly, however, grazing increases the rate of species replacement.

The studies of Sousa (1979) and Lubchenco (1983) unambiguously support Farrell's (1991) prediction that under the inhibition model of succession, selective grazing damage to earlier species should increase the rate of succession. The other studies in Table 19.1 are also consistent with this prediction, assuming that the presence of grazers *per se* is not facilitating the establishment of later species. While it may seem far-fetched, it is possible that mucus (Connor and Quinn 1984) or metabolic waste produced by the grazer could selectively fertilize later successional species, thereby increasing their rate of establishment, independent of any grazing damage incurred by earlier species.

Underwood *et al.* (1983) investigated the effects of different

Table 19.1. Successional species transitions in which grazing damage to earlier species accelerates their replacement by later species.

Habitat	Earlier species	Later species	Grazer	Reference
Temperate rocky intertidal	Green alga: *Ulva* sp.	Red algae: *Gigartina* spp.	Crab: *Pachygrapsus crassipes*	Sousa (1979)
	Green and red algae: *Urospora* sp. and *Bangia* sp.	Green and red algae: *Enteromorpha* spp., *Ulva* sp., and *Porphyra* sp.	Larval diptera: *Paraclunio* spp.	Robles and Cubit (1981)
	Green and red algae: *Ulva lactuca* *Enteromorpha* spp., and *Porphyra* spp.	Brown alga: *Fucus vesiculosus*	Snail: *Littorina littorea*	Lubchenco (1983)
	Red alga: *Halosaccion glandiforme*	Red alga: *Endocladia muricata*	Limpets and a chiton: *Lottia* spp. and *Lepidochitona dentiens*	Johnson (1989), pers. comm.
	Diatoms and the green alga: *Ulva* sp.	Barnacles: *Chthamalus* spp.	Limpets: *Collisella* spp. and *Notoacmaea fenestrata*	Sousa (1979)

Habitat	Early colonizers	Later colonizers	Grazer	Reference
	Green algae: *Ulva lactuca* and *Enteromorpha intestinalis*	Barnacle: *Tesseropora rosea*	Limpet: *Cellana tramoserica* (mid-shore: at densities of 2 and 4 per 400 cm^2; low shore: at all densities)	Underwood et al. (1983)
	Diatoms and the green algae: *Ulva* sp. and *Enteromorpha* sp.	Barnacles: *Chthamalus* spp.	Limpets: *Collisella* spp.	Van Tamelen (1987)
Tropical coral reef (subtidal)	Diatoms and the green algae: *Enteromorpha* spp. and *Derbesia vaucheriaeformis*	Red algae: *Polysiphonia* spp.	Damselfish: *Eupomacentrus planifrons*	Irvine (1983)
Tropical coral reef (laboratory microcosm)	Mixed turf of filamentous green, brown, and red algae	Red alga: *Hypnea spinella*	Amphipod: *Ampithoe ramondi*	Brawley and Adey (1981)

densities of limpets on the successional transition from green algae to barnacles at low and mid-shore levels. On the low shore, where algae grow more rapidly than at higher levels, algal cover declined and barnacle recruitment increased monotonically with increasing limpet density. At mid-shore heights, barnacle recruitment was similarly enhanced by low densities of limpets (2 or 4 limpets 400 cm^{-2}) which differentially removed the cover of green algae. However, as discussed below, higher densities of limpets at mid-shore levels reduced the abundance of both green algae and barnacles, and the rate of establishment of barnacles was no different from that in the absence of limpets. All these results agree with Farrell's (1991) preditions for the inhibition model (Fig. 19.1).

2. Cases in which grazing slows or stops succession

We have identified three examples of grazers slowing or stopping successional replacement of species. Sousa *et al.* (1981) examined the effects of grazing by the urchin *Strongylocentrotus purpuratus* on algal succession in the low-intertidal zone of southern California. When urchins were experimentally removed, the successional sequence from early → later on the newly opened space was: mixed assemblage of the green alga *Ulva* sp., filamentous red algae, and crustose corallines → turf of perennial red algae including *Gigartina canaliculata* and *Laurencia pacifica* → large, perennial, brown algae primarily *Egregia laevigata* and *Cystoseira osmundacea*. Urchin-grazing maintained the assemblage in the early successional stage, with a gradual increase over time in the cover of grazing-resistant coralline algae. Laboratory feeding trials showed that *E. laevigata* is relatively highly preferred by urchins over all other fleshy algal species in the sequence; there is little obvious discrimination among the other algal species. Experimental manipulations of algal cover in the absence of urchin-grazing showed that inhibition characterized each step of the sequence. Thus, this example appears to be consistent with the prediction by Farrell (1991) that under an inhibition model of succession, selective damage to later species will slow or stop succession (Fig. 19.1). Sea urchins appear to have a similar effect on algal species composition in many rocky subtidal habitats (Lawrence 1975; Harrold and Reed 1985; Chapman and Johnson 1990), but the phenomenon has rarely been examined in a successional context.

Farrell (1991) found that barnacles (*Chthamalus dalli* and *Balanus glandula*) increased the rate of establishment of later successional, perennial brown (*Pelvetiopsis limitata* and *Fucus distichus*) and red (*Mastocarpus papillatus* and *Endocladia muricata*) algae by providing them a spatial refuge from limpet grazing. This case represents an example of facilitation succession in which the grazer selectively damages the later successional species. In accord with his own predictions (Fig. 19.1), Farrell found the rate of succession, expressed as the rate of macroalgal establishment, to be lower in the presence than in the absence of limpets. Other workers have documented similar barnacle *refugia* for perennial algae (Hawkins 1981; Lubchenco 1983), but the successional context of these relationships was not discussed.

Harris *et al.* (1984) documented an analogous case of facilitation in a temperate subtidal kelp forest. Early successional, filamentous brown algae provide young kelp sporophytes with a refuge from fish-grazing. Their experiments manipulated the cover of filamentous algae, but not the abundance of grazers, so it is not known if the net effect of grazers is to slow the establishment of kelp as Farrell (1991) would predict.

Irvine's (1983) study of algal succession within damselfish territories on a coral reef provides another example of selective grazing damage to later species slowing succession. When algal mats are protected from fish-grazing, the middle successional stage dominated by the red alga *Polysiphonia* is invaded by blue-green algae and perennial red and brown algae including *Dictyota* and *Jania*. Damselfish prevent this from occurring by selectively weeding out these later successional species. They are plucked from the turf (but not consumed), leaving behind earlier successional species which are preferred by the fish as food. In this case, while the mechanism of succession that leads to the establishment of the later successional species when fish are excluded is unknown, their selective removal by the fish stalls succession. This pattern would be predicted under any of successional models (Fig. 19.1). Several other studies of the effects of damselfishes on coral-reef algal succession have shown that these grazers retard the establishment of later successional species by differentially weeding them (Lassuy 1980) or by consuming them, selectively or not, at a faster rate than they can regrow (Montgomery 1980; Hixon and Brostoff 1983).

3. Cases in which grazing has little or no effect on the rate of succession

Finally, there are some studies that have manipulated the abundance of grazers during a succession and found little effect on the rate or pattern of species replacement. Turner (1983*a*, *b*) experimentally demonstrated that both inhibition and facilitation occurred over the course of a low intertidal plant succession in Oregon. The early successional brown alga *Phaeostrophion irregulare* inhibited the establishment of other early successional species, including a species of the green alga *Ulva* and filamentous diatoms. In contrast, the middle successional red algae

Rhodomela larix and *Odonthalia floccosa* facilitated the establishment of the surfgrass *Phyllospadix scouleri*. Manipulations of limpet or turban snail densities demonstrated a modest negative impact on the cover of the *Ulva* species but no obvious effect on the rate of succession. Turner (1983a) suggested that physical stress, especially seasonal desiccation stress, and variable recruitment play a more significant role in species replacement than does grazing, especially in the early and middle stages of succession. The late successional establishment and dominance of surfgrass is effected by specific morphological features of its seeds that enhance recruitment into the red algal turf (Turner 1983b). Grazing seems to play little role at this stage of succession.

Similarly, Jernakoff (1985) studying algal succession on a temperate rocky shore in Australia found little effect of grazing by the limpet *Patelloida latistrigata* on the rate or pattern of succession. Earlier successional species (diatoms and ephemeral green and red algae) were negatively affected by limpet grazing, while later species (blue-green and encrusting brown algae) were unaffected. However, the rate of establishment of later species did not differ in the presence vs absence of limpets. The model of succession was not identified. Both this system and that studied by Turner were characterized by large spatial variation in algal recruitment.

Foster (1982) monitored the re-establishment of cover by the red algae *Iridaea flaccida* and *Iridaea cordata* in experimental clearings within and below the zone dominated by *I. flaccida*. At the higher tidal height, recovery of *Iridaea* cover was temporarily slowed in control plots compared with plots in which the densities of molluscan grazers (limpets and turban snails) were experimentally reduced, but this difference persisted for less than one year, after which no significant difference could be detected among treatments. At the lower tidal height, no difference ever developed among treatments. Overall, grazers had little effect on the rate or course of succession. Foster (1982) did not investigate the mechanism of succession or the species-specific susceptibility of the algae to grazing.

Sousa (1979) found that limpet-grazing indirectly enhanced barnacle recruitment by removing the inhibiting influence of early successional algal sporelings that pre-empt the space (Table 19.1). However, this grazing had little effect on the rate of establishment of later stages dominated by perennial red algae. The densities of limpets gradually declined on control plots as these asexually reproducing red algal turfs gradually dominated the space.

As mentioned above, Underwood *et al.* (1983) found that at mid-shore levels high densities of the limpet *Cellana tramoserica* (6 and 8 per 400 cm^2) strongly reduce the recruitment and standing abundance of both early successional green algae and later successional barnacles. The rate of succession at high limpet densities, judged by the rate of establishment of barnacles, was not different from that in the treatment with no limpets. In other words, under an inhibition model, when both stages are equally negatively affected by grazers, the rate of succession is unaffected, as predicted in Fig. 19.1.

Discussion

A rich variety of successional mechanisms has been experimentally documented in assemblages of marine algae. Inhibition of late species by early colonists is common, more so than instances of the traditional facilitation mechanism. Of the three demonstrated cases of facilitation, none involved amelioration of harsh physical conditions *per se*. Harris et al. (1984) and Farrell (1991) found that early species provided later colonists with a refuge from grazing, and Turner (1983b) showed that earlier species provided a required substrate for the attachment of propagules of a later species.

Farrell's (1991) predictions concerning the impact of different patterns of grazing on rates of succession seem to be generally supported by the available experimental evidence. The number of studies that bear on the problem is, however, surprisingly few. Even in those that do, the successional mechanisms and the influence of grazers are rarely evaluated independently. Future progress towards understanding the interaction between these processes will require both kinds of information.

Of the three possible effects of grazing on the rate of succession, positive, negative, or none, cases of the latter require especially careful investigation. Such a neutral effect could occur in spite of substantial grazing pressure if the successional model is:

(1) tolerance, and grazing damage falls mainly on an early species, or
(2) inhibition, and grazing, is indiscriminate (Fig. 19.1).

Alternatively, grazers may have little or no effect on successional rate simply because they have little impact on algal populations. The density of grazers might be too low, due to poor recruitment, high predation, or competition with sessile species, including the algae themselves (e.g., Underwood and Jernakoff 1981), to have much effect. Other possibilities are that grazers may be too small relative to the algal thalli to remove much tissue, or algal productivity (recruitment or growth) may be too high for existing grazers to affect standing algal cover.

With regard to the impact of grazers on successional dynamics, the results of Underwood et al. (1983) suggest that it will be important to evaluate the effects of different densities of grazers across a range of environments that vary in algal productivity (also see Underwood and Denley 1984; Underwood 1985). Grazer density and the intensity of their impact on algal assemblages will often vary in space and time, as well as interactively with the patch structure of the environment (Sousa 1984, 1985). Finally, we need to pay closer attention to the role that variation in the availability of propagules plays in successional dynamics (Sousa 1984; Underwood and Denley 1984).

Acknowledgements

We wish to thank B. Mitchell, D. M. John, and an anonymous reviewer for critical comments on early versions of the paper. Discussions with G. Branch, S. Brawley, M. Foster, B. Hatcher, S. Hawkins, L. Johnson, L. McCook, A. Underwood, and R. Vadas helped to clarify our ideas and provided additional examples. The support of NSF grants OCE 80-08530 to WPS and OCE 88-22930 to JHC are gratefully acknowledged.

References

Brawley, S. H. and Adey, W. H. (1981). The effects of micrograzers on algal community structure in a coral reef microcosm. *Marine Biology*, **61**, 167-77.

Breitburg, D. L. (1985). Development of a subtidal epibenthic community: factors affecting species composition and the mechanisms of succession. *Oecologia* (Berlin), **65**, 173-84.

Chapman, A. R. O. and Johnson, C. R. (1990). Disturbance and organization of macroalgal assemblages in the North west Atlantic. *Hydrobiologia*, **192**, 77-121.

Connell, J. H. and Keough, M. J. (1985). Disturbance and patch dynamics of subtidal marine animals on hard substrata. In *The ecology of natural disturbance and patch dynamics*, (ed. S. T. A. Pickett and P. S. White), pp. 125-51. Academic Press, Orlando, Florida.

Connell, J. H. and Slatyer, R. O. (1977). Mechanisms of succession in natural communities and their role in community stability and organization. *American Naturalist*, **111**, 1119-44.

Connell, J. H., Noble, I. R., and Slatyer, R. O. (1987). On the mechanisms producing successional change. *Oikos*, **50**, 136-7.

Connor, V. M. and Quinn, J. F. (1984). Stimulation of food species growth by limpet mucus. *Science*, **225**, 843-4.

Day, R. W. and Osman, R. W. (1981). Predation by *Patiria miniata*

(Asteroidea) on bryozoans: prey diversity may depend on the mechanism of succession. *Oecologia* (Berlin), **51**, 300-9.

Dayton, P. K. (1985). Ecology of kelp communities. *Annual Review of Ecology and Systematics*, **16**, 215-45.

Dean, T. A. and Hurd, L. E. (1980). Development in an estuarine fouling community: the influence of early colonists on later arrivals. *Oecologia* (Berlin), **46**, 295-301.

Farrell, T. M. (1991). Models and mechanisms of succession: an example from a rocky intertidal community. *Ecological Monographs*, **61**, 95-113.

Foster, M. S. (1982). Factors controlling the intertidal zonation of *Iridaea flaccida* (Rhodophyta). *Journal of Phycology*, **18**, 285-94.

Harris, L. G., Ebling, A. W., Laur, D. R., and Rowley, R. J. (1984). Community recovery after storm damage: a case of facilitation in primary succession. *Science*, **224**, 1336-8.

Harrold, C. and Reed, D. C. (1985). Food availability, sea urchin grazing, and kelp forest community structure. *Ecology*, **66**, 1160-9.

Hawkins, S. J. (1981). The influence of season and barnacles on the algal colonization of *Patella vulgata* exclusion areas. *Journal of the Marine Biological Association of the United Kingdom*, **61**, 1-15.

Hixon, M. A. and Brostoff, W. N. (1983). Damselfish as keystone species in reverse: intermediate disturbance and diversity of reef algae. *Science*, **220**, 511-13.

Huston, M. A. (1985). Patterns of species diversity on coral reefs. *Annual Review of Ecology and Systematics*, **16**, 149-77.

Irvine, G. V. (1983). Fish as farmers: an experimental study of herbivory by a coral reef damsel fish. Unpublished Ph.D. Thesis. University of California, Santa Barbara.

Jernakoff, P. (1985). An experimental evaluation of the influence of barnacles, crevices and seasonal patterns of grazing on algal diversity and cover in an intertidal barnacle zone. *Journal of Experimental Marine Biology and Ecology*, **88**, 287-302.

Johnson, C. R. and Mann, K. H. (1988). Diversity, patterns of adaptation, and stability of Nova Scotian kelp beds. *Ecological Monographs*, **58**, 129-54.

Johnson, L. E. (1989). Spatial and temporal influences on the recruitment of intertidal red algae. Unpublished Ph.D. Thesis. University of Washington.

Lawrence, J. M. (1975). On the relationships between marine plants and sea urchins. *Oceanography and Marine Biology Annual Review*, **13**, 213-86.

Lassuy, D. R. (1980). Effects of 'farming' behavior by *Eupomacentrus lividus* and *Hemiglyphidodon plagiometopon* on algal community structure. *Bulletin of Marine Science*, **30**, 304-12.

Lubchenco, J. (1983). *Littorina* and *Fucus*: effects of herbivores, substratum heterogeneity, and plant escapes during succession. *Ecology*, **64**, 1116-23.

Montgomery, W. L. (1980). The impact of non-selective grazing by the giant blue damselfish, *Microspathodon dorsalis*, on algal communities in the Gulf of California, Mexico. *Bulletin of Marine Science*, **30**, 290-303.

Paine, R. T. (1977). Controlled manipulations in the marine intertidal zone, and their contributions to ecological theory. *Academy of Natural Sciences of Philadelphia, Special Publication*, **12**, 245-70.

Pickett, S. T. A., Collins, S. L., and Armesto, J. J. (1987). Models, mechanisms and pathways of succession. *Botanical Review*, **53**, 335-71.

Quinn, J. F. and Dunham, A. E. (1983). On hypothesis testing in ecology and evolution. *American Naturalist*, **122**, 602-17.

Robles, C. J. and Cubit, J. (1981). Influence of biotic factors in an upper intertidal community: dipteran larvae grazing on algae. *Ecology*, **62**, 1536-47.

Sousa, W. P. (1979). Experimental investigations of disturbance and ecological succession in a rocky intertidal algal community. *Ecological Monographs*, **49**, 227-54.

Sousa, W. P. (1984). Intertidal mosaics: patch size, propagule availability, and spatially variable patterns of succession. *Ecology*, **65**, 1918-35.

Sousa, W. P. (1985). Disturbance and patch dynamics on rocky intertidal shores. In *The ecology of natural disturbance and patch dynamics*, (ed. S. T. A. Pickett and P. S. White), pp. 101-24. Academic Press, Orlando, Florida.

Sousa, W. P., Schroeter, S. C., and Gaines, S. D. (1981). Latitudinal variation in intertidal algal community structure: the influence of grazing and vegetative propagation. *Oecologia* (Berlin), **48**, 297-307.

Tamelen, P. G. van (1987). Early successional mechanisms in the rocky intertidal: the role of direct and indirect interactions. *Journal of Experimental Marine Biology and Ecology*, **112**, 39-48.

Turner, T. (1983a). Complexity of early and middle successional stages in a rocky intertidal surfgrass community. *Oecologia* (Berlin), **60**, 56-65.

Turner, T. (1983b). Facilitation as a successional mechanism in a rocky intertidal community. *American Naturalist*, **121**, 729-38.

Underwood, A. J. (1985). Physical factors and biological interactions: the necessity and nature of ecological experiments. In *The ecology of rocky coasts*, (ed. P. G. Moore and R. Seed), pp. 372-90. Hodder and Stoughton, London.

Underwood, A. J. and Denley, E. J. (1984). Paradigms, explanations, and generalizations in models for the structure of intertidal communities on rocky shores. In *Ecological communities: conceptual issues and the evidence*, (ed. D. R. Strong Jr., D. Simberloff, L. G. Abele, and A. B. Thistle), pp. 151-80. Princeton University Press, Princeton, New Jersey, USA.

Underwood, A. J. and Jernakoff, P. (1981). Effects of interactions between algae and grazing gastropods on the structure of a low-shore intertidal algal community. *Oecologia* (Berlin), **48**, 221-33.

Underwood, A.J., Denley, E.J., and Moran, M.J. (1983). Experimental analyses of the structure and dynamics of mid-shore rocky intertidal communities in New South Wales. *Oecologia* (Berlin), **56**, 202-19.

Walker, L.R. and Chapin, F.S. III. (1987). Interactions among processes controlling successional change. *Oikos*, **50**, 131-5.

20. Competition and marine plant–animal interactions

A.J. UNDERWOOD
Institute of Marine Ecology, Zoology Building, University of Sydney, NSW, Australia

Abstract

Competitive interactions for food are widespread amongst marine invertebrate grazers. The common occurrence of such interactions is due to the unpredictable increases in density of grazers that can occur as a result of large fluctuations in the timing and intensity of settlement or recruitment from planktonic stages. In addition, the supplies of algal foods, particularly microalgae that are consumed by many intertidal grazing gastropods, are also variable. Thus, a superfluity of grazers, or shortage of resource, may happen in any area without regard to any previous abundances of the food or its consumers. Despite the widespread nature of competitive interactions for food, general synthesis is not yet possible about many aspects of this process. These include the nature of symmetry or asymmetry of competition, particularly where grazers have different modes of feeding. It is not yet clear whether competition will necessarily lead to local exclusion of species from patches of a habitat. These areas of uncertainty are discussed with respect to suitable designs of experimental analyses of this widespread process in benthic systems.

Introduction

Competitive interactions in shallow coastal habitats have received considerable attention both for original experimental studies and as topics for reviews. Competition is generally considered to be widespread in these habitats. There are several reasons why this might be so and competitive interactions are of undoubted importance as one of the processes determining the structure of assemblages of species in this type of habitat. Firstly, competitive interactions for space are known to be important, even between organisms of quite different phyla, or even kingdoms (e.g., algae and Crustacea, algae and molluscs). This sort of competitive interaction for space is critical in many marine habitats, because space, as a two-dimensional resource, is often in short supply (see reviews by Connell 1972; Underwood 1979, 1986; Branch 1984). Space, as a resource, is not often easily renewed — renewal requires some sort of disturbance or other process to remove the existing occupants. In some studies, disturbance is relatively frequent (e.g., Dayton 1971; Paine 1979). In other situations, organisms that acquire space may require serious large-scale disturbances, or a long period of time to elapse before they are removed and the space is again free for colonization (e.g., Sousa 1979a, b).

In addition, organisms in many shallow coastal habitats, particularly rocky intertidal ones, have widespread dispersive stages of their life history, and there is little opportunity for regulation of numbers in local populations (Underwood and Fairweather 1989). As a result, there is considerable decoupling between the numbers of adults in any area at one time and the numbers of offspring that may arrive there at some subsequent time. Therefore, situations in which numbers of organisms increase beyond the carrying capacity of some local habitat tend to be widespread in organisms with dispersive propagules. For this reason, competitive interactions have often been studied in such organisms on rocky intertidal shores.

Finally, the types of organisms that are in these habitats tend to be numerous, widespread, have life histories that are of the scale of years to tens of years (rather than minutes or decades) and do not have the emotional appeal of many vertebrate organisms in terrestrial habitats. These features make them very suitable for experimental manipulations, particularly ones that can be well replicated and controlled. The spatial scales on which many of the organisms operate are very small. A large majority of organisms that have been studied experimentally have been sessile (algae, barnacles) or relatively sedentary (limpets, chitons, gastropods, starfish). As a result, experimental manipulations can be done in relatively small areas and be replicated easily in space and time. This is a much harder proposition where organisms roam over many metres, hundred of metres, or even larger distances during their normal activities.

The purpose of the present review is not to go over all the examples of competitive interactions that occur in these habitats, since they have been widely reviewed in recent years (Connell 1972, 1983; Underwood 1979, 1986; Schoener 1983; Branch 1984). Here, the purpose is to examine situations which might affect the interactions between animals and plants, and to determine in what ways competitive interactions might affect plant–animal interactions. Throughout, however, it will be necessary to discuss some of the problems of the design, analysis, or interpretation of experimental studies. There are still far too many problems for complacency and passive acceptance of the types of experimental procedures that are used (see particularly, Hurlbert 1984; Underwood 1986).

Throughout this chapter competitive interactions are considered to be those as defined by Birch (1957). He considered that competition would be found when any two or more organisms that required the same resources occurred together, and the resources were in short supply such that the organisms in some way harmed or impaired each other while trying to use them. This is an effective operational definition. It has the features of focusing on the necessities to identify the resources, the reasons why they are in short supply, and the harm that organisms might do to each other under these circumstances. There are always difficulties with definitions, and this one is no exception. Welden and Slauson (1986) have objected to the simple use of the term 'resources'. They preferred 'resource items' to draw attention to the fact that, for most organisms, competitive interactions are for particular items, rather than some long list of potential resources. This may or may not be an improvement, but it is recognized here that, as with Welden and Slauson's (1986) requirements, the particular resources for which competition may be occurring must be identifiable and identified.

The other problem with definitions of competition is that there are many ways in which competitive interactions can be brought about (Schoener 1983). Because there are already so many problems in the design of a number of experimental field studies (Underwood 1986), it is probably not yet possible to be sure exactly what process is operating in each case or even that the process operating is competitive (e.g., Benke 1978).

The circumstances under which competition is likely to occur in ways that might affect interactions between marine animals and plants are threefold. Firstly, competition will occur when resources of food are in short supply for grazers. Under these circumstances, competitive interactions within species and among species of grazers are inevitable. Many studies (review by Branch 1984) have indicated that grazing gastropods and chitons are subject to extreme competition, leading to mortality when food supplies are short. Secondly, resources of space might intrinsically be in short supply. This is particularly the case on rocky shores, where much of the space is already occupied by sessile organisms, including plants, that otherwise do not share similar resources with the grazers. There are essentially two forms of competitive interaction when space is in short supply. First, there is pre-emption — one organism already occupies space, making it unavailable for another (Underwood and Denley 1984). Second, but much more often described, are overgrowth interactions where organisms manage to settle and grow on the shore but subsequently some organisms smother others (Connell 1961a, b). These two processes of competition for space are quite different. In the first case, organisms arriving at a site where the resource of space is already occupied may, under some circumstances, be able to move away and thus will not be affected (unless the increased time in the plankton causes increased mortality) by the competitive interaction. Competition will then dictate aspects of the structure of the local community where space is occupied, but does not particularly have any effect on the populations of the species which cannot invade. This assumes, of course, that free space somewhere else will be available. In the other case, not only do competitive interactions leading to overgrowth of one species by another result in changes in structure of a local assemblage, but the individuals that are smothered are also killed. Under these circumstances, there is an effect not only on the local assemblage, but also directly on the size of the population. Underwood (1978) and Branch (1984) have discussed other differences between competition for food and competition for space.

The third set of circumstances under which competition is going to be evident in marine habitats are those where the resources are not intrinsically in short supply. Thus, at one time, there may be sufficient food and space for populations of grazers living on a shore. Competition is then caused by the arrival of a large number of consumers of resources from the pelagic stage of their life history. Thus, if an inordinate number of barnacles (or limpets) happen to settle, metamorphose, survive, and grow in an area, they will remove the space (or food) that was previously available to the grazers. Clearly, therefore, shortage of, and problems in the use of resources may be caused by quite different processes. There may be declines in food supply that are not directly related to the populations of grazers. There may be increases in numbers of grazers that are unrelated to the food supply.

The emphasis in what follows is on the role of experimental studies of competitive interactions among plants, among grazers, and between grazers and plants. This is not intended to be a complete review because of the large number of previous reviews that are available.

Competition among plants

Competitive interactions among seaweeds have been reviewed by Denley and Dayton (1985). In general, the consensus is that the plants tend to compete for space (Branch 1984). It is extremely unlikely that plants would compete for nutrients (they are all bathed equally in nutrients), but they might be competing for light. This would be manifested, in most cases, by competition for space. Space on which to live and grow is an absolute prerequisite for gaining access to light. If some plants grow taller than others, thus depriving the shorter ones of access to light, the effect is essentially that there is no space on which the taller plants are absent, and competition for space is the ultimately observable result of competitive interactions for light.

Pre-emption is a common phenomenon in interactions between algae (Foster 1982). For example, Lubchenco (1980) removed the encrusting alga *Chondrus crispus* from areas low on the shore, below the lower boundary of species of *Fucus*. Provided that the *Chondrus* remained clear, *Fucus* could become established at lower levels than normal. Thus, she was able to demonstrate, by direct manipulation, that competition for space at low levels on the shore eliminated the higher-shore species and prevented them from moving down-shore.

In some other similar studies, however, removal of a lower-shore species allowed an increase in abundance of a higher-shore species, but the effect of the lower-shore species was insufficient to account for the lower boundary of the upper species (e.g., Schonbeck and Norton 1978; see Underwood 1991, for details). Because such experiments have produced different results, it is always necessary to do the experimental work before invoking competition as the mechanism which dictates the appearance of boundaries between species of plants (see also Chapman 1973).

Dayton (1975) has also demonstrated competitive interactions not brought about by pre-emption. He removed canopy species from experimental areas and demonstrated a rapid decrease and disappearance of the obligate understorey species, coincident with an increase in the cover of fugitive species. The canopy was thus essential as a resource for the understorey species. What was not clear, however, was that the understorey species were responsible for eliminating fugitive species by competition for space. It is also possible that the fugitive species would not live under the canopy because of some effect of the canopy itself. The appropriate experimental test would be to remove the understorey species while leaving the canopy intact.

Sousa (1979*a*, *b*) demonstrated competitive interactions of a different sort in a succession of algae on the tops of intertidal boulders.

Early colonists like *Ulva* tend to inhibit later arrivals by pre-emption of the space, but are very vulnerable to being consumed by animals. Their removal makes space free for the arrival of various red species which are typical of boulders which have not been disturbed for some time. These species themselves inhibit other later arrivals. When boulders are rolled around by waves, red algae (including the latest arrivals such as *Gigartina canaliculata*) are scraped off their surfaces, but remnant parts of the *Gigartina* are able to recover very quickly and re-establish themselves by vegetative growth. As a result, as time progresses, more and more of a boulder becomes covered by *Gigartina*, which can retain its space.

Intraspecific competition within marine pants is also known (Black 1977). Plants at large densities do not, however, always compete (Schiel and Choat 1980). Again, as discussed above, experimental analyses are always necessary before the presence of competition is known.

Interestingly, there are no, or no widely cited, experimental studies on competition among microalgae. Whether this is because of the intrinsic problems of working with such small plants, or because they are often overgrazed (leading to competition among microalgal grazers, see below) is not clear. MacLulich (1987) has demonstrated a series of successional stages of colonization of cleared surfaces by microalgae. Such successions are often indicative of competitive interactions (Connell and Slatyer 1977; Sousa 1979a, b), suggesting that there may be circumstances under which microalgae compete.

Competition among species of grazers

1. Competition for food

There have been many experimental investigations of competitive interactions for food among, particularly, intertidal grazers. Some of these studies indicate behavioural changes when animals are in sufficient densities to deplete their supplies of food. For example, Ebert (1977) found that when urchins were in sufficient local abundance to eliminate their food, they tended to disperse over large distances. Thus, intraspecific density altered the behaviour of the urchins. Similarly, Mackay and Underwood (1977) demonstrated that the limpet *Cellana tramoserica* would change behaviour from homing to wandering around at random when local densities were increased. In this case, there was evidence that homing was related to the local availability of microalgal foods (Mackay and Underwood 1977). In many cases of competitive interactions between grazers, one or other of the competing species shows increased mortality when at increased density of the other species (e.g., Underwood 1976, 1978, 1984a; Creese and Underwood 1982). In New South Wales, there have been repeated demonstrations that food is nearly always in critically short supply for intertidal snails. As a result, even relatively small increases in local density can cause increased mortality (see also Creese 1982; Fletcher 1984; Fletcher and Creese 1985).

McKillup (1983) demonstrated some interesting consequences of interactions for food among mud snails, *Nassarius pauperatus*. These animals live on mud-flats, some of which provide point sources of food in the form of carrion, while others provide dispersed sources of food in the form of algae in and on the surface of the mud. *Nassarius* are able to graze on algae when these are available, but algae did not provide a particularly good source of food. Thus, when an index of hunger of the animals was measured (McKillup and Butler 1983), snails feeding on algae generally tended to be more hungry than those that also had carrion available. When snails were packed around the point sources of food, trying to consume the carrion, they showed two different behavioural traits. Some twisted from side to side as they fed, which prevented other snails from gaining access to the food and prevented the twisters from being dislodged from positions close to the supply of food. Thus, the twisting behaviour was a fairly clear advantage to the animals when densities of feeding snails were large. Twisting behaviour was an inherited characteristic and was more prevalent amongst populations where there were point sources of food (McKillup 1983).

In many of the studies on competitive interactions for food between marine grazers, there is scant information about the nature, quantities, variety and potential shortage of the food resource. On rocky shores in New South Wales, it has been demonstrated that the supply of microalgal foods is much sparser at higher than lower levels (MacLulich 1987; Underwood 1984a). From this information, Underwood (1984b) predicted that competitive interactions among limpets (*Cellana tramoserica*) and snails (*Nerita atramentosa*) would be more intense at higher than lower levels on the shore. This prediction was borne out and it was demonstrable that shortage of food was the most likely explanation for the results (Underwood 1984b, 1985). There are few similar accounts of situations where the supply of food has been measured while evaluating competitive interactions (but see Quinn 1988). Because these studies have been reviewed so often before (Underwood 1979, 1986; Branch 1981, 1984; Connell 1983), they will not be discussed further here. Some of the problems in the design of these experiments will, however, be considered later.

2. Competition for space

In addition to studies demonstrating direct competitive interactions for food, there are others which demonstrate interference competition between grazing species for space. The ultimate resource over which the competitive struggles are fought is actually food, but the organisms involved need to keep potential competitors away from areas of space so that the food will grow. An example is provided by the owl limpet *Lottia gigantea* in California (Stimson 1970, 1973). *Lottia* are found surrounded by a thin film of algae which does not seem to grow in areas where there are no limpets. Each territory is approximately 900 cm^2 in area, but territories increased with the size of limpets. Stimson (1973) demonstrated that densities of other grazing limpets, notably smaller species of the genus *Acmaea*, were very small in the regions around each *Lottia*, compared with other parts of the shore. Stimson experimentally removed *Lottia* from some areas and introduced them into areas where they had previously been absent. He hypothesized that, if competition were the appropriate mechanism altering the densities of *Acmaea*, the *Acmaea* should decline in density where he introduced *Lottia*. Where he removed *Lottia*, the *Acmaea* would increase to match those surrounding regions. Both predictions were corroborated by the experiments. In addition, Stimson (1973) observed that when *Acmaea* arrived in an area previously occupied by *Lottia*, the algal film disappeared. Stimson (1973) also demonstrated that *Lottia* actively defended their territories against intruders, including other limpets of the same species. Stimson's experiments involved disturbances of the *Lottia* but, subsequently, Wright (1982) has demonstrated that when the *Lottia* are not removed from one area and placed into the territory of another limpet, but are introduced as they move along naturally, they do not fight, but the intruders tend to retreat. This study demonstrates a direct interference process of competition where one species actively and aggressively pushes another out of parts of the habitat. Similar processes have been described for South African limpets (*Patella longicosta, Patella tabularis*) by Branch (1971, 1975, 1984). Again, these studies have been discussed in detail elsewhere and will not be considered further here. They do, however, indicate the potential for competitive interactions for food and the interactions between the resources of food and space that typify much of what happens, at least in intertidal habitats.

Competition between grazers and plants

Because grazers require space over which to feed and because plants must have space on which to grow, there is the potential for competitive interactions between these types of organisms. At low levels on the shore in New South Wales, it has been demonstrated that plants grow very quickly even in the presence of grazing limpets. Limpets from high levels on the shore were able to survive in areas low down that were cleared of foliose plants (Underwood and Jernakoff 1981). The plants were, however, able to recolonize such experimental areas very rapidly. As a result, unless densities of grazers were very large, algae occupied all of the space, making it impossible for microalgal grazing species, such as *Cellana*, to continue to survive there. The *Cellana* were unable to feed on the algae after they had grown; increasing growth of the plants decreased the amount of space over which the limpets could feed. Siphonarian limpets, in contrast, were able to feed on macroalgae but were unable to reduce the cover that the algae occupy, because the limpets snipped off the growing tips of the algae, but never removed them down to the substratum. Underwood and Jernakoff's (1981) experiments also demonstrated that *Cellana* would migrate away from areas that were occupied by mature stands of algae. On the low-shore experimental plots, the only way that limpets were able to keep the area free from algae was at artificially large and continuously maintained densities. Thus, lower areas on the shore are dominated by foliose plants, which exclude microalgal grazing gastropods. These gastropods at higher levels are sufficiently voracious and active that they can eliminate all of the propagules of the lower-shore species of plants (Underwood 1985; Underwood and Jernakoff 1981, 1984).

Such interactions between plants and grazers seem to be widespread. This has resulted in a number of investigations of mutualism, positive associations among species, such that one species is present or in enhanced numbers because of the presence or activities of the other species. This is particularly common where large grazing species (notably large chitons or echinoids) are able to remove well-established, fully grown foliose algae which otherwise prevent the persistence of smaller grazing limpets and chitons (Paine 1980; Ayling 1981; Choat and Schiel 1982; Dethier and Duggins 1984; Choat and Andrew 1986).

The story is not always as simple as this, because in some areas there may be increased mortality due to the presence of the urchins themselves (Fletcher 1987). Nevertheless, there seems to be a widespread phenomenon that some smaller or microalgal grazers are outcompeted for space by plants in areas where the plants rapidly grow too big to be consumed.

Competition between grazers and other organisms

The other plant–animal interaction in shallow coastal marine assemblages is that other organisms, particularly sessile species, may compete with or be outcompeted for space by the grazers. Connell (1961a, 1961b) demonstrated that the limpet *Patella vulgata*, while wandering around grazing, deleteriously affected barnacles. This 'bulldozing' has been observed in many species of limpets and barnacles (e.g., Dayton 1971). In contrast, there are situations where barnacles prevent grazing by large molluscan grazers (Choat 1977; Jernakoff 1983; Lubchenco 1983; Underwood et al. 1983).

Barnacles can also interfere with processes of competition between grazers. Thus, Creese (1982) demonstrated that competition between larger limpets (*Cellana*) and the small limpet (*Patelloida latistrigata*) was intense on open rock surfaces, leading to increased mortality of the smaller limpet. Where barnacles were present, however, the smaller limpets (*Patelloida*) were able to survive because the larger limpets were unable to graze (Underwood et al. 1983). This interaction is more complicated, because where grazers were active, foliose plants were unable to colonize the rock surfaces. Under these circumstances, barnacle larvae could settle and metamorphose (Denley and Underwood 1979; Underwood et al. 1983). Thus, there was some positive effect of the limpet *Cellana* on barnacles where algal growth was fast and profuse; where algae were not able to grow fast, limpets only had a deleterious effect on the barnacles (see review in Underwood 1985).

Clearly, competitive interactions for space can be complex and mediated by a number of species (see also later). In addition to barnacles, other sessile species may encroach on to the space needed by a grazer. In Stimson's (1973) removal of the owl limpet *Lottia gigantea* from experimental areas, the gradual encroachment of mussels was significantly greater than in similar areas where the limpets were left intact. Patrolling the edges of the territory by *Lottia* on a frequent basis kept the mussels at bay. Without this competitive interaction between the grazer and the sessile mussels which needed the space, there would be no bare space over which limpets could graze.

Competition between plants and other organisms

Plants may also have deleterious effects on occupiers of space on rocky shores. Some foliose species of macroalgae are washed around by the waves in such a way that their fronds are continuously brushed over the surface of the rock. This removes newly settled species of sessile animals and has been referred to as a 'whiplash' effect (Dayton 1971). In areas where there are such large plants, there will be fewer barnacles or mussels, and space under the fronds of the macroalgae will therefore be available for limpets to graze (Southward 1964).

Plants may also provide shelter for predatory animals. Menge's (1978a, 1978b) experiments indicated that predatory whelks (*Nucella* [as *Thais*] *lapillus*) were much more active as consumers of mussels in areas where algae provided moisture and shelter from desiccatory conditions during low tide. Under these circumstances, the algae provide a shelter for predatory whelks, which in turn can keep the substratum free from sessile organisms and, provided that the whelks do not also eat grazers, will therefore make surfaces available for grazing by other animals. In this way, algae can provide habitat for grazers through the indirect means of providing shelter for predatory organisms.

In addition to this indirect effect, barnacles and possibly other sessile organisms can also provide direct shelter for the algae from their grazers. Jernakoff (1985a, 1985b) investigated why foliose algae were found at higher levels on the shore where there were barnacles than where there were no such sessile species. Jernakoff found that barnacles provided a refuge from grazing by the larger molluscan grazers on the shore. Although barnacles harboured a number of smaller grazers, these were less able to eliminate species of algae than were the larger grazers on the open shore. Jernakoff (1985a) also investigated whether the barnacles provided a direct alteration of habitat that would enhance the persistence of growth of algae, but concluded that the major effect on the barnacles was to eliminate grazing by the larger molluscs.

In this way, the existence of sessile species of animals will alter the structure of the local assemblage and is, in effect, an indirect result of competition for space between the sessile animals and the grazers. Where the sessile species such as barnacles are very abundant, they reduce grazing (they compete for space with the grazers) and thereby enhance the cover, diversity, and abundance of algae. In some areas, barnacles provide the only refuges from grazing the propagules of species of algae, so that, for example, *Fucus* only grows in areas where propagules are safe from grazing by snails (Lubchenco 1983).

Inter- and intraspecific competition among grazers

The net effect of competitive interactions between species of grazers is not always to cause elimination of one species by a superior competitor although this is usually the outcome of interference competition for space (e.g., among barnacles, Connell 1961a, b). As had been foreshadowed by Darwin (1882), intraspecific competition is often more likely to be more intense than is the competitive interaction between species. Individuals of the same species will usually tend to have a closer requirement for identical resources than is the case between individuals of different species. Under these circumstances, the densities of one species that are required to increase mortality, leading to eventual local extinction of competitively inferior species, are large, but unlikely to be maintainable. Intraspecific competition (e.g., for food) is likely to cause decreases in the density of the superior competitor, perhaps even at a faster rate than the decreases in density of the inferior species. Whatever the relative rate of disappearance of the two types of organisms, the fact that intraspecific competitive interactions will reduce the density of a superior competitive species, leads inevitably to the conclusion that the superior competitor cannot continue at sufficient densities to eliminate all members of the inferior species.

This was demonstrated experimentally by Creese and Underwood (1982) for competition between the prosobranch limpet *Cellana tramoserica* and pulmonate limpets, *Siphonaria* spp. *Cellana* was a superior competitor in areas where both types of limpets occurred. Increased densities of *Cellana* caused decreased densities (due to increased mortality) of the *Siphonaria*. Nevertheless, some siphonarians always stayed alive in experimental plots because the density of *Cellana* also declined and was reduced below that necessary to exceed the carrying capacity, and thereby to restrict the availability of food for siphonarians. Similar results have subsequently been obtained for other species of limpets (Ortega 1985).

Here, again there are complications. Although both *Siphonaria* and *Cellana* require the same foods, *Cellana* requires microscopic, early stages in the life history of the plants and will therefore consume the plants long before they are of a size suitable to be eaten by *Siphonaria*. Thus, the competition must be asymmetrical in the sense that *Cellana* will exploit the food resource long before it is available to be consumed by *Siphonaria*. It is, however, also the case that *Cellana* will not allow many other microalgal grazers on their backs. As a general rule, therefore, except at the very highest levels on the shore, *Cellana* carry around a growth of foliose plants on the surfaces of their shells (Creese and Underwood 1982) which provide a source of food for *Siphonaria* even though they are being outcompeted on the surface of the rock by the *Cellana*. It is entirely possible, therefore, that even were *Cellana* able to eliminate all the food from the surface of rocks so rapidly that none was ever available for *Siphonaria*, the latter, pulmonate limpets would always be able to survive simply by foraging on the backs of the prosobranch microalgal grazers.

It remains to be seen whether other examples of this type of asymmetrical competition will be found. At the moment, it is clear that competitive exclusion of one species by another is extremely unlikely for shallow coastal marine grazers. The shortage of food resources is so acute that intraspecific competition for food will always reduce the densities of the potentially superior competitor below those necessary to keep on causing such deleterious effects on the inferior competitor that the latter eventually has a density reduced to zero.

Experimental evaluations of the effects of intraspecific competition on the behaviour and dispersal of grazers can be problematic. Several studies have shown that limpets disperse more rapidly from areas of increased density (Aitken 1962; Breen 1971; Mackay and Underwood 1977). The experiments are, however, subject to the problem that limpets moved into areas to cause increased density are disturbed, handled, and placed in unfamiliar surroundings, in addition to being at increased density. Unless there are adequate (and complex) controls for these necessary artefacts of the experimental manipulation, the results are confounded and uninterpretable. Appropriate designs for these experiments are discussed in detail in Underwood (1988).

Experimental designs for detecting competition

Appropriate designs for investigating competitive interactions in field experiments have been discussed in detail in Underwood (1986). There have been very many problems in field experiments, including difficulties with 'natural experiments' which are not experiments at all (Underwood 1986). They are comparisons of areas with and without some putative competitor. The notion is that these can be compared in order to investigate the effects of the presence of the putative competitor on some species. Unfortunately, many differences may exist between the areas with and naturally without the putative competing species, which may not be related specifically to its presence or absence. Thus, comparisons of some species in the natural presence and natural absence of some other species and the discovery that there are differences in the growth, reproductive output, or mortality, are not indicative of competitive effects due to the other species; it is irrational to conclude that they are. Examples are discussed in Underwood (1986), but such experiments have been considered to be valid in a review by Diamond (1986). There is no possible validity of such 'experimental' procedures. At best, if competition occurs, it is necessary that there should be differences in the mortality, growth, weight, and similar, of organisms with and without the putative competitor. Discovery of such differences is, however, not sufficient to demonstrate that competition is occurring.

There have also been numerous experimental studies where treatments have not been replicated, leading to problems and irrationality in interpretations (see also Hurlbert 1984). Some 16 per cent of the studies examined in Underwood (1986) had no replication, or inadequate replication of some treatments.

One of the areas, however, that has caused the most difficulty in experimental studies of competitive interactions, is that of confounded designs of the experiment (Underwood 1986), but there has been some misunderstanding about the importance of this point (e.g., Hawkins and Hartnoll 1985). The problem arises (see Table 20.1) when experiments are designed to compare the survival, or growth, or reproductive output, or tissue weight, of some species, for example A, in the presence and absence of some potentially competing species B. Under natural conditions, A may be found alone at a density of 10 per unit area, and B may be found alone, also, for example, at a density of 10 per unit area. These control conditions are then contrasted against experimental conditions where A and B are placed together. Unfortunately, the experimental treatments are often constructed so that there are five A and five B per unit area. For example, Bertness (1981) compared rock-pools with 200 hermit crabs (*Clibanarius*) and pools with 200 crabs (*Calcinus*) with pools in which the two potentially competing species were together, but at a density of 100 of each species. Under such circumstances, there are two differences between the areas with species A alone and those with species A and B together. One is the presence of species B, which is the treatment about which hypotheses of competition are made, and which the experiment is supposedly designed to test. The second difference is that the density of species A has, itself, been altered between the two types of area. Thus, there is an intrinsic confounding between the two types of plot, which cannot be separated in the experiment. If intraspecific interactions within species A occur at a similar intensity to interspecific effects of species B on A, there will be no observable difference in the growth, survival, etc., of species A in the presence and absence of species B. This will occur despite competitive interactions from B. Only if there were no intraspecific competition at all within species A, would it be possible to detect, unambiguously, the effects of species B. A better design is that indicated in Table 20.1, where the density of one species is fixed and the other species is added to it. This is exactly how an experiment would be done to detect the effect of adding, for example, a chemical, or some other non-biological experimental treatment to an area. It is irrational to change the design simply because the item being added is some number of individuals of another species.

Another aspect of designing experiments to detect competitive interactions is the use of asymmetrical analyses to cover the situation where there might be a single control plot and a number of experimental densities of one or more species. Because there is considerable evidence

Table 20.1. Experimental treatments necessary to determine influences of inter- and intraspecific competition on density, mortality, growth, weight, or reproductive output of each of two species at several densities.

For Species A:

	Control	Increased densities		
		(+10)	(+20)	
	1	2	3	
	10A	20A	30A	(+Species A)
		(=10A + 10A)	(=10A + 20A)	
		4	5	
		10A + 10B	10A + 20B	(+Species B)

For species B:

	Control	Increased densities		
		(+10)	(+20)	
	6	7	8	
	10B	10B + 10A	10B + 20A	(+Species A)
		9	10	
		20B	30B	(+Species B)
		(=10B + 10B)	(=10B + 20B)	

Contrasts of treatments:	Interaction estimated:
1 vs 2 vs 3	Intraspecific competition, A on A
1 vs 4 vs 5	Interspecific competition, B on A
6 vs 7 vs 8	Interspecific competition, A on B
6 vs 9 vs 10	Intraspecific competition, B on B
(8–7) vs (5–4)	Symmetry or asymmetry: A on B vs B on A
(3–2) vs (8–7)	Symmetry or asymmetry: A on A vs A on B
(3–2) vs (5–4)	Symmetry or asymmetry: A on A vs B on A
(10–9) vs (5–4)	Symmetry or asymmetry: B on B vs B on A
(10–9) vs (8–7)	Symmetry or asymmetry: B on B vs A on B

that inter- and intraspecific competitive interactions need to be investigated simultaneously (e.g., Pontin 1960; Creese and Underwood 1982; reviewed by Underwood 1986), experiments must be designed such that controls consist of a chosen density of one species, which is then contrasted to experimental plots with different densities of that species and with that species in combination with a second one. This design has been described in full elsewhere (Underwood 1978, 1984b, 1986) and is summarized in Table 20.1.

The use of asymmetrical analyses simplifies the design of such experiments. In the simplest case, where there are two species (A and B), intraspecific competition among individuals of species A can be determined by comparing a single control treatment with some density of A (e.g., 10A in Table 20.1) and experimental plots with increased density of A (e.g., 20A and 30A in Table 20.1). At the same time, if there are plots with similar densities of species B (as described in Table 20.1), the effects of interspecific competition can also be measured. Each of these treatments needs to be replicated independently a number of times and the results can then be analysed by an asymmetrical analysis of variance as described in Table 20.2. This design has been used successfully in a number of studies (summarized, for example, from studies of limpets by Fletcher and Creese, 1985). There has been a long history of recommendation that such designs are necessary for interpretations of any competitive interaction (e.g., Connell 1961a; Pontin 1960; Reynoldson and Bellamy 1970; Underwood 1978, 1986). Because this design and its ramifications have been described in full elsewhere (Underwood 1986), they will not be considered further here.

Symmetry and asymmetry

Use of experimental designs as described in the previous section leads directly to the possibilities of interpreting interactions between species in different ways. There has been some discussion in the literature of the relative importance of competition from one species on another, v. the reciprocal effect. Thus, there is often an asymmetry between the competitive effects of species B on species A, compared with the effects of A on B. There have been several major reviews of this topic (Lawton and Hassell 1981; Connell 1983; Schoener 1983). There are theoretical constructs in the use of Lotka–Volterra equations for determining competition coefficients in studies of populations (e.g., Williamson 1972; Krebs 1978; Lawton and Hassell 1981). As has been argued in full elsewhere (Underwood 1986), there are problems with the design of experiments to determine whether competitive interactions are symmetrical or asymmetrical. Unless two competing organisms happen

Table 20.2. Asymmetrical analysis of variance of experimental treatments (as in Table 20.1) to determine influences of inter- and intraspecific competitive interactions on one species (A) by members of its own and another species (B).

(A) Analysis of proportional mortality (or density) or other variable measured per experimental plot; there are r replicate plots per treatment

Source of variation	Degrees of freedom
Among all treatments (1–5)	4
Control (1) vs Others (2–5)	1
Among other treatments (2–5)	3
Between increased densities [(2 + 4) vs (3 + 5)]	1
Between species [(2 + 3) vs (4 + 5)]	1
Interaction: density × species	1
Residual	$5(r - 1)$
Total	$5r - 1$

(B) Analysis of size, weight, growth, etc., measured for n replicate individuals of Species A in each of r replicate plots of each treatment (as in Table 20.1)

Source of variation	Degrees of freedom
Among all treatments (1–5)	4
Control (1) vs others (2–5)	1
Among other treatments (2–5)	3
Between increased densities [(2 + 4) vs (3 + 5)]	1
Between species [(2 + 3) vs (4 + 5)]	1
Interaction: density × species	1
Among replicate plots within treatments	$5(r - 1)$
Residual	$5r(n - 1)$
Total	$5rn - 1$

---

to occur in similar densities (or biomasses, or whatever currency of abundance or measures of abundance are appropriate), then experiments to determine the effect of one species on another and the reciprocal interaction, will be rather difficult. This topic is not pursued further here; a complete analysis was provided by Underwood (1986).

It is important to realize that a complete study of competitive interactions requires that all possible forms of symmetry and asymmetry be considered. Underwood (1986) argued that there are three major forms of symmetrical/asymmetrical interaction. First is the classic one that symmetry or asymmetry will occur in the effect of one species on the second. Thus, if the effect on growth, reproduction, mortality, etc., of species A by addition of some number of species B is estimated and found to be equal to the effect on growth, reproduction, etc., of species B when that number of species A is added, the interaction is entirely symmetrical. If, in contrast, the addition of one or more A causes a different effect, per capita on B than the corresponding effect of adding one or more B to A, the interaction is asymmetrical. This is illustrated in Table 20.1. There are, however, two other forms of symmetrical or asymmetrical interaction. The second form is the comparison of the effect of addition of the first species A to some number of its own species A, in contrast to the effect of adding them to some number of B. If this is investigated experimentally, it could reveal whether or not there is symmetry or asymmetry between the intraspecific effect of A on A, vs the interspecific effect of B on A. This would be important for predicting the outcome of competitive interactions between species in which intraspecific competition is at least as intense as interspecific competition (e.g., Creese and Underwood 1982; Ortega 1985). This also is illustrated in Table 20.1.

Finally, the third type of interaction that might be symmetrical or asymmetrical is the effect of one species A on itself, compared with the interspecific effect of another species B. This is again indicated in Table 20.1. These have practical importance for understanding the effects of competitive interactions (Underwood, 1986). Consider competition for the resource of space on a rocky shore where space in one area was occupied by species A and that in another area by species B. The first type of symmetry or asymmetry (the effect of A on B and the effect of B on A) would be useful to determine whether future arrivals of species A into an area occupied by B, or of species B into an area already occupied by A, is likely to lead to a change in the composition of the species, and therefore the structure of the assemblage in that locality. In contrast, knowledge of whether the second process of competition is symmetrical or asymmetrical may be useful to determine what happens if members of species A start to arrive in each of the two areas. If the interactions are symmetrical then the arrival of A into areas already occupied by A will cause the same changes in numbers of individuals as the arrival of A into areas occupied by species B. If the interaction is not symmetrical, there can be greater turnover of individuals in one patch compared with the other.

Finally, if the patch containing A is considered, the future occupancy of the patch will be affected quite differently if new arrivals are of species A or of species B. Prediction or interpretation of events depends on knowledge of the degree to which individuals of A or B are more likely to invade a patch of A (i.e., the degree of asymmetry between the effects of A on A and those of B on A).

Several studies of competitive interactions within and between species of intertidal gastropods have been sufficiently complete to examine the three types of interaction (reviewed by Fletcher and Creese 1985). In the cases that they discussed, there were many complexities in the relationships among and within the species. Some species had weak, symmetrical inter- and intraspecific interactions; others had strong symmetrical interactions with interspecific competition being less intense than intraspecific effects. It is clear that there is no general pattern of interaction among grazers even though some of the processes are very similar for different pairs of species. This type of study needs to be done in many more cases, so that general patterns, if they exist, will emerge.

Processes affecting the outcome of competition

1. Disturbances

As discussed earlier, physical disturbances of the habitat will alter the outcome of competition between plants, between animals or between plants and animals. Sousa (1979a, b) demonstrated that the disturbance of intertidal boulders clearly influenced the outcome of competitive processes. Where boulders were overturned commonly, successional processes among the algae were disrupted and the early colonizing species (*Ulva*) tended to predominate. These were able to pre-empt the space, preventing it from becoming available for other colonists. At the other end of the successional series on these boulders were 'grab and hold' algae such as *Gigartina canaliculata*, which thrived by vegetative growth from remnants after the plants had been abraded when boulders were overturned.

Connell (1975) described a general model for the processes of colonization of patches of a habitat that are occupied by dominant species. He indicated that when the dominant organisms and any associated species that are dependent upon them are removed, early colonization and subsequent events would depend on the physical harshness of the environment. In general, in physically harsh environments, organisms that settled would often be killed by physical components of the environment. Only when, for whatever reason, the physical harshness of the environment was ameliorated, would colonists be able to thrive and survive. At the opposite end of the physical spectrum, in physically benign habitats, organisms that settled would tend to be removed by their predators or grazers. Thus, early succession would normally be reset back to bare surfaces and the assemblage would consist only of the juveniles of early colonizing species. Occasionally, however, there would be periods when the grazers or predators were absent because of some change in environmental circumstances (e.g., Dayton 1971; Crisp 1964). Under these circumstances, Connell (1975) proposed that the colonists would thrive.

Under either of the conditions that allowed colonists to thrive, densities of colonizing species should be sufficiently large for competition between them to determine which ones survived and, eventually, the competitively dominant species would occupy the space, restoring the system back to its original starting position. It is interesting to note that all these pathways lead, inexorably, to competitive interactions as the only means by which the original species in some area could eventually become re-established.

There have been relatively few experimental investigations of plant–animal competitive interactions in the sea under different physical environment conditions. One example was the experiments on the competitively dominant snail *Nerita atramentosa* which caused increased mortality of the limpet *Cellana tramoserica* whenever densities were sufficiently large (Underwood 1978, 1984*b*). Underwood (1984*a*) had already demonstrated that the available supplies of microalgal foods were much smaller at high, rather than low, levels on the shore and also during summer compared with winter. Competition was more intense at high levels on the shore than at low levels and more intense during summer than winter (Underwood 1984*b*, 1985). The results revealed that physical features of the environment (period of immersion and season of the year) could directly influence the outcome of competitive processes by influencing the available supplies of food and, presumably, by influencing the activities of the grazers.

2. Competition can affect competition

Kastendiek's (1982) almost unique investigation of competitive interactions among subtidal species of macroalgae revealed that competition between a canopy and an understorey species influenced the competitive interactions among the understorey species living closer to the surface of the rock. The canopy species, *Eisenia arborea*, prevented growth on the surface of the rock by *Halidrys dioica*. The mechanism of this competitive interaction is not clear, but *Eisenia* seemed to shade *Halidrys* from the light. In experimental plots where *Eisenia* plants were removed, there was a marked increase during the next few months, in the percentage cover of *Halidrys*. Simultaneously, another species, *Pterocladia*, was reduced in cover. Kastendiek (1982) also demonstrated that, unless *Eisenia* plants were present, *Halidrys* was unable to grow in these areas. In experiments in which he removed *Eisenia* and also manually removed *Halidrys* as they encroached into the plots, *Pterocladia* were able to survive. In control plots, into which *Halidrys* could invade once the competitors (*Eisenia*) were removed, *Pterocladia* declined in cover. Thus, competition from one species (*Eisenia* competing with *Halidrys*) can prevent a competitive interaction between other species (*Halidrys* outcompeting *Pterocladia* for space). This is a very interesting study because it demonstrates the complexity of competitive interactions and the unpredictability of their outcome in situations where more than two species compete, but in different ways, for the same resources.

3. Grazing and preferences for different types of food

In some areas, because of the preferences for particular types of plants, grazing animals may influence the outcome of competition among the species of plants. For example, Paine and Vadas (1964) investigated grazing by sea-urchins on sublittoral species of algae. Where sea-urchins were relatively rare, the alga *Chondrus* outcompeted other species of plants and therefore, gradually, diversity of algae decreased. Where urchins were abundant, however, they consumed *Chondrus* preferentially and thereby made space available for other less competitive species of plants. In general, in such regions, diversity of algae increased (see also Lubchenco and Menge 1978). This interaction of grazers influencing the outcome of competitive interactions is akin to that more commonly known now as 'keystone predation' (Paine 1974), when a predator (or grazer), by concentrating its attention on more competitive species among those available as food, eliminates competition among the prey thus making resources available for a diversity of other species. The grazer or predator is considered a 'keystone' because its interactions with its prey influence the outcome of a series of competitive processes, resulting in increased diversity of species which would otherwise be eliminated by competition. There can be doubts about the validity of this particular case (Underwood and Denley 1984), but the principle is an interesting one and Paine and Vadas (1964) appear to have demonstrated a similar phenomenon of grazing by urchins on diverse species of plants.

The whole issue of whether grazing or predation can increase or decrease the number of species of plants has been addressed by Lubchenco (1978) and, in more theoretical terms, by Lubchenco and Gaines (1981). Lubchenco (1978) described the situation where algae were relatively diverse in rock-pools which had grazing snails (*Littorina littorea*), as compared with pools that had no such grazers. In the pools without grazers, soft, fleshy species such as *Enteromorpha* tended to predominate. Lubchenco introduced grazing snails into one pool from which they were naturally absent, where they rapidly consumed the *Enteromorpha*, making space available for species such as *Chondrus crispus*. Diversity of species of algae was reduced in another pool from which *Littorina* were removed; the surfaces of the pools became occupied by *Enteromorpha*. The replication of these experiments was very small and, for some treatments, there were no replicates. Lubchenco (1978) interpreted the results to be due to the fact that *Littorina* were preferentially grazing on *Enteromorpha*, which was acting as the superior competitor among the species of algae. Where *Littorina* were consuming *Enteromorpha*, the alga was no longer able to outcompete and eliminate some species of algae, therefore grazing had the effect of allowing increases in diversity of species.

By contrast, in areas out of pools, grazing snails (*Littorina*) continued to consume *Enteromorpha* preferentially. Here, however, *Enteromorpha* was no longer the competitively dominant species. According to Lubchenco (1978), other species such as *Fucus* spp. and *Ascophyllum nodosum* were superior competitors, but were relatively unaffected by grazing snails. In contrast to the situation in the pools, grazing by the snails could decrease the number of species of algae on the surface of the rocks. Where grazing was sufficiently intense, the grazers could actually eliminate *Enteromorpha*.

Lubchenco (1978) contrasted these results and suggested that the outcome of grazing on competitive interactions among the species of food plants would be dependent on whether the grazers were preferentially feeding on a superior competitive species or on a competitively inferior one. In the former case, grazing would have the effect of reducing the intensity of competitive interactions, thereby freeing resources for other species, thus having the effect of increasing species diversity. This is a 'keystone' effect. Alternatively, where grazers were focusing on competitively inferior species, the only possible effect they could have on the number of species present was elimination of the preferred plant. Here, there would be no 'keystone' effect, but species diversity could decline as a result of intense overgrazing. It is a pity that similar studies have not been pursued elsewhere, because these ideas are very useful ones in any synthesis of results from a range of different studies with different outcomes. It is also a pity that no evidence was provided to demonstrate that *Enteromorpha* was, in fact, outcompeted by the fucoid algae, as suggested by Lubchenco (1978). This study would repay further development and experimentation done with appropriate replication, both in individual pools and at a number of sites, to determine the generality of the results.

The relative importance of competition

Welden and Slauson (1986) have considered the problem of the relationship between the measurable intensity of competitive interactions and the determination of whether this is important to the species concerned. The issue is that, even though competitive interactions between two species may be demonstrably large or statistically significant, it does not follow that competitive interactions actually matter much to the overall size of populations, the well-being of individuals in populations, or the reproductive output of individuals. For example, competition between species of barnacles may be very intense locally, but it matters very little because predators would have consumed most of the individuals anyway. Under these circumstances, competition can be intense and can be demonstrated in relatively simple experiments, but has little importance to the eventual occupancy of space on the shore or to the populations of either of the competing species.

Welden and Slauson (1986) therefore suggested that intensity of competition between species B and species A could be measured, for species A, as the difference in density, or whatever relevant variable, between areas where A is alone and where A is present with B. This would provide some index of intensity, but it would depend on the prevailing condition of the individuals of species A. Those might, for example, be more or less susceptible to competition from B because of the current state of supply of resources (e.g., Schoener 1983; Underwood 1984b), density (e.g., Underwood 1986), size (e.g., Underwood 1976) or previous history of stress (e.g., Peterson and Black 1988). Thus, Welden and Slauson (1986) originally proposed that intensity of competition should be measured as the difference from the optimal physiological state induced by competition. Their proposed measure is not independent of existing state and assumes that the condition of individuals of species A is optimal wherever B is absent.

The importance of competition is much more difficult to document. Welden and Slauson (1986) presumed that it would be estimated in an analysis of variance by the proportion of the total variability attributable to differences in mean abundance (or size, weight, etc.) of A in the presence or absence of B. This measure assumes that it is independent of the size of the experiment. As discussed in a different context (Underwood and Petraitis 1991), this measure is, among other things, a function of the number of experimental densities compared in the analysis. This is illustrated in Table 20.3, for two experiments on the effects of different densities of species B on density of A. In the first experiment, there are only 2 treatments, one with 10A alone per experimental area and the other with 10A and 10B together. The

Table 20.3. Problems with using proportions of total variation as a measure of relative importance of competition (see Welden and Slauson 1986). Relative importance is measured as proportion of total variation attributable to experimental treatments (i.e., different densities) in an analysis of variance of mean weights of species A. Expected values of mean squares are as in Winer (1971), Underwood (1981); for further details, see Underwood and Petraitis (1990).

(A) Experiment 1: 2 treatments, $n = 5$ replicates, residual mean square = 20

Treatments:	10A + 0B	10A + 10B
Mean Weight of A (grams):	100	80

Source of variation	Sum of squares	Degrees of freedom	Mean square
Between treatments	1020	1	1020
Residual	160	8	20
Total	1180	9	

Importance of competition = 1020/1180 = 0·86

(B) Experiment 2: 4 Treatments, $n = 5$ replicates, residual mean square = 20

Treatments:	10A + 0B	10A + 10B	10A + 20B	10A + 30B
Mean Weight of A (grams):	100	80	60	40

Source of variation	Sum of squares	Degrees of freedom	Mean square
Among treatments	10 060	3	3353
Residual	720	36	20
Total	10 780	39	

Importance of competition = 10 060/10 780 = 0·93

---

treatments are replicated 5 times. The percentage total of variation attributable to competition is 86 per cent (Table 20.3). In the second experiment, there is an extra treatment of 10A with 20B, again replicated five times. The percentage of total variation attributable to competition is now 93 per cent (Table 20.3). The perceived importance of competition is influenced by the size of the experiment used to detect it.

A better method of assessing the relative importance of some ecological processes such as competition would be through experiments on other processes (such as predation and the effects of physical harshness and disturbances) in conjunction with the study of competition. Under these circumstances, the relative importance of each process should be assessable (e.g., Menge 1978*a*, *b*) for factors influencing rates of predation by whelks; Underwood et al. (1983) for interactions of predation, competition, recruitment and different habitats).

Another approach that is of great value in attempting to determine the importance of some processes such as competition is the repeated investigation of a given process at a large number of sites or at many times. For example, Foster (1990) has investigated patterns in the structure of assemblages at numerous sites along the coast of California. Of particular relevance, he examined the vertical distribution of several species of algae which were known to compete for space. Foster's (1990) extensive analysis revealed that competition among the algae could not explain, or be relevant to, the patterns of distribution of the plants. From studies such as this, the importance of competition could be assessed as the proportion of those sites examined in which competition influences the distribution of a species.

Widespread comparative investigations of this sort are needed for many marine habitats. Other examples are those on grazing of intertidal algae by Sousa et al. (1981) and on the effects of predation on the grazing snail *Tegula* (Fawcett 1984).

Conclusions

Competition can have important consequences for interactive processes between plants and animals, the resulting population dynamics of the participating species and the structure of marine assemblages. There can be competition, largely for space, between plants, which often influences their patterns of distribution. Little is known of competition among plants that are the food of grazers. No experimental studies have shown competition among microalgal foods; few studies on macroalgae have been on species that are the major food of grazers. Where competition is among species of algae that are major components of grazers' diets, there are complex interactions between preferences by the grazers and the physical environment (Lubchenco 1978).

Grazers, particularly microalgal grazers, are very likely to compete for food in some habitats. Overexploitation of food is a widespread phenomenon on many rocky shores. This reflects the fact that, even though resources of food are, at least partially, renewed without removal of the grazers (Underwood 1978; MacLulich 1986), they are distributed over restricted areas of space. Competition for space over which to graze, coupled with exploitative overgrazing, is then intense. Perhaps also the widespread incidence of competition among intertidal grazers reflects the lack of predation on these organisms in many habitats. Predators do not regulate abundances of those microalgal grazers that compete.

There are several examples of plants that can exclude other species from areas of habitat, thus causing limits to distribution of the inferior competitors. By contrast, competition among grazers tends to regulate abundances of populations and to influence local patchiness of the diversity. There have been no unambiguous demonstrations that one species of, for example, intertidal grazer can limit the vertical distribution of another species. Partly, this must be because intraspecific competition for food is at least as intense as interspecific effects, making it generally unlikely that one species could eliminate another in any part of their potentially overlapping habitat.

A wide variety of processes can affect the intensity or outcome of competitive interactions (including physical factors in the environment, disturbances, predators, etc.). It is not possible to predict accurately the outcome of a competitive interaction unless the processes affecting the supply of the resources and the non-competitive processes affecting the abundances of the competitors are also investigated. Studies of competition cannot be successful in isolation from studies of other aspects of the ecology of the participants.

Despite the broad variety of studies on competitive interactions, more accurate information is needed before it will be possible to propose useful general models of the role of these processes in plant–animal interactions. Better attention to the requirements of experimental designs is needed. Too many studies are poorly-designed, illogical, or lacking in replication. In many instances, there is inadequate information about the nature, rate of supply, etc., of the resources.

Very few studies have been repeated at sufficiently varied spatial or temporal scales, or with a wide enough range of starting densities of the participating species, for any conclusion to be reached about the general effects of competition. How often does a competitive interaction influence the density of a grazer, or the distribution of a plant? In how many patches of habitat does competition influence the abundance or reproductive output of an organism? Any attempt to determine the importance of competition as a selective mechanism influencing the evolution of marine species must be based on knowledge of the spatial and temporal likelihood that individuals in a species will be, or have been, affected by the processes involved.

This brief review has indicated some of the main features of competition in plants and animals, in addition to aspects of their interactions. The processes are complex and variable, warranting much more experimental analysis. The problems underlying these and the range of potentially related processes affecting competition have been considered here to assist in better planning of future studies.

Acknowledgements

The preparation of this review was supported by funds from the Australian Research Council and the Institute of Marine Ecology of the University of Sydney. Discussions with J. H. Connell, P. Jernakoff, K. A. McGuinness, W. P. Sousa, P. A. Underwood and, particularly, M. G. Chapman clarified many issues. The preparation of the manuscript was much aided by K. Astles and M. G. Chapman.

References

Aitken, J. J. (1962). Experiments with populations of the limpet *Patella vulgata*. *Irish Naturalists Journal*, **14**, 12-15.

Ayling, A. M. (1981). The role of biological disturbance in temperate subtidal encrusting communities. *Ecology*, **62**, 830-47.

Benke, A. C. (1978). Interactions among coexisting predators — a field experiment with dragonfly larvae. *Journal of Animal Ecology*, **47**, 335-50.

Bertness, M. D. (1981). Competitive dynamics of a tropical hermit crab assemblage. *Ecology*, **62**, 751-61.

Birch, L. C. (1957). The meanings of competition. *American Naturalist*, **91**, 5-18.

Black, R. (1977). Population regulation in the intertidal limpet *Patelloida alticostata* (Angas, 1865). *Oecologia* (Berlin), **30**, 9-22.

Branch, G. M. (1971). The ecology of *Patella* Linnaeus from the Cape Peninsula, South Africa. 1. Zonation, movements and feeding. *Zoologica Africana*, **6**, 1-38.

Branch, G. M. (1975). Mechanisms reducing intraspecific competition in *Patella* spp.: migration, differentiation and territorial behaviour. *Journal of Animal Ecology*, **44**, 575-600.

Branch, G. M. (1981). The biology of limpets: physical factors, energy flow, and ecological interactions. *Oceanography and Marine Biology Annual Review*, **19**, 235-80.

Branch, G. M. (1984). Competition between marine organisms: ecological and evolutionary implications. *Oceanography and Marine Biology Annual Review*, **22**, 429-593.

Breen, P. A. (1971). Homing hebavior and population regulation in the limpet *Acmaea (Colisella) digitalis*. *Veliger*, **14**, 177-83.

Chapman, A. R. O. (1973). A critique of prevailing attitudes towards the control of seaweed zonation on the sea shore. *Botanica Marina*, **41**, 80-2.

Choat, J. H. (1977). The influence of sessile organisms on the population biology of three species of acmaeid limpets. *Journal of Experimental Marine Biology and Ecology*, **26**, 1-26.

Choat, J. H. and Andrew, N. L. (1986). Interactions amongst species in a guild

of subtidal benthic herbivores. *Oecologia* (Berlin), **68**, 387–94.
Choat, J. H. and Schiel, D. R. (1982). Patterns of distribution and abundance of large brown algae and invertebrate herbivores in subtidal regions of northern New Zealand. *Journal of Experimental Marine Biology and Ecology*, **60**, 129–62.
Connell, J. H. (1961a). The influence of interspecific competition and other factors on the distribution of the barnacle *Chthamalus stellatus*. *Ecology*, **42**, 710–23.
Connell, J. H. (1961b). Effects of competition, predation by *Thais lapillus* and other factors on natural populations of the barnacle *Balanus balanoides*. *Ecological Monographs*, **40**, 49–78.
Connell, J. H. (1972). Community interactions on marine rocky intertidal shores. *Annual Review of Ecology and Systematics*, **3**, 169–92.
Connell, J. H. (1975). Some mechanisms producing structure in natural communities: a model and evidence from field experiments. In *Ecology and evolution of communities*, (ed. M. S. Cody and J. M. Diamond), pp. 460–90. Harvard University Press, Cambridge, Massachusetts.
Connell, J. H. (1983). On the prevalence and relative importance of interspecific competition: evidence from field experiments. *American Naturalist*, **122**, 661–96.
Connell, J. H. and Slatyer, R. O. (1977). Mechanisms of succession in natural communities and their role in community stability and organization. *American Naturalist*, **111**, 1119–44.
Creese, R. G. (1982). Distribution and abundance of the acmaeid limpet *Patelloida latistrigata*, and its interaction with barnacles. *Oecologia* (Berlin), **52**, 85–96.
Creese, R. G. and Underwood, A. J. (1982). Analysis of inter- and intraspecific competition amongst limpets with different methods of feeding. *Oecologia* (Berlin), **53**, 337–46.
Crisp, D. J. (1964). An assessment of plankton grazing by barnacles. In *Grazing in terrestrial and marine environments*, (ed. D. J. Crisp), pp. 151–64. Blackwell, Oxford.
Darwin, C. (1882). *The origin of species*, (6th edn). Murray, London.
Dayton, P. K. (1971). Competition, disturbance and community organization: the provision and subsequent utilization of space in a rocky intertidal community. *Ecological Monographs*, **41**, 351–89.
Dayton, P. K. (1975). Experimental evaluation of ecological dominance in a rocky intertidal algal community. *Ecological Monographs*, **45**, 137–59.
Denley, E. J. and Dayton, P. K. (1985). Competition among macroalgae. In *Handbook of phycological methods. 4. Ecological field methods: macroalgae*, (ed. M. M. Littler and D. S. Littler), pp. 511–30. Cambridge University Press, Cambridge.
Denley, E. J. and Underwood, A. J. (1979). Experiments on factors influ-

encing settlement, survival and growth of two species of barnacles in New South Wales. *Journal of Experimental Marine Biology and Ecology*, **36**, 269-93.

Dethier, M. N. and Duggins, D. O. (1984). An 'indirect commensalism' between marine herbivores and the importance of competitive hierarchies. *American Naturalist*, **124**, 205-19.

Diamond, J. M. (1986). Overview: laboratory experiments, field experiments and natural experiments. In *Community ecology*, (ed. J. M. Diamond and T. J. Case), pp. 3-22. Harper and Row, New York.

Ebert, T. A. (1977). An experimental analysis of sea urchin dynamics and community interactions on a rock jetty. *Journal of Experimental Marine Biology and Ecology*, **27**, 1-22.

Fawcett, M. H. (1984). Local and latitudinal variation in predation on an herbivorous marine snail. *American Naturalist*, **131**, 257-70.

Fletcher, W. J. (1984). Intraspecific variation in the population dynamics and growth of the limpet, *Cellana tramoserica*. *Oecologia* (Berlin), **63**, 110-21.

Fletcher, W. J. (1987). Interactions among subtidal Australian sea urchins, gastropods and algae: effects of experimental removals. *Ecological Monographs*, **57**, 89-109.

Fletcher, W. J. and Creese, R. G. (1985). Competitive interactions between co-occurring herbivorous gastropods. *Marine Biology*, **86**, 183-92.

Foster, M. S. (1982). Factors controlling the intertidal zonation of *Iridaea flaccida* (Rhodophyta). *Journal of Phycology*, **18**, 285-94.

Foster, M. S. (1990). Organization of macroalgal assemblages in the Northeast Pacific: the assumption of homogeneity and the illusion of generality. *Hydrobiology*, **192**, 21-33.

Hawkins, S. J. and Hartnoll, R. G. (1985). Factors determining the upper limits of intertidal canopy-forming algae. *Marine Ecology Progress Series*, **20**, 265-71.

Hurlbert, S. J. (1984). Pseudoreplication and the design of ecological field experiments. *Ecological Monographs*, **54**, 187-211.

Jernakoff, P. (1983). Factors affecting the recruitment of algae in a midshore region dominated by barnacles. *Journal of Experimental Marine Biology and Ecology*, **67**, 17-31.

Jernakoff, P. (1985a). An experimental evaluation of the influence of barnacles, crevices and seasonal patterns of grazing on the algal diversity and cover in an intertidal barnacle zone. *Journal of Experimental Marine Biology and Ecology*, **88**, 287-302.

Jernakoff, P. (1985b). Interactions between the limpet *Patelloida latistrigata* and algae on an intertidal rock-platform. *Marine Ecology Progress Series*, **23**, 71-8.

Kastendiek, J. (1982). Competitor-mediated coexistence: interactions among three species of benthic macroalgae. *Journal of Experimental Marine Biology and Ecology*, **62**, 201–10.

Krebs, C.J. (1978). *Ecology: the experimental analysis of distribution and abundance.* Harper and Row, New York.

Lawton, J. H. and Hassell, M. P. (1981). Asymmetrical competition in insects. *Nature*, **289**, 793–5.

Lubchenco, J. (1978). Plant species diversity in a marine intertidal community: importance of herbivore food preference and algal competitive abilities. *American Naturalist*, **112**, 23–39.

Lubchenco, J. (1980). Algal zonation in the New England rocky intertidal community: an experimental analysis. *Ecology*, **61**, 333–44.

Lubchenco, J. (1983). *Littorina* and *Fucus*: effects of herbivores, substratum heterogeneity and plant escapes during succession. *Ecology*, **64**, 1116–23.

Lubchenco, J. and Gaines, S. D. (1981). A unified approach to marine plant-herbivore interactions. I. Population and communities. *Annual Review of Ecology and Systematics*, **12**, 405–37.

Lubchenco, J. and Menge, B. A. (1978). Community development and persistence in a low rocky intertidal zone. *Ecological Monographs*, **48**, 67–94.

Mackay, D. A. and Underwood, A. J. (1977). Experimental studies on homing in the intertidal patellid limpet *Cellana tramoserica* (Sowerby). *Oecologia* (Berlin), **30**, 215–38.

MacLulich, J. H. (1986). Colonization of bare rock surfaces by microflora in a rocky intertidal habitat. *Marine Ecology Progress Series*, **32**, 91–6.

MacLulich, J. H. (1987). Variations in the density and variety of intertidal epilithic microflora. *Marine Ecology Progress Series*, **40**, 285–93.

McKillup, S. C. (1983). A behavioural polymorphism in the marine snail *Nassarius pauperatus*: geographic variation correlated with food availability and differences in competitive ability between morphs. *Oecologia* (Berlin), **56**, 58–66.

McKillup, S. C. and Butler, A. J. (1983). Measurement of hunger as a relative estimate of food available to populations of *Nassarius pauperatus*. *Oecologia* (Berlin), **56**, 16–22.

Menge, B. A. (1978a). Predation intensity in a rocky intertidal community: relation between predator foraging activity and environmental harshness. *Oecologia* (Berlin), **34**, 1–16.

Menge, B. A. (1978b). Predation intensity in a rocky intertidal community: effect of an algal canopy, wave action and desiccation on predator feeding rates. *Oecologia* (Berlin), **34**, 17–36.

Ortega, S. (1985). Competitive interactions among tropical intertidal limpets. *Journal of Experimental Marine Biology and Ecology*, **90**, 21–5.

Paine, R. T. (1974). Intertidal community structure: experimental studies on the relationship between a dominant competitor and its principal predator. *Oecologia* (Berlin), **15**, 93-120.

Paine, R. T. (1979). Disaster, catastrophe, and local persistence of the sea palm *Postelsia palmaeformis*. *Science*, **205**, 685-6.

Paine, R. T. (1980). Food-webs: linkage, interaction strength and community infrastructure. *Journal of Animal Ecology*, **49**, 667-85.

Paine, R. T. and Vadas, R. L. (1964). The effects of grazing by sea urchins, *Strongylocentrotus* spp. on benthic algal populations. *Limnology and Oceanography*, **14**, 710-19.

Peterson, C. H. and Black, R. (1988). Density-dependent mortality caused by physical stress interacting with biotic history. *American Naturalist*, **131**, 257-70.

Pontin, A. J. (1960). Population stabilization and competition between the ants *Lasius flavus* (F.) and *L. niger* (L.). *Journal of Animal Ecology*, **30**, 47-54.

Quinn, G. P. (1988). Ecology of the intertidal pulmonate limpet *Siphonaria diemenensis* Quoy et Gaimard. I. Population dynamics and availability of food. *Journal of Experimental Marine Biology and Ecology*, **117**, 115-36.

Reynoldson, T. B. and Bellamy, L. S. (1970). The establishment of interspecific competition in field populations, with an example of competition in action between *Polycelis nigra* (Mull.) and *P. tenuis* (Ijima) (Turbellaria, Tricladida). In *Proceedings of the advanced study institute on dynamics of numbers in populations, Oosterbeck*, (ed. P. J. den Boer and G. R. Gradwell), pp. 282-97. Netherlands Centre for Agricultural Publication and Documentation, Wageningen.

Schiel, D. R. and Choat, J. H. (1980). Effects of density on monospecific stands of marine algae. *Nature*, **285**, 324-6.

Schoener, T. W. (1983). Field experiments on intraspecific competition. *American Naturalist*, **122**, 240-85.

Schonbeck, M. and Norton, T. A. (1978). Factors controlling the upper limits of fucoid algae on the shore. *Journal of Experimental Marine Biology and Ecology*, **31**, 303-14.

Sousa, W. P. (1979a). Experimental investigations of disturbance and ecological succession in a rocky intertidal algal community. *Ecological Monographs*, **49**, 227-54.

Sousa, W. P. (1979b). Disturbance in marine intertidal boulder fields: the nonequilibrium maintenance of species diversity. *Ecology*, **60**, 1225-39.

Sousa, W. P., Schroeter, S. C., and Gaines, S. D. (1981). Latitudinal variation in intertidal algal community structure — the influence of grazing and vegetative propagation. *Oecologia* (Berlin), **48**, 297-307.

Southward, A. J. (1964). Limpet grazing and the control of vegetation on rocky shores. In *Grazing in terrestrial and marine environments*, (ed. D. J. Crisp), pp. 265-73. Blackwell, Oxford.

Stimson, J. (1970). Territorial behavior of the owl limpet *Lottia gigantea*. *Ecology*, **51**, 113–18.

Stimson, J. (1973). The role of the territory in the ecology of the intertidal limpet *Lottia gigantea* (Gray). *Ecology*, **54**, 1020–30.

Underwood, A.J. (1976). Food competition between age-classes in the intertidal neritacean *Nerita atramentosa* Reeve (Gastropoda: Prosobranchia). *Journal of Experimental Marine Biology and Ecology*, **23**, 145–54.

Underwood, A.J. (1978). An experimental evaluation of competition between three species of intertidal prosobranch gastropods. *Oecologia* (Berlin), **33**, 185–208.

Underwood, A.J. (1979). The ecology of intertidal gastropods. *Advances in Marine Biology*, **16**, 111–210.

Underwood, A.J. (1981). Techniques of analysis of variance in experimental marine biology and ecology. *Oceanography and Marine Biology Annual Review*, **19**, 513–605.

Underwood, A.J. (1984a). The vertical distribution and seasonal abundance of intertidal microalgae on a rocky shore in New South Wales. *Journal of Experimental Marine Biology and Ecology*, **78**, 199–220.

Underwood, A.J. (1984b). Vertical and seasonal patterns in competition for microalgae between intertidal gastropods. *Oecologia* (Berlin), **64**, 211–22.

Underwood, A.J. (1985). Physical factors and biological interactions: the necessity and nature of ecological experiments. In *The ecology of rocky coasts*, (ed. P.G. Moore and R. Seed), pp. 171–90. Hodder and Stoughton, London.

Underwood, A.J. (1986). The analysis of competition by field experiments. In *Community ecology: pattern and process*, (ed. J. Kikkawa and D.J. Anderson), pp. 240–68. Blackwells, Melbourne, Australia.

Underwood, A.J. (1988). Design and analysis of field experiments on competitive interactions affecting behaviour of intertidal animals. In *Behavioural adaptation to intertidal life*, (ed. G. Chelazzi and M. Vannini), pp. 333–58. Plenum Press, New York.

Underwood, A.J. (1991). The logic of ecological experiments: a case history from studies of the distribution of macro-algae on rocky intertidal shores. *Journal of the Marine Biological Association of the United Kingdom*. **71**, 841–866.

Underwood, A.J. and Denley, E.J. (1984). Paradigms, explanations and generalizations in models for the structure of intertidal communities on rocky shores. In *Ecological communities: conceptual issues and the evidence*, (ed. D.R. Strong, D. Simberloff, L.G. Abele, and A. Thistle), pp. 151–80. Princeton University Press, New Jersey.

Underwood, A.J. and Fairweather, P.G. (1989). Supply-side ecology and benthic marine assemblages. *Trends in Ecology and Evolution*, **4**, 16–20.

Underwood, A.J. and Jernakoff, P. (1981). Interactions between algae and grazing gastropods in the structure of a low-shore algal community. *Oecologia* (Berlin), **48**, 221–33.

Underwood, A.J. and Jernakoff, P. (1984). The effects of tidal height, wave-exposure, seasonality and rock-pools on grazing and the distribution of intertidal macroalgae in New South Wales. *Journal of Experimental Marine Biology and Ecology*, **75**, 71–96.

Underwood, A.J. and Petraitis, P.S. (1991). Structure of intertidal assemblages in different locations: how can local processes be compared? In *Historical and geographical determinants of community diversity*, (ed. R. Ricklefs and D. Schluter). University of Chicago Press, Chicago. (In press).

Underwood, A.J., Denley, E.J., and Moran, M.J. (1983). Experimental analyses of the structure and dynamics of mid-shore rocky intertidal communities in New South Wales. *Oecologia* (Berlin), **56**, 202–19.

Welden, C.W. and Slauson, W.L. (1986). The intensity of competition versus its importance: an overlooked distinction and some implications. *Quarterly Review of Biology*, **61**, 23–44.

Williamson, M. (1972). *The analysis of biological populations*. Edward Arnold, London.

Winer, B.J. (1971). *Statistical principles in experimental design* (2nd edn). McGraw-Hill Kogakusha, Tokyo.

Wright, W.G. (1982). Ritualized behaviour in a territorial limpet. *Journal of Experimental Marine Biology and Ecology*, **60**, 245–52.

21. Plant–herbivore coevolution: a reappraisal from the marine realm and its fossil record

ROBERT S. STENECK

Department of Botany and Plant Pathology and Department of Oceanography, University of Maine, Darling Marine Center, Walpole, USA

Abstract

Stable coexistence and interdependencies between herbivores and their food plants are often interpreted as resulting from coevolution. An untested assumption is that such intimate associations reflect an 'evolutionary arms-race' over geological time. An examination of the fossil record of apparently 'coevolved' species of marine limpets and their algal hosts indicated that these mutualistic associations evolved recently and the timing of specific adaptations and coadaptations does not support the arms-race model. Alternatively, several recent ecological studies support the hypothesis that trophic specialization and stable coexistence will be most prevalent among herbivores that are small, have low metabolic demands, reduced mobility and are predator-susceptible. They will commonly be associated with algae that are relatively large, well-defended, of poor food quality, and have great thallus longevity and regenerative ability. Among the pool of herbivores capable of stable coexistence, it is possible that unilateral adaptations to generalized anti-herbivore defences occasionally evolve and in rare instances species-specific algal coadaptations result in mutualisms. The significance of mutualisms may be exaggerated if they are interpreted as being 'coevolved' when they may represent only an end point along an adaptive continuum which has been channelled by phyletic constraints.

Plant–Animal Interactions in the Marine Benthos, (ed. D.M. John, S.J. Hawkins, and J.H. Price), Systematics Association Special Volume No. 46, pp. 477-91. Clarendon Press, Oxford, 1992. © The Systematics Association 1992.

Introduction

'We hold that plants and phytophagous insects have evolved in part in response to one another, and that the stages we have postulated have developed in a stepwise manner' Ehrlich and Raven (1964).

Plant-animal interactions often impact the fitness of the interacting species (Vadas 1977; Lubchenco and Gaines 1981) and thus have the potential to be strong agents of selection for each other. The asymmetrical outcome of most plant-herbivore interactions is thought to provide the evolutionary 'push' for plants (i.e., directional selection) to evolve more effective anti-herbivore defences, which in turn become the evolutionary 'pull' for coadaptation of herbivores dependent upon those newly adapted plants. When trophically specialized herbivores coexist with chemically or structurally defended plants the interaction is often interpreted to be the result of an evolutionary arms race called coevolution (e.g., Ehrlich and Raven 1964).

The topic of coevolution has been extensively studied by terrestrial entomologists (e.g., see Bernays and Graham 1988) who note that most herbivorous insects are trophically specialized and the plants on which they feed often have identifiable deterrents making them unpalatable or inedible to other herbivores. Plant-herbivore trophic specialization in the marine realm is less common (Lubchenco and Gaines 1981; Hay *et al.* 1989) but it does exist and offers the advantage that some of the interacting groups have a relatively complete fossil record. This provides more than an opportunity to examine coevolution in the fossil record, it provides an opportunity to achieve some insight into the evolutionary mechanisms of adaptation in general.

Specifically, I will explore whether chronic ecological interactions such as those between plants and herbivores have created evolutionary pressures that resulted in reciprocal adaptations and change over time. If so, coevolved mutualisms should have a long evolutionary history of association during which time evolutionary change within both interacting phyletic lineages should be evident. If not, what plausible alternatives are there for the examples of species-specific ('pairwise') trophic specialization of herbivores on chemically or structurally defended plants?

Coevolution: definition, expectations, and evidence

Coevolution as originally defined is a 'stepwise reciprocal selective response' between 'ecologically closely linked organisms' (Ehrlich and Raven 1964). It was assumed that 'a trait of one species has evolved in response to a trait of another species which itself has evolved in response to the trait in the first' (Futuyma and Slatkin 1983).

Mitter and Brooks (1983) suggested that if trophically specialized phytophagous organisms are coevolved with their prey then they must be the result of a reciprocal adaptive radiation, in which the timing of the evolution of an herbivore clade closely follows the timing of major evolutionary events within the plant prey clade. As a result, groups of related plants are consumed by groups of related herbivores. Unfortunately, the question of how long and under what conditions did such associations evolve could not be addressed for insect–plant interactions because neither group has an adequate fossil record.

I will re-examine the question of coevolution for encrusting coralline red algae (Rhodophyta, Corallinales) and their herbivores because:

(1) this plant group is commonly associated with herbivores;
(2) close, relatively stable coralline–herbivore interactions have been documented that are species-specific (e.g., Steneck 1986);
(3) this association has the most complete fossil record of any plant–animal association in the world (Steneck 1983, 1985, 1986).

1. Pairwise (species-specific) mutualism: a case study

The association between the limpet *Tectura* (= *Acmaea*) *testudinalis* and the coralline *Clathromorphum circumscriptum* (Steneck 1982a) is an apparently pairwise mutualism. The limpet preferentially recruits to, occupies, and eats this coralline species over all others using a coralline-grazing radula. The limpet's survivorship against sea star predation is greater on this species than it is on other corallines (Steneck 1990). If the limpet is excluded from *C. circumscriptum* the coralline dies (Steneck 1982a). *Clathromorphum circumscriptum* is protected from limpet grazing by possessing a thick multilayered epithallus which overlies and protects the meristem. Epithallial cell replacement rate matches limpet ingestion rate. Reproductive structures are sunken beneath the thallus surface minimizing the damage from rasping herbivores. *C. circumscriptum* is the competitively dominant coralline in this region only when grazed by this limpet (Steneck 1990). Competitive dominance reverses to thinner, faster-growing corallines when simulated grazing from limpets is reduced (Steneck 1982b, 1990). Both species are circumglobal in their distribution (Steneck 1982b).

2. Evolutionary histories: searching for evidence of reciprocal adaptations

The evolutionary history of both *T. testudinalis* and *C. circumscriptum* suggests that the association is recent and major evolutionary events are neither tightly linked nor strictly stepwise (Fig. 21.1). The genus

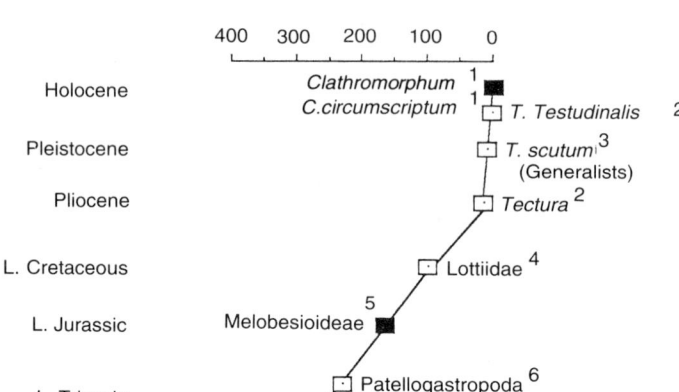

Fig. 21.1. Evolutionary history of a limpet–coralline association. The origins of *Clathromorphum circumscriptum* and related higher taxa are plotted using the black square and labelled to the left of the solid line. The origins of *Tectura testudinalis* and its related higher taxa are plotted with the white squares and are identified to the right of the solid line. References: 1. Adey and Johansen 1972, 2. Lindberg 1988, 3. Lindberg, pers. comm., 4. Vermeij 1987, 5. Johansen 1981, 6. Moore 1964, 7. Wray 1964.

Clathromorphum is not known prior to the Holocene (Johnson 1961; Adey and Johansen 1972) whereas the limpet genus *Tectura* is Pliocene in origin (Lindberg 1988). Although all the subtidal species of *Clathromorphum* are associated with herbivores, only *C. circumscriptum* appears to have such a close association with a single limpet species (Lebednik 1977; Steneck, personal observations). The most closely related limpet, *T. scutum* (once thought to be a subspecies of *T. testudinalis*, Test 1945), is known from the Pleistocene. Although *Tectura scutum* can consume corallines (Steneck *et al.* 1991), it is thought to be a trophic generalist (Padilla 1985). The closest coralline relative to *Clathromorphum*, the genus *Lithothamnion*, which is one of the oldest coralline genera, is identified in fossils from the Jurassic period (Johnson 1961). The earliest coralline-like crust, *Archaeolithophyllum*, is from the Pennsylvanian (Wray 1964). *Tectura* is phyletically close to *Lottia* of the Lottiidae family, a relatively new patellogastropod clade (Lindberg 1988) which is Pliocene in origin. The origin of the true limpets is Middle Triassic

(Moore 1964) but they did not become abundant until the Jurassic (reviewed in Steneck 1983).

If the two interacting groups evolved in conformity with the reciprocal adaptive radiation model for coevolution (Mitter and Brooks 1983), then the higher taxa of related herbivores should be closely associated with higher taxa of related plants. However, there is no evidence of this at any level of higher taxa within the coralline or limpet clades (Fig. 21.2). Instead, coralline grazing has evolved repeatedly within each major limpet clade. Convergent radular and shell morphologies characteristic of coralline associations have been identified in several distantly related limpet groups (Lindberg 1988). Similarly, several characteristics determined to be adaptive under limpet grazing have evolved convergently among several distantly related corallines (Steneck 1986, 1990).

At a coarser scale, did calcified algal crusts evolve in response to evolutionary pressure from limpets or any major herbivore group? Apparently not: calcified algal crusts first appear in the fossil record in the Late Proterozoic of the Precambrian (Grant et al. 1991) and thus evolved prior to most metazoans and all known herbivores. They were

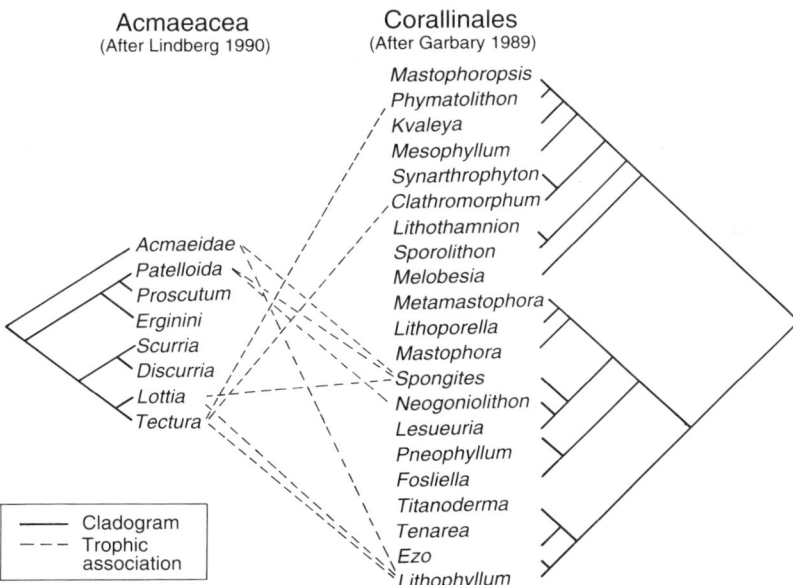

Fig. 21.2. Cladograms for acmaeid limpets (Lindberg 1990) and coralline algae (Garbary, unpublished data). Major groups of coralline grazing limpets (Acmaeidae, Patelloida, *Lottia*, and *Tectura*) show no association with any particular higher taxa (e.g., subfamilies) of coralline algae.

thin calcified crusts, with evident hypobasal calcification and possibly with raised conceptacles. Thus they possess characters found in peyssonnelid and coralline crusts today. The Solenoporaceae are thought to be early ancestors to the corallines which evolved in the Cambrian (Steneck 1983). Solenopores never radiated significantly and went extinct during a period of major diversification of coralline grazing herbivores in the early Tertiary (Steneck 1983). During the Palaeozoic, *Archaeolithophyllum* possessed several coralline characters such as a calcified thallus, a high degree of tissue differentiation, and conspicuous raised conceptacles (Wray 1964). Both the Precambrian and Palaeozoic corallines usually grew over sediment in relatively herbivore-free environments (e.g., Steneck 1983, 1985; Grant *et al.* 1991). It is possible that the calcified thallus and prostrate growth form of corallines and their relatives provided a successful means of invading sedimentary habitats which were a relatively unoccupied adaptive zone for algae. There is no evidence that the genesis of these early crusts was a 'response' to herbivores, since most of them evolved prior to or in environments lacking intense herbivory (Steneck 1983).

Were calcareous algae the 'cause' of the evolution of calcium carbonate excavating limpets? Again probably not: true docoglossate limpets evolved during the late Triassic (Vermeij 1987), when calcareous algae (particularly encrusting forms) were exceedingly rare (Flügel *et al.* 1989). Their first radiation was during the Jurassic and graze marks on calcium carbonate first appear at that time (Voigt 1977; Boucot 1990). The earliest graze marks, however, are on bivalve and ammonite shells (Voigt 1977; Akpan *et al.* 1982) and may have been the result of exploiting epiphytes or endophytes (Akpan *et al.* 1982). Limpet graze marks on corallines are relatively common in the Cainozoic (Bosence, pers. comm.). It is probable that limpets of the Mesozoic were grazing on corallines (Steneck 1983), but there is no evidence to suggest, that limpet evolution was a 'response' to the presence of corallines.

Trophic specialization and stable coexistence: a consideration of mechanisms

Today the association between limpets and corallines is strong (Fig. 21.3), although the fossil record does not apparently support a direct evolutionary stepwise pattern of mutual decent. In a review on the diets of 37 species of limpets, over 30 per cent were identified as grazers of encrusting corallines (Steneck and Watling 1982). However other strong associations with fleshy crusts (11 per cent), kelp, fucoids (19 per cent), and marine angiosperms (11 per cent) (Fig. 21.3;

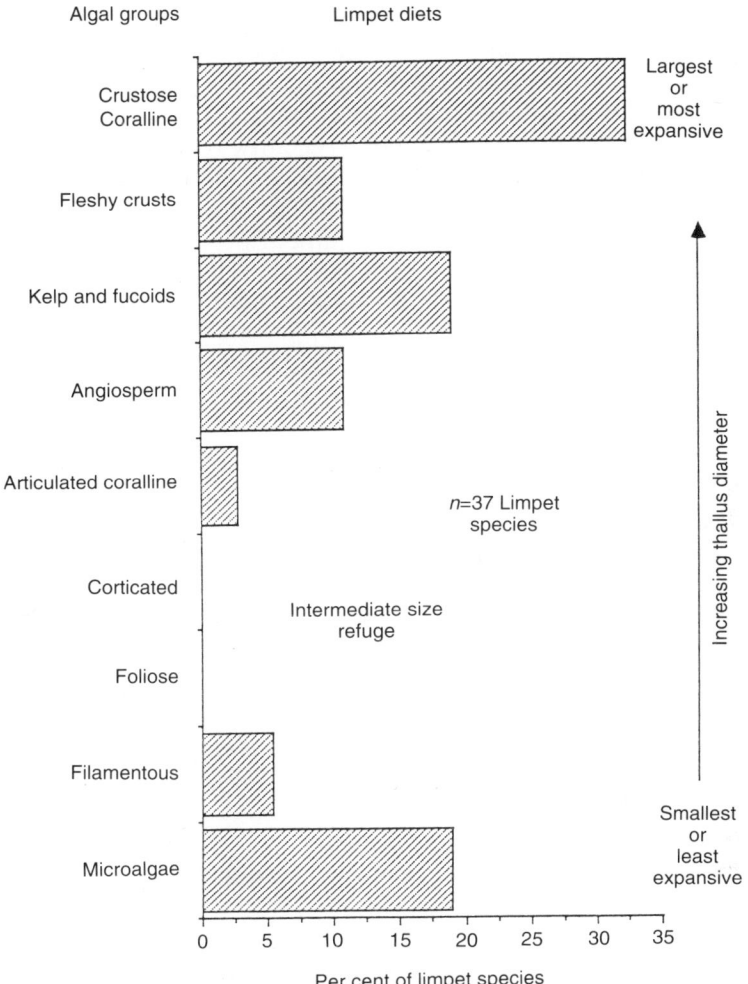

Fig. 21.3. The per cent of limpet species feeding on nine distinctly different morphological groupings of benthic algae and seagrass (from Steneck and Watling 1982). Algal groups are arranged in order of increasing thallus diameter from the bottom bar to the top.

Lindberg 1990) are also well known. One pairwise association between the limpet, *Lottia incessa*, and the kelp, *Egregia menziesii*, is said to be coevolved (Black 1976). Other pairwise associations have been reported for limpets and angiosperms such as *Lottia alveus* on *Zostera marina* and *Tectura paleacea* on *Phyllospadix* (reviewed in Steneck and Watling 1982). Obligate pairwise associations risk being on an evolutionary limb in that the survival of one species becomes dependent upon the survival of the other. Case in point is *L. alveus*, which became locally extinct throughout the western North Atlantic following the rapid near-complete decline in *Zostera* beds in the 1930s (Lindberg 1990).

Limpet–kelp associations were recently reviewed by Estes and Steinberg (1988). They reported the work of Lindberg and others, who indicate that limpets specializing on kelp often have saddle-shaped shells conforming to the stipe's cylindrical shape. Specialists on angiosperms are elongate and have relatively parallel sides to their aperture conforming to strap-like seagrass blades. Reviewing the fossil record, Estes and Steinberg (1988) conclude that limpet–kelp associations are very recent having appeared only in the last 3 million years. Thus it appears that other relatively stable, pairwise associations may have evolved suddenly and rather recently.

It is significant that a relatively high proportion of limpets feed on diverse assemblages of microalgae (Fig. 21.3). The pattern suggests that limpets tend to graze minute algae off the substrata they occupy or they occupy the alga that they graze. Where specific species or groups of algae have been identified as the food plant on which limpets specialize, they are commonly large or expansive forms (e.g., corallines, kelp, and angiosperms). This was supported by Branch (1986) who found a correlation between dietary specialization among 11 southern African *Patella* species and the ratio of food plant to limpet size. Among herbivorous gastropods, limpets have a relatively large perimeter (aperture). Since the primary defensive behaviour against perceived threats is to clamp down, the absence of suitably wide substrata may explain the virtual absence of significant consumption of foliose and corticated macroalgae on which limpet occupation is difficult (Fig. 21.3).

Phyletic constraints: channelling evolutionary patterns

Most examples of coexistence between trophically specialized herbivores and algae are among small, relatively slow-moving invertebrates on relatively large poor-quality food (Steneck 1982a; Hay et al. 1989, 1990a, b). This pattern transcends all taxonomic and phylogenetic categories, since excellent examples can be found among groups as diverse as decapods and green algae (Hay et al. 1989), amphipods and brown algae (Hay et al. 1990a), ascoglossan molluscs and green algae (Hay et al. 1990b), as well as the examples of limpets and their associations with angiosperms, kelp and corallines discussed above. In most cases, the herbivores are associated with algae possessing structural or chemical defences which have been demonstrated to reduce generalized herbivory.

There are limited variations in structure or function for the body plan of any related group of plants or animals. For animals, body size, mechanics of movement, and feeding tend to be conservative within phyletic lineages (i.e., they are phyletically constrained, see Gould 1989). Usually there are developmental, anatomical, biomechanical, or other physical reasons why organisms attain a certain maximum size, or can move only so rapidly, or can feed only on certain plants and not on others. Among herbivorous gastropods, the foot and radula constrain the latitude of their mobility and feeding, respectively. Thus there are some general body-size patterns among herbivores, in that most amphipods are smaller than molluscs which are smaller than most regular echinoids, which in turn tend to be smaller than most herbivorous fishes. The same is true for algae since large size is linked to practical problems of translocating photosynthates throughout the thallus. All massive algal forms of all three major divisions have convergently evolved mechanisms for translocation and wound-healing. Except for species in the order Siphonales (which are coenocytic), most of the highly diverse Chlorophyta (green algae) have no means of translocation and tend to be relatively diminutive in size and limited in their morphological variety.

Important evolutionary events such as the origin of cell fusion allow translocation to occur in coralline algae (Steneck 1983). This may aid growth and facilitate wound healing by physiologically integrating the algal thallus (Steneck 1986). As a result, coralline algae have a superior capacity for regeneration following wounds (Steneck 1986; Steneck and Dethier, unpublished data). However, in the past massive calcareous algae (i.e., Solenoporaceae) may have been incapable of translocation and thus had a limited ability to regenerate lost tissue following wounding (Steneck 1983). Possibly as a result, the solenopore clade declined steadily as the herbivore groups capable of grazing calcium carbonate diversified. Patterns of continually increasing frequency of unrepaired wounds were evident in the fossil record during the period of solenopore decline (Steneck 1983). Therefore it seems likely that strong selective pressures did not cause solenopores to adapt, they just went extinct. The ability of corallines to diversify during the escalation in herbivory during the late Mesozoic and Cainozoic may have been due to their ability to translocate, which appears to have evolved in the Palaeozoic Era prior to the herbivore radiation (Steneck 1983, 1985).

Small, slow-moving organisms are most vulnerable to predators (Peters 1983). For them habitat selection to refuges or habitats where their predator-susceptibility is low is critical to their survival. Within this subset of habitats in which they can live is the subset of algae

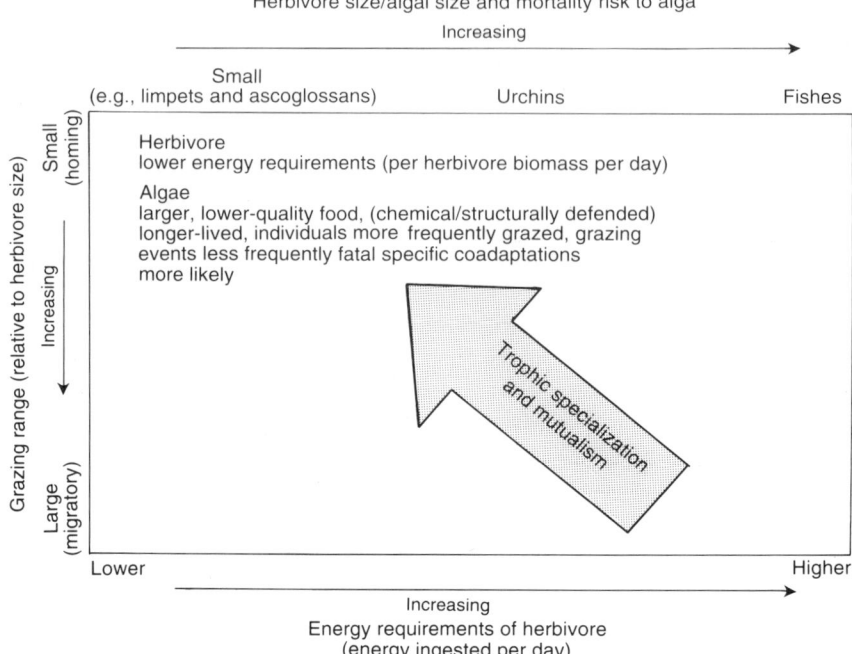

Fig. 21.4. A model predicting the functional characteristics under which trophic specialization and mutualism will be most likely (modified from Steneck 1982a).

on which they can feed. These constraints limit trophic latitude and contribute to trophic specialization. Steneck (1982a) proposed a general model of algal–herbivore interactions (Fig. 21.4), which suggested that trophic specialization would be most prevalent among herbivores that are small relative to their food plant and have small grazing ranges. Further it was suggested that these herbivores would have lower metabolic requirements due to their reduced vagility and the algae would be relatively large, of low food quality (i.e., chemically or structurally defended) and long-lived. Under such conditions, individual herbivore impact is relatively low and the probability of algal coexistence great (Fig. 21.5). Among small, slow-moving herbivores, their habitat is their food plant and thus species-specific herbivore adaptations (i.e., high digestive efficiency or detoxification or tolerance of an anti-herbivore compound) and stable coexistance are more likely. Conversely, larger herbivores are not similarly constrained. They are less susceptible to predators and are capable of migrating greater distances to obtain higher-quality foods (Vadas 1977). Herbivory from

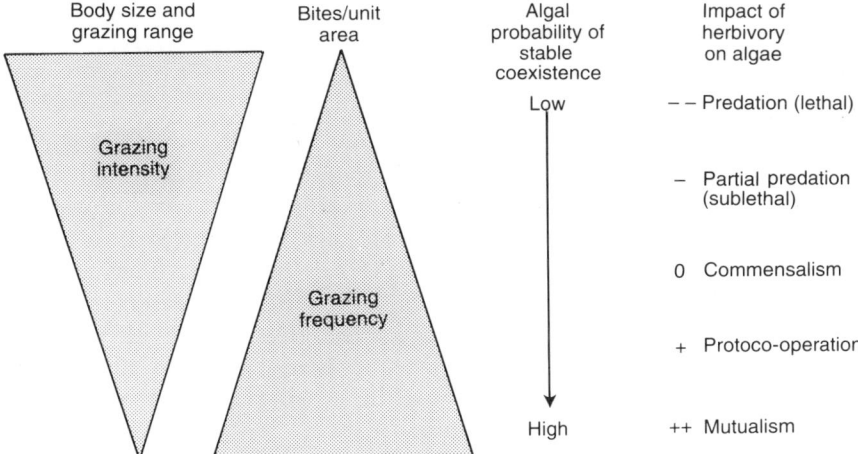

Fig. 21.5. A model suggesting the relationship between herbivore body size and grazing range as it affects grazing intensity and frequency. It suggests that the likelihood of stable coexistence and algal–herbivore mutualism is greatest for herbivores having a small grazing range and impact algae with grazing of low intensity but high frequency.

larger, more deeply grazing urchins and fishes make stable coexistence with their algal prey less likely to occur and thus make mutualisms unlikely. Trophic specialization is most common among urchins that are most mobile (Vadas 1990) so depleting their preferred food is of little consequence to their survival.

It seems that the prevalence of trophic specialization and herbivore coexistence on algal species possessing anti-herbivore defences is most common among herbivore groups that are phyletically constrained in terms of their body size, body plan, mobility, and feeding ability. Thus the functional aspects of plant–herbivore trophic interactions may determine the groups within which 'intimate associations' occur. From this pool of interacting species, occasional, species-specific, pairwise mutualisms have a higher probability of evolving. In this way, phyletic constraints in functional latitude (i.e., dietary or habitat breadth) channel evolutionary change. As a result, tight pairwise associations may appear suddenly without a long phyletic association between the interacting groups.

It is now widely recognized that pairwise associations are rare (Futuyma and Slatkin 1983). Much more common are unilateral adaptations and those involving adaptations of a species to a suite of often phylogenetically diverse species to which it holds a similar ecological relationship (i.e., 'diffuse coevolution'). However, Vermeij (1987) maintains that for both pairwise and diffuse modes, 'the word coevolution should be restricted to cases of adaptive reciprocity' and the 'evolution of mutualism cannot be taken as evidence of reciprocity'. My examination of the evolution of corallines and limpets revealed no adaptive reciprocity at any level. The evolution of corallines appears to be much more influenced by the sudden evolution of major herbivore groups which collectively possessed greatly increased abilities to excavate calcium carbonate substrata (Steneck 1983, 1985, 1986). This conforms with the findings of Boucot (1990), who concluded a major study of coevolution in the fossil record by saying: 'I have encountered no examples of gradual change on a geological time scale between behaviors or coevolutionary relations . . . Major changes in coevolution and behavior are suggested by the fossil record to occur very rapidly, and to be very spasmodic in time.'

Herbivory is ubiquitous and often intense, but the assumption that this process creates strong selective pressures on algal prey, to which they may adapt, is not supported by the fossil record. The non-random distribution of trophically specialized herbivores does not necessarily suggest long-term mutual descent (e.g., a reciprocal adaptive radiation). It could be the result of a complex suite of body plan, size, mobility, and feeding characteristics, that limit the potential pool of algal resources to only a few growth forms. The algal growth forms tend to be large, and capable of regeneration, thereby facilitating stable coexistence. From such trophically limited plant–animal interactions, occasional pairwise specializations or even mutualisms have evolved. However, to date, there is no evidence that they are the result of the evolutionary arms race of coevolution.

Acknowledgements

Many colleagues have contributed to the ideas presented here. In particular I wish to thank W. Adey, G. Branch, M. Dethier, S. Hacker, J. Jackson, R. Paine, C. Pfister, M. Hay, S. Stanley, L. Watling, R. Vadas, and G. Vermeij. Mark Hay and Les Watling read and made valuable comments on this chapter. The research was funded by grants from the National Science Foundation (Grant Nos. OCE-8315136 and OCE-8600262 with additional funding from NOAA's UM/UNH Sea Grant (R/FM-169 and R/FMD-297, NURC-UCAP awards 1984–1988). This is Darling Center Contribution No. 238.

References

Adey, W. H. and Johansen, H. W. (1972). Morphology and taxonomy of Corallinaceae with special respect to *Clathromorphum Mesophyllum* and

Neopolyporolithon gen. nov. (Rhodophyceae, Cryptonemiales). *Phycologia*, **11**, 159-80.

Akpan, E. B., Farrow, G. E., and N. Morris. (1982). Limpet grazing on Cretaceous algal-bored ammonites. *Paleontology*, **25**, 361-67.

Bernays, E. and Graham, M. (1988). On the evolution of host specificity in phytophagous arthropods. *Ecology*, **69**, 886-92.

Black, R. (1976). The effects of grazing by the limpet, *Acmaea insessa*, on the kelp, *Egregia laevigata*, in the intertidal zone. *Ecology*, **557**, 265-77.

Boucot, A. J. (1990). *Evolutionary paleobiology of behavior and coevolution*. Elsevier, Amsterdam.

Branch, G. M. (1986). Limpets: their role in littoral and sublittoral community dynamics. In *The ecology of rocky coasts*, (ed. P. G. Moore and R. Seed), pp. 97-116. Columbia University Press, New York.

Ehrlich, P. R. and Raven, P. H. (1964). Butterflies and plants: a study in coevolution. *Evolution*, **18**, 586-608.

Estes, J. A. and Steinberg, P. D. (1988). Predation, herbivory, and kelp evolution. *Paleobiology*, **14**, 19-36.

Flügel, E., Senowbari-Daryan, B., and Stanley Jr., G. D. (1989). Late Triassic Dasycladacean alga from Northeastern Oregon: significance of first reported occurrence in western North America. *Journal of Paleontology*, **63**, 374-81.

Futuyma, D. J. and Slatkin, M. (1983). Epilogue: the study of coevolution. In *Coevolution*, (ed. D. J. Futuyma and M. Slatkin), pp. 459-64. Sinauer Associates Inc., Sunderland, Massachusetts.

Gould, S. J. (1989). A developmental constraint in *Cerion*, with comments on the definition and interpretation of constraint in evolution. *Evolution*, **43**, 516-39.

Grant, S. W. F., Knoll, A. H., and Germs, G. J. B. (1991). Probable calcified metaphytes in the Latest Proterozoic Nama group, Namibia: origin, diagenesis and implications. *Journal of Phycology*, **65**, 1-18.

Hay, M. E., Duffy, J. E., and Fenical, W. (1990*a*). Host-plant specialization decreases predation on a marine amphipod: an herbivore in plant's clothing. *Ecology*, **71**, 733-43.

Hay, M. E., Pawlik, J. R., Duffy, J. E., and Fenical, W. (1989). Seaweed-herbivore-predator interactions: host-plant specialization reduces predation on small herbivores. *Oecologia* (Berlin), **81**, 418-27.

Hay, M. E., Duffy, J. E., Paul, V. J., Renaud, P. E., and Fenical, W. (1990*b*). Specialist herbivores reduce their susceptibility to predation by feeding on the chemically-defended seaweed *Avrainvillea longicaulis*. *Limnology and Oceanography*, **35**, 1734-43.

Johansen, H. W. (1981). *Coralline algae, a first synthesis*. CRC Press, Florida.

Johnson, J. H. (1961). *Limestone building algae and algal limestones*. Johnson Publishing Co., Boulder, Colorado.

Lebednik, P. A. (1977). The Corallinaceae of northwestern North America. 1. *Clathromorphum* Foslie emend Adey. *Syesis*, **9**, 59-112.

Lindberg, D. R. (1988). The Patellogastropoda. *Malacological Review*, Supplement **4**, 35-63.

Lindberg, D. R. (1990). Morphometrics and the systematics of marine plant limpets (Mollusca: Patellogastropoda). *Proceedings of the Michigan Morphometrics Workshop*, (ed. F. J. Rohlf and F. L. Bookstein), pp. 301-10. The University of Michigan Museum of Zoology, Ann Arbor, Michigan.

Lubchenco, J. and Gaines, S. D. (1981). A unified approach to marine plant-herbivore populations. I. Populations and communities. *Annual Review of Ecology and Systematics*, **12**, 405-37.

Mitter, C. and Brooks, D. R. (1983). Phylogenetic aspects of coevolution. In *Coevolution*, (ed. D. J Futuyma and M. Slatkin), pp. 65-98. Sinauer Associates Inc., Sunderland, Massachusetts.

Moore, R. C. (1964). *Treatise on invertebrate paleontology*; Part I. *Mollusca*. Vol. 1. Geological Society of America and University of Kansas Press, Boulder, Colorado.

Padilla, D. K. (1985). Structural resistance of algae to herbivores: a biomechanical approach. *Marine Biology*, **90**, 103-109.

Peters, R. H. (1983). *The ecological implications of body size*. Cambridge University Press, Cambridge.

Steneck, R. S. (1982*a*). A limpet-coralline alga association: adaptations and defences between a selective herbivore and its prey. *Ecology*, **63**, 507-22.

Steneck, R. S. (1982*b*). Adaptive trends in ecology and evolution of crustose coralline algae (Corallinaceae, Rhodophyta). Unpublished Dissertation. The Johns Hopkins University, Baltimore, Maryland, USA.

Steneck, R. S. (1983). Escalating herbivory and resulting adaptive trends in calcareous algal crusts. *Paleobiology*, **9**, 44-61.

Steneck, R. S. (1985). Adaptations of crustose coralline algae to herbivory: patterns in space and time. In *Paleoalgology; contemporary research and applications*, (ed. D. Toomey and M. Nitecki), pp. 352-66. Springer-Verlag, Berlin.

Steneck, R. S. (1986). The ecology of coralline algal crusts: convergent patterns and adaptive strategies. *Annual Review of Ecology and Systematics*, **17**, 273-303.

Steneck, R. S. (1990). Herbivory and the evolution of nongeniculate coralline algae (Rhodophyta, Corallinales) in the North Atlantic and North Pacific. In *Evolutionary biogeography of the marine algae of the North Atlantic*, (ed. D. J. Garbary and G. R. South), NATO Advanced Research Workshop Series G, **22**, 107-29.

Steneck, R. S. and Watling, L. (1982). Feeding capabilities and limitations of herbivorous molluscs: a functional group approach. *Marine Biology*, **68**, 299-319.

Steneck, R. S., Hacker, S. D., and Dethier, M. N. (1991). Mechanisms of competitive dominance between crustose coralline algae: an herbivore-mediated competitive-reversal. *Ecology*, **72**, 938–50.

Test, A. R. (1945). Ecology of California *Acmaea*. *Ecology*, **26**, 395–405.

Vadas, R. L. (1977). Preferential feeding: an optimization strategy in sea urchins. *Ecological Monographs*, **47**, 337–71.

Vadas, R. L. (1990). Comparative foraging behavior of tropical and boreal sea urchins. In *Behavioral mechanisms of food selection*, (ed. R. N. Hughes), NATO Advanced Research Workshop Series Vol. G, pp. 479–5. Springer-Verlag, Berlin.

Vermeij, G. J. (1987). *Evolution and escalation, an ecological history of life.* Princeton University Press, Princeton, New Jersey.

Voigt, E. (1977). On grazing traces produced by the radula of fossil and Recent gastropods and chitons. In *Trace fossils*, Geological Journal Special Issue, 9, (ed. T. P. Crimes and J. C. Harper), pp. 335–46.

Wray, J. L. (1964). *Archaeolithophyllum*, an abundant calcareous alga in limestones of the Lansing group (Pennsylvanian), southeastern Kansas. *Kansas Geological Survey Bulletin*, **170**, 1–13.

22. Bioerosion and biogeomorphology

T. SPENCER
Department of Geography, University of Cambridge, Downing Place, Cambridge, England, UK

Abstract

Bioerosion is a complex set of physical and chemical processes of rock removal, driven by the interrelationships between (1) 'biological corrosion', a solutional process (accounting for 10–30 per cent of total erosion) largely engineered by Cyanobacteria, and (2) 'biological abrasion', the removal of these micro-organisms and particulate debris, notably by boring sponges and bivalve molluscs, burrowing echinoids, and grazing amphineurids and gastropods (70–90 per cent of total erosion). Rates of biological abrasion, although largely derived from either live coral surfaces or unconsolidated reef sediments and thus not readily transferable to lithified substrates, have been established through studies of *in situ* surface modification, experimentation with block transplants and through the collection of sedimentary products. Mean rates of general surface-lowering characteristically range between 0.5–1.0 mm yr^{-1}. As cyanobacterial zonation, and thus erosion rate, varies with position on the littoral profile, and as this zonation changes with exposure, it is possible to explain the characteristic notched form to the littoral profile and its evolution on rocky shores. However, bioerosion rates do not necessarily increase on progressively higher energy coasts: under microtidal regimes, encrustations of coralline algae and vermetid gastropods on surf platforms act as protective structures, sealing the underlying bedrock from bioerosional processes.

Introduction

One of the consequences of plant–animal interactions in marine ecosystems is that on rocky shorelines such interactions may drive bioerosion, a complex set of biologically mediated chemical and physical processes of *in situ* weakening and subsequent removal of consolidated substrate. Two sets of processes, linked by a 'strict trophic relationship' (Schneider and Torunski 1983, p. 51), are involved: 'biological corrosion', or rock dissolution resulting from the action of Cyanobacteria, and 'biological abrasion', the removal of these micro-organisms and the associated production of lithoclasts, notably by sponge and bivalve boring activity, burrowing echinoids, and grazing amphineurids and gastropods (Trudgill 1985; Spencer 1988). In the northern Adriatic Sea, total erosion appears to be partitioned: 10 to 30 per cent biological corrosion, and 70 to 90 per cent biological abrasion (Torunski 1979).

Unfortunately, many reports of bioerosion remain anecdotal and few rates of bioerosion are directly comparable as they refer, within sites, to only a small sample of individuals at a site without taking the age structure or population dynamics of the species into account and, across sites, to varying lithology and degree of coastline exposure. Only on limestone coasts (curiously, because traditionally limestones have been regarded as the product of chemical dissolution and thus unique coastal landforms) has it been possible to quantify these processes in sufficient detail across a range of organisms through studies of *in situ* surface modification, experimentation with block transplants, and through the collection of faecal sedimentary products. Such research has allowed for the construction of coastal biogeomorphologies (Trudgill 1987; Viles 1988) involving generalizations about the spatial pattern of substrate modification by macro-organisms (e.g., within-coral reef (Kiene 1988) and across-reef (Sammarco and Risk 1990) patterns of bioerosion) and explanations as to how these patterns may determine the evolution of the characteristic intertidal notches of many rocky shore profiles. This review, therefore, concentrates upon carbonate substrates.

Both long-term estimates of coastal retreat, and accurate but short-term measurements of surface lowering in subtidal and intertidal environments on tropical and temperate coasts, suggest erosion rates of between 0·5 and 2 mm yr^{-1} (Trudgill 1985; Spencer 1988). Clearly lithology exerts a strong overiding control: thus heterogeneous, weakly cemented tropical limestones tend to be characterized by mean erosion rates in excess of 1 mm yr^{-1}, while older limestones with a history of considerable diagenesis on temperate coasts retreat at less than 0·5 mm yr^{-1}. However, within this control, the similarity of erosion rates from different localities and determined by different methods, has been used as evidence for a balance between the rock removal potential of grazing organisms on the one hand, and the regenerative ability of substrate-penetrating endolithic micro-organisms on the other (Spencer

1988). Similar linkages have been well illustrated in reef environments where shifts in grazing pressure have led to changes in the dominant bioeroding community and the type of sediment produced (Sammarco et al. 1987).

Biological corrosion

The diverse micro-organic community on rocky shores is often dominated by Cyanobacteria, but also includes chlorophytes, rhodophytes, heterotrophic fungi, lichens, and other bacteria. Conspicuous zonation among these lithophytic communities is well known (e.g., Le-Campion Alsumard 1970; Golubic et al. 1975; Schneider 1976; Torunski 1979; Whitton and Potts 1982), and results from both the presence or absence of indicator species and the changing importance of different 'lithobiontic' niches (Golubic et al. 1981) relating to environmental gradients of insolation, desiccation and illumination, and ecological interactions with macro-organisms. The degree of substrate modification, particularly by within-rock endoliths, can be considerable. The greatest number of boreholes are to be found in the upper intertidal zones: along the Adriatic coast (Torunski 1979) densities reach 4460 to 8140 boreholes mm^{-2} with up to 33 per cent by volume of the substrate occupied by borings (Le-Campion Alsumard 1979). Examination of rock samples by scanning electron microscopy has shown that endolithic borings are not randomly distributed, but organized into intersecting trends, producing lines of preferential fracture which predetermine the size and shape of grain which can be removed by a grazing organism; the modal size appears to fall in the range 20–60 μm in indurated limestones (Torunski 1979), but shows a greater range (< 10–400 μm in diameter) on weakly cemented Pleistocene limestones (Rasmussen and Frankenberg 1990). In addition, borers may preferentially select impurities and components of varying solubility within the host substrate as erosional pathways (Whitton and Potts 1982).

Biological abrasion

Substrate removal during grazing activity has long been established (e.g., Lowenstam 1962). Erosional efficiency is related to foraging behaviour, radula structure and strength (Schneider and Torunski 1983; Trudgill et al. 1987), and the size distribution of individuals and population densities (e.g., Bak 1990 for Pacific echinoids; Taylor and Way 1976; Trudgill 1983a, b, for *Acanthopleura* spp.). Most of the quantitative information on biological abrasion rates comes from coral-reef environments, and follows McLean's (1967a, b, 1974) classic experiments which showed how this process could be quantified by the collection of faecal pellets and the analysis of their carbonate content. Comparisons of bioabrasion between species groups suffer from the difficulties, in some cases, of extrapolating short-term laboratory results to the field and of the need to control for rock density. Nevertheless, Tables 22.1 and 22.2 show the rock removal potential of chitons, with their large, magnetite-strengthened radulae, and echinoids, which employ basal spines and teeth in burrow excavation.

By comparison, rock-boring bivalves and barnacles, which internally weaken rather than externally lower rock substrates, appear to be much less effective erosion agents (Table 22.3), although it is difficult to evaluate known rates of boring in the absence of information on variation in boring between individuals of different sizes and at different life stages, on the areal coverage of borings, and on the percentage occupancy of these boreholes (Trudgill et al. 1987; Trudgill and Crabtree 1987). However, on tropical coasts, boring sponges (and locally polychaete worms) are clearly highly important bioerosional agents (Hutchings 1986; Table 22.4), particularly in the early stages of substrate infestation (e.g., Neumann 1966; Rützler 1975). The detrital 'chips' which result from sponge excavation processes often contribute 0.1 to 6.0 per cent and, exceptionally, in excess of 30 per cent, of the total sediment (for the silt fraction: 10–30 percent, 98 per cent respectively) in tropical lagoons and fore-reef environments (Young and Nelson 1985).

Biogeomorphology

On Aldabra Atoll, Indian Ocean (Trudgill 1976), and Grand Cayman Island, Caribbean Sea (Spencer 1985), the use of controlled exclusion experiments has permitted the partitioning of total littoral erosion between physical, chemical, and biological processes. In these environments, grazing bioabrasion accounts for 34 to 64 per cent of total coastal retreat on undercut shore profiles. Further extension of this approach shows that it is possible to link cyanobacterial zonation, distribution of macro-organisms, and physical exposure to measurements of erosion rate to give a model for littoral profile variation (Fig. 22.1a). These interrelationships may be extrapolated firstly, temporally to predict long-term profile evolution (Fig. 22.1b) and secondly, spatially to show how the littoral profile varies with changing coastline exposure. Such explanations can only be partial ones because of the strong confounding influences of rock structure (and thus coastal slope angle) and those physical processes unrelated to biological activity. While on high-

Table 22.1. Quantitative estimates of substrate removal and/or surface down-wearing by grazing amphineurids and gastropods.

Agent	Rate of rock removal	Substrate	Locality	Author
Chiton sp.	8·0 cm^3 yr^{-1}	Beachrock	Barbados	McLean (in Trudgill, 1983b)
Acanthopleura sp.	13·0 cm^3 yr^{-1}	Beachrock	Barbados	McLean (in Trudgill, 1983b)
Acanthozostera gemmata	18·0 cm^3 yr^{-1} chiton^{-1}	Beachrock	Heron Is., Great Barrier Reef	McLean (1974)
	0·5–2·9 mm yr^{-1}			
Acanthozostera gemmata	0·2–2·9 mm yr^{-1}	Boulder rock	One Tree Is., Great Barrier Reef	Trudgill (1983a)
Acanthopleura brevispinosa	5·4 cm^3 yr^{-2} chiton^{-1}	Reef limestone	Aldabra Atoll	Taylor and Way (1976)
Acanthopleura granulata	22·8 cm^3 yr^{-1} chiton^{-1}	Reef limestone	San Salvador	Rasmussen and Frankenberg (1990)
	0·12 mm yr^{-1}			
Acanthopleura granulata	0·21 mm yr^{-1}	Reef limestone	Andros Is.	Donn and Boardman (1988)
Acanthopleura granulata	38·8 cm^3 yr^{-1} chiton^{-1}	Coral rubble	Puerto Rico	Glynn (1973)
Chiton tuberculatus	11·6 cm^3 yr^{-1} chiton^{-1}	Coral rubble	Puerto Rico	Glynn (1973)
Acanthopleura granulata	36·5 cm^3 yr^{-1} chiton^{-1}	Reef limestone	Grand Cayman	Benn (in Spencer and Benn, in press)
Acanthopleura granulata	38·6 cm^3 yr^{-1} chiton^{-1}	Beachrock		
Chiton squamosus	34·6 cm^3 yr^{-1} chiton^{-1}	Reef limestone		
Chiton squamosus	34·1 cm^3 yr^{-1} chilton^{-1}	Beachrock		
Chiton marmoratus	6·8 cm^3 yr^{-1} chiton^{-1}	Reef limestone		
Chiton marmoratus	6·5 cm^3 yr^{-1} chiton^{-1}	Beachrock		

Table 22.1. cont'd

Agent	Rate of rock removal	Substrate	Locality	Author
Patella coerulea	0·51–0·76 mm yr^{-1}	Limestone	Northern Adriatic	Torunski (1979)
Littorina neritoides	0·07–0·13 mm yr^{-1}			
Littorina ziczac	0·40 cm^3 yr^{-1}	Beachrock	Barbados	McLean (1967b)
Littorina meleagris	0·15 cm^3 yr^{-1}			
Nodolittorina tuberculata	0·60 cm^3 yr^{-1}			
Littorina planaxis	0·93 cm^3 yr^{-1}			
	2·5 mm yr^{-1}	Sandstone	California	North (1954)
Littorina scutulata	3·8 mm yr^{-1}			
Cittarium pica	1·30 cm^3 yr^{-1}	Beachrock	Barbados	McLean (1967b)
Nerita versicolor	0·80 cm^3 yr^{-1}			
Nerita tesselata	0·40 cm^3 yr^{-1}			
	154 gm^{-2} yr^{-1}			
Acmaea sp.	0·99 cm^3 yr^{-1}	Beachrock	Barbados	McLean (in Trudgill, 1983b)
	1·5 mm yr^{-1}			
	2·4 g yr^{-1}			
Fissurella sp.	5·0 cm^3 yr^{-1}			

Table 22.2. Substrate removal by burrowing and grazing echinoids. Full key to sources in Spencer (1988) and additional references in this chapter.

Agent	Sediment removal	Substrate	Locality	Author
Tropical and sub-tropical environments				
Echinometra lucunter	24 g yr^{-1} 9·96–14·0 cm^3 yr^{-1}	Beachrock	Barbados	McLean (1967a)
Echinometra lucunter	0·77 g d^{-1} urchin^{-1} (7·0 kg m^{-2} yr^{-1})	Aeolianite	Bermuda	Hunt (1969)
Echinometra lucunter	0·11 g d^{-1} urchin^{-1} (3·9 kg m^{-2} yr^{-1})	Algal ridge	Virgin Is.	Odgen (1977)
Eucidaris thouarsii	0·40–0·84 g d^{-1} urchin^{-1} 0·47–0·77 g d^{-1} urchin^{-1} (1·7 kg m^{-2} yr^{-1})	Coralline algae Live coral Overall, reef edge	Galapagos	Glynn et al. (1979)
Diadema savignyi	1·90 g d^{-1} urchin^{-1} (3·4 kg m^{-2} yr^{-1})			
Echinometra mathaei	0·14 g d^{-1} urchin^{-1} (0·4 kg m^{-2} yr^{-1})	Coral rubble	Moorea, Society Is.	Bak (1990)
Echinothrix diadema	3·30 g d^{-1} urchin^{-1} (0·8 kg m^{-2} yr^{-1})			
Echinometra mathaei	0·9–1·4 g d^{-1} urchin^{-1}	*Porites* reef	Kuwait	Downing and El-Zahr (1987)
Echinometra mathaei	0·11–0·13 g d^{-1} urchin^{-1}	Lagoonal coral knoll and reef platform	Enewetak Atoll	Russo (1980)

Table 22.2. cont'd

Agent	Sediment removal	Substrate	Locality	Author
Echinostrephus aciculatus	0.18–0.40 g d^{-2} urchin^{-1}	Lagoonal coral and reef platform	Enewetak Atoll	Russo (1980)
Echinometra mathaei and *E. oblonga*	0.11–0.82 g d^{-1} urchin^{-1}	Limestone veneer on basalt platform	Hawaii	Russo (1977, 1980)
Echinometra mathaei	0.5 g d^{-1} urchin^{-1}	Dead *Acropora*	Persian Gulf	Shinn (in Hughes Clarke and Keij 1973)
Diadema antillarum	1.85 g d^{-1} urchin^{-1} 5.0–5.6 kg m^{-2} yr^{-1}	Coral reef	Barbados	Scoffin *et al.* (1980)
Diadema antillarum	2.07 g d^{-1} urchin^{-1}	Coral reef	Barbados	Lewis (1974)
Diadema antillarum	4.6 kg m^{-2} yr^{-1}	Coral reef	Virgin Islands	Ogden (1977)
Diadema antillarum	5.8 kg m^{-2} yr^{-1}	Coral reef	Barbados	Hunter (1977)
Temperate environments				
Paracentrotus lividus	0.25–1.5 cm yr^{-1} (exposed coast) 0.00–1.0 cm yr^{-1} (sheltered coast)	Limestone	Co. Clare, Eire	Trudgill *et al.* (1987)
Paracentrotus lividus	1.1 mm yr^{-1} surface lowering 18 g yr^{-1} urchin^{-1}	Limestone	Northern Adriatic	Torunski (1979)
		Limestone	Northern Adriatic	Torunski (1979)

Table 22.3. Burrow extension rates, substrate removal, and surface lowering by bivalve molluscs.

Agent	Rate	Substrate	Locality	Author
Lithophaga nasuta	0·91 cm yr^{-1} (0·87 cm^3 yr^{-1})	Reef limestones	Aldabra Atoll	Trudgill (1976)
Lithophaga sp.	0·25 cm yr^{-1}	Reef limestones	Oman	Vita-Finzi and Cornelius (1973)
Lithophaga sp.	1·5 cm yr^{-1}	Beachrock	Great Barrier Reef	Otter (1937); McLean (1974)
Lithophaga lithophaga	0·4–1·3 cm yr^{-1}	Various limestones		Kleeman (1973)
Hiatella arctica	0·50–0·67 cm yr^{-1}	Limestones, mudstones, sandstones	Cumbrae, Plymouth UK	Hunter (1949)
Hiatella gallicana				
Hiatella arctica	0·125–1·00 cm yr^{-1}	Limestone	Co. Clare, Eire	Trudgill and Crabtree (1987)
Lithophaga/Gastrochaena	0·18 kg m^{-2} yr^{-1} (0·05–4·4 cm^3 yr^{-1})	Coral reef	Florida	Hein and Risk (1975); calculation by Davies (1983)
Lithophaga/Gastrochaena	0·13–0·15 kg m^{-2} yr^{-1} 0·01–0·02 kg m^{-2} yr^{-1} 0·04–0·05 kg m^{-2} yr^{-1}	Reef slope Flat reef Reef lagoon	Lizard Is., Great Barrier Reef	Kiene (1985)
Lithotrya sp.	0·84 cm yr^{-1} (0·78 cm^3 yr^{-1})	Reef limestones	Aldabra Atoll	Trudgill (1976)
Tridacna crocea	0·14 kg m^{-2} yr^{-1}	Reef flat	Great Barrier Reef	Hamner and Jones (1976)
Saccostrea amasa	0·02–0·38 cm yr^{-1}	Boulder rock	One Tree Island, Great Barrier Reef	Trudgill (1983a)

Table 22.4. Bioerosion by the cryptofaunal coral reef community: selected sponges and polychaetes.

Carbonate removal ($kg\,m^{-2}\,yr^{-1}$)	Locality	Method of estimation and substrate type	Author
Clionid sponges			
20·0–25·0 (1·0–1·4 $cm^3\,yr^{-1}$)	Bermuda	Small block transplant; 100 days	Neumann (1966)
0·25–0·30	Bermuda	Small block transplant; 12 months	Rützler (1975)
1·35 (0·21–3·50)	St Croix, Virgin Is.	Thin-section of algal ridge cores	Moore and Shedd (1977)
0·96 (0·21–1·81)	Jamaica	Deep-reef sediment traps	Moore and Shedd (1977)
2·6–3·3	Netherlands Antilles	Experimental blocks	Bak (1976)
8·0	Grand Cayman	Changes in specific gravity between bored, unbored blocks	Acker and Risk (1985)
1·35 (1·12–1·60)	Barbados	Coral head; X-ray section (1980)	Scoffin et al. (1980)
3·0–13·4	Florida Keys	Coral head; X-ray section	Hudson (1977) (calculation from Davies 1983)
0·17	Florida Keys	Coral head; X-ray section	Hein and Risk (1975) (calculation from Davies 1983)
0·01–0·11 (reef-slope)	Lizard Is., Great Barrier Reef	Small block transplants; 2–3 years	Kiene (1985)
0·009–0·02 (reef-flat)			
0·03–0·43 (lagoon)			
0·35 (lagoon)	Davies Reef, Great Barrier Reef	Maufactured carbonate sand grains	Tudhope and Risk (1985)

Polychaetes			
0·18	Florida Keys	Coral head; polydorid polychaetes	Hein and Risk (1975)
0·694 (reef-front) 0·843 (reef-flat) 1·788 (patch-reef)	Lizard Is., Great Barrier Reef	Coral block transplants; cirratulids, polydorids, and sabellids	Davies and Hutchings (1983)
4·08–4·72 (reef-flat) 13·23–6·40 (leeward reef) 0·33 (patch reef)	Lizard Is., Great Barrier Reef	Coral block transplants; polydoids and fabricinids	Hutchings and Bamber (1985)
0·36–0·93 (reef-slope) 0·15–0·38 (reef-flat) 0·10–0·37 (lagoon)	Lizard Is., Great Barrier Reef	Coral blocks; polychaetes, then replaced by sipunculids	Kiene (1985)

a
Height (m)
Zones
GR
BS
DB
GB
WE
GN
2.0
1.75
1.5
1.25
1.0
0.75
0.5
0.25
0
EHWS
MHWS
MHWN
MSL
MLWN
MLWS
ELWS
winter
annual mean
summer
0
0.25
0.5
0.75
1.0
1.25
Erosion rate (mm yr⁻¹)
exposed coasts
overlap
sheltered coasts
1
2
3
4
5
6
b
Initial surface
EHWS
MSL
1000
2000
3000 (years)
(m)
0
1
2
3
Initial surface
1000
2000
3000
0
2
4
6

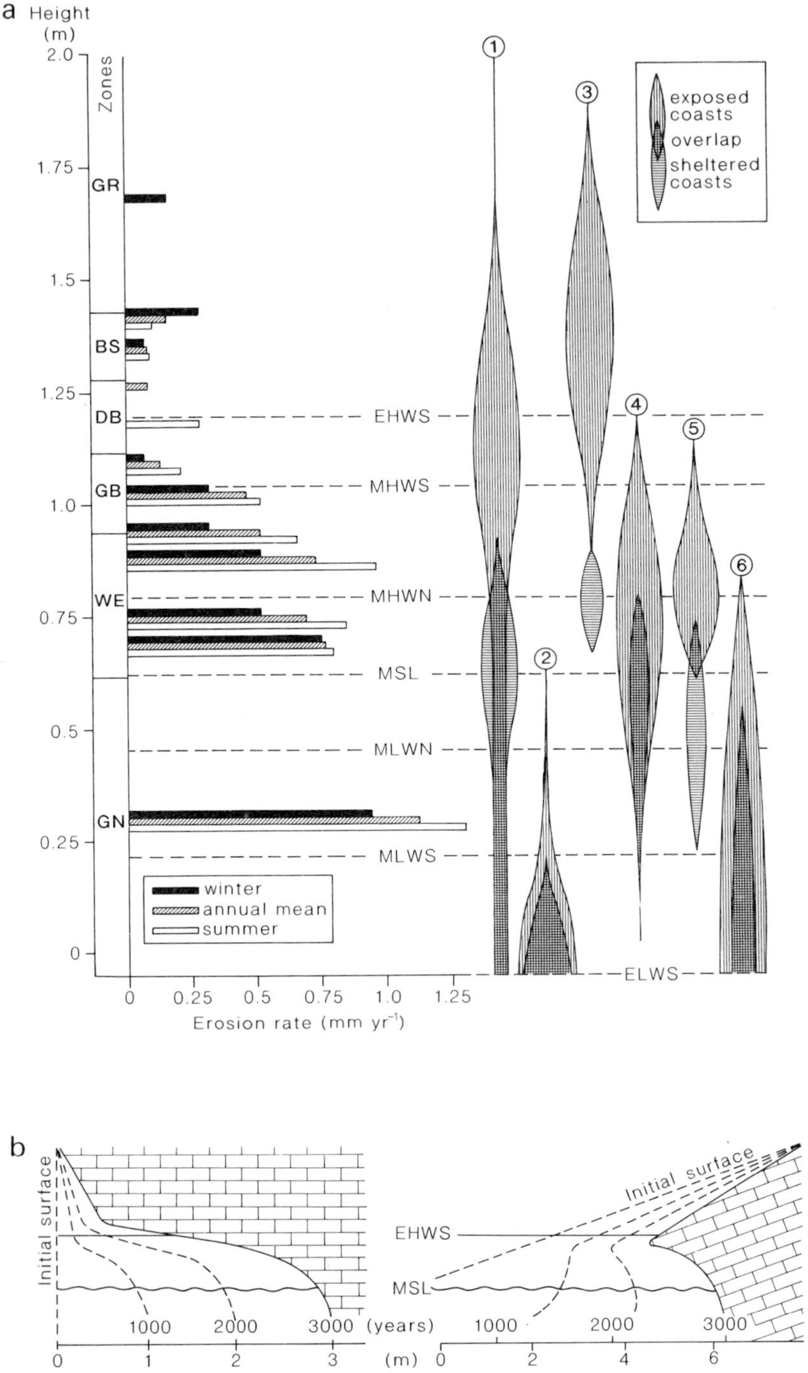

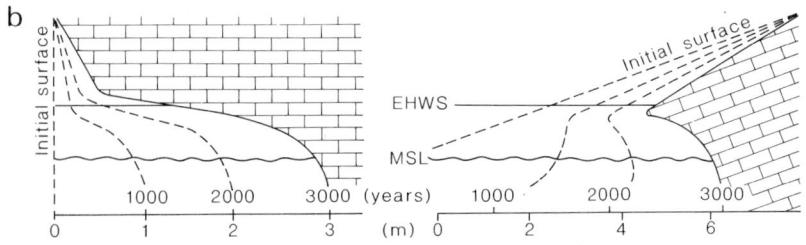

energy coasts, the processes of wetting and drying, salt weathering, and direct wave attack may cause the replacement of the intertidal notch by a planar coastal ramp (Trudgill 1976), under microtidal regimes the lithified encrustations of coralline algae and vermetid gastropods may build surf bench structures near mean sea-level as exposure increases. (Föcke 1978; Spencer 1985). Furthermore, there has been no assessment of the bioerosional consequences of the periodic major restructuring of shallow marine ecosystems, e.g. the 1983–1984 die-off of *Diadema antillarum*, a major grazer on Caribbean reefs (Lessios *et al*. 1984; Liddell and Ohlhorst 1986). Nevertheless, the explicit consideration of plant–animal interactions in the study of coastal erosion processes, and the attempt to link these relationships to known rates of littoral weathering, offers the possibility of constructing much more comprehensive models of long-term and large-scale rocky shore erosion dynamics than has hitherto been achieved.

References

Acker, K. L. and Risk, M. J. (1985). Substrate destruction and sediment production by the boring sponge *Cliona*. *Journal of Sedimentary Petrology*, **55**, 705–11.

Bak, R. P. M. (1976). The growth of coral colonies and the importance of crustose coralline algae and burrowing sponges in relation with carbonate accumulation. *Netherlands Journal of Sea Research*, **10**, 285–337.

Bak, R. P. M. (1990). Patterns of echinoid bioerosion in two Pacific coral reef lagoons. *Marine Ecology Progress Series*, **66**, 267–72.

Davies, P. J. (1983). Reef growth. In *Perspectives on coral reefs*, (ed. D. J. Barnes), pp. 69–106. Clouston, Canberra.

Davies, P. J. and Hutchings, P. A. (1983). Initial colonization, erosion and accretion on coral substrate: experimental results. Lizard Island, Great Barrier Reef. *Coral Reefs*, **2**, 27–35.

Donn, T. F. and Boardman, M. R. (1988). Bioerosion of rocky carbonate coastlines on Andros Island, Bahamas. *Journal of Coastal Research*, **4**, 381–94.

Downing, N. and El-Zahr, C. R. (1987). Gut evacuation and filling rates in the rock-boring sea urchin, *Echinometra mathaei*. *Bulletin of Marine Science*, **41**, 579–84.

Fig. 22.1. Bioerosion and biogeomorphology, northern Adriatic Sea (after Torunski 1979). (a) Interrelationships between erosion rates, Cyanobacteria (zone classification from Schneider 1976), and macro-organisms, with profile position and exposure. 1: Cyanobacteria, 2: *Paracentrotus lividus*, 3: *Littorina neritoides*, 4: *Monodonta turbinata*, 5: *Patella coerulea*, 6: *Cliona* sp. (b) Intertidal profile evolution from extrapolation of erosion rates.

Föcke, J. W. (1978). Limestone cliff morphology on Curaçao (Netherland Antilles) with special attention to the origin of notches and vermetid/coralline algal surf benches ('cornices', 'trottoirs'). *Zeitschrift für Geomorphologie*, **22**, 329-49.

Glynn, P. W. (1973). Aspects of the ecology of coral reefs in the western Atlantic region. In *Biology and geology of coral reefs*, Volume II, Biology 1, (ed. O. A. Jones and R. Endean), pp. 271-324. Academic Press, New York.

Glynn, P. W., Wellington, G. M., and Birkeland, C. (1979). Coral reef growth in the Galapagos: limitation by sea urchins. *Science*, **203**, 47-9.

Golubic, S., Perkins, R. S., and Lukas, K. J. (1975). Boring microorganisms and microborings in carbonate substrates. In *The study of trace fossils*, (ed. R. W. Frey), pp. 229-59. Springer-Verlag, Berlin.

Golubic, S., Friedmann, I., and Schneider, J. (1981). The lithobiontic ecological niche with special reference to micro-organisms. *Journal of Sedimentary Petrology*, **51**, 475-79.

Hamner, W. M. and Jones, M. S. (1976). Distribution, burrowing and growth rates of the clam *Tridacna crocea* on interior reef flats. *Oecologia* (Berlin), **24**, 207-27.

Hein, F. J. and Risk, M. J. (1975). Bioerosion of coral heads: inner patch reefs, Florida reef tract. *Bulletin of Marine Science*, **25**, 133-8.

Hudson, J. H. (1977). Long-term bioerosion rates on a Florida reef: new method. In *Proceedings of the Third International Coral Reef Symposium*, 2, pp. 491-98. Atlantic Reef Committee, Miami.

Hughes Clarke, M. W. and Keij A. J. (1973). Organisms as producers of carbonate sediment and indicators of environment in the southern Persian Gulf. In *The Persian Gulf: Holocene carbonate sedimentation and diagenesis in a shallow epicontinental sea*, (ed. B. H. Purser), pp. 33-56. Springer-Verlag, Berlin.

Hunt, M. (1969). A preliminary investigation of the habits and habitat of the rock-boring urchin *Echinometra lucunter* near Devonshire Bay, Bermuda. In *Seminar on organism sediment relationships*, Bermuda Biological Research Station, Special Publications, 2, (ed. R. N. Ginsburg and P. Garrett), pp. 35-40. Bermuda Biological Research Station.

Hunter, I. G. (1977). Sediment production of *Diadema antillarum* on a Barbados fringing reef. In *Proceedings of the Third International Coral Reef Symposium*, 2, pp. 105-109. Atlantic Reef Committee, Miami.

Hunter, W. R. (1949). The structure and behaviour of *Hiatella gallicana* (Lamarck) and *H. artica* (L), with special reference to the boring habit. *Proceedings of the Royal Society of Edinburgh*, **B63**, 271-89.

Hutchings, P. A. (1986). Biological destruction of coral reefs. A review. *Coral Reefs*, **4**, 239-52.

Hutchings, P. A and Bamber, L. (1985). Variability of bioerosion rates at

Lizard Island, Great Barrier Reef: preliminary attempts to explain these rates and their significance. In *Proceedings of the Fifth International Coral Reef Congress*, 5, pp. 333-8. EPHE, Tahiti.

Kiene, W. E. (1985). Biological destruction of experimental coral substrates at Lizard Island, Great Barrier Reef, Australia. In *Proceedings of the Fifth International Coral Reef Congress*, pp. 339-44. EPHE, Tahiti.

Kiene, W. E. (1988). A model of bioerosion on the Great Barrier Reef. In *Proceedings of the Sixth International Coral Reef, Symposium*, 3, pp. 449-54. Townsville, Queensland, Australia.

Kleeman, K. (1973). *Lithophaga lithophaga* (L.) (Bivalvia) in different limestones. *Malacologia*, 14, 345-7.

Le-Campion Alsumard, T. (1970). Cyanophycées marines endoliths colonisant les surfaces rocheuses denudées (Étages supralittoral et mediolittoral de la region de Marseille). *Schweizerische Zeitschrift für Hydrologie*, 32, 552-8.

Le-Campion Alsumard, T. (1979). Les cyanophycées endoliths marines. Systématique, ultrastructure et biodestruction., *Oceanologia Acta*, 2, 143-56.

Lessios, H. A., Robertson, D. R., and Cubit, J. D. (1984). Spread of *Diadema* mass mortality through the Caribbean. *Science*, 226, 335-37.

Lewis, J. R. (1974). *The ecology of rocky shores*. English Universities Press, London.

Liddell, W. D. and Ohlhorst, S. L. (1986). Changes in benthic community composition following the mass mortality of *Diadema* at Jamaica. *Journal of Experimental Marine Biology and Ecology*, 95, 271-78.

Lowenstam, H. A. (1962). Magnetite in denticle capping in Recent chitons (Polyplacophora). *Bulletin of the Geological Society of America*, 73, 435-8.

McLean, R. F. (1967a). Erosion of burrows in beachrock by the tropical sea urchin *Echinometra lucunter*. *Canadian Journal of Zoology*, 45, 586-8.

McLean, R. F. (1967b). Measurement of beachrock erosion by some tropical marine gastropods. *Bulletin of Marine Science*, 17, 551-61.

McLean, R. F. (1974). Geologic significance of bioerosion of beachrock. In *Proceedings of the Second International Coral Reef Symposium*, 2, pp. 401-8. Great Barrier Reef Committee, Brisbane, Australia.

Moore, C. H. and Shedd, W. W. (1977). Effective rates of sponge bioerosion as a function of carbonate production. In *Proceedings of the Third International Coral Reef Symposium*, 2, pp. 499-505. Atlantic Reef Committee, Miami, Florida.

Neumann, A. C. (1966). Observations on coastal erosion in Bermuda and measurements of the boring rate of the sponge *Cliona lampa*. *Limnology and Oceanography*, 11, 92-108.

North, W. J. (1954). Size, distribution, erosive activities and gross metabolic efficiency of the marine intertidal snails *Littorina planaxis* and *L. scutulata*. *Biological Bulletin*, **106**, 185–97.

Ogden, J. (1977). Carbonate sediment production by parrotfish and sea urchins on Caribbean reefs. In *Reefs and related carbonates — ecology and sedimentology*, American Association of Petroleum Geologists, Studies in Geology, 4, (ed. S.H.J. Frost, M.J. Weiss, and J.B. Saunders), pp. 281–7. AAPG, Tulsa.

Otter, G.W. (1937). Rock-destroying organisms in relation to coral reefs. *Scientific Reports, Great Barrier Reef Expedition 1928-29, British Museum (Natural History)*, **1**, 323–52.

Rasmussen, K.A. and Frankenberg, E.B. (1990). Intertidal erosion by the chiton *Acanthopleura granulata*: San Salvador, Bahamas. *Bulletin of Marine Science*, **47**, 680–95.

Russo, A.R. (1977). Water flow and the distribution and abundance of echinoids (Genus: *Echinometra*) on an Hawaiian reef. *Australian Journal of Marine and Freshwater Research*, **28**, 693–702.

Russo, A.R. (1980). Bioerosion by the two rock boring echinoids (*Echinometra mathaei* and *Echinostrephus aciculatus*) on Enewetak Atoll, Marshall Islands. *Journal of Marine Research*, **38**, 99–110.

Rützler, K. (1975). The role of burrowing sponges in bioerosion. *Oecologia (Berlin)*, **19**, 203–16.

Sammarco, P.W. and Risk, M.J. (1990). Large-scale patterns in internal bioerosion of *Porites*: cross continental shelf trends on the Great Barrier Reef. *Marine Ecology Progress Series*, **59**, 145–56.

Sammarco, P.W., Risk, M.J., and Rose, C. (1987). Effects of grazing and damselfish territoriality on internal bioerosion of dead corals: indirect effects. *Journal of Experimental Marine Biology and Ecology*, **112**, 185–99.

Schneider, J. (1976). Biological and inorganic factors in the destruction of limestone coasts. *Contributions to Sedimentology*, **6**, 1–112.

Schneider, J. and Torunski, H. (1983). Biokarst on limestone coasts, morphogenesis and sediment production. *Marine Ecology*, **4**, 45–63.

Scoffin, T.P. *et al.* (1980). Calcium carbonate budget of a fringing reef on the west coast of Barbados. *Bulletin of Marine Science*, **30**, 475–508.

Spencer, T. (1985). Marine erosion rates and coastal morphology of reef limestones on Grand Cayman Island, West Indies. *Coral Reefs*, **4**, 59–70.

Spencer, T. (1988). Limestone coastal morphology: the biological contribution. *Progress in Physical Geography*, **12**, 6–101.

Spencer, T. and Benn, J.R. Limestone topography and rock weathering in the Cayman Islands. In *The biogeography and ecology of the Cayman Islands*, (ed. D.R. Stoddart, J.E. Davies, and M.A. Brunt). Junk, The Hague. (In press).

Taylor, J.D. and Way, K. (1976). Erosive activities of chitons at Aldabra Atoll. *Journal of Sedimentary Petrology*, **46**, 974–77.

Torunski, H. (1979). Biological erosion and its significance for the morphogenesis of limestone coasts and for nearshore sedimentation. *Senckenbergiana maritima*, **11**, 193–265.

Trudgill, S. T. (1976). The marine erosion of limestones on Aldabra Atoll, Indian Ocean. *Zeitschrift für Geomorphologie Supplementband*, **26**, 164–200.

Trudgill, S. T. (1983a). Preliminary estimates of intertidal limestone erosion, One Tree Island, Southern Great Barrier Reef, Australia. *Earth Surface Processes and Landforms*, **8**, 189–93.

Trudgill, S. T. (1983b). Measurements of rates of erosion of reefs and reef limestones. In *Perspectives on coral reefs*, (ed. D. J. Barnes), pp. 256–62. Clouston, Canberra, Australia.

Trudgill, S. T. (1985). *Limestone geomorphology*. Geomorphology Texts, No. 8. Longman, London.

Trudgill, S. T. (1987). Bioerosion of intertidal limestone, Co. Clare, Eire — 3: Zonation, process and form. *Marine Geology*, **74**, 111–21.

Trudgill, S. T. and Crabtree, R. W. (1987). Bioerosion of intertidal limestone, Co. Clare, Eire — 2. *Hiatella artica*. *Marine Geology*. **74**, 99–109.

Trudgill, S. T., Smart, P. L., Friederich, H., and Crabtree, R. W. (1987). Bioerosion of intertidal limestone, Co. Clare, Eire — 1: *Paracentrotus lividus*. *Marine Geology*, **74**, 85–98.

Tudhope, A. W. and Risk, M. J. (1985). Rate of dissolution of carbonate sediments by microboring organisms, Davies Reef, Australia. *Journal of Sedimentary Petrology*, **55**, 440–7.

Viles, H. A. (ed.) (1988). *Biogeomorphology*, Blackwell, Oxford.

Vita-Finzi, C. and Cornelius, P. F. S. (1978). Cliff sapping by molluscs in Oman. *Journal of Sedimentary Petrology*, **43**, 31–2.

Whitton, B. A. and Potts, M. (1982). Marine littoral. In *The biology of cyanobacteria*, (ed. N. G. Carr and B. A. Whitton), pp. 515–42. Blackwell, Oxford.

Young, H. R. and Nelson, C. S. (1985). Biodegradation of temperate-water skeletal carbonates by boring sponges on the Scott shelf, British Columbia, Canada. *Marine Geology*, **65**, 33–45.

23. Endosymbiosis in marine cnidarians

P. SPENCER DAVIES
Department of Zoology, The University, Glasgow, Scotland, UK

Abstract

Endosymbiotic relationships with either prokaryote or algal symbionts are found in the marine benthos, in members of the Protozoa, Porifera, Cnidaria, Turbellaria, and Mollusca. This review will concentrate upon the most widespread association, that of the dinoflagellate *Symbiodinium microadriaticum* (zooxanthellae) and its anthozoan hosts. During the daytime, photosynthetically fixed carbon compounds, mainly glycerol and lipids, pass from the symbiont to the host tissue. The magnitude of this flux has now been quantified and related to the nutritional requirements of the host. Both carbon and energy budget studies indicate that in shallow water and on clear sunlit days, the total photosynthetic production exceeds the requirements of symbiont and host. The excess is excreted, mainly as mucus-lipid and is thereby available to the food chain. Zooxanthellae also play a role in the uptake of dissolved inorganic nitrogen and in the recycling of nitrogenous excretory products back to the host as amino acids.

The host is thought to regulate the number of zooxanthellae in its tissues, possibly by limiting the supply of nutrients passing through the vacuolar membrane to the symbiont. Perturbation of environmental factors, including light, temperature, and salinity, can affect the regulatory mechanisms, resulting in the expulsion of the symbionts by the host. Coral reefs in several parts of the world are currently displaying this phenomenon, usually referred to as 'bleaching', which may be related to global warming of seawater.

Introduction

Endosymbiotic associations are widespread in benthic marine environments. By far the most important involve members of the Dinophyceae (dinoflagellates) and are commonly referred to as zooxanthellae (Smith and Douglas 1987). They are found in a phyletically diverse range of host groups, including members of the Porifera, Cnidaria, Platyhelminthes, and Mollusca (McLaughlin and Zahl 1966; Taylor 1974). The most widespread of the dinoflagellate symbioses are those with members of the Cnidaria, particularly with the class Anthozoa. These associations are found from temperate to tropical regions and find their maximum expression in shallow-water coral reefs where symbioses involving octocorals, actinians, zoanthids and scleractinian corals abound. It is the 'benefit' conferred by these mutualistic associations, resulting from a tight nutrient recycling, which allows these ecosystems to flourish in what are otherwise nutrient-poor waters. This review will examine the relationship between zooxanthellae and their cnidarian hosts, and their trophic interactions in particular.

Characteristics of zooxanthellae in symbiosis

Cnidarian zooxanthellae, irrespective of host, have a superficial morphological similarity which has led them to be grouped into a single species *Symbiodinium microadriaticum* (Freudenthal 1962). They are yellow-brown in colour, spherical or coccoid in form, and range in size (according to host species) from approximately 7 to 12 μm in diameter (Wilkerson *et al.* 1988). The alga is invested in a thin outer theca or amphiesma or periplast, composed of several closely apposed membranes (Taylor 1968; Kevin *et al.* 1969; Schoenberg and Trench 1980*b*). By contrast, zooxanthellae grown in culture have a thick fibrous or granular pellicle, up to 0·2 μm in thickness which develops beneath the outermost limiting membrane. Flagella, although normally absent, have been observed occasionally in zooxanthellae *in situ* (Schoenberg and Trench 1980*b*). This raises the possibility that flagellated zoospores which develop in cultured algae may also be produced within the host, and may represent the dispersal and reinfection phases (Steele 1975, 1977).

Within the host, the zooxanthellae are generally located only in the gastrodermal cell layer. However in the zoanthid genus *Palythoa* they are found, in addition, in the mesogloea and epidermal cells (Trench 1971*a*), and in the medusa *Cassiopeia* they are found only in the mesogloea (Kevin *et al.* 1969). The number of zooxanthellae varies between host species from one to several per cell. In each case, the alga is confined within a phagocytic vacuole, and therefore separated from the host cytoplasm by the vacuolar fluid and the vacuolar membrane.

The gross distribution of the zooxanthellae is not uniform throughout the polyp. In general, they tend to be concentrated in areas which are directly exposed to light. Thus in corals, which have the tentacles exposed in daylight, they are evenly dispersed throughout the tentacles. Conversely, in corals whose tentacles are retracted during the day, the tentacles are devoid of algae, which are then concentrated in the polyp wall and in the coenosteum between adjacent polyps. In these locations, the zooxanthellae are most abundant in the outer of the two gastrodermal layers (Johnston 1980). There is a limit to the overall size of the algal population: in corals it is normally in the region of $1-2 \times 10^6$ zooxanthellae cm^{-2} of colony surface (summary in Muscatine 1980). At higher concentrations than this, the algae would become self-shading and photosynthetic efficiency would decline (Crossland and Barnes 1977; Hoegh-Guldberg and Smith 1989).

Zooxanthellae divide within their vacuole, by binary fission. The frequency of division is in some way related to the rate of division of the host cells. In some of the fast-growing branching corals such as *Acropora cervicornis* there is a phase lag in cell division, which results in the terminal polyps always being devoid of algae. The generation time of dividing zooxanthellae *in situ*, of 13 different species of anthozoa, has a mean of approximately 26 days, compared with about 6 days for zooxanthellae in culture (from Wilkerson *et al.* 1988). Clearly the algae within the host have a reduced rate of growth and division. The trophic implications of this will be considered in a later section.

Despite the superficial similarity between zooxanthellae from different hosts, there is growing evidence of genetic differences. Examination of cultured algae from 17 different host species revealed differences in isoenzyme and soluble-protein patterns (Schoenberg and Trench 1980*a*), dimensions of the unseparated two-cell stage following division (Schoenberg and Trench 1980*b*) and specificity for reinfection when presented to an aposymbiotic host. These observations suggest that the conservation of superficial morphology might result from the relative uniformity of the intravacuole environment, while the differences are indicative of genotypic separation. Differences in chromosome number and chromosome volume may be detected from morphometric analysis of three-dimensional reconstructions of the nuclei of algae from different hosts (Blank and Trench 1985). Therefore it now seems likely that *Symbiodinium microadriaticum* in reality encompasses a large complex of species, and that each host species may have a separate species of symbiont, as first envisaged by Loeblich and Sherley (1979). Until the genetics of these algae are better understood, it has been suggested that the name *Symbiodinium microadriaticum* should only be used for the symbionts from *Cassiopeia frondosa* and *C. xamachana*. All other zooxanthellae should be referred to as *Symbiodinium* sp. (Blank and Trench 1986).

Differences are also observed in the mode of transmission of zooxanthellae from one generation of host to another. In some host species there is a closed system, and transmission takes place via the egg and planula larva. In others, the planula is devoid of algae (see summary table in Trench, 1987) and reinfection therefore has to take place before or after settlement. The mechanism of this infection by free-living zooxanthellae is not known. Furthermore, there is only one record, itself somewhat questionable, of free-living *Symbiodinium* sp. in the sea (Taylor 1983). However, both corals and sea anemones are known to continually expel apparently healthy zooxanthellae to the surrounding water (Trench 1974; Steele 1975, 1977; Hoegh-Guldberg et al. 1987; Gates, 1990a) and populations of the free-living form must presumably be present.

Infection by free-living zooxanthellae could take place following random contact with the host; by chemotaxis to attractants liberated by the host (Kinzie 1974; Fitt 1984) or by ingestion and digestion of a zooplankter containing viable zooxanthellae in its gut (Trench 1979; Fitt 1984). Whatever the route into the animal, final entry into the host tissue takes place by phagocytosis by gastrodermal cells (Colley and Trench 1983; Fitt and Trench 1983). Initial formation of the phagocytic vesicle is dependent upon surface recognition of the symbiont, possibly by the recognition of specific surface antigens (Muscatine et al. 1975). In heterologous infections of aposymbiotic *Aiptasia tagetes* with cultured zooxanthellae, the success of uptake is related to the genetic similarity to the homologous symbiont. When a heterologous infection is successful, the foreign symbiont fails to be recognized by a second intracellular recognition mechanism and is then expelled by exocytosis (Trench 1987). Following homologous infection, the alga is thought to evade digestion by inhibiting the fusion of lysosomes with the vacuolar membrane (Fitt and Trench 1983). Again the molecular mechanisms involved are not known.

In symbiotic associations, 'benefits' are, by definition, thought to accrue to both partners. The symbiont may appear to be disadvantaged when its rates of growth and division are compared with those of the same species in culture. However, there is, as yet, no knowledge of the growth and division rates of the free-living forms which in the oligotrophic waters of coral reefs may be less than those of the symbiont. The enormous difference in biomass of symbionts compared with free-living forms in the coral reef ecosystem provides strong evidence for 'benefit' at a population level. 'Benefit' to the host, on the other hand, is more readily observed. Since the rate of division of symbiotic zooxanthellae is restricted, the products of photosynthesis, instead of being metabolized in growth, could diffuse into the animal cytoplasm and thereby contribute to its nutrition. In addition, nitrogenous excretory products may be resynthesized into amino acids and translocated back to the host, thereby minimizing its dependence upon an exogenous source of nitrogen.

In addition to the morphological changes in zooxanthellae when grown in culture, already noted, there are concomitant changes in trophic physiology. It is not yet clear how soon these changes appear after zooxanthellae are experimentally removed from host tissue. For this reason, interpretations of the functioning of the symbionts *in situ*, made from observations on either 'freshly isolated' or cultured algae, have to be made with caution.

Carbon fluxes between symbiont and host

Evidence for the translocation of metabolites from symbiont to host has come primarily from experiments in which either freshly isolated zooxanthellae or the intact symbiotic animal is incubated in the light with a ^{14}C-labelled precursor, usually $H^{14}CO_3^-$, as a source of carbon dioxide. Early experiments (Von Holt and Von Holt 1968a; Muscatine and Cernichiari 1969; Trench 1971a) showed that the ^{14}C label became incorporated primarily into lipid and protein within the host. Freshly isolated zooxanthellae, when illuminated in seawater containing $H^{14}CO_3^-$, release only very small amounts of labelled metabolites (Von Holt and Von Holt 1968b; Trench 1971b). However, if they are incubated in a homogenate of host tissue, a range of ^{14}C-labelled compounds including glucose, glycerol, and alanine are released into the medium (Muscatine and Cernichiari 1969; Trench 1971c). Of these compounds, glycerol, a small water-soluble molecule which forms the critical structural element of triglycerides and phosphoglycerides, is the major form of carbon translocated (Muscatine 1967; Trench 1971b; Schmitz and Kremer 1977; Hofman and Kremer 1981). Corroborative evidence came from 'inhibition' experiments in which corals were incubated in the light in seawater containing both $H^{14}CO_3$ and unlabelled glycerol. Some of the translocated labelled glycerol was recovered from the seawater to which it passed by exchange diffusion (Lewis and Smith 1971).

It was suggested that, *in vivo*, zooxanthellae produce a surfeit of glycerol which is then translocated to the host cytoplasm where metabolism to lipid occurs. When ^{14}C-labelled host lipid was de-acylated to yield its constituent glycerol and fatty acids, the label appeared to be associated with the glycerol moiety only (Muscatine and Cernichiari 1969; Trench 1971a), thus substantiating the experiments with isolated zooxanthellae. However, this raised the problem of accounting for the source of the carbon forming the long-chain fatty acids of the triacylglycerols. This was resolved by Patton *et al.* (1983) using a more comprehensive methodology, which showed that ^{14}C is in fact rapidly incorporated into both the glycerol and fatty acid residues of the triacylglycerols of both zooxanthellae and host tissues.

Patton *et al.* (1977) were the first to propose that the major part of photosynthetically fixed carbon is translocated not as glycerol but as lipid. Their argument was based on the observation that in *Pocillopora capitata* almost identical fatty acid residues are found in triacylglycerols from both algal and host tissue, and electron micrographs showed lipid globules clustered around zooxanthellae *in situ*. Electron micrographs of zooxanthellae of the coral *Acropora acuminata* showed small lipid droplets both within the chloroplast and adjacent to the amphiesma (Crossland *et al.* 1980) (Fig. 23.1). In freshly isolated zooxanthellae from the tentacles of *Condylactis gigantea* (Kellogg and Patton 1983) and *Stylophora pistillata* (Patton and Burris 1983) the lipid is more prominent than in the *in situ* electron micrographs, and can be seen with Nomarsky interference or phase-contrast optics to form large 'blebs' distending the amphiesma. The lipid blebs may be up to $1-5\,\mu m$ in diameter. No actual release of lipid from the zooxanthellae has yet been observed.

Lipogenesis is confined to daytime periods when zooxanthellae are actively photosynthesizing. Dark incorporation of ^{14}C-labelled acetate results in labelling of structural lipids only, whereas during photosynthesis the label is incorporated into storage triacylglycerols and wax esters (Blanquet *et al.* 1979). In *A. acuminata* the maximum rate of incorporation into triacylglycerols occurs in the early afternoon, and into wax esters around midday (Crossland *et al.* 1980). Similarly, the proportion of freshly isolated zooxanthellae displaying lipid blebs varies from 5–7 per cent in the early morning to 11–13 per cent in the early afternoon (Patton and Burris 1983).

The lipids within zooxanthellae are mainly in the form of neutral lipids or triacylglycerols. Very little wax ester (1 per cent or less) appears to be synthesized (Kellogg and Patton 1983). It seems therefore that lipid is translocated mainly as triacylglycerols. However, lipid droplets isolated from gastrodermal cells contain up to 83 per cent of wax ester. It has been suggested therefore that, after liberation into the host cytoplasm, triacylglycerols are transformed into wax esters by enzymes in the membrane of the lipid droplet (Kellogg and Patton 1983). It is still not clear whether free fatty acids are also translocated, as proposed by Blanquet *et al.* (1979) and Schlichter *et al.* (1984). If this is the case, the fatty acids could be reduced in the host cytoplasm to fatty alcohols before esterification to wax esters.

In addition to lipid translocation, some free glycerol has been

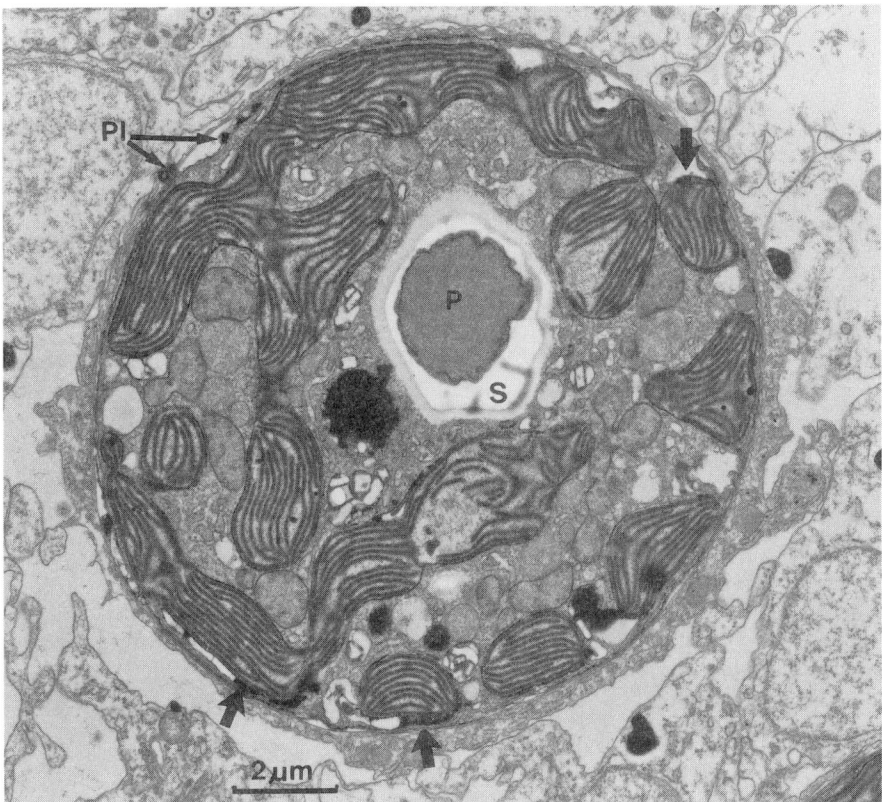

Fig. 23.1. Electron micrograph of zooxanthella of *Acropora acuminata* showing lipid deposits. Arrow heads: lipid deposits on chloroplast envelope; P, pyrenoid; S, starch; Pl, lipid deposits near plasmalemma (from Crossland et al. 1980; courtesy of M. Borowitzka).

shown to pass across the algal membranes to the host, where it may account for about 2 per cent of the ^{14}C incorporation. While acylation to triacylglycerols cannot be ruled out, it appears that most of the glycerol is used as a substrate for cell respiration (Battey and Patton 1984, 1987). Since glycerol is converted to CO_2, while lipid is stored, it has not been possible to determine the relative proportions of the two groups of compounds which are translocated. A summary of the main routes of photosynthetic carbon fluxes is shown in Fig. 23.2.

The excess lipids resulting from photosynthesis are stored. Sea anemones and corals can accumulate prodigious amounts of lipid, as much as 30–40 per cent of the body dry weight (Patton *et al.* 1977; Kellogg and Patton 1983; Stimson 1987). Up to 97 per cent of this

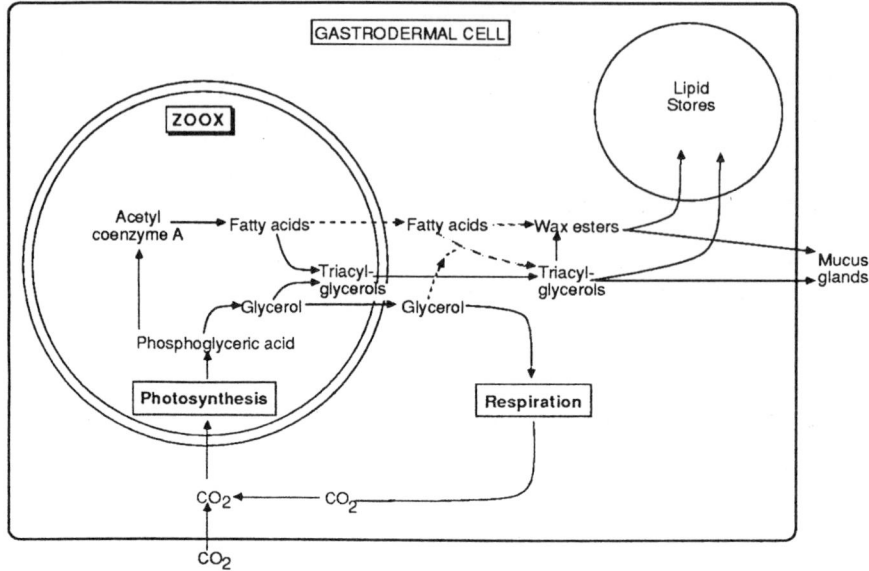

Fig. 23.2. Probable routes of carbon metabolism in a gastrodermal cell of a symbiotic anthozoan in the light. Some other small water-soluble molecules including glucose, alanine, succinate, malate, and citrate may also pass out of the algae. Some triacylglycerol and small amounts of wax ester may be stored within the zooxanthellae. When the animal cell lipid stores are replete, triacylglycerols and wax esters pass to mucus glands to be excreted. Broken lines indicate routes for which present evidence is less certain.

is stored in the animal tissues, mainly in the mesenteries of the polyps (Kellogg and Patton 1983). When lipid stores are replete, the excess photosynthetically fixed carbon is excreted as mucus–lipid. Crystalline inclusions of wax ester may be seen in electron micrographs of mucus-producing cells (Crossland *et al.* 1980). The mucus secretions of corals contain both triacylglycerols and wax esters (Benson and Muscatine 1974) which account for 60–90 per cent of the ^{14}C incorporated into 'mucus' during maximum rates of photosynthesis (Crossland *et al.* 1980).

Carbon and energy fluxes: quantitative aspects

The mass of carbon passed to the host depends upon the rate of photosynthesis, which may be measured from the rate of oxygen production of the whole organism when exposed to light. The rate measured is net photosynthesis, since some of the oxygen produced is immediately used for the respiration of both zooxanthellae and animal tissues. Gross photosynthesis is usually estimated by adding to the net photosynthesis the rate of oxygen uptake by respiration measured in darkness, although it is becoming clear that this may not be totally justified (see later). The relationship between rate of photosynthetic carbon production (P) and light or irradiance (I) takes the form of an asymptotic curve for which the best fit is given by the hyperbolic tangent function (Chalker 1981) (Fig. 23.3).

Photosynthetic saturation occurs at relatively low irradiance levels (see summary table in Trench 1987) with I_k values in the region of 200–300 μE m^{-2} s^{-1}. However, where comparisons of I_k values have been made on freshly isolated zooxanthellae and on intact organisms, the latter have significantly higher values. This suggests that *in situ* there is some degree of shelf-shading of the algae (Crossland and Barnes 1977). In the branching coral *Pocillopora damicornis* the value of I_k increases with increasing colony size, probably again as a result of progressive shading of zooxanthellae in the deeper recesses between the branches (Jokiel and Morrissey 1986).

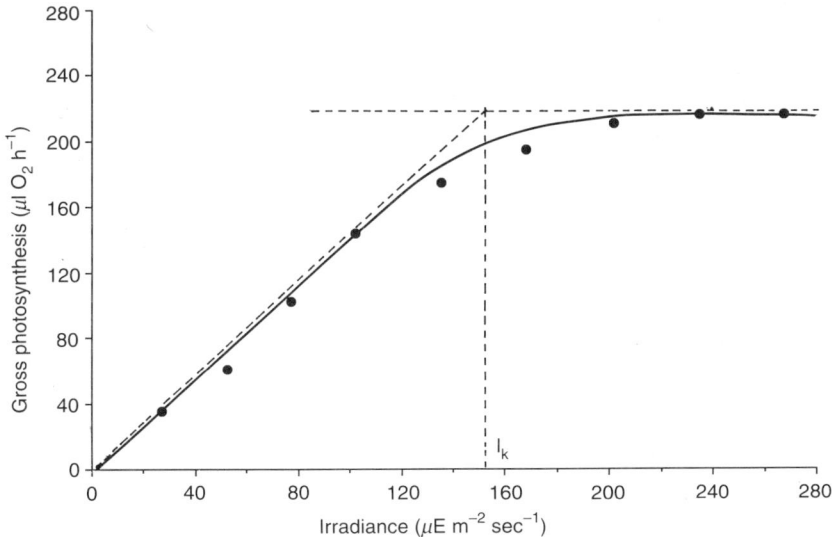

Fig. 23.3. Plot of gross photosynthesis v. irradiance for *Montastrea annularis* at 28°C. The curve was fitted to the hyperbolic tangent function equation $p_{max}^{gross} = p^{gross} \tan h \, (I/I_k)$, where p^{gross} = net photosynthesis + dark respiration rate, and I_k is the irradiance at which the initial gradient of the curve intercepts the horizontal asymptote. Low values of I_k indicate enhanced efficiency of photosynthesis at low light intensities.

Zooxanthellae from corals living under reduced light conditions undergo photoadaptation, during which the shape of the P vs I curve changes so that saturation occurs at a lower irradiance level (Wethey and Porter 1976; Porter *et al* 1984; Gattuso 1985). The mass of chlorophyll *a* in each algal cell increases up to fourfold, resulting in an increase in the maximum rate of photosynthesis (P_{max}) per unit of coral surface area (Falkowski and Dubinsky 1981). With increasing depth on the reef, photoadaptation therefore results in more efficient photosynthetic utilization of the available light. As a result, in *Acropora* spp., Chalker and Dunlap (1983) showed that the diel rates of photosynthesis remained approximately constant from depths of 1–35 m where the light flux was only 5·8 per cent of the surface irradiance. In *Montastrea annularis*, however, carbon fixation fell progressively from 0·5 m to 50 m depth (Battey and Porter 1988).

Several attempts have been made to determine the quantitative importance of photosynthetic carbon fixation to the symbiotic organism. The simplest quantitative measure is the daily ratio of photosynthetic carbon fixation to carbon metabolized in respiration. This is usually determined from measures of gross oxygen production during daytime (P) to 24-hour oxygen consumption (R) (extrapolated from dark respiration measurements) to give a P/R ratio. If this exceeds a value 1 of then excess photosynthate, over and above the needs of symbiont and host for respiration, is available to meet their combined (and minor) requirements for growth and reproduction. P/R values of 1·08 and 1·02 were recorded in shallow water for the massive corals *Montastrea annularis* and *M. cavernosa* respectively. At greater water depths, P/R is less than unity, suggesting that heterotrophic nutrition is required to supplement the energy or carbon budget (Davies 1977). By contrast, *Acropora* spp. showed daily P/R ratios above 1 to depths of 60 m, close to the lower limits of coral distribution (Chalker and Dunlap 1983). This may reflect fundamental differences in effectiveness of light gathering between massive corals with low relative surface area and large polyps, and the small-polyped branching corals which may be more dependent upon photosynthetic carbon for their nutrition (Porter 1976).

The P/R ratios, however, do not provide information on the extent to which translocated carbon meets the nutritional requirements of the host. In corals living in shallow waters, approximately 90–95 per cent of the carbon fixed in photosynthesis during a 24-hour period is estimated to be made available to the animal tissues (Davies 1984; Muscatine *et al*. 1984; Edmunds and Davies 1986). Of this, the contribution of the zooxanthellae to animal respiration requirements (CZAR) for *Stylophora pistillata* was calculated to be 157 per cent and 143 per cent in shallow water (3 m), while in shade (Muscatine et al. 1984) or at 35 m depth (McCloskey and Muscatine 1984) the values fell to 78 per cent and 58 per cent respectively, again indicating a requirement for a heterotrophic contribution to the nutrition.

A more detailed analysis of the quantitative utilization of the photosynthetically fixed carbon can be obtained from examination of the energy budgets of symbiont and host. The energy budget for the corals *Pocillopora damicornis* and *Porites porites* on a cloudless day at depths of 3 m and 10 m, respectively, shows that the utilization of energy for algal growth and respiration is low, so that 91 per cent and 78 per cent respectively of photosynthetically fixed energy is translocated (Edmunds and Davies 1986; Davies 1991). The major allocation of this energy is for respiration by the animal tissues, with only minor utilization for either growth or reproduction. Assuming that the lipid stores were replete, 20–45 per cent of the energy fixed in photosynthesis was predicted to be lost, probably as mucus–lipid. In shallow waters with abundant light, the nutrition of these corals is analogous to that of certain aphids which take in large quantities of carbon and subsequently excrete most of it.

Little information is yet available on the ways in which the energy budget changes with environmental change. However, reponses to short-term variations in light are different from long-term adaptive changes. *Porites porites* living in water with a high load of suspended sediment show typical photosynthetic adaptation to the reduced illumination with I_k values of $215\,\mu\text{E}\,\text{m}^{-2}\,\text{s}^{-1}$, compared with $416\,\mu\text{E}\,\text{m}^{-2}\,\text{s}^{-1}$ in a control group living in clear water. Carbon fixation increases as a result of photoadaptation, while growth and respiration decrease. This results in the predicted surplus of energy increasing from 45 per cent to 67·3 per cent of that fixed in photosynthesis (Edmunds and Davies 1989).

With acute reductions in light level during the course of the day, the total photosynthetic input into the organism falls. Thus in a 10 g skeletal weight *Pocillopora damicornis*, the predicted daily energy excess of 71 J on an 'ideal' cloudless day would be reduced to 39 J on an 'average' day with occasional scattered clouds. However, on a heavily overcast day, the reduced levels of photosynthesis would result in a deficit of 59 J, which would have to be met from the coral's lipid reserves (Davies 1991) (Fig. 23.4). From knowledge of the magnitude of the stored lipid of *P. damicornis* (Stimson 1987), it is possible to calculate that the reserves would last for 28 consecutive cloudy days (Davies 1991).

In none of the energy budgets investigated so far has the loss of mucus–lipid been measured directly. Furthermore, it now seems likely that estimates of daytime respiration rates may be too low, since

Porites porites
10 m depth, Jamaica

Zooxanthellae

	Photosynthesis	=	Respiration	+	Growth	+	Excess
Ideal day	281.9 (100%)		59.7 (21.2%)		2.2 (0.8%)		220.0 (78.0%)

Animal

	Respiration	+	Growth	+	Reproduction	+	Losses
	74.2 (26.3%)		17.9 (6.3%)		1.1 (0.4%)		126.8 (45.0%)

Pocillopora damicornis
3 m depth, Hawaii

	Photosynthesis	=	Respiration	+	Growth	+	Excess
Ideal day	355.2 (100%)		29.5 (8.3%)		1.2 (0.3%)		324.5 (91.4%)
Overcast day	224.7 (100%)		29.5 (13.1%)		1.2 (0.5%)		191.1 (86.4%)

	Respiration	+	Growth	+	Reproduction	+	Losses
Ideal day	241.1 (67.9%)		12.4 (3.5%)		a		71.0 (20.0%)
Overcast day	241.1 (107.3%)		12.4 (5.5%)		a		[-59.4] (-26.4%)

(^{a} Not measured)

Fig. 23.4. Twenty-four hour energy budget (in Joules) for 10 g skeleton weight nubbins of the corals *Porites porites* and *Pocillopora damicornis*. The excess energy, after utilization for respiration and growth by zooxanthellae, is translocated to the host. On overcast days the budget of *P. damicornis* is predicted to be in deficit, and the coral would then draw upon lipid reserves (from Edmunds and Davies 1986; Davies 1991).

respiration measured immediately following a period of photosynthesis is elevated relative to early morning levels (Edmunds and Davies 1988). One result of this is that mucus–lipid surpluses would be lower than predicted. However, direct measurements of mucus–lipid excretion have been made in *Acropora acuminata* and calculated to represent approximately 40 per cent of the net carbon fixation (Crossland *et al.* 1980). Shallow-water colonies of *Stylophora pistillata* are estimated to lose 26–52 per cent (Muscatine *et al.* 1985), or approximately 20 per cent (Crossland 1987) of the carbon fixed, as mucus and mucus–lipid.

Nitrogen fluxes between symbiont and host

The majority of studies on the role of zooxanthellae in the nitrogen metabolism of cnidarian symbioses have been carried out on corals. The nutrient intake from heterotrophic feeding on zooplankton is extremely small (Johannes *et al.* 1970; Johannes and Tepley 1974; Porter 1974; Edmunds and Davies 1986) and no quantitative experimental determinations of the nitrogen input from this source have been made. Uptake of dissolved inorganic nitrogen (DIN) as ammonia, and perhaps nitrate, does occur but is limited by the very low levels of these nutrients in the waters surrounding coral reefs (see table 1 in Muscatine 1980). Zooxanthellae might be expected therefore to fulfil an important role in recycling the animal excretory nitrogen and in scavenging DIN from the seawater. The nitrogen fluxes can be considered as comprising three separate stages:

(1) uptake of DIN from seawater to the animal cytoplasm;
(2) assimilation of DIN by zooxanthellae from the animal cytoplasm;
(3) translocation of dissolved organic nitrogen products from zooxanthellae to animal cytoplasm.

1. Uptake of dissolved inorganic nitrogen

All cnidarian symbioses examined have the ability to take up ammonia from seawater. Most, but not all, corals appear to take up nitrate, while sea anemones do not (summary in Wilkerson and Trench 1986).

(a) Uptake of ammonia Uptake of ammonia has invariably been measured indirectly from the rate of depletion of the nutrient from seawater. Because of the low inherent levels of ammonia in seawater it is usually experimentally enriched by the addition of NH_4Cl to initial levels of 10 or 20 μg atoms NH_4-N. ($=10$–$20\ \mu M\ NH_4^+$). These experiments have shown unequivocally that symbiotic cnidarians can deplete ammonia while exposed to light (Kawaguti 1953; Muscatine

and D'Elia 1978; Wilkerson and Trench 1986). Conversely, aposymbionts (Cates and McLaughlin 1976; Muscatine et al. 1979; Wilkerson and Muscatine 1984; Szmant-Froelich and Pilson 1977) and nonsymbiotic ahermatypic corals (Muscatine and D'Elia 1978; Burris 1983) excrete ammonia.

As with free-living microalgae (Syrett 1981), light is normally required for ammonia uptake by zooxanthellae *in situ*. In *Anemonia viridis*, the rate of ammonia uptake is proportional to irradiance over the range of $5-300\,\mu E\,m^{-2}\,s^{-1}$ (Davies 1988). Conversely, in darkness ammonia is normally released, except for the period immediately following exposure to light. Thus when *Pocillopora damicornis* was exposed to light for 1 h following an 18 h dark period, ammonia uptake continued for about 2 h. After a 4 h light exposure following a 16 h dark period, uptake lasted about 3 h before ammonia release took place (Muscatine and D'Elia 1978). It seems likely, therefore, that, as in the free-living *Chlamydomonas* (Syrett 1981), zooxanthellae accumulate carbon compounds during photosynthesis, and ammonia uptake continues in darkness until these stores are depleted. In *P. damicornis* it was predicted that ammonia uptake will continue during the night-time period following a normal daylight period, so that there will be no net loss of excretory nitrogen (Muscatine and D'Elia 1978). This has now been confirmed experimentally in *A. viridis*, which displays a diminished rate of ammonia uptake during the night, but no net loss (Davies 1988).

The rate of uptake increases in proportion to the external concentration of ammonia, and it has been suggested (D'Elia 1977; D'Elia et al. 1983) that uptake conforms to a diffusion–depletion model, i.e. that depletion of nutrients from the animal cytoplasm by the zooxanthellae creates a concentration gradient for diffusion from the external medium. In a number of symbiotic Cnidaria, the uptake rate is linear with increase in external concentration at least up to $10-20\,\mu M\,NH_4^+$ (Burris 1983; Wilkerson and Muscatine 1984; Wilkerson and Trench 1986).

Pocillopora capitata (Muscatine and D'Elia 1978) and the anemone *Aiptasia pulchella* (Wilkerson and Muscatine 1984) also deplete ammonia from unenriched seawater of 0.75 and $2.7\,\mu M$ concentration, respectively. Since uptake is a result of diffusion processes, this suggests that the internal concentration of ammonia is low. No reliable measurements of intracellular ammonia concentration are yet available. The relatively high concentration values of $5-50\,\mu M$ estimated by Crossland and Barnes (1977) and Wilkerson and Muscatine (1984) may result from degradative deamination during preparative procedures (Wilkerson and Muscatine 1984).

(b) Uptake of nitrate The literature on uptake of nitrate by symbiotic Cnidaria is equivocal. A number of corals appear to remove nitrate from enriched seawater (e.g., D'Elia and Webb 1977; Franzisket 1973, 1974; Webb and Wiebe 1978; Wilkerson and Trench 1986) while other anthozoans do not (table 3, Wilkerson and Trench 1986; Davies 1988). In reviewing the literature, Miller and Yellowlees (1989) argued that nitrate uptake would require the presence of a specific carrier mechanism at the outer membranes of the host cells. The critical experiments of uptake of ^{15}N-labelled nitrate do not appear to have been performed. Observations of nitrate uptake may be artefacts arising from bacterial contamination, since in two series of experiments (Cates and McLaughlin 1979; Webb and Wiebe 1978) nitrate depletion was recorded in the absence of zooxanthellae.

2. Assimilation of DIN by zooxanthellae

The fate of ammonium taken up by the gastrodermal cells is the subject of current debate. Probably all animal cells contain enzymes to assimilate some of the ammonia from catabolism of proteins to synthesize amino acids. Algal cells also contain enzymes for assimilation of exogenous ammonia for the synthesis of their amino acids. In the *Hydra/ Chlorella* symbiosis, Rees (1987) proposed that exogenous ammonia was primarily assimilated via the animal assimilation pathway, thereby maintaining the algae in a state of nitrogen deficiency which limited their growth and division. This assimilation route has been recently suggested by Miller and Yellowlees (1989) to apply to marine symbioses involving zooxanthellae.

The above authors argue that the lack of ammonium uptake in prolonged darkness is due to the lack of availability of carbon skeletons to act as amino-group acceptors in transaminase reactions in the host tissue. However, an alternative hypothesis is that a proportion of the ammonium from animal cell catabolism is recycled to glutamate by the route used in all animal cells (Fig. 23.5). The remainder is taken up by the zooxanthellae during photosynthesis, and uptake continues in darkness until the availability of carbon skeletons for ammonium assimilation is exhausted. During further darkness ammonium would be excreted. This hypothesis suggests therefore that carbon limitation of assimilation occurs in the zooxanthellae rather than in the animal tissue.

In animal cell assimilation, ammonia reacts with 2-oxoglutarate to form the amino-acid glutamate, the reaction being catalysed by NADP-glutamate dehydrogenase (NADP–GDH). Miller and Yellowlees (1989) point to the high specific activity of NADP–GDH in animal cytoplasm of symbiotic Anthozoa, which is approximately two

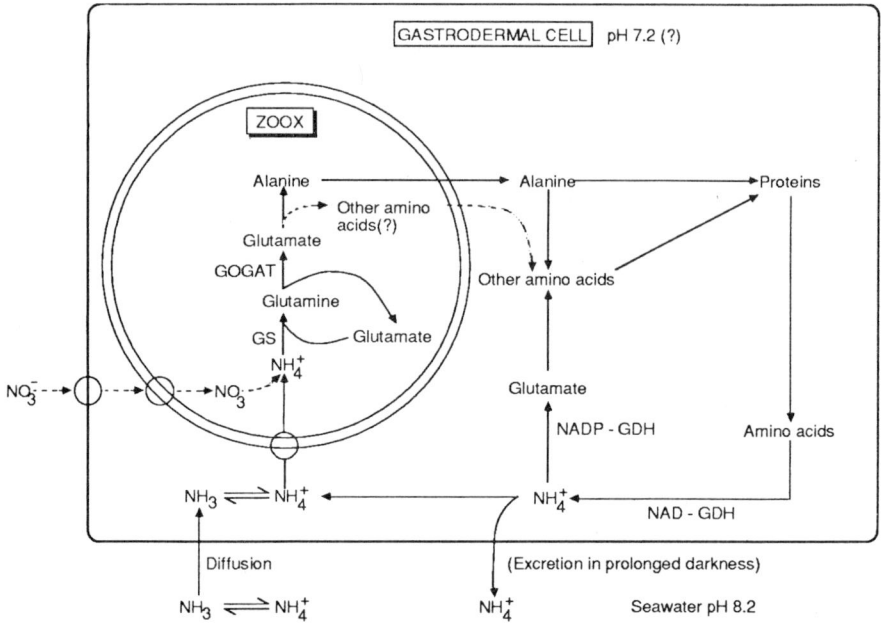

Fig. 23.5. Nitrogen fluxes in a gastrodermal cell of a symbiotic anthozoan in the light. In prolonged darkness, ammonia which would otherwise pass to the zooxanthellae is excreted. Evidence for the nitrate uptake route is equivocal. Abbreviations: GS glutamine synthetase; GOGAT glutamine 2-oxoglutarate amido transferase; GDH glutamate dehydrogenase. ○ specific membrane transport mechanism. For clarity, the protein synthesis and catabolism paths of the zooxanthellae have been omitted.

orders of magnitude greater than that of freshly isolated zooxanthellae (Catmull *et al.* 1987; Dudler and Miller 1988) when expressed on a unit-protein basis. However this low activity of GDH in zooxanthellae is not unexpected since there is now evidence that the algae use a different assimilatory pathway.

The most likely assimilation route in zooxanthellae is the linked glutamine synthetase/glutamine 2-oxoglutarate amido transferase (GS/GOGAT) pathway, which is commonly used in plant cells. Ammonia reacts with glutamate to produce the amide glutamine, catalysed by GS. The amido group of glutamine is then transferred to 2-oxoglutarate by GOGAT yielding 2 moles of glutamate, one of which is recycled to react with ammonia again. GS has been identified in zooxanthellae by Wilkerson and Muscatine (1984) and Anderson and Burris (1987). GOGAT has not yet been positively identified. Nevertheless patterns of incorporation of the stable isotope ^{15}N suggest the activity of the GS/GOGAT pathway (Summons and Osmond 1981; Summons et al. 1986). In *Stylophora pistillata*, inhibition of GOGAT was achieved by incubation of the coral in the light with the specific inhibitor azaserine. This resulted in total suppression of NH_4 uptake, and excretion of NH_4 at a rate similar to that of corals kept in prolonged darkness (Rahav et al. 1989). This appears to confirm that the major route for assimilation of both the excess catabolic ammonia and the exogenous ammonia is the GS/GOGAT pathway in zooxanthellae.

On the basis of present evidence it appears that ammonia assimilation in gastrodermal cells involves both the animal NADP-GDH and the GS/GOGAT pathway of zooxanthellae. However, it is likely that, as in other animal cells, the GDH pathway is normally fully saturated, and both excess excretory ammonium and ammonium entering the cytoplasm by diffusion are assimilated in light-driven reactions in the zooxanthellae. A summary of this interpretation is given in Fig. 23.5.

3. Nitrogen translocation

Very little information is available on the fate of nitrogen assimilated into the zooxanthellae. However, Trench (1971a) observed that after incubating various symbiotic Anthozoa in labelled $NaHCO_3$ a significant proportion of the ^{14}C was subsequently incorporated into animal protein. Freshly isolated zooxanthellae were shown to liberate ^{14}C-labelled alanine (Trench 1971b), while Lewis and Smith (1971), using an inhibition technique, showed that ^{14}C-labelled alanine from zooxanthellae could pass from the host tissue into alanine-enriched seawater. It appears therefore that the amino acid alanine is the main route by which organic nitrogen is made available to the host. It is likely (but this does not appear to have been tested) that alanine reacts with 2-oxoglutarate in the animal cytoplasm to yield glutamate, which could in turn be transaminated to provide other acids for protein synthesis.

Nitrogen budgets

It is clear from the foregoing that the host in symbiotic Anthozoa obtains benefit both from the recycling of excretory ammonium and from the effective conversion of ambient dissolved inorganic nitrogen into amino acids for protein synthesis. The importance of this in the overall nitrogen budget is only just beginning to be investigated. Calculations from the data of Rahav et al. (1989) would suggest that 80 per cent of the daily flux of nitrogen from the zooxanthellae in *Stylophora pistillata* is recycled (Fig. 23.6). In *Acropora palmata*, Bythell (1988) calculated that 30 per cent of the annual nitrogen intake is from dissolved

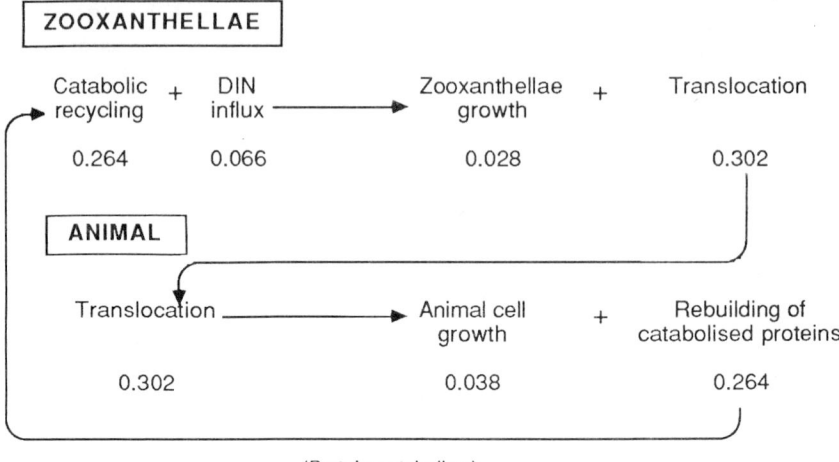

Fig. 23.6. Daily nitrogen budget for *Stylophora pistillata*, adapted from data of Rahav *et al.* (1989). All values are in units of μg at N d^{-1} cm^{-2} of coral surface. Note that the influx value of 0·066 is a derived value to balance the budget and may comprise dissolved inorganic nitrogen (DIN) which passes to the zooxanthellae and particulate organic nitrogen which passes to the animal cytoplasm. The value of 0·066 is probably too low, since the model does not allow for the nitrogen lost as mucus secretion. The value of 0·264 for catabolic recycling is the excretory nitrogen which passes into the zooxanthellae in the light. In addition to this, some as yet unquantified amount will be recycled via the NADP-GDH pathway of the animal cytoplasm.

inorganic nitrogen, while the remainder comes from particulate food. On the expenditure side of the budget, 50 per cent is lost as mucus, 40 per cent is used for tissue growth and 11 per cent for gamete production.

Regulation of symbiont populations

The number of zooxanthellae harboured within the gastrodermal tissue is closely regulated, probably to optimize photosynthesis by the available surface of host exposed to light. Regulation implies the maintenance of a constant ratio of biomass of zooxanthellae to that of the host during growth, and the growth rates of zooxanthellae and animal tissue would have to be the same. Trench (1987) suggested three ways in which this could be achieved:

(1) by a mechanism, under either host or algal control, which promotes synchrony of division of host and algal cells;

(2) by exocytosis and expulsion of excess algal cells produced by cell division when the biomass ratio is exceeded;
(3) by host control of algal division rates by limitation of nutrient availability.

The first of these hypotheses does not appear to have been investigated. The second hypothesis has some experimental evidence in its favour. Steele (1976) and Hoegh-Guldberg et al. (1987) demonstrated that apparently healthy and viable zooxanthellae were continuously expelled from several different species of host. In *Stylophora pistillata* the specific growth rate of the algae is greater than that of the animal tissue, and expulsion of zooxanthellae would therefore be necessary in order to maintain a constant biomass ratio (Muscatine et al. 1985). This provides reasonable support for the second hypothesis, but leaves unanswered the mechanism by which the symbiosis detects the set point or population density above which expulsion would take place.

The third hypothesis was outlined in some detail by Cook and D'Elia (1987) who proposed that the biomass of zooxanthellae could be controlled by limiting the availability of nutrients, particularly nitrogen, to the zooxanthellae. Miller and Yellowlees (1989) reviewed the evidence for this, and concluded that zooxanthellae do not show the usual characteristics of nitrogen-limited plant cells such as high NH_4^+ uptake activity, high levels of NH_4^+ assimilation enzymes and high nitrate reductase enzyme activity. Nevertheless several recent studies have indicated that if zooxanthellae are not nitrogen-limited they may be nitrogen impoverished. One indication of this in *Aiptasia pallida* is that algae freshly isolated from sea anemones from the field have a total C:N ratio of 9·4 compared with a value of 7·5 for algae from hosts which had been fed to repletion and about 16 for algae from starved hosts (Cook et al. 1988). Furthermore, the growth of zooxanthellae *in situ* can be promoted by increasing the level of NH_4^+ in the seawater. The mitotic index of the depleted population of zooxanthellae in *Aiptasia pallida*, which had been experimentally starved in nutrient-free seawater, increased when they were placed in NH_4^+/PO_4^{2-} enriched seawater (Cook et al. 1988). In *Stylophora pistillata* the algal population density doubled after 14 days incubation in seawater enriched to 20 μM NH_4^+ (Muscatine et al. 1989). Similar results were obtained in *S. pistillata* and *Seriatopora hystrix* by Hoegh-Guldberg and Smith (1989). At the end of the incubation period, in both studies, the mitotic index was almost identical to that of the control group, although Stimson (1988) in an apparently similar experiment reported that addition of NH_4^+ produced a higher rate of expulsion of zooxanthellae. One explanation for these observations is that, when living in nutrient-poor waters, the whole symbiotic association is nitrogen impoverished. At higher ambient nitrogen levels, the nitrogen flux to the zooxanthellae and thence to the animal cytoplasm is enhanced, resulting in a higher growth rate of both zooxanthellae and animal tissue. The increased mass of protein cm^{-2}, and the relatively unchanged biomass ratio at the end of the incubation in ammonia-enriched seawater, as observed by Muscatine et al. (1989), provide some evidence for this. Although the mitotic index remained the same, the expulsion rate would have to increase, as observed by Stimson (1988), in order to maintain the constancy of the biomass ratio.

These experiments suggest that control of zooxanthellae numbers may be determined by environmental levels of nitrogen, rather than by host-controlled nutrient limitation. Nevertheless, there would still need to be an endogenous control mechanism to prevent overgrowth of host tissue by algae under conditions of high nitrogen availability.

Stress-related expulsion of zooxanthellae

When exposed to a range of environmental stresses, symbiotic organisms commonly respond by 'bleaching', becoming pale and ultimately white or translucent. In the case of corals which have been adapted to low light intensity and then exposed to full sunlight, the bleaching results from loss of photosynthetic pigment from the zooxanthellae (Hoegh-Guldberg and Smith 1989). In all other cases examined, bleaching results from a breakdown of the symbiotic association and the expulsion of zooxanthellae, which in extreme cases may be seen issuing from the mouths of polyps in mucous strands. This response has been observed after exposure to prolonged darkness (Yonge and Nicholls 1931b; Franzisket 1970), elevated temperatures (Yonge and Nicholls 1931a; Jokiel and Coles 1977; Coles and Jokiel 1978; Hoegh-Guldberg and Smith 1989), low temperatures (Steen and Muscatine 1987) and low salinities (Goreau 1964).

In recent years there have been numerous reports of spontaneous bleaching by symbiotic organisms, mainly corals, on reefs in many parts of the world (Brown 1987). Severe bleaching, resulting in deaths of coral colonies, was reported in the eastern Pacific in 1983 (Glynn 1983, 1984; Brown and Suharsono 1990) (Fig. 23.7) and in the Caribbean in 1987–1988 (Williams et al. 1987). The 1983 bleaching event was correlated with increased seawater temperatures of 2–3°C resulting from the 1982–1983 'El Niño', which produced the strongest warming of the equatorial Pacific in this century (Kerr 1983). The bleaching of Caribbean corals in 1987–1988 was likewise attributed to localized seawater warming (Williams et al. 1987; Gates 1990b).

Fig. 23.7. Bleaching as a result of loss of zooxanthellae in *Acropora* species on a reef-flat in the Pulau Seribu Islands, Indonesia, during April 1983 (Suharsono, Institute of Oceanology, Jakarta).

Experimental work has shown that Pacific corals normally live within 2–3°C of their upper lethal limit, and that physiological changes, notably in respiration and photosynthetic rates, occur below the lethal limits (Coles and Jokiel 1977).

Despite the correlation of bleaching with elevated temperature, it is not clear whether temperature has a direct effect or whether there is mediation by a pathological organism (Brown 1987). Corals may show a mosaic response, with only localized areas of loss of zooxanthellae (Gates 1990*b*) which recover their full complement of algae several months later (Hayes and Bush 1990). During partial bleaching in *Montastrea annularis* the coral tissue biomass is reduced (Szmant and Gassman 1990), possibly as a result of autolytic catabolism, and the structural integrity of the vacated gastrodermal cells is disrupted, the cells filling with mucoid material (Hayes and Bush 1990). The remaining zooxanthellae are vacuolated and agranular (Hayes and Bush 1990) and become concentrated in the gastroderm at the base of the polyps — an area normally devoid of algae (Szmant and Gassman 1990). The bleached areas cease growth (Goreau and MacFarlane 1990) and are unable to complete gametogenesis (Szmant and Gassman 1990).

These histopathological studies may indicate changes associated with early stages of cell death. The incipient changes may have occurred at an earlier stage, possibly involving a breakdown in the mechanism of recognition of the symbiont by the host. It is clear that more research is needed in order to unravel the intricate nature of stress-related expulsion of symbionts, not least because there may be more frequent and widespread bleaching events on coral reefs if the 'greenhouse effect' is found to induce localized areas of elevated seawater temperatures in tropical regions.

References

Anderson, S. L. and Burris, J. E. (1987). Role of glutamine synthetase in ammonia assimilation by symbiotic marine dinoflagellates (zooxanthellae). *Marine Biology*, **94**, 451-8.

Battey, J. F. and Patton, J. S. (1984). A re-evaluation of the role of glycerol in carbon translocation in zooxanthellae-coelenterate symbiosis. *Marine Biology*, **79**, 27-38.

Battey, J. F. and Patton, J. S. (1987). Glycerol translocation in *Condylactis gigantea*. *Marine Biology*, **95**, 37-46.

Battey, J. F. and Porter, J. W. (1988). Photoadaptation as a whole organism response in *Montastrea annularis*. In *6th International Coral Reef Symposium*, **3**, pp. 79-87. Townsville, Queensland, Australia.

Benson, A. A. and Muscatine, L. (1974). Wax in coral mucus: energy transfer from corals to reef fishes. *Limnology and Oceanography*, **19**, 810-4.

Blank, R. J. and Trench, R. K. (1985). Speciation and symbiotic dinoflagellates. *Science*, **229**, 656-8.

Blank, R. J. and Trench, R. K. (1986). Nomenclature of endosymbiotic dinoflagellates. *Taxon*, **35**, 286-94.

Blanquet, R. S., Nevenzel, J. S., and Benson, A. A. (1979). Acetate incorporation into the lipids of the anemone *Anthopleura elegantissima* and its associated zooxanthellae. *Marine Biology*, **54**, 185-94.

Brown, B. E. (1987). Worldwide death of corals — natural cyclidal events or man-made pollution? *Marine Pollution Bulletin*, **18**, 9-13.

Brown, B. E. and Suharsono (1990). Damage and recovery of coral reefs affected by El Niño related seawater warming in the Thousand Islands, Indonesia. *Coral Reefs*, **8**, 163-70.

Burris, R. H. (1983). Uptake and assimilation of $^{15}NH_4^+$ by a variety of corals. *Marine Biology*, **75**, 151-5.

Bythell, J. C. (1988). A total nitrogen and carbon budget for the elkhorn coral *Acropora palmata* (Lamarck). In *6th International Coral Reef Symposium* 2, pp. 535-40. Townsville, Queensland, Australia.

Cates, N. and McLaughlin, J. J. A. (1976). Differences in ammonia metabolism in symbiotic and aposymbiotic *Condylactis* and *Cassiopeia* spp.

Journal of Experimental Marine Biology and Ecology, **21**, 1-5.

Cates, N. and McLaughlin, J.J.A. (1979). Nutrient availability for zooxanthellae derived from physiological activities of *Condylactis* spp. *Journal of Experimental Marine Biology and Ecology*, **37**, 31-41.

Catmull, J., Yellowlees, D., and Miller, D.J. (1987). NADP$^+$ dependent glutamate dehydrogenase from *Acropora formosa*: purification and properties. *Marine Biology*, **95**, 559-63.

Chalker, B.E. (1981). Simulating light-saturation curves for photosynthesis and calcification by reef-building corals. *Marine Biology*, **63**, 135-41.

Chalker, B.E. and Dunlap, W.C. (1983). Primary production and photoadaptation by corals on the Great Barrier Reef. In *Proceedings of the Great Barrier Reef Conference*, (ed. J.T. Baker et al.), pp. 294-8. James Cook University, Queensland, Australia.

Coles, S.L. and Jokiel, P.L. (1977). Effects of temperature on photosynthesis and respiration in hermatypic corals. *Marine Biology*, **43**, 209-16.

Colley, N.J. and Trench, R.K. (1983). Selectivity in phagocytosis and persistence of symbiotic algae by the scyphistoma stage of the jellyfish *Cassiopeia xamachana*. *Proceedings of the Royal Society, London*, Series B, **219**, 61-82.

Cook, C.B. and D' Elia, C.F. (1987). Are natural populations of zooxanthellae ever nutrient-limited? *Symbiosis*, **4**, 199-212.

Cook, C.B., D' Elia, C.F., and Muller-Parker, G. (1988). Host feeding and nutrient sufficiency for zooxanthellae in the sea anemone *Aiptasia pallida*. *Marine Biology*, **98**, 253-62.

Crossland, C.J. (1987). In situ release of mucus and DOC-lipid from the corals *Acropora variabilis* and *Stylophora pistillata* in different light regimes. *Coral Reefs*, **6**, 35-42.

Crossland, C.J. and Barnes, D.J. (1977). Gas exchange studies with the staghorn coral *Acropora acuminata* and its zooxanthellae. *Marine Biology*, **40**, 185-94.

Crossland, C.J., Barnes, D.J., and Borowitzka, M.A. (1980). Diurnal lipid and mucus production in the staghorn coral *Acropora acuminata*. *Marine Biology*, **60**, 81-90.

D'Elia, C.F. (1977). The uptake and release of dissolved phosphorus by reef corals. *Limnology and Oceanography*, **22**, 301-15.

D'Elia, C.F. and Webb, K.L. (1977). The dissolved nitrogen flux of reef corals. In *3rd International Coral Reef Symposium*, 1, pp. 325-30. Atlantic Reef Committee, Miami, Florida.

D'Elia, C.F., Domotor, S.L., and Webb, K.L. (1983). Nutrient uptake kinetics of freshly isolated zooxanthellae. *Marine Biology*, **75**, 157-67.

Davies, L.M. (1988). Nitrogen flux in the symbiotic sea anemone *Anemonia viridis* (Forskal). Unpublished Ph.D. Thesis. University of Glasgow.

Davies, P.S. (1977). Carbon budgets and vertical zonation of Atlantic reef corals. In *3rd International Coral Reef Symposium*, 1, pp. 391-6. Atlantic Reef Committee, Miami, Florida.

Davies, P.S. (1984). The role of zooxanthellae in the nutritional energy requirements of *Pocillopora eydouxi*. *Coral Reefs*, **2**, 181-6.

Davies, P.S. (1991). Effects of daylight variations on the energy budgets of shallow-water corals. *Marine Biology*, **108**, 137-44.

Dudler, N. and Miller, D.J. (1988). Characterization of two glutamate dehydrogenases from the symbiotic microalga *Symbiodinium microadriaticum* isolated from the coral *Acropora formosa*. *Marine Biology*, **97**, 427-30.

Edmunds, P.J. and Davies, P.S. (1986). An energy budget for *Porites porites* (Scleractinia). *Marine Biology*, **92**, 339-47.

Edmunds, P.J. and Davies, P.S. (1988). Post-illumination stimulation of respiration rate in the coral *Porites porites*. *Coral Reefs*, **7**, 7-9.

Edmunds, P.J. and Davies, P.S. (1989). An energy budget for *Porites porites* (Scleractinia), growing in a stressed environment. *Coral Reefs*, **8**, 37-43.

Falkowski, P.G. and Dubinsky, Z. (1981). Light-shade adaptation of *Stylophora pistillata*, a hermatypic coral from the Gulf of Eilat. *Nature*, **289**, 172-4.

Fitt, W.K. (1984). The role of chemosensory behaviour of *Symbiodinium microadriaticum*, intermediate hosts and host behaviour in the infection of coelenterates and molluscs with zooxanthellae. *Marine Biology*, **8**, 9-17.

Fitt, W.K. and Trench, R.K. (1983). Endocytosis of the symbiotic dinoflagellate *Symbiodinium microadriaticum* Freudenthal by endodermal cells of the scyphistomae of *Cassiopeia xamachana* and resistance of the algae to host digestion. *Journal of Cell Science*, **64**, 195-212.

Franzisket, L. (1970). The atrophy of hermatypic reef corals maintained in darkness and their subsequent regeneration in light. *Internationale Revue Gesamten Hydrobiologie*, **55**, 1-12.

Franzisket, L. (1973). Uptake and accumulation of nitrate and nitrite by reef corals. *Naturwissenschaften*, **60**, p. 552.

Franzisket, L. (1974). Nitrate uptake by reef corals. *International Review of Hydrobiology*, **59**, 1-7.

Freudenthal, H. (1962). *Symbiodinium* gen. nov. and *Symbiodinium microadriaticum* sp. nov., a zooxanthellae; taxonomy, life cycle and morphology. *Journal of Protozoology*, **9**, 45-52.

Gates, R.D. (1990a). Seawater temperature and algal–cnidarian symbiosis. Unpublished Ph.D. Thesis. University of Newcastle upon Tyne.

Gates, R.D. (1990b). Seawater temperature and sublethal coral bleaching in Jamaica. *Coral Reefs*, **8**, 193-8.

Gattuso, J.-P. Features of depth effects on *Stylophora pistillata*, an hermatypic coral in the Gulf of Aqaba (Jordan, Red Sea). In *5th International Coral Reef Symposium*, 6, pp. 95–100. EPHE, Tahiti.

Glynn, P.W. (1983). Extensive bleaching and death of reef corals on the Pacific coast of Panama. *Environmental Conservation*, 10, 149–54.

Glynn, P.W. (1984). Widespread mortality and the 1982–83 El Niño warming event. *Environmental Conservation*, 11, 133–46.

Goreau, T.F. (1964). Mass expulsion of zooxanthellae from Jamaican reef communities after hurricane Flora. *Science*, 145, 383–6.

Goreau, T.J. and MacFarlane, A.H. (1990). Reduced growth rate of *Montastrea annularis* following the 1987–1988 coral-bleaching event. *Coral Reefs*, 8, 211–16.

Hayes, R.L. and Bush P.G. (1990). Microscopic observations of recovery in the reef-building scleractinian coral, *Montastrea annularis* after bleaching on a Cayman reef. *Coral Reefs*, 8, 203–10.

Hoegh-Guldberg, O. and Smith, G.J. (1989). Influence of the population density of zooxanthellae and supply of ammonium on the biomass and metabolic characteristics of the reef corals *Seriatopora hystrix* and *Stylophora pistillata*. *Marine Ecology Progress Series*, 57, 173–86.

Hoegh-Guldberg, O., McCloskey, L.R., and Muscatine, L. (1987). Expulsion of zooxanthellae by symbiotic cnidarians from the Red Sea. *Coral Reefs*, 5, 201–4.

Hofman, D.K. and Kremer, B.P. (1981). Carbon metabolism and strobilation in *Cassiopeia andromedea* (Cnidaria: Scyphozoa): significance of endosymbiotic dinoflagellates. *Marine Biology*, 65, 25–33.

Von Holt, C. and Von Holt, M. (1968a). The secretion of organic compounds by zooxanthellae isolated from various types of *Zoanthus*. *Comparative Biochemistry and Physiology*, 24, 83–92.

Von Holt, C. and Von Holt, M. (1968b). Transfer of photosynthetic products from zooxanthellae to coelenterate hosts. *Comparative Biochemistry and Physiology*, 24, 73–81.

Johannes, R.E. and Tepley, L. (1974). Examination of feeding of the reef coral *Porites lobata in situ* using time lapse photography. In *2nd International Coral Reef Symposium*, 1, pp. 127–31. Great Barrier Reef Committee, Brisbane, Australia.

Johannes, R.E., Coles, S.L., and Kuenzel, N.T. (1970). The role of zooplankton in the nutrition of some scleractinian corals. *Limnology and Oceanography*, 15, 579–86.

Johnston, I.S. (1980). The ultrastructure of skeletogenesis in hermatypic corals. *International Review of Cytology*, 67, 171–214.

Jokiel, P.L. and Coles, S.L. (1977). Effects of temperature on the mortality and growth of Hawaiian reef corals. *Marine Biology*, 43, 201–8.

Jokiel, P.L. and Morrissey, J.I. (1986). Influence of size on primary production in the reef coral *Pocillopora damicornis* and the macroalga *Acanthophora spicifera*. *Marine Biology*, **91**, 15-26.

Kawaguti, S. (1953). Ammonium metabolism of the reef corals. *Biological Journal of Okayama University*, **1**, 171-6.

Kellogg, R.B. and Patton, J.S. (1983). Lipid droplets, medium of energy exchange in the symbiotic anemone *Condylactis gigantea*: a model coral polyp. *Marine Biology*, **75**, 1-49.

Kerr, R.A. (1983). Fading El Niño broadening scientist's view. *Science*, **221**, 940-1.

Kevin, M.J., Hall, W.T., McLaughlin, J.J.A., and Zahl, P.A. (1969). *Symbiodinium microadriaticum* Freudenthal, a revised taxonomic description, ultrastructure. *Journal of Phycology*, **5**, 341-50.

Kinzie, R.A. (1974). Experimental infection of aposymbiotic gorgonian polyps with zooxanthellae. *Journal of Experimental Marine Biology and Ecology*, **15**, 335-45.

Lewis, D.H. and Smith, D.C. (1971). The autotrophic nutrition of symbiotic marine coelenterates with special reference to hermatypic corals. I. Movement of photosynthetic products between the symbionts. *Proceedings of the Royal Society of London*, Series B, **178**, 111-29.

Loeblich, A.R. and Sherley, J.L. (1979). Observations on the theca of the motile phase of the free-living and symbiotic isolates of *Zooxanthella microadriatica* (Freudenthal) comb. nov. *Journal of the Marine Biological Association of the United Kingdom*, **59**, 195-205.

McCloskey, L.R. and Muscatine, L. (1984). Production and respiration in the Red Sea coral *Stylophora pistillata* as a function of depth. *Proceedings of the Royal Society of London*, Series B, **222**, 215-20.

McLaughlin, J.J.A. and Zahl, P.A. (1966). Endozoic algae. In *Symbiosis*, (ed. S.M. Henry), pp. 257-97. Academic Press, New York.

Miller, D.J. and Yellowlees, D. (1989). Inorganic nitrogen uptake by symbiotic marine cnidarians: a critical review. *Proceedings of the Royal Society of London*, Series B, **237**, 109-25.

Muscatine, L. (1967). Glycerol excretion by symbiotic algae from corals and *Tridacna* and its control by the host. *Science*, **156**, 516-19.

Muscatine, L. (1980). Productivity of zooxanthellae. In *Primary productivity in the sea*, (ed. P.G. Falkowski), pp. 381-402. Plenum Press, New York.

Muscatine, L. and Cernichiari, E. (1969). Assimilation of photosynthetic products of zooxanthellae by a reef coral. *Biological Bulletin (Woods Hole, Massachusetts)*, **137**, 506-23.

Muscatine, L. and D'Elia, C.F. (1978). The uptake, retention, and release of ammonium by reef corals. *Limnology and Oceanography*, **23**, 275-34.

Muscatine, L., Pool, R.R., and Trench, R.K. (1975). Symbiosis of algae and invertebrates: Aspects of the symbiont surface and the host-symbiont interface. *Transactions of the American Microscopical Society*, **94**, 450-69.

Muscatine, L., Masuda, H., and Burnap, R. (1979). Ammonium uptake by symbiotic and aposymbiotic reef corals. *Bulletin of Marine Science*, **29**, 572-5.

Muscatine, L., Falkowski, P. G., Porter, J. W., and Dubinsky, Z. (1984). Fate of photosynthetic fixed carbon in light- and shade-adapted colonies of the symbiotic coral *Stylophora pistillata*. *Proceedings of the Royal Society, London*, Series B, **222**, 181-202.

Muscatine, L., McCloskey, L., and Loya, Y. (1985). A comparison of the growth rates of zooxanthellae and animal tissue in the Red Sea coral *Stylophora pistillata*. In *5th International Coral Reef Symposium*, 6, pp. 119-23. EPHE, Tahiti.

Muscatine, L., Falkowski, P. G., Dubinsky, Z., Cook, P. A., and McCloskey, L. R. (1989). The effect of external nutrient resources on the population dynamics of zooxanthellae in a reef coral. *Proceedings of the Royal Society of London*, Series B, **236**, 311-24.

Patton, J. S. and Burris, J. E. (1983). Lipid synthesis and extrusion by freshly isolated zooxanthellae (symbiotic algae). *Marine Biology*, **75**, 131-6.

Patton, J. S., Abraham, S., and Benson, A. A. (1977). Lipogenesis in the intact coral *Pocillopora capitata* and its isolated zooxanthellae: evidence for a light driven carbon cycle between symbiont and host. *Marine Biology*, **44**, 235-47.

Patton, J. S., Battey, J. F., Rigler, M. W., Porter, J. W., Black, C. C., and Burris, J. E. (1983). A comparison of the metabolism of bicarbonate ^{14}C and acetate l- ^{14}C and the variability of species lipid compositions in reef corals. *Marine Biology*, **75**, 121-30.

Porter, J. W. (1974). Zooplankton feeding by the Caribbean reef-building coral *Montastrea cavernosa*. In *2nd International Coral Reef Symposium*, pp. 111-25. Great Barrier Reef Committee, Brisbane, Australia.

Porter, J. W. (1976). Autotrophy, heterotrophy and resource partitioning in Caribben reef-building corals. *American Naturalist*, **110**, 731-42.

Porter, J. W., Muscatine, L., Dubinsky, Z., and Falkowski, P. G. (1984). Primary production and photoadaptation in light and shade-adapted colonies of the symbiotic coral, *Stylophora pistillata*. *Proceedings of the Royal Society of London*, Series B, **222**, 161-80.

Rahav, O., Dubinsky, Z., Achituv, Y., and Falkowski, P. G. (1989). Ammonium metabolism in the zooxanthellate coral, *Stylophora pistillata*. *Proceedings of the Royal Society of London*, Series B, **236**, 325-37.

Rees, T. A. V. (1987). The green hydra symbiosis and ammonium. I. The role of the host in ammonium assimilation and its possible regulatory significance. *Proceedings of the Royal Society of London*, Series B, **229**, 299-314.

Schlichter, D., Kremner, B. P., and Svoboda, A. (1984). Zooxanthellae providing assimilatory power for the incorporation of exogenous acetate in

Heteroxenia fuscescens. Marine Biology, **83**, 277–86.

Schmitz, K. and Kremer, B.P. (1977). Carbon fixation and analysis of assimilates in a coral–dinoflagellate symbiosis. *Marine Biology*, **42**, 305–13.

Schoenberg, D.A. and Trench, R.K. (1980a). Genetic variation in *Symbiodinium* (= *Gymnodinium*) *microadriaticum* Freudenthal, and specificity in its symbiosis with marine invertebrates. I. Isoenzyme and soluble protein patterns of axenic cultures of *Symbiodinium microadriaticum*. *Proceedings of the Royal Society, London*, Series B, **207**, 405–27.

Schoenberg, D.A. and Trench, R.K. (1980b). Genetic variation in *Symbiodinium* (= *Gymnodinium*) *microadriaticum* Freudenthal, and specificity in its symbiosis with marine invertebrates. II. Morphological variation in *Symbiodinium microadriaticum*. *Proceedings of the Royal Society of London*, Series B, **207**, 429–44.

Smith, D.C. and Douglas, A.E. (1987). *The biology of symbiosis*. Edward Arnold, London.

Steele, R.D. (1975). Stages in the life history of a symbiotic zooxanthella in pellets extruded by its host *Aiptasia tagetes* (Duch and Mich). (Coelenterata, Anthozoa). *Biological Bulletin (Woods Hole, Massachusetts)*, **149**, 590–600.

Steele, R.D. (1976). Light intensity as a factor in the regulation of the density of symbiotic zooxanthellae in *Aiptasia tagetes* (Coelenterata, Anthozoa). *Journal of Zoology, London*, **179**, 387–405.

Steele, R.D. (1977). The significance of zooxanthella-containing pellets extruded by sea anemones. *Bulletin of Marine Science*, **27**, 591–4.

Steen, G.R. and Muscatine, L. (1987). Low temperature evokes rapid exocytosis of symbiotic algae by a sea-anemone. *Biological Bulletin (Woods Hole, Massachusetts)*, **172**, 246–63.

Stimson, J.S. (1987). Location, quantity and rate of change in quantity of lipids in tissue of Hawaiian hermatypic corals. *Bulletin of Marine Science*, **41**, 889–904.

Stimson, J.S. (1988). The rate and diel pattern of release of zooxanthellae by undisturbed colonies of *Pocillopora damicornis* at two levels of dissolved nitrogen. In *6th International Coral Reef Symposium*, Abstract, p. 382.

Summons, R.E. and Osmond, C.B. (1981). Nitrogen assimilation in the symbiotic marine alga *Gymnodinium microadriaticum*. Direct analysis of ^{15}N incorporation by GC–MS methods. *Phytochemistry*, **20**, 575–8.

Summons, R.E., Boag, T.S., and Osmond, C.B. (1986). The effect of ammonium on photosynthesis and the pathway of ammonium assimilation in *Gymnodinium microadriaticum in vitro* and in symbiosis with tridacnid clams and corals. *Proceedings of the Royal Society of London*, Series B, **227**, 147–59.

Syrett, P.J. (1981). Nitrogen metabolism of microalgae. In *Physiological bases* of *phytoplankton ecology*, Bulletin 210, Department of Fisheries and Oceans, Ottawa, (ed. T. Platt), pp. 182-210.

Szmant, A.M. and Gassman, N.J. (1990). The effects of prolonged 'bleaching' on the tissue biomass and reproduction of the reef coral *Montastrea annularis*. *Coral Reefs*, **8**, 217-24.

Szmant-Froelich A. and Pilson, M.E.Q. (1977). Nitrogen excretion by colonies of the temperate coral *Astrangia danae* with and without zooxanthellae. In *3rd International Coral Reef Symposium*, 1, pp. 417-23. Atlantic Reef Committee, Miami, Florida.

Taylor, D.L. (1968). *In situ* studies on the cytochemistry and ultrastructure of a symbiotic marine dinoflagellate. *Journal of the Marine Biological Association of the United Kingdom*, **48**, 349-66.

Taylor, D.L. (1974). Symbiotic marine algae: taxonomy and biological fitness. In *Symbiosis in the sea*, (ed. W.B. Vernberg), pp. 245-62. University of South Carolina Press.

Taylor, F.J.R. (1983). Possible free living *Symbiodinium microadriaticum* (Dinophyceae) in tide pools in southern Thailand. In *Endocytobiology III*, (ed. Schenk, H.E.A. and W. Schwemmler), pp. 1009-14. De Gruyter, Berlin.

Trench, R.K. (1971a). The physiology and biochemistry of zooxanthellae symbiotic with marine coelenterates. I. The assimilation of photosynthetic products of zooxanthellae by two marine coelenterates *Proceedings of the Royal Society of London*, Series B, **177**, 225-35.

Trench, R.K. (1971b). The physiology and biochemistry of zooxanthellae symbiotic with marine coelenterates. II. Liberation of fixed ^{14}C by zooxanthellae *in vitro*. *Proceedings of the Royal Society of London*, Series B, **177**, 237-50.

Trench, R.K. (1971c). The physiology and biochemistry of zooxanthellae symbiotic with marine invertebrates. III. The effects of homogenates of host tissues on the excretion of photosynthetic products *in vitro* by zooxanthellae from two marine coelenterates. *Proceedings of the Royal Society of London*, Series B, **177**, 251-64.

Trench, R.K. (1974). Nutritional potentials in *Zoanthus sociatus*. *Helgoländer Wissenschaftliche Meeresunters*, **26**, 174-216.

Trench, R.K. (1979). The cell biology of plant-animal symbiosis. *Annual Reviews of Plant Physiology*, **30**, 485-531.

Trench, R.K. (1987). Dinoflagellates in non-parasitic symbioses. In *The biology of dinoflagellates*, (ed. F.J.R. Taylor), pp. 530-70. Blackwell, Oxford.

Webb, K.L. and Wiebe, W.J. (1978). The kinetics and possible significance of nitrate uptake by several algal-invertebrate symbioses. *Marine Biology*, **47**, 21-7.

Wethey, D.S. and Porter, J.W. (1976). Sun and shade differences in

productivity of reef corals. *Nature*, **262**, 281–2.
Wilkerson, F.P. and Muscatine, L. (1984). Uptake and assimilation of dissolved inorganic nitrogen by a symbiotic sea anemone. *Proceedings of the Royal Society of London,* Series B, **211**, 71–86.
Wilkerson, F.P. and Trench, R.K. (1986). Uptake of dissolved inorganic nitrogen by the symbiotic clam *Tridacna gigas* and the coral *Acropora* sp. *Marine Biology*, **93**, 237–46.
Wilkerson, F.P., Kobayashi, D., and Muscatine, L. (1988). Mitotic index and size of symbiotic algae in Caribbean reef corals. *Coral Reefs*, **7**, 29–36.
Williams, E.H., Goenaga, C., and Vicente, V. (1987). Mass bleachings on Atlantic coral reefs. *Science*, **237**, 877–8.
Yonge, C.M. and Nicholls, A.G. (1931*a*). Studies on the physiology of corals. IV. The structure, distribution and physiology of the zooxanthellae. *Scientific Reports of the Great Barrier Reef Expedition*, **1**, 135–76.
Yonge, C.M. and Nicholls, A.G. (1931*b*). Studies on the physiology of corals V. The effects of starvation in light and in darkness on the relationship between corals and zooxanthellae. *Scientific Reports of the Great Barrier Reef Expedition*, **1**, 177–211.

24. Summary and future prospects for plant–animal interactions

A.J. UNDERWOOD

Institute of Marine Ecology, Zoology Building, University of Sydney, NSW, Australia

---

The topics discussed in this volume make it obvious that there is a wide range of approaches and concepts used in the study of interactions between animals and plants. As a result, a considerable body of knowledge can be assembled on many types of processes. There are, however, two major themes that could be explored profitably either in their own right or as part of the planning and implementation of future studies. Both would greatly aid future comparative assessments of the nature of interactions within and between marine animals and plants.

First, there are great difficulties in any attempt to compare patterns and processes in different parts of the world. Efforts to generalize about the importance of different aspects of interactions are completely dependent on the types of information available from biogeographical regions. There is, for example, a world of disparity between the long-term detailed observations on small-scale intertidal habitats on the north-eastern coast of the USA (Vadas and Elner, Chapter 2 this volume) and the shorter-term, but wide-scale, comparative quantitative descriptions on the Californian coast (Foster 1990 and Chapter 3 this volume).

Regions more removed from traditional centres of marine ecology pose special difficulties, in that there are few studies and smaller total scientist-hours have been spent in such regions (see the information on West Africa by John *et al.*, Chapter 4 this volume). Other parts of the world have been more productive of small-scale experimental analyses of plant–animal interactions (e.g., Chapman and Johnson 1990; Underwood 1985). These, however, have not always been

Plant–Animal Interactions in the Marine Benthos, (ed. D.M. John, S.J. Hawkins, and J.H. Price), Systematics Association Special Volume No 46, pp. 541–4. Clarendon Press, Oxford, 1992. © The Systematics Association, 1992.

accompanied by large-scale descriptions of the distributions and abundances of the component animals and plants (e.g., the Australian coastline, Underwood and Kennelly 1990).

As a consequence, meaningful biogeographical comparisons of the patterns of distribution and diversity of marine flora and fauna are almost impossible. The processes causing or maintaining such patterns may be known, to a greater or lesser extent (Underwood and Denley 1984; Foster 1990), for any particular region. Metamodels that seek to explain patterns in several places, with different faunas and floras, will continue to be inadequate until more attention is paid to the collection of comparable, commensurable data from the different regions. There will remain numerous problems (reviewed by Underwood and Petraitis 1991). Nevertheless, comparable information should include estimates of seasonal and annual variation in abundances and distributions of animals and plants and, as much as is practicable, information about patchiness of the populations and of the richness or diversity of the species, at several spatial scales. These scales should range from very small (the sort of scale often used in detailed experimental studies) to large (hundreds of kilometres represent the scale of many broad biogeographical descriptions, Stephenson and Stephenson 1972). The contributions to this volume indicate the spatial and temporal scales which ecologists currently find interesting. Addition of some extra components will greatly increase the comparative worth of many extant and future studies.

The second area that would greatly improve the assessment of interactions between animals and plants is the integration of processes themselves. There is a tendency, for example, to investigate processes of competition for microalgal foods by intertidal grazers without any concomitant investigation of the effects of grazing on the structure, distribution and behaviour of the microalgae (although there are exceptions; see Underwood, Chapter 20 this volume). All sorts of components of foraging behaviour by grazers tend to be examined in isolation of each other (Chapman and Underwood, Chapter 13 this volume). Most studies have little historical or evolutionary underpinning (Branch 1984; Steneck, Chapter 21 this volume). Detailed studies of the chemistry of the plants (Hay, and Fenical, Chapter 14 this volume) are crucial for the eventual determination of details of processes affecting the grazers. They should, however, be more tightly integrated, in future, into studies on the abundance, distribution, and diversity of the plants. For example, the following questions might be asked: does this chemical defence really affect patterns of species richness? What happens if all species are well-defended? Also, the effects on the animals are largely unknown. Does consumption of potentially toxic algae, or learning to avoid them, really affect fitness of individuals or sizes of populations of fish? Again, many of the contributions in this volume indicate the areas of connectedness among studies and topics that are likely to be most usefully explored if there are slight changes of emphasis in future projects.

There will, inevitably, continue to be new directions, scopes, and scales of work on the interactions between animals and plants in marine habitats. These will transcend, underpin, and encompass the existing themes and concepts. Increased attention will have to be directed to the needs of experimental and sampling designs and the necessity to adopt sophisticated procedures to extract messages and coherence from variable and 'noisy' biological systems. The experimental tools needed for this are constantly being improved (there are examples throughout this volume). Increased rigour in their logical use will enhance work on many apsects of these topics.

Finally, it is impossible to predict with certainty what may lie ahead for the study of these interactions. New techniques, such as recent developments in molecular methodologies and in large-scale satellite mapping, will undoubtedly be brought into the studies, changing our understanding of, and our ability to understand, these fields. No one should make confident predictions about the future of ecological research. Nevertheless, these proceedings will have served an admirable purpose if they generate new insights, syntheses, and integration into what is clearly a vibrant and diverse field of marine research.

References

Branch, G. M. (1984). Competition between marine organisms: ecological and evolutionary implications. *Oceanography and Marine Biology Annual Review*, **22**, 429-593.

Chapman, A. R. O. and Johnson, C. R. (1990). Disturbance and organization of macroalgal assemblages in the Northwest Atlantic. *Hydrobiologia*, **192**, 21-34.

Foster, M. S. (1990). Organization of macroalgal assemblages in the Northeast Pacific: the assumption of homogeneity and the illusion of generality. *Hydrobiologia*, **192**, 21-34.

Stephenson, T. A. and Stephenson, A. (1972). *Life between tide marks of rocky shores*. W. H. Freeman and Company, San Francisco.

Underwood, A. J. (1985). Physical factors and biological interactions: the necessity and nature of ecological experiments. In *The ecology of rocky coasts*, (ed. P. G. Moore and R. Seed), pp. 171-90. Hodder and Stoughton, London.

Underwood, A. J. and Denley, E. J. (1984). Paradigms, explanations and generalizations in models for the structure of intertidal communities on rocky shores. In *Ecological communities: conceptual issues and the evidence*, (ed. D. R. Strong, D. Simberloff, L. G. Abele, and A. Thistle), pp. 151-80. Princeton University Press, Princeton, New Jersey.

Underwood, A.J. and Kennelly, S.J. (1990). Ecology of marine algae on rocky shores and subtidal reefs in temperate Australia. *Hydrobiologia*, **192**, 3-20.

Underwood, A.J. and Petraitis, P.S. (1991). Structure of intertidal assemblages in different locations: how can local processes be compared? In *Historical and geographical determinants of community diversity*, (ed. R. E. Ricklefs and D. Schluter). University of Chicago Press, Chicago. (In press).

Index

Standard benthic topics that do not appear as expected in this index may be taken to be so frequently mentioned in the text that page-listing is impracticable. Bracketed parts of multipage references indicate especially important pages.

abalone 299, 300, 304, 394
  Pacific Red 388
abiotic factors 2-3, 12, 34-41, 45,
    71-3, 87, 89-91, 93, 95, 137,
    139, 265-81, 326, 331-2, 377-8,
    395, 407, 456, 461-2, 466-8,
    495-6; *see also specific entries*
  desiccation 65-74, 95, 106, 453, 495
  in coral reefs 101-25, 213-30,
    319-33, 405-20
  in expulsion of zooxanthellae 530-2
  in 'gardening' 407, 411-14
  in *Posidonia* epiphytes 168 *et seq*
  perturbation of 511, 528-32
  Posidonia resistance to 167 *et seq*
  salinity 104, 193, 511
  space 34, 386, 394, 444, 445-6,
    447-8, 450-2, 467-8
  space in reefs 411-14
  substrate *passim*
  temperature 196-7, 511
  tides 90, 102-4, 110-2, 120, 293,
    296-7, 301-2, 436
  wave exposure *passim*, but
    especially 2-24, 195, 201, 240,
    306, 389
absorption efficiency (Ae) 172, 174,
    215, 217, 218-19, 224, 270, 272,
    273-4, 275, 327, 363-[372-3]-80
  see also assimilation efficiency
abundance *passim*
  effect of barnacle larval settlement on
    adult barnacles 386
  enhancement by barnacles 453
Acanthaster planci 124, 387, 396
Acanthopleura 495, 497
  brevispinosa 497
  granulata 497
Acanthozostera gemmata 115, 497
Acanthuridae 92, 105, 218-20, 222,
    341, 342, 343, 346-8, 350-1,
    355, 357, 414
Acanthurus 341, 342, 351
  dussumieri 341
  nigrofuscus 341, 346-8, 355, 356
  nigroris 341
acetogenins, in seaweeds 319-33
Acmaea 196, 450, 498
  digitalis 298
  insessa 295
  scutum 294
  testudinalis 295, 396, 479
  virginea 20
  see also ***Tectura***
Acmaeacea 481
Acmaeidae 481
Acrochaetium 370
Acropora 500, 520, 531
  acuminata 516, 523
  cervicornis 513
  palmata 527
Actinia equina 16
actinians 512
adhesion
  of algal propagules 376-7
  of faecal pellets 363-[376-8]-380
Agaricia 396
  agaricites 387, 392
  humilis 387, 392
  tenuifolia 387, 392
Agariciidae 392
Agarum cribrosum 36, 43, 44

Agarum cribrosum (cont.)
  coexisting with urchins 46
Aiptasia
  pallida 529
  pulchella 524
  tagetes 514
alanine 515, 518, 526, 527
Alaria
  esculenta 3, 43-4
  marginata 75
Alcyonidium 193, 203
  hirsutum 195, 200, 202
Alcyonium siderium 387, 392, 396
Algae
  as group 443-68
  blooms 39, 42, 51, 140, 341
  gardens 92, 133-41, 379-80, 405-20
  mats, *see* turf [algal]
  plains 94
  ridge 499, 502
algivory, *see specific organisms*
alkaloids
  absent from seaweeds 319-33
allelochemicals 375
allelopathic agents
  secondary metabolites as 331
Allorchestes 248
  compressa 243
  compta 245
amino acids 280, 511, 515, 525-6; *see also* nitrogenous excretion; protein
  derived metabolites in algae 319-33
ammonia/ammonium
  absorption 196
  concentrations 139, 141
  excretion 269, 277-9, 523-4
  in N_2 fluxes in symbiosis 523-7
  under urchins 410-11
  uptake of DIN as 523-4
Amphibola crenata 139
amphineurids 493-4, 497
Amphipoda/amphipods 36-7, 39, 42, 44, 87, 91, 151, 153, 157, 165, 170-1, 174-5, 192-3, 196, 198, 203, 223, 235-43, 245-6, 248-50, 252, 290, 295, 297, 299, 324, 368, 369-70, 375-6, 379-80, 387, 433, 484-5
  and seaweed defences 319-33
Ampithoidae 248
Ampithoe* 198, 236, 248, 396
  humernalis 242
  lacertosa 237, 247
  longimana 153-5, 247, 249, 321-2
  marcuzii 247, 249
  mea 247
  ramondi 237, 247, 433
  simulans 241
  valida 247, 322
Anemonia viridis 524
annuals 45, 49, 50, 113
Anthozoa 511-12, 518, 524-5, 527
antigens 514
  see also biochemical cues
Antithamnion 370
apical cells 366
  see also meristematic cells of algae
Aplodactylidae 342, 343
Aplysia
  californica 392, 396
  winneba 94
Aplysiidae 393
Apolemichthys xanthopunctatus 348
apparency theory, *see* plant apparency model
Arbacia 321
  punctulata 322
Archaeolithophyllum 480, 482
Archosargus probatocephalus 348
Arenicola 135
Argopecten irradians 268
aromatic compounds
  in seaweeds 319-33
Arthropoda 235, 246-8, 253
ascidians 153
ascoglossans 324, 484, 486
Ascophyllum 11, 21, 195, 253
  nodosum 3, 23, 35, 36-41, 43, 50, 192, 199, 464
  recruitment 52
  assemblages 42, 45, 88, 89, 61-78, 93, 395-6, 425-38, 467
  composition model in

*Note: The genus *Amphithoe* has hitherto been rendered with and without the first 'h', although the correct form is now accepted as in this note. Both forms occurred throughout relevant submitted texts for this volume; we have accepted the majority rendering without the initial 'h', since there is no confusion as to the organisms concerned.

'gardens' 410-14
diatoms 135
epiphytes on *Posidonia* 168-9
  in disturbances 461-2
  in gardening 405-20
  in shallow coastal habitats 443-4
assimilation efficiency *passim*, but,
  of grazers 215, 363-80
  experiments on 272, 327, 372-3
  see also absorption efficiency
Asterias 43, 444
  rubens 36-7
asteroids 290, 387
asymmetrical analyses 456-61
ATP, in protein turnover 280-1
attachment, of algal propagules 363-80
attractants, *see* biochemical cues
Audouinella purpurea 366, 375-6
Aulacomya ater 272

Bachelotia antillarum 97
bacteria/bacterial films 136-9, 156, 168,
    176, 178, 200, 265-[270]-81,
    349, 355-6, 388, 406, 495, 525
  agar-digesting bacteria 356
  Vibrio 356
  see also microbial effects; *individual species*
Balanus
  glandula 4-5, 434
  improvisus 153-5
Balistidae 342
Bangia 37, 432
  fuscopurpurea 370
Bangiales 367-8
barnacles 2-3, 7, 9-17, 39-40, 42-3,
    88, 94, 103-6, 109, 114, 117-20,
    122, 124, 298, 386, 432-7, 444,
    446, 452-3, 465, 496
  as refugia 43
  larvae 452
barrens 45, 47-8, 50, 391
  detrital shores 133
  in grazed reefs 416
beachrock 497-9, 501
Bembicium auratum 139
Bifurcaria bifurcata 7
biochemical cues (including attractants,
    inducers, repellants, stimuli in
    general) 154-6, 199-201, 268,
    294, 304, 386-9, 392-6

to larval settlement 385-96
polysaccharide inducer 392
in zooxanthellae infection 514
bioerosion 493-505
  rates of 494, 504-5
biological abrasion 493-6
biological corrosion [493-5]-505
  see also blue-greens
biomass 2, 5-9, 14, 18-19, 21, 48, 68,
    75, 93, 109, 112, 138, 140, 150,
    156-8, 165, 177, 191, 240,
    242-3, 249, 251, 265-81,
    363-80, 458-68
  algae on coral reefs 213-30
  animals in *Posidonia* 177-8
  fish on coral reefs 213-30
  fish in *Posidonia* 170-1
  in 'gardening' 405-[414-7]-420
  of symbionts in coral reefs 514
  partitioning 165
  photosynthetic 165
  phytoplankton 265-81
  Posidonia 166-7, 171-2, 177
  subsurface 165
  zooxanthellae, biomass
    control 528-530
  see also primary production, standing
    crops
biotic factors *passim*, but especially 12,
    41-5, 87, 89, 91, 139, 167,
    213-30, 265-81, 319-[332]-33,
    363-[373-4]-80,
    405-[411-14]-20, 495
  arrival of excessive numbers 446
  in *Posidonia* epiphytes 168
bioturbating fauna *passim*, but
  especially 139, 141, 214, 319-33,
  363-80
bivalves, *see* Mollusca
'bleaching'
  expulsion of, or pigment loss from,
    symbionts due to host 511,
    530-2
Blenniidae 341, 342, 343, 348, 355
  in reefs 218-9, 220, 222, 238, 297
Blidingia 9, 13, 109
  minima 115
blue-greens 2, 37, 39, 87, 91, 93,
    114-15, 134-5, 140, 349, 366,
    375, 388, 408, 411, 435-6,
    493-5, 496, 504-5
Boleophthalmus 137

Bossiella plumosa 70–1
Brachyura 248
bream 170–1
brittle stars 217
brominated phenols 23
bryozoans 45, 94, 170, 175, 193, 195–7, 200–3
buccal morphology
  of grazers 363–[372–3]–80
  in 'gardening' fish 408
  see also feeding apparatus
Bullia digitalis 299
burrowing habit 94, 133–41, 501
butter-fish 343, 353

C/N ratios
  in 'gardens' 408
  Mollusca 279–81
  of algae from symbioses 529
  Posidonia 167, 171–3, 176
caesionids 215
calcareous/calcified algae, *see* corallines
Calcinus 456
Callianassa kraussi 139
Callinectes sapidus 156
Calliopiidae 248
Calliopius 248
Callithamnion halliae 393
callus, of frondose Bangiales and other algae 367–8
Calothrix 115
Cancer 43–4
canopy 3, 9, 11, 16–18, 23, 39–41, 45, 65, 68, 74–5, 89, 103, 105, 110, 113, 116, 119–20, 149, 154, 427
  competition 462–3
  Posidonia 166, 170, 177
  removal 447
  Ulva 375–6
Caprella 248
Caprellidae/Caprellidea 237, 240, 247, 248
carbohydrates
  structural 171–3
  as energy sources 277, 281
carbon budgets 135, 197, 265–81, 326–7, 511–32
  carbon excretion 521
  DOC (dissolved organic carbon) 201–2, 217
  fluxes in coral reefs 213–30

  fluxes symbiont/host 515–23
  POC (particulate organic carbon) 217
  symbiotic metabolism 518
  see also energetics
carbon compounds 493–[494–6]–505, 511, 515–18
  see also specific substances – alanine, lipid, phosphoglycerides, protein, triglycerides
Carcinus maenas 36–7, 41, 51
Caridea 248
Carpophyllum 343
carpospores 241, 374–6
Cassiopeia 512
  frondosa 513
  xamachana 513
Cebidichthys violaceus 342, 347–8, 356
Cellana 451–2
  tramoserica 108, 294, 433, 436, 448–9, 454, 462
Celleporella hyalina 201
Centroceras 407
  clavulatum 370
Centropyge flavissimus 348
Centrostephanus
  coronatus 301
  rodgersii 296, 301
Ceramium 407
  rubrum 370
Cerastoderma edule 266
cerithoid snails 137
Chaetomorpha 365, 369
  antennina 95–6
Chama 106, 117
Chanidae 342, 348
Chanos chanos 348
Chione stutchburyi 271
Chiton 373, 497
  granosus 369–71
  marmoratus 497
  squamosus 497
  tuberculatus 497
chitons 65–7, 69–70, 73, 75, 115, 290, 294, 387, 432, 444–5, 451, 496
  feeding 372–3, 496
Chlamydomonas 524
Chlorella 525
chlorophyll content 6
  in reduced-light zooxanthellae 520
  in sediment 137, 139
Chlorophyta 191, 238, 341–2, 346, 348, 366–70, 386–8, 393, 432–7,

484-5, 495
chloroplasts 516-7
Chondria 248, 329
Chondrococcus hornemanni 393
Chondrus 36-7, 39-43, 47, 248, 393, 463
  crispus 39, 107, 447, 464
Choromytilus meridionalis 272
Chthamalus 11, 432-3
  dalli 434
  dentatus 95-7
  stellatus 105
ciliates 136
cirratulids 503
Cittarium pica 498
Cladophora 202, 369
clams 42, 265-81
Clathromorphum 43, 196, 480-1
  circumscriptum 479-80
clearance experiments, *see* exclusion experiments
Clibanarius 456
climate/climatic seasons/shifts
  effects on biota 43, 87, 95, 97, 104, 108-109, 111-2, 119-23, 135, 177, 204, 265 *et seq.*, 296
climax vegetation 93
  of/on *Posidonia* 166, 168
Cliona 504-5
clionid sponges 502
Cnidaria 511-40
cobbles, *see* rhodoliths
cockles 266
cod 48, 51
Codium dimorphum 369
coevolutionary processes 62, 76-7, 118, 174-5, 204, 324-8, 379-80, 406, 477-91
  evolving digestive mechanisms 356
  evolving herbivory 357
  evolving predator/prey relations 249-50
  in 'gardening' 417-20
Colisella 365, 432-3
  ceciliana 369-71
  purpureus 303
  scabra 299
  zebrina 369-70
community structures 2, 4, 8-9, 14-15, 23, 34, 45, 47, 50-1, 63, 114-16, 117, 120-1, 149, 158, 191, 196, 203, 235-6, 240, 244, 246,

252-3, 344, 385, 395
animals on *Posidonia* 165
coral reef fish 214, 405-420
evolution of feeding specialization 324
  in 'gardens' 411-14
  planktonic 364
  Posidonia system 165-[168]-78, 173-4
stability of 51
competition *passim*, but especially 443-68
  in coral reef 'gardening' 405-[411-14, 417-20]-20
  importance of 456-7
  influencing lower limit of *Fucus* 107
  in macroalgae 462-3
  in microalgae 448
  in *Posidonia* epiphytes 168
  intensity of 465-7
  interference by barnacles 452-3
  intraspecific 453-61, 468
  limitations on data 443-68
  philosophy of 443-68
  symmetry/asymmetry 454-5, 458-61
Condylactis gigantea 516
copepods 136, 178, 217, 223, 235, 387
*Corallina (officinalis/*spp.) 11, 251, 297, 306, 370, 479
corallines 21, 22, 36, 43-5, 87, 91, 93-4, 96, 107, 113-14, 196, 214, 220, 342, 351, 386-92, 394, 480-2, 484-5, 488, 499
  articulated 392, 483; *see also Corallina, Jania*
  coralline barrens 45, 48-9
  crustose corallines 20-1, 49, 64, 66-7, 74-6, 93, 105, 122, 190, 214, 220, 295, 370, 372, 409, 412, 416-17, 479-80, 482-3, 493, 505
  on *Posidonia* 168
  see also barrens, *specific entries*
corals irregularly *passim*, but see 241, 511-32
  as abrasion surfaces 493
  agariciids 392
  ahermatypic 524
  block transplants 502, 563
  expelling zooxanthellae 514, 530-2
  hard corals 213-[214]-30
  heads 410, 502-3

corals irregularly *passim* (*cont.*)
  killed by damselfish 411
  lagoonal 499
  live 499
  mucus secretions 518
  removing N_2 from seawater 525
  rubble in erosion 497, 499
  settlement inhibited 414
  soft corals 387
  spat mortality 414
  storing lipid 517-18
  symbiosis in 511-40
  temperature effects on 530-2
  see also specific entries
coral reefs 87-8, 92, 102, 124, 136, 213-30, 240, 289, 319-33, 344, 405-[410-14]-20, 426, 431, 433, 435, 493-501, 511-14
  cryptofauna 502
  flats 501-3
  front 503
  lagoon 501-2
  leeward reef 503
  patch reef 503
  slopes 501-3
  symbiosis in 511-40
Corbicula fluminea 273
Corophiidae 248
Corophium
  arenarium 139
  volutator 139-40
Coryne uchidai 393-4
crabs 17, 42, 48, 66, 69, 94, 106, 112, 115, 122, 156, 235, 238-9, 241, 247, 249-50, 290, 296, 425, 432
  ocypodids 137
  and seaweed defences 319-33
  see also genera/species, hermit crabs, green crabs
Crassostrea
  gigas 273
  virginica 269, 272
Crepidula convexa 153-5
Crustacea 43, 139, 170, 223, 225, 235-6, 290, 444
crustose algae 19, 22, 65, 69, 71, 75, 103-7, 111-15, 117, 119-22, 237, 289-90, 295-6, 304, 342, 434, 436, 479-80, 482-3; *see also* corallines, peyssonellids, *Corallina*
cryptic habitats

crypsis in epiphytes on algae 196
cryptofauna in algae 190-1
cryptofauna in coral reef 502
cryptofauna in 'gardens' 411, 413
  in *Posidonia* 177
Cryptopleura 66
Ctenidia 269
Ctenochaetus striatus 343, 349-51
cues, *see* biochemical cues
cunner 44, 48
Cyanobacteria, *see* blue-greens
Cymadusa 248
  compta 153-5, 246
  crassicornis 237
cymopol 320
Cymopolia 320
cystocarps, of red algae 241, 367, 372, 374
Cystoseira 21
  osmundacea 68, 434

damselfish
  giant blue 407
  see also Pomacentridae, *specific organisms*
Dasya 393
Decapoda 152, 165, 170-2, 248, 251, 484
defences
  against digestion 363-80
  biochemical & morphological in algae 14, 23, 42, 49, 50, 75-7, 93, 113-4, 140, 171-2, 198-201, 204-5, 243, 246, 252, 319-37, 339, 357, 374-5, 393, 477-8, 484, 486-7, 542
  compounds 319-37
  costs 326-7
  deterrents in *Posidonia* 167, 171-2
  in 'gardening' model 414-7
  see also strategies, secondary metabolites
density
  passim, but *see* 443-68
  in populations 495-6
  as concept 448-9
  of *Cellana* 451
  of recruits 386, 391
  intraspecific 453-4
  experiments 456-8
  determining effect of

competition 465–7
  in experimental design 458–61
  of colonizers 462
deposit feeders 133–[137, 139]–41, 165, 265–81
  subsurface 178
  surface 178
Derbesia vaucheriaeformis 433
dessication 69, 72–4, 95, 106–7, 110–13, 120, 156, 194, 196–7, 201, 240, 296–7, 301, 431, 436, 495
  shelter by algae from 453
  spore resistance to 376–7
Desmarestia 36, 43–4, 67, 75–6, 91
deterrents, *see* defences
detrital feeders, *see* detritivores
detritivores 175–8, 221, 242–6, 248, 343, 351, 355, 406; *see also* detritus
detritus 5, 38, 94, 133–41, 148–9, 190, 201, 213–30, 242–6, 248, 251, 265–81, 343, 351, 355
  high suspended load on reefs 521
  in coral reefs 213–20
  in *Posidonia* ecosystem 165–6, 174–6
  production of 224
  see also detritivores
Devalerea 45
Dexamine spinosa 171, 174
Diadema
  antillarum 94, 251, 321–2, 407–8, 410, 414, 500, 505
  savignyi 499
diatoms 9, 13–14, 66, 108, 120, 133–41, 238, 248, 267, 290, 303, 343, 349, 351, 388, 406, 432–3, 436
  dimensions 135
  epipsammic 133–[140]–41
  epipelic 133–[140]–41
  filamentous 435
  growth 139
  numbers 139, 140
  on *Posidonia* 168, 171, 174, 178
dictyols 321, 322, 328, 330
dictyopterenes
  A/B 328
Dictyopteris 328–30
  delicatula 328–30
Dictyota 105, 192, 198, 241, 295, 321, 324, 326–31, 435

  bartayresii 328–9
  cervicornis 328
  dichotoma 249
  divaricata 328–9
  furcellata 328
  indica 328
  linearis 328
Dictyotaceae 319
Dictyotales 320, 322, 330
  in mimicry 328–9, 330–1
dietary changes ('Andean plane crash') 244–6
digestion
  seaweed spore/fragment survival of 363–80
  see also absorption, assimilation
Dilophus 328–9, 330
  alternans 329
DIN (Dissolved Inorganic N_2) 523–8
dinoflagellates 10, 135, 511–12
Diplodus holbrooki 322
Diptera 235–63; *see also* larvae
Discurria 481
dispersal 395
  as a life process 444
  of *Haliotis* larvae 388–90
  of seaweed propagules 77–8, 375–6
  of zooxanthellae 512
DOC (Dissolved Organic Carbon)
  in *Posidonia* 167
distributions *passim*, especially 2–24, 33–[35–9]–51, 62–78, 88–97, 107–8, 112, 148
  barnacle recruits 386
disturbance *passim*, for concept refer to 33–[34–8, 42]–52, 76, 78, 119–21, 443–[444, 461–2]–68
diterpene alcohols 321
diversity 2, 21, 88, 93, 177, 204, 236, 463, 468
  $\alpha/\beta/\gamma$ in 'gardens' 405–[412–13]–20
  enhancement by barnacles etc. 453
  in model of 'gardening' effects 414–17
  of diatoms 137
  of *Posidonia* epiphytes 169
division
  algal division under N_2-deficiency 525
  binary fission of zooxanthellae in hosts 513
  of symbionts 514

Docoglossa 372-3, 482; *see also specific organisms*, e.g *Collisella*
dog whelks 12, 15, 42-3, 119, 298; *see also Nucella, Thais*
dolastanes 328-9
drift 42, 45, 50, 74, 248, 299-300, 303-4
Dynamene 202
   bidentata 245

Echinaster sepositus 171
echinoderms 38, 170-2, 178, 387, 391
echinoids 19, 165, 290-1, 304, 451, 485
   basal spines and teeth 496
   burrowing 493-4, 496, 499
   grazing 499
   Pacific 495
   see also sea urchins
Echinometra
   lucunter 91, 107, 499
   mathaei 499-500
   oblonga 500
   viridis 414
Echinostrephus
   aciculatus 500
   molaris 300
Echinothrix diadema 499
Echinus 19
   esculentus 19-20
Ecklonia 194, 243, 295, 303-4
   maxima 377
   radiata 343
ecotrophic efficiency, *see* consumption efficiency
Ectocarpaceae 366
Ectocarpus 369
Egregia 66, 194, 295
   laevigata 434
   menziesii 483
Eisenia arborea 462-3
El Niño (1982-3) 242
   coral bleaching from 530
elatol 320
Electra pilosa 193, 195, 197, 202-3
Endocladia 66
   muricata 68, 432, 434
endophytism 89, 482; *see also specific organisms*
endozoism 89; *see also specific organisms*
energetics 1-9, 47, 272, 279-81
   in algal 'gardens', 405-[411-17]-20
   coral reefs 213-30, 319-33
   costs of defence energetics 326-7
   in symbiosis 518-23
   net energy balance 275-7
   nutritional values 173
   Posidonia 165-7, 171-2, 177
Enteromorpha 9, 13, 21, 36, 38, 41, 96, 106, 120, 245, 247, 249, 251, 365-6, 369, 371, 374, 377, 432-3, 464
   compressa 365, 376
   intestinalis 433
Entocladia 369
environmental stimuli 386, 392
   cue-associated settlement substrata 394-6
enzymes 319-33, 345-6, 363-[373-4]-80
   acetyl coenzyme A 518
   in assimilation of ammonia 525-7, 529
   cellulolytic enzymes 345-6, 373
   fuconases 373
   lipases 373
   NADP-glutamate dehydrogenase 525-8
   proteases 373
   ribonucleases 373
   transaminase activity 525
   transformation triacylglycerol-wax ester 516
   zooxanthellae isoenzymes 513
Eogammarus confervicolus 243
ephemerals 2, 12-14, 17, 21, 23, 35-40, 42, 47, 50, 64, 69, 71, 113, 120, 204, 238, 240-1, 244, 247-8, 299, 386-7, 436, 447
epifauna/epiflora *passim*, especially
   intertidal 34-43
   sublittoral 43-9
epipelic microalgae 139-40
epiphytes/epiphytism 89, 118-119, 148, 156-8, 172-3, 175-8, 189-205, 240-2, 244-6, 249, 254, 303, 482
   zonation and presence on seagrasses 167, 174, 237
epipsammic microalgae 140
epizoism 89
Epulopiscium fishelsoni 355
Erichsonella attenuata 153-5

Index

Erichthorius braziliensis 299
Erginini 481
erosion
  littoral 496, 505
  rates 497–504
Erythocladia 370
Erythrotrichia 370
escape mechanisms, *see* defences, defensive compounds
Eucidaris thouarsii 499
Eudiplogaster pararmatus 137
euglenophytes 135
Eupomacentrus
  acapulcoensis 414
  lividus 409, 414
  nigricans 348, 411
  planifrons 348, 411, 414, 433
Eusiridae 248
Evechinus chloroticus 300–1, 304
evolution of relationships, *see* coevolutionary processes
exclusion experiments 9–10, 40–1, 71, 107, 109, 139–41, 238–40, 250–2, 408–14, 436, 452, 496
  clearance experiments 19–20, 46, 67–9, 96–7, 120, 447–8, 452
Exogoninae 175
experimental studies
  on bioerosion etc 493
  carbon isotope analysis 246
  Caribbean coral recruitment 392
  conceptualized 445
  in competition 443–68
  density etc 456
  design 23–4, 34, 244–6, 305–7, 390, 396, 455–61, 467–8, 542–3
  displacement experiments 305–7
  macroalgal competition 462–3
  on numbers/densities of limpets 451, 455
  on Pacific coral temperature limits 530–2
  in regulation of symbiont populations 528–30
  replication 455–6
  reproductive output 455–61
  symbiont/host relationships 515–30
  transplantation 393–4, 493–4
  see also clearance/exclusion experiments
extinction of solenopores 485

Ezo 481

fabricinids 503
faecal pellets 140, 172, 174, 224, 363–[368]–80
  as sedimentary products 494
  contents 365–7, 369–70, 371–9, 496
faeces 269–81; *see also* pseudofaeces
faeces feeders 216–17; *see also* pseudofaeces
fatty acids
  in phytoplankton 279
  in symbiosis 515–6, 518
  VFAs (volatile fatty acids) 353–4, 356
fatty alcohols 516
feeding apparatus
  structure in grazers 372–3
  see also buccal morphology
feeding rates
  in molluscs 274–7, 449
filamentous algae (all groups) 36, 39–40, 93–4, 103, 105–6, 242, 247–8, 251, 341–2, 366, 407–8, 433–5, 483
  in 'gardening' 405–20
  on *Posidonia* 168–9
  in regeneration 366
  radula effects on 372–3
fish
  algivorous 87–8, 91–3, 137, 368–70
  and seaweed defences 319–33, 364
  'beaks' of 106, 339
  digestion by 339, 345–8, 363–80
  distribution in Atlantic 92–4
  generalists 165
  microalgal feeders 137
  of *Posidonia* beds 170–1
  omnivorous 92, 214–30
  on coral reefs 213–30, 319–33, 405–20
  teeth 339
  teleosts 431 *et seq.*
  territorial, in 'gardening' 93, 405–20
  see also specific organisms
Fissurella 498
  crassa 365, 369–70
  limbata 369–70
flagella
  on green algal spores/protoplasts 367–8

flagella (cont.)
  in zooxanthellae 512
flagellates 133-41, 355-6
flounder 44
Flustrellidra 193, 201, 203
  hispida 195, 197, 200, 202
foliose algae 3, 39, 87, 93-4, 366, 451-2, 454
  radula effects on 372-3
food web/chains *passim*, but *see*
  coral reefs, especially 224-7
  'food-driven self thinning' 268
  nutrient supply, *passim* but especially 72, 109
  searching, visual/non-visual 319-33
  small 137-8
foraging behaviour, *passim* but especially 291-307
Foslietta 168, 481
free space 36-7, 42, 50, 88-9
  competition for 89
  recolonization 96
Fucales, defences in 325-6
fucoids 1, 2, 4-17, 21, 23, 37, 39, 41-3, 103-5, 119, 190, 199, 201, 203, 242, 295, 343, 387, 464, 482-3
Fucus 2, 7, 9, 11-17, 21, 39-40, 66, 75, 107, 113, 202, 237, 245, 247-8, 373, 447, 453, 464
  distichus 35, 37, 41, 198, 332, 434
  *d. evanescens, see also **F. evanescens*** 35
  *d. edentatus, see also **F. edentatus*** 35, 41
  *edentatus, see also **F. distichus*** 192
  *evanescens, see also **F. distichus*** 35, 37, 41
  serratus 3, 11, 16-7, 192-3, 195-8, 200-3, 393
  spiralis 3, 7, 11, 16, 21, 35, 37, 39, 41, 193
  vesiculosus 7-9, 11-12, 16, 35-9, 77, 192-3, 239, 244, 432
    v. var. *evesiculosus* 3, 10
fungal effects 89, 176, 495

GABA (gamma-aminobutyric acid) 389-92
gametes (algae)
  on coral reefs 213-30
  in digestion survival 363-[366-7, 371-2, 377-8]-80
  production in symbiosis 528
gammarids 236-7, 245, 248
Gammarus 248
  lawrencianus 245
  mucronatus 322
  salinus 375
gardening, *see* algal gardens
garfish 353
Gastrochaena 501
gastropods 8, 15, 17, 20, 41-2, 64, 75, 94, 101, 106-8, 110, 115, 123, 137-40, 153, 165, 173-4, 178, 198, 203, 290, 292-3, 295, 297, 303, 305-6, 322, 332, 387, 444-5, 451, 461, 484-5, 493-4, 497; *see also specific organisms*
Gelidiopsis 407
Gelidium 66, 109, 118
  chilense 370
  lingulatum 371
  micropterum 406
  rex 193
germination, of algal propagules 366-7
Geukensia demissa 267, 269, 271-2
Gibbula 173-4
  ardens 165, 174
  umbilicaris 165, 174
Gigartina 66-7, 121, 432
  canaliculata 434, 448, 461
Girella
  fimbriata 345, 350-1, 358
  nebulosa 345, 351
  nigricans 356
  tricuspidata 345, 348-9, 357
Girellidae 341, 342, 343, 345, 350-1, 356-7
global warming 511, 532
Glossophora 330
  kunthii 369
glutamate 525-7
  dehydrogenase 526
glutamine 526-7
  Glutamine 2-oxoglutarate amido transferase 526-7
  Glutamine sythentase 526-7
Gobiidae 341, 342
'grab and hold' algae 461
Gracilaria 109, 133
  asiatica 245
Grandidierella 248
Grapsidae 65, 248

Grapsus grapsus 91
green crabs 41–3
  recent invaders (NW Atlantic) 43
  see also **Carcinus**
'greenhouse effect', *see* global warming
growth *passim*, but especially 443–68
  of algae under N_2 deficiency 525
  of algae with season 108–109
  experiments 455–61
  shallow depth energy required for 521
  of symbionts 514, 528–30
  of zooxanthellae in hosts 513, 528
growth efficiencies
  of algal propagules 366–7
  of partly-digested algal tissues 364–6
  of reef fish 221
guilds
  feeding 174–5
  functional 177
gut analyses, *see* stomach contents
Gymnogongrus 374
  furcellatus 370
Gyrodinium aureolum 10

habitat *passim*, especially
  complexity 177
  fidelity 174
haemulid fish 410
halfmoon fish 356
Halichoeres bivattatus, *see* wrasse
Halidrys dioica 462–3
Halimeda 326, 331, 333
Haliotis 373, 389–90, 396
  cracherodii 300, 389
  discus hannai 300, 389
  laevigata 387
  rubra 389
  rufescens 387–92, 394
  scalaris 389
  sieboldi 300
Halodule wrightii 157
halogens, in seaweed secondary compounds 319–33
Halosaccion
  glandiforme 432
  see also **Devalerea**
harpacticoids 178
harvesting seaweeds 47
Helcion (Patina) pellucidum 20, 192, 295

Hemichordata 137
Hemiglyphododon plagiometopon 409
Hemiramphidae 342
herbivory, *see specific organisms*
hermit crabs 94, 290, 324, 456; *see also* crabs
Hermosilla azurea 356
Herposiphonia 370
  heringii 406
Hiatella
  arctica 501
  gallicana 501
Himanthalia 3, 200
Holothuria tubulosa 171
holothurians 170
  importance in detritus recycling 176
Homarus americanus 43
horse mussel 46; *see also* **Modiolus**
Hyale 193, 237, 241, 247–8, 369–70, 396
  frequens 237
  hirtipalma 241, 374–5
  macrodactyla 322
  media 367, 374
  nilssoni 245–6, 295
Hyalidae 248
Hydra 525; *see also* hydroids
Hydrobia 136
  ulvae 8, 136, 139–40
  effects on topography 140
hydrobiid snails 139
hydroids 170, 175, 387, 393–4; *see also* **Hydra**
Hypnea 109, 241, 297
  spinella 240, 433
Hyporhamphus melanochir 353

ice-scour 34–5, 38, 44, 51, 120
Idotea 36–7, 44, 192–3, 202, 244, 248, 297
  baltica 242, 245, 247
  baltica basteri 173
  hectica 173
  wosnesenskii 241
Idoteidae 173, 248
Ilyanssa
  obsoleta 139–40
inducers, *see* biochemical cues
insects 66, 321, 323
insecticides 238, 240

Iridaea 66, 297
  boryana 113
  cordata 241, 436
  cornucopiaea 297
  flaccida 436
  laminarioides 237, 241, 367, 370-1, 374, 375-6
  splendens 70-1
iridescence blue, in Dictyotales 330-1
irradiance, *see* light, abiotic factors
Ischyroceridae 248
isolaurinterol 320
Isopoda 8, 39, 42, 153, 157, 165, 170, 173-4, 192, 235-7, 239, 241-3, 245, 247-50, 253, 290, 297

Jania 118, 435
Jassa 248
jellyfish 249

Katharina tunicata 73, 387, 390
KCl, mimicry of inducer 393
kelp (incl. kelp forest) 1, 3, 20, 23, 37, 43-6, 48-9, 63-6, 68, 74-5, 103-4, 136, 192, 200-4, 243, 245, 269, 272, 295, 300, 304-5, 325-6, 389, 391, 435, 482-4
  competition by 45
  defences in 325-6
  downward extensions 46
  importance to lobster 48
  refuge beds 45
kelp fish 242
keystone predation 34, 46-7, 463-4
Kvaleya 481
Kyphosidae 341, 342, 343, 353-7
Kyphosus
  cornelii 353-5
  sydneyanus 343, 353-5

Labridae 170, 217
Lacuna
  pallidula 203
  vincta 37, 41, 44, 192-3
Lagodon rhomboides 151, 322
lagoons
  brackish 269
  coral 499-500, 501, 503
  tropical 496

Laminaria 4, 9, 19, 20, 44, 66, 190-4, 295, 303, 343, 373
  digitata 3, 17, 43, 46
  hyperborea 3, 18-20, 195
  longicruris 43
  pallida 377
  saccharina 3, 43, 46, 192, 201
larvae (including insect larvae) 18-19, 65-6, 119, 124, 147, 149, 152-4, 156, 158, 199, 200, 385-403
  dipteran larvae 235-6, 238-9, 247, 253, 290, 304, 425, 432
  swimming behaviour 388-9
  see also metamorphosis
late successionists, marine algae 363-[369-72]-80
Laurencia 94, 319-20, 326, 329, 331, 393
  pacifica 393-4, 434
  poitei 393
Lepidochitona dentiens 432
Leitoscoloplos fragilis 139
Lembos 248
Lessonia 194
  nigrescens 369
Lesueuria 481
life histories
  heteromorphic 113
  of algae 38, 41, 363-80
  of shallow water marine organisms 444
light 18-19, 21, 66, 68, 73, 75-6, 158, 250, 392, 427, 495, 511, 518-24, 528
  affecting regulatory mechanisms 511-40
  as competitive requirement 447-8
  illumination 495
  insolation 495
  photoadaptation in zooxanthellae 520
  relation to photosynthesis 519-20
  as zonation factor 45
  see also abiotic factors
Ligia 247-8
  dilatata 243
  gracilipes 91
  oceanica 237
Ligiidae 248
limiting resources, *passim*, especially 443-68

limpets 2–17, 21, 39, 64–9, 71, 73–4,
      91, 106, 108, 110, 113–6, 123,
      192–4, 196, 203, 238, 291–2,
      294–9, 301–4, 425, 432–7, 444,
      446, 451–2, 454–5, 477, 479–84,
      486, 488
  in 'gardening' 405–20
lipids 23, 511, 515–18, 521–2
  as energy sources 277
  lipogenesis 516
Lithophaga 501
  lithophaga 501
  nasuta 501
Lithophyllum 389, 481
Lithoporella 481
Lithothamnia 11, 103, 105, 120
Lithothamnion 43, 104, 304, 389–91,
      480–1
Lithotrya 501
Littorina 3, 5–6, 8, 10, 13, 68, 137
  granosa 91
  irrorata 296
  littorea 7, 10, 12, 17, 23, 33, 35–44,
      50–1, 107, 110, 293, 303, 432,
      463–4
    distribution NW Atlantic 35
  mariae 198, 203
  meleagris 498
  neglecta 7, 13
  neritoides 498, 504–5
  obtusata 4, 36–9, 192, 198–9, 295
  peruviana 369–71
  planaxis 498
  punctata 91, 95–6
  saxatilis 36–9
  scutulata 198, 303, 498
  sitkana 198
  unifasciata 293, 302, 306
    upper intertidal foraging 40
  ziczac 498
littorinids 2, 4, 16–7, 23–4, 37, 64–6,
      73, 75, 87, 91, 114, 137, 139,
      199, 303
  pathogens of 50
lobsters 43–4, 46–9, 51
  as keystone predators 46
  kelp relations 48
  long-term changes 51
  predation of urchins hypothesis 47
  production 48
  seaweeds as cover for 49
lobster (rock) 243

Lottia 432, 450, 480–1
  alveus 483–4
  gigantea 71, 298–9, 406, 450, 452
  incessa 483
Lottiidae 480
luderick 345, 348; *see also* **Girella**
Lysianassa pilicornis 171
Lytechinus anamesus 68

Macoma balthica 273
Macrocystis 67, 373, 377
  angustifolia 377
  pyrifera 68, 193, 242
Macrophthalmus 137
Majiidae 171, 173, 248
maldanids 407
mangroves 102, 136, 139
  red mangrove 243
Margarites 44
Mastocarpus 37, 39–40, 66
  papillatus 68, 77, 434
  stellatus 39
Mastophora 481
Mastophoropsis 481
mats
  Posidonia ('matte') 177–8; *see also*
      algal mats, turf (algal)
mechanical stimuli 386; *see also* cues
Medialuna californiensis 356
Melichthys niger 92–3
Melita 248
Melobesia 481
Melobesioideae 480
Membranipora 193, 195, 201
  membranacea 203
Mercenaria mercenaria 269, 274
Merismopedia 140
meristematic cells
  in digestion 364–6, 371–2
  see also apical cells
Mesophyllum 481
  insigna 304
Metamastophora 481
metamorphosis 388, 390–1 *et seq.*; *see*
      also larvae
  of invertebrate larvae 386–7
microbial effects 4–7, 23–4, 89, 137,
      200, 217, 265, 272–3, 346, 348,
      353–7
  degradation of faeces 217, 265
  epigrowths on detritus 176

microbial effects (*cont.*)
  Posidonia resistance to
    decomposition 167
microscopic algae/stages 133–41, 174,
    292, 300, 303, 305, 388, 488–9,
    454, 462, 484, 542
  algal film 5–6, 8
  grazed 451, 461–2, 467–8
  in coral reefs 214, 220, 524
  productivity 419
  see also unicellular algae
Microspathodon dorsalis 407, 409
Mictyris 137
migration 177
mimicry Batesian, Mullerian
    319–[328–31]–33
Mithrax 248
  sculptus 240
mobility, meso-herbivores 236–8, 477
models
  algae-herbivore trophic specialization
    486
  'arms race' 477
  behaviour patterns 306–7
  body size vs grazing range 487
  community factor predictions 121–3,
    125
  diffusion – depletion of ammonia 524
  extinction model 120
  fauna in seagrass 152
  of fish production 224–7
  general model of 'gardening' 414–7
  of colonization by dominants in
    patches 461–2
  of grazing pressure in 'gardens' 410
  of herbivore impact 111–3
  of interactions and defences 324–8
  macrophyte production 157–8
  pattern explaining metamodels 542
  ration-based model 48
  reciprocal adaptive radiation
    model 481
  resource availability model 325–6
  successional models 426–41, 435, 437
  variation in grazer importance 72
  see also trophodynamic modelling
Modiolus modiolus 46
  communities 46
Molgula manhattensis 153–5
Mollusca 43, 87, 91, 105–14, 121–2,
    165, 170–1, 173–5, 218–9,
    222–3, 235, 253, 265–81, 290–2,

  368–77 *et seq.*, 387, 436, 444,
    485, 511–2
  bivalves 15, 20, 103, 105, 117–19, 482
  rock-boring bivalves 493–4, 496, 501
  see also specific groups/organisms
monkeyface prickleback 342, 348, 356
Monodonta turbinata 504–5
Montastrea
  annularis 519–20, 531
  cavernosa 520
moon snails 42
Mopalia muscosa 387, 390
mortality 4, 13, 63, 66, 73, 104,
    116–117, 119–20, 152, 158, 194,
    200–3, 251, 295, 297–8, 386,
    388, 396, 427–9, 486, 531–2
  experiments 455–61
mucopolysaccharides, *see* mucus
mucus 4, 133–41, 274, 431, 518
  glands 518
  loss of fixed carbon as 523
  secretion 528
mucus-lipid 511, 518, 521, 523
mudskipper 137
mud snails 449; *see also* ***Nassarius***
Mugil (mullet) 137, 351
  cephalus 344, 349–51
Mugilidae 342, 343, 350, 357
mullet, *see* ***Mugil***
Mullus surmuletus 171
mussels 2, 3, 9–10, 12, 15, 17, 24, 40,
    42, 44–6, 49, 69, 103–5,
    118–119, 122, 195, 202, 265–81,
    298, 304, 452
  beds 269, 278
  dislodgement 46
Mustela macrodon 43
mutualisms 46, 197, 204–5, 451, 477–8
    et seq., 486 *et seq.*, 512
  commensalism 487
Myrionema orbicularis on ***Posidonia*** 168
Mytilus 17, 36–7
  edulis 202, 266, 269, 271–3, 277–9

Nassarius
  obsoletus 33, 35
  pauperatus 305, 449
Navicula
  pygmaea 137
  salinarum 137
nematodes 136–8, 178

algivorous 136-7
Neogoniolithon 481
Neorhodomela larix 241-2; see also
    Rhodomela larix
Nereididae 175, 235, 407, 411
Nereis vexillosa 407
Nereocystis 67, 303, 377
    luetkana 201
Nerita 109-110, 116-117
    atramentosa 449, 462
    funiculata 113-114, 296
    scabricosta 113-114, 296
    tesselata 498
    versicolor 498
nitrate
    reductase activity 529
    uptake in N_2 fluxes 523
    uptake in symbiotic Cnidaria in
        general 525
nitrogen demands/budgets *passim*, but
        especially 265-81
    in 'gardens' 410-2, 414, 416, 511-40
    in zooxanthellae symbioses 515,
        523-8
nitrogenous excretion 511, 515, 523-4;
    see also amino-acids
nocturnal activity 91-4, 172, 249-50
    C-product incorporation 516
    lack of N_2 uptake 525
    O_2 uptake by rock 218
    release of ammonia by
        zoxanthellae 524, 526
Nodolittorina tuberculata 498
non-predatory keystone species 414
Nothogenia fastigiata 370
Notoacmaea 194
    fenestrata 432
Nucella 5-8, 11-12, 15, 202
    lapillus 3-4, 16, 36-7, 40, 50
    see also *Thais*
nudibranchs 196, 203, 304
numbers, see density
Nuttallina
    californica 69-71
    kata 65

ocean port 44
octocorals 392, 512
Ocypode 137
ocypodid crabs 137, 140
Odacidae 341, 342, 352-3

Odax pullus 343, 352-3, 357
Odonthalia floccosa 436
opal-eye 356
opisthobranchs 175, 392-3
opportunistic agariciid corals 392
opportunistic algae 37-8, 44, 96,
        363-[366 et seq.]-80, 448, 461-2
    in 'gardens' 407 et seq.
    in 'turfs' 416
opportunistic feeders 172, 265-[271]-81
Orchestia 242, 248
Ostrea edulis 273
owl limpets 450, 452
oysters 94, 105, 117, 156,
    265-[265-9]-81

pachydictyol – A 320-2, 327, 330
Pachydictyon 330
Pachygrapsus 66, 248
    crassipes 238, 432
Padina 332
pagurids 94
Palaemonetes 248
    pugio 239, 243
Palaemonidae 248
Palmaria palmata 245
Palythoa 512
Paracentrotus lividus 19, 171-2, 500,
    504-5
Paraclunio 432
parrot fish, see Scaridae, *specific organisms*
PIM (Particulate Inorganic Matter) 267
POM (Particulate Organic Matter)
    176, 267
Patella 1, 4-6, 8-10, 12-14, 17, 106,
        194, 419, 484
    barbara 407, 419
    cochlear 74, 108, 112, 406, 411-3
    coerulea 498, 504-5
    compressa 295, 303
    granularis 301
    longicosta 298, 304, 406-7, 410-1,
        417-8, 450
    miniata 301, 407, 419
    safiana 87
    tabularis 406, 450
    vulgata 4, 12, 14, 16, 23, 302, 306,
        452
Patellogastropoda 480
Patelloida latistrigata 298, 366, 375,
    436, 452, 481

pathogens/pathogen inhibitors 50, 76-7, 331
Patina 194, 203
  pellucida, see **Helcion**
Pecten maximus 273
Pelagophycus 377
Pelvetia 21, 66
  canaliculata 3, 193
Pelvetiopsis limitata 434
peptides, small, complexed to proteins 388, 391-2
peracarids 178
Perciformes 341
perennials (algae) 38-41, 45, 49-50, 71
periphyton 173-5, 177
periwinkles 4, 17, 42, 198-9, 292, 302-3; *see also* **Littorina**
Perumytilus purpuratus 367, 369-70
Petalonia fascia 369
peyssonelid crusts 482
Phaeophyta 21-3, 191, 341-2, 345, 353, 355, 357, 369, 386-8, 432, 434-6, 484
  encrusting 436
Phaeostrophion irregulare 435
phenolics 321, 324-8
phlorotannins 23, 320, 323, 325-8, 332; *see also* polyphenolic compounds
phosphoglyceric acid 518
phosphoglycerides 515
phosphorus 269
photosynthesis
  algae on coral reefs 213-30
  in coral symbioses 511-40
Phragmatopoma californica 395
Phycodrys 44
Phyllophora truncata 47
Phyllospadix 66, 483
  scouleri 436
Phymatolithon 43, 481
Pilayella 369
pipe fish 151
'pipette' feeders 137
Pisa 173
  muscosa 173
  nodipes 173
Pisaster 122
piscivorous biota, coral reef fish 213-30
Placopecten magellanicus 271
planktivores, in coral reef fish 213-30, 265-[271]-81

plankton 201, 238, 265-81, 295, 389, 394, 523
planktonic larvae (incl. veligers) 14, 15, 153, 203, 295, 386, 388, 392-3, 395; *see also* larvae
plant apparency model 324-8
planula larva, of sponges 514
Platyhelminthes 136, 138, 512
Platynereis
  bicanaliculata 407
  dumerilii 175, 321-2
Plectroglyphidodon lacrymatus 221, 411
Plocamium 393
  pacificum 370
Pneophyllum 168, 481
Pocillopora
  capitata 516, 524
  damicornis 519, 521-2, 524
polychaetes 139, 153, 165, 170-1, 174-5, 178, 192, 217, 223, 235, 237, 249, 290, 324, 395, 406-7, 496, 502-3
  and seaweed defences 319-33
  in 'gardening' 405-20
Polydora ligni 153-5
polydorids 503
polynoids 175
polyphenolic compounds 23, 75, 198, 200, 243
  in *Posidonia* 167, 171-3
  in seaweeds 320-[324-8]-33, 377, 417-18
  see also phlorotannins
Polyplacophora 372-3
polyps 512-40
Polysiphonia 295, 370, 393, 407, 433, 435
  lanosa 246
Pomacanthidae 342, 346, 348
Pomacentridae 92-3, 215, 218-20, 222, 238-9, 251, 299, 341, 342, 343, 346, 348, 355, 406-20, 431, 433, 435
  gut pH in 408
  see also specific organisms
Pomatocentrus lividus 299
Pomatoceros 18
Pontogeneia 248
  rostrata 237, 245
populations *passim*, especially abundance 468

dynamics 443-[467-8]-68, 494
experiments 455-61
processes 268
regulation of symbionts in 528-30
trends 95
Porifera 511-12
Porites 499
  porites 241, 521-2
Porphyra 9, 13-66, 432
  'conchocelis' phase 20
  columbina 370
  umbilicalis 3
Posidonia 237
  oceanica 165-78
  prairies 166, 176
  roots 166-7
Postelsia palmaeformis 118
potamidid snails 137
Prasiola meridionalis 73
prediction of events 52, 460-1
pre-emption, in shore-space 446-8
primary productivity, *passim*, but
    especially 108-110, 112, 133-41,
    148, 156, 241, 269, 299, 437-8
  excretion of excess 511
  6B exposed rocky shore 5
  in coral reef symbiosis 514-15
  in 'gardens' 405-[414-7]-20
  microalgal 135-7, 139
  reef fish 213-[219-21]-30
  see also standing crops, yields
propagules 18, 238, 241, 253, 426,
    437-8, 444, 451
  of algae 363-80, 453
Proscutum 481
prosobranchs 174, 454
proteins
  ammonia assimilation 526-7
  as energy source 277, 515
  content in 'gardens' 408
  in N_2 excretion 279
  patterns in zooxanthellae/hosts 513,
    515, 525
  synthesis 527-9
  whole-body turnover 280-1
  protoplasts of algae 363-[367-8, 373,
    379-80]-80
Protozoa 156, 354, 511
Psammechinus microtuberculatus 172
Pseudamphithoides incurvaria 252, 295,
    321

pseudofaeces 269 *et seq.*; *see also* faeces
Pterocladia 462-3
Pterosiphonia dendroidea 370
Pterothamnion 370
Pterygophora 67
Ptilota 44
  serrata 46
Ptychodera bahamensis 137
Pugettia 66, 248
pulmonate snails/limpets 139, 454
pycnogonids 196

queen conch 393

rabbit fish 348
radulae *passim*, but especially 106,
    372-3
  in chitons,
    magnetite-strengthened 496
  strength 495
  structure 290-1, 495
Ralfsia 112, 298, 304, 369, 417-18
  verrucosa 406-7, 410, 417-18
recruitment 14-15, 34, 47-52, 115,
    117, 120, 124-5, 147, 149,
    153-4, 158, 236, 249, 304,
    385-96, 427, 430, 434, 436-7,
    446-7, 479
redlip parrotfish 344, 353
reefs 107, 109, 250, 289, 296-7, 341,
    353, 392
  back 235
  edge 499
  fish 299, 344
  rock 88, 92-4
  windward 213-30
  see also coral reefs
refugia 93, 101, 116-19, 147, 149-52,
    158, 240, 252, 295, 306, 385,
    434-5, 437, 483, 485
  artificial surfaces 88
  barnacles as 453
  from 'grazing' on reefs 412
  kelp beds 45
  from predation 331
  from urchin grazing 45-7
  wave-exposed 49
regeneration 363-80
  by endolithic microorganisms 494-5

regeneration (*cont.*)
    of partially-digested algal
        tissues 364-6, 371-5, 379-80
repellents, *see* biochemical cues
replication, *see* experimental studies
resource partitioning, *passim*, but
        especially 443-68
    algae 371, 443-68
    availability 465-7
    concept 445
    among macroalgae 462-3
'reverse ramp function' 268
rhizomes, ***Posidonia*** 165-7, 169
Rhizophora mangle 243
Rhodochorton purpureum 376; *see also*
        Audouinella purpurea
rhodoliths 43, 87-8, 93-4
Rhodomela 196
    larix 195, 436; *see also* ***Neorhodomela***
        larix
Rhodophyta 18, 21-3, 191, 238, 341-2,
    346, 348, 355, 386-8, 393-4,
    432-6
Rissoa ventricosa 171

Sabella microphthalma 153-5
sabellids 503
Saccorhiza 20, 295
    dermatodea 43-4
    polyschides 19, 21
Saccostrea amasa 501
salinity, *see* abiotic factors
salt marsh 267, 269, 271
Sargassum 21, 93, 103, 105, 238, 249,
    387, 394
    tortile 393-4
Sarpa salpa 165, 170, 172
scallops bay 268
Scaridae 92, 94, 105, 108, 110, 218-20,
    222-4, 229, 330, 342, 343-5,
    352-3, 355, 414; *see also specific*
    organisms
Scarus
    frenatus 221
    rubroviolaceus 344, 352-3
    sordidus 221, 344
    see also parrotfish
Schottera nicaeensis 370
sciaphilic macroalage 178
scleractinians 387, 512

Scomberomorus 228
Scopimera 137
Scorpididae 342, 356-7
Scurria 194, 481
sea anemones
    algae from 529
    expelling zooxanthellae 514
    nitrate uptake 523
    storing lipid 517-18
sea grasses 42, 88, 147 *et seq*, 165-78,
        228, 237, 239-40, 243, 248-9,
        296, 320, 340, 344, 353, 356,
        483-4; *see also specific organisms*
sea mink 43, 51; *see also* ***Mustela***
sea-stars 43-4, 63, 479; *see also* starfish,
        Asterias
sea urchins 3, 19, 41-3, 45-9, 51,
        63-9, 74-6, 87, 91, 93-4, 105-7,
        165, 214-5, 238, 251, 253, 293,
        296-7, 299-301, 303-4, 364,
        368-80, 425, 434, 451, 463,
        486-7
    on coral reefs 214-30, 405-20
    dominating zones 46, 91
    fronts formation 47
    'gardening' 406-7
    green sea urchin 391
    low densities 49
    manipulation/removal
        experiments 46, 412, 414
    mortalities 45, 47, 49
    parasites of 50
    pathogens of 48, 50
    population explosions 45, 47, 49-50
    population size 52, 448-9
    recolonization 46
    and seaweed defences 319-33
seasonality, *see* climate
secondary metabolites/production 139,
        198-9, 203, 331-2
    see also specific substances, defences
    as antifouling agents 331-2
    in defence mechanisms 319-33
    in ***Posidonia*** 167
    on coral reefs 215 *et seq.*
sediment, *see* detritus
selectivity, by grazing in 'gardens'
        405-[410-1]-20
Semibalanus balanoides 4-6, 8, 11-12,
        14-15, 36-7, 39
Seriatopora hystrix 529

Serpulidae 171, 396, 414
Sertularella miurensis 393
seston 270, 272-5, 281
   concentration 268
   in salt marsh 267
   see also PIM, POM
sestonophagous species 176, 265-81
settlement 11, 13, 19, 71, 96-7, 118, 152-4, 156, 158, 190, 193, 197, 199-205, 240, 251, 294, 299, 304, 383-96, 425-7, 429-30, 437
   cues 386-96
   reattachment 366-80
   selective 386-96
   velocities 267 *et seq.*, 374 *et seq.*
   see also biochemical cues; includes colonization
shrimps 15, 235, 237, 239, 243, 246-7, 249; *see also specific entries*
Sicyases sanguineus 369-70, 373
Siganidae 341, 342, 348, 355
Siganus doliatus 322
silver drummer 343
Siphonales 485
siphonarians 106, 294, 451, 454
Siphonaria 87, 373, 451, 454
   alternata 291, 302, 306
   capensis 301-2
   denticulata 294
   gigas 113
   lessonii 365, 369-70, 374, 377
   normalis 291, 302, 306
   pectinata 91, 95-6, 301
   thersites 297
   virgulata 294
sipunculids 178, 503
sit-and-wait grazers 299-300
size
   in competition analyses 455-67
   distribution 495
   as limiting factor 485-6
snails 109-10, 114, 117-18, 157, 192, 196, 198, 291, 295-6, 299, 301, 303, 432
   coiled snails 425
   mud nails 305
   turban snails 436
Solenoporaceae 482, 485
Solieria chordalis 366
space *passim, see particular aspects*, abiotic factors

Sparidae 170, 172, 341, 342, 348
Spartina 242, 271-2, 296
   alterniflora 38
Sphacelaria 373
Sphaeroma peruvianum 243
Spirorbis 389, 396
   rupestris 387
   spirorbis (borealis) 199, 393
spirorbids 202, 387
sponges 45, 175, 502
   boring 493-4, 496
Spongites 481
spores/sporangia/sporelings 106, 115, 119, 124, 196, 240, 290, 299, 436
   of seaweeds 363-[364-8, 371-2, 376-9]-80
   sori 372
   tetrasporangia 374
   see also carpospores
Sporolithon 481
sporophylls 372, 377
standing crops 166-7
   see also yields, primary products, biomass of *Posidonia*
starfish 304, 444
   see also sea-stars, *Asterias*
Stegastes
   apicalis 221, 411
   fasciolatus 409
steroid-like compounds, epoxide of δ-tocotrienol as inducer 394
Stichaeidae 341, 342, 347
Stichaster australis 304, 387, 396
stimuli, *see* biochemical cues, environmental stimuli, mechanical stimuli
stomach contents/gut analyses 94, 136, 171-2, 176, 367, 369-70
strategies anti-grazing 140, 173; *see also* defences
stress *passim*, but especially 443-75
   desiccation 69
   effect of, on competitive ability 465-7
   in expulsion of zooxanthellae 530-2
   of summer conditions 68
   on tropical shores 110
   physiological 73
striped bristlemouth 343, 349
Strombus gigas 393
Strongylocentrotus 65, 396

Strongylocentrotus (*cont.*)
   droebachiensis 20, 33, 36, 41, 43-4, 46, 50, 303, 387, 391
    (***Modiolus*** interaction) 46
   franciscanus 300, 303-4, 391
   purpuratus 73, 387, 391, 434
Stylophora pistillata 516, 520-1, 523, 527-9
Stypopodium 320
stypotriol 320
sublittoral algae/plants 9, 17, 18-20, 43-9, 64-8, 74-5
   adaptations 93
   distribution under urchin-grazing 45, 74
   dominating zones 46; *see also* sublittoral zonation
   as refuge for fish/invertebrates 48, 51
   at record depths 46
sublittoral zonation
   Baltic 242
   Mediterranean 165-87
   NE Atlantic 17-20
   NE Pacific 61-85
   NW Atlantic 43-9
   seagrasses 147-64
   tropical E Atlantic 91-4
subsurface feeders 139, 178
successional stages 67-9, 119-20, 174, 205, 236, 241-2, 396, 417-20, 425-38, 461-2
surfgrass, *see* ***Phyllospadix***
surgeon fish 105, 107, 109, 341, 351, 355-6, 358
   brown surgeon 341, 346
   see also Acanthuridae
suspension feeding 265-81
sweeping (fronds), *see* whiplash effect
syllids 175
Symbiodinium 513
   free-living 514
   microadriaticum 511-13
   a species complex 513
   see also zooxanthellae
symbiosis *passim*, but especially 214, 511-40
   Hydra/Chlorella 525
   regulation of symbiont populations 528-30
   see also endosymbiosis
symmetry analyses, experimental design 456-8

Synarthrophyton 481
synchrony, in symbiont cell division 528-30

Talitridae 248
Talitroidea 242
Talorchestia 248
tannins 200, 320, 324-8; *see also* phlorotannins
Tautoglabrus 37
Tectura 480-1
   paleacea 483
   scutum 480
   testudinalis 36-7, 39, 44, 479-80
   see also ***Acmaea***
Tegula (turban snails) 66, 68, 396, 467
   funebralis 75, 293
tellinid bivalves 137
temperature, *see* abiotic factors
Tenarea 481
terpenes in seaweeds 320-33
territory/territorial animals *passim*, but especially 298-9, 304, 435
Tesseropora rosea 433
Tetraodontidâe 342
Tetrapygus niger 369-70
Thais 119, 122
   lapillus 453
   see also ***Nucella***
Thalassia 320
thalassinidean prawns 139
Thalassoma bifasciatum, *see* wrasse
Titanoderma 481
tocotrienol, epoxide of, as inducer 394
Tonicella 44, 390
   lineata 387, 389
Tozeuma carolinense 151
triacyglycerols 516-18
Tricholia speciosa 171
Tridacna crocea 501
triglycerides 515
trochids 21
Trochus niloticus 387, 390
trophodynamic modelling reef fish 213-30
turban snails, *see* ***Tegula***
Turbellaria 511
Turbo undulatum 297, 306
turf (algal) (including algal mats) 1, 3, 11, 16-17, 21, 23, 41, 63, 65,

93, 103, 105, 109, 117, 190, 220, 238, 241, 251, 297, 345, 391, 433-6
  algal mats 133, 135, 137, 140
  in 'gardens' 410, 414-7
  in intermediate grazing 416
  in larval settlement 391
  mats in coral reefs 214, 412
  outside 'gardens' 412
  restriction by urchins 47
Turritella 94
turtles 119

Uca 137
Ulothrix 13, 37
  flacca 38
Ulva 21, 66, 69-71, 96, 106, 120, 238, 241, 247-8, 297, 299, 366, 371, 374, 393, 432-6, 448, 461
  fasciata 96
  lactuca 432-3
  lobata 348
  rigida 369, 375-6
Ulvella 369
ulvoids 366, 375-6
unapparent plants 324-8
understorey algae/spp. 18-19, 47, 447, 462-3
  guilds 47
unicellular algae 133-41; *see also* microscopic algae
Urospora 66, 369, 432

vacuoles 511-40
vermetids 493, 505
VFAs, *see* fatty acids
Vibrio 356; *see also* bacteria

wax esters 516, 518
weight
  effect on competitive ability 465-7
  experiments 455-8
whelks 453, 466-7; *see also specific organisms*
whiplash effect, of algal fronds 1-[12, 17]-32, 119, 452
wolffish (Anarchichatidae) 44
worms and seaweed defences 319-[322]-33
wrackbeds (*Posidonia*) 175-6
Wrangelia argus 95-6
wrasse 252, 301
  Halichoeres bivattatus 252
  Thalassoma bifasciatum 252

yields
  coral reef fish 213-30
  see also standing crops, primary productivity

Zanclidae 355
zebra-perch 356
zoanthids 512
zooxanthellae 511-20, 523-4, 526, 528
  in coral (hard) 214
  chemical attractants 514
  distribution of light 512-13
  genetic distinctions 513-14
  in gastrodermal layers, mesogloea and epidermis 512-13
  limits to numbers 512-3, 528-30
  specificity to hosts 513
  transmission to hosts 514
  see also **Symbiodinium**
Zostera 243, 272, 484
  marina 149-50, 153, 156-7, 175, 483

Systematics Association Publications

1. Bibliography of key works for the identification of the British fauna and flora, *3rd edition* (1967)
   Edited by G.J. Kerrich, R.D. Meikle, and N. Tebble
2. Function and taxonomic importance (1959)
   Edited by A.J. Cain
3. The species concept in palaeontology (1956)
   Edited by P.C. Sylvester-Bradley
4. Taxonomy and geography (1962)
   Edited by D. Nichols
5. Speciation in the sea (1963)
   Edited by J.P. Harding and N. Tebble
6. Phenetic and phylogenetic classification (1964)
   Edited by V.H. Heywood and J. McNeill
7. Aspects of Tethyan biogeography (1967)
   Edited by C.G. Adams and D.V. Ager
8. The soil ecosystem (1969)
   Edited by H. Sheals
9. Organisms and continents through time (1973)†
   Edited by N.F. Hughes

Published by the Association (out of print)

Systematics Association Special Volumes

1. The new systematics (1940)
   Edited by J.S. Huxley (Reprinted 1971)
2. Chemotaxonomy and serotaxonomy (1968)*
   Edited by J.G. Hawkes
3. Data processing in biology and geology (1971)*
   Edited by J.L. Cutbill
4. Scanning electron microscopy (1971)*
   Edited by V.H. Heywood
   Out of print
5. Taxonomy and ecology (1973)*
   Edited by V.H. Heywood

6. The changing flora and fauna of Britain (1974)*
   Edited by D. L. Hawksworth
   Out of print
7. Biological identification with computers (1975)*
   Edited by R. J. Pankhurst
8. Lichenology: progress and problems (1976)*
   Edited by D. H. Brown, D. L. Hawksworth, and R. H. Bailey
9. Key works to the fauna and flora of the British Isles and north-western Europe, *4th edition* (1978)*
   Edited by G. J. Kerrich, D. L. Hawksworth, and R. W. Sims
10. Modern approaches to the taxonomy of red and brown algae (1978)*
    Edited by D. E. G. Irvine and J. H. Price
11. Biology and systematics of colonial organisms (1979)*
    Edited by G. Larwood and B. R. Rosen
12. The origin of major invertebrate groups (1979)*
    Edited by M. R. House
13. Advances in bryozoology (1979)*
    Edited by G. P. Larwood and M. B. Abbot
14. Bryophyte systematics (1979)*
    Edited by G. C. S. Clarke and J. G. Duckett
15. The terrestrial environment and the origin of land vertebrates (1980)*
    Edited by A. L. Panchen
16. Chemosystematics: principles and practice (1980)*
    Edited by F. A. Bisby, J. G. Vaughan, and C. A. Wright
17. The shore environment: methods and ecosystems (2 Volumes) (1980)*
    Edited by J. H. Price, D. E. G. Irvine, and W. F. Farnham
18. The Ammonoidea (1981)*
    Edited by M. R. House and J. R. Senior
19. Biosystematics of social insects (1981)*
    Edited by P. E. Howse and J.-L. Clément
20. Genome evolution (1982)*
    Edited by G. A. Dover and R. B. Flavell
21. Problems of phylogenetic reconstruction (1982)*
    Edited by K. A. Joysey and A. E. Friday
22. Concepts in nematode systematics (1983)*
    Edited by A. R. Stone, H. M. Platt, and L. F. Khalil
23. Evolution, time and space: the emergence of the biosphere (1983)*
    Edited by R. W. Sims, J. H. Price, and P. E. S. Whalley
24. Protein polymorphism: adaptive and taxonomic significance (1983)*
    Edited by G. S. Oxford and D. Rollinson

25. Current concepts in plant taxonomy (1983)*
    Edited by V. H. Heywood and D. M. Moore
26. Databases in systematics (1984)*
    Edited by R. Allkin and F. A. Bisby
27. Systematics of the green algae (1984)*
    Edited by D. E. G. Irvine and D. M. John
28. The origins and relationships of lower invertebrates (1985)‡
    Edited by S. Conway Morris, J. D. George, R. Gibson, and H. M. Platt
29. Infraspecific classification of wild and cultivated plants (1986)‡
    Edited by B. T. Styles
30. Biomineralization in lower plants and animals (1986)‡
    Edited by B. S. C. Leadbeater and R. Riding
31. Systematic and taxonomic approaches in palaeobotany (1986)‡
    Edited by R. A. Spicer and B. A. Thomas
32. Coevolution and systematics (1986)‡
    Edited by A. R. Stone and D. L. Hawksworth
33. Key works to the fauna and flora of the British Isles and north-western Europe, 5th edition (1988)‡
    Edited by R. W. Sims, P. Freeman, and D. L. Hawksworth
34. Extinction and survival in the fossil record (1988)‡
    Edited by G. P. Larwood
35. The phylogeny and classification of the tetrapods (2 Volumes) (1988)‡
    Edited by M. J. Benton
36. Prospects in systematics (1988)‡
    Edited by D. L. Hawksworth
37. Biosystematics of haematophagous insects (1988)‡
    Edited by M. W. Service
38. The chromophyte algae: problems and perspectives (1989)‡
    Edited by J. C. Green, B. S. C. Leadbeater, and W. Diver
39. Electrophoretic studies on agricultural pests (1989)‡
    Edited by Hugh D. Loxdale and J. den Hollander
40. Evolution, systematics, and fossil history of the Hamamelidae (2 Volumes) (1989)‡
    Edited by Peter R. Crane and Stephen Blackmore
41. Scanning electron microscopy in taxonomy and functional morphology (1990)‡
    Edited by D. Claugher
42. Major evolutionary radiations (1990)‡
    Edited by P. D. Taylor and G. P. Larwood
43. Tropical lichens: their systematics, conservation, and ecology (1991)‡
    Edited by D. J. Galloway

44. Pollen and spores: patterns of diversification[‡]
   Edited by S. Blackmore and S. H. Barnes
45. The biology of free-living heterotrophic flagellates[‡]
   Edited by D. J. Patterson and J. Larsen
46. Plant–animal interactions in the marine benthos[‡]
   Edited by D. M. John, S. J. Hawkins, and J. H. Price

[*] Published by Academic Press for the Systematics Association
[†] Published by the Palaeontological Association in conjunction with the Systematics Association
[‡] Published by the Oxford University Press for the Systematics Association